SOLUTIONS TO EXERCISES

Organic Chemistry

SECOND EDITION

THE DAILY GRIND   (watercolor on paper, 30 x 40 cm)
Thomas N. Sorrell, 1998

The Daily Grind is the name of a coffee shop on the campus of the University of North Carolina where students and professors often congregate to socialize and relax. It is also a good place to study, especially on the clear, warm days during the fall and spring that make Chapel Hill such a pleasant place to live.

The title of this painting also refers to the daily discipline required to learn organic chemistry. Rodger Griffin, Jr., my undergraduate organic chemistry instructor, made the point like this: "Unlike history or philosophy, chemistry cannot profitably be read chapter by chapter but must be vigorously attacked with a dozen sharp pencils and a ream of inexpensive paper close at hand. When information is used to solve problems, it rapidly becomes part of your knowledge."

SOLUTIONS TO EXERCISES

Organic Chemistry

SECOND EDITION

Thomas N. Sorrell

THE UNIVERSITY OF NORTH CAROLINA AT CHAPEL HILL

University Science Books
Sausalito, California

University Science Books
www.uscibooks.com

Design Consultant: *Robert Ishi*
Compositor: *Thomas N. Sorrell*
Printer & Binder: *Von Hoffmann Corporation*

This book is printed on acid-free paper.

University Science Books
Sausalito, California
©1999, 2006
All rights reserved.
First edition 1999. Second edition 2006.

Library of Congress Control Number: 2005935137

ISBN 10: 1-891389-40-8
ISBN 13: 978-1-891389-40-5

Printed in the United States of America
10 9 8 7 6 5 4 3 2

CONTENTS

To learn organic chemistry, you must see and understand the recurring patterns that correlate the thousands of facts that will be presented during your studies. Working the exercises in the textbook is one way to learn this factual material, and many of the exercises will comprise a familiar repetition of problems for which the solution has been presented and described in the Examples appearing throughout the text. Some of the exercises, however, especially those at the ends of the chapters, will force you to recall information from earlier chapters. Others will ask you to recall the facts in a different format. Still others will require you to draw analogies with what you know, and then to predict the solutions to problems that you have never seen previously. I cannot stress strongly enough that you should work problem after problem after problem if you want to master this subject.

This *Solutions to Exercises* book has the answer to every exercise that appears in the textbook. In many instances, the approach needed to work toward the answer is also included along with the factual solution. Note that many of the synthesis and spectroscopy problems either have multiple solutions or can be solved by alternate approaches. Just because your answer or strategy is different from the one presented in this book, do not assume that you have made a mistake. Talk to your instructor and find out if your solution is equally valid.

The first edition of *Solutions to Exercises* was originally prepared with significant help from Julius Beau Lucks, an undergraduate student at UNC, who received a Churchill Scholarship after graduation and is currently a graduate student in Chemistry at Harvard University. Beau worked every exercise in the first edition and checked their solutions. Subsequent groups of students during the past five years have worked many of the same exercises, so most of the errors have been corrected. This second edition was copyedited and checked by Christine Cleveland, an undergraduate English major who worked as a professional copy editor for several years before she returned to school to take organic chemistry so that she could attend medical school. (She started at the UNC–CH School of Medicine in the fall semester of 2005.)

A mistake in the solution to an exercise is such a frustrating matter when you are trying to learn a new subject, and the last thing I want to do is to add uncertainty to your growing understanding. So as with the textbook itself, if you find factual errors or discussions that are confusing, please let me know (sorrell@unc.edu) so that I can post the corrections and make changes before subsequent printings.

Enjoy your study of organic chemistry—it is a fascinating subject!

—Tom Sorrell
Chapel Hill, 2005

SOLUTIONS TO EXERCISES

Organic Chemistry

SECOND EDITION

THE STRUCTURES OF ORGANIC MOLECULES

1.1. The functional groups in a molecule are easily recognized because they consist of heteroatoms (atoms other than C or H) or multiple bonds. The identities of the common functional groups are summarized in Table 1.1 of the text. Functional groups normally include the carbon atom to which a heteroatom is attached.

a.

carboxylic acid
alcohol

b.

amine
nitrile

c.

carboxylic acid ester

d.

alkene
thiol

1.2. To draw condensed formulas, express the hydrocarbon units with the designations CH, CH_2, or CH_3. Functional groups are included by using the condensed formulas given in Table 1.1 of the text.

a. $HO-CH_2-CH_2-CH_2-COOH$

b. $H_2N-CH_2-CH_2-CN$

c.

$$\begin{array}{c} H_2 \\ C \\ H_2C \qquad CH-COOCH_3 \\ H_2C \qquad CH_2 \\ C \\ H_2 \end{array}$$

d.

$$\begin{array}{c} SH \\ H_2 \qquad | \\ C \qquad CH_2 \\ HC \qquad CH \\ HC-CH_2 \end{array}$$

1.3. When drawing a full structural formula from a condensed one, focus first on creating the molecule's carbon skeleton. (Heteroatoms—particularly O, N, and S—may have to be included if they are part of a ring or chain.) Make certain to include all of the carbon atoms in this first step.

For acyclic molecules, draw the backbone atoms in the chain, including the substituents that contain carbon atoms. Then attach the atoms needed to establish the functional groups. Finally attach the hydrogen atoms and non-carbon substituents and include multiple bonds in order to give the correct numbers of bonds to each atom—four bonds to each carbon atom, three bonds to each nitrogen atom, and two bonds to each oxygen and sulfur atom.

For cyclic compounds, draw the ring and attach any substituents that contain carbon atoms, including heteroatoms needed to create the molecular skeleton. Then attach the atoms needed to form the functional groups. Finally attach the hydrogen atoms and non-carbon substituents and include multiple bonds in order to give the correct numbers of bonds to each atom.

1.3.   (continued)

a. CH_3CH_2CHO ⟶ C—C—C ⟶ [aldehyde structure] ⟶ [expanded structure]

  aldehyde   **aldehyde**

b.   [cyclopentane with CHOH structure] ⟶ [ring skeleton] ⟶ [ring with C—O] ⟶ [expanded structure]

  alcohol   **alcohol**

oxygen atom in the chain

c. $HSCH_2CH_2OCH_3$ ⟶ C—C—O—C ⟶ [S-C-C-O-C structure] ⟶ [expanded structure]

  thiol   ether   **ether**   **thiol**

oxygen atom in the chain

d. $CH_3CHBrCH_2COOCH_3$ ⟶ C—C—C—C—O—C ⟶ [C-C-C(=O)-O-C structure] ⟶

  carboxylic acid ester   **carboxylic acid ester**

[expanded structure with H, Br, O]

1.4.   Follow the procedure given in the solution to Exercise 1.3.

a.   [cyclopentane with CH₃ and amide structure] ⟶ [ring skeleton with C—N] ⟶ [ring with N, C=O] ⟶ [expanded structure]

  Amide (carboxamide)   **Amide (carboxamide)**

b.   oxygen atom in the chain

  [ether/aldehyde line structure] ⟶ [C-O-C structure] ⟶ [C-O-C-C-H with aldehyde] ⟶ [expanded structure]

  ether   **ether**   **aldehyde**

  aldehyde

1.5.   To convert from a structural formula to a condensed line structure, represent the carbon skeleton first as a polygon (cyclic molecule) or a zigzag line (acyclic molecule) and include any double and triple bonds that are present. Then attach the heteroatoms to the carbon atoms in the ring or chain. If a heteroatom is attached via a multiple bond to the carbon atom at the end of a chain (for example in the nitrile, aldehyde, carboxylic acid, and carboxylic acid ester functional groups), include the carbon atom(s) for that functional group, using its condensed notation (Table 1.1). Make certain that the total number of carbon atoms is correct.

1.5.   (continued)

a.

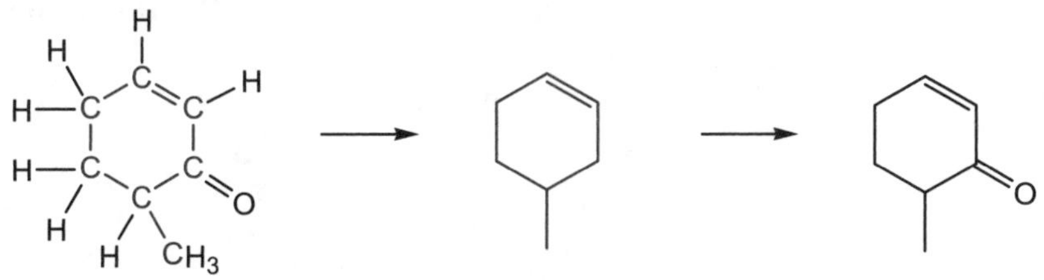

7 carbon atoms
6-membered ring

7 carbon atoms
6-membered ring

b.

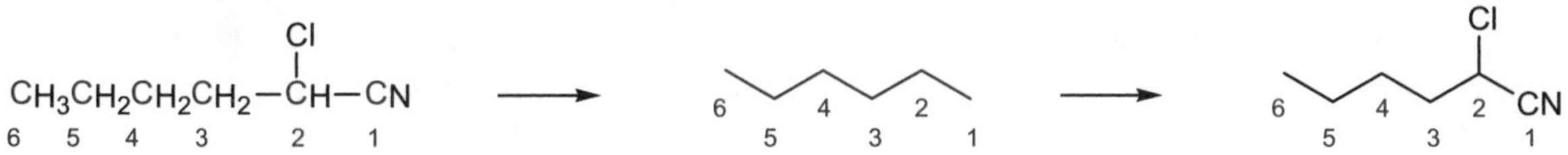

$$CH_3CH_2CH_2CH_2-\underset{\underset{2}{|}}{\overset{\overset{Cl}{|}}{CH}}-\underset{1}{CN}$$
6   5   4   3

6 carbon atoms

6 carbon atoms

6 carbon atoms

1.6.   For the compounds in this exercise, convert the name of the compound root to its representation as a chain or ring, and then include any double or triple bonds between the appropriate carbon atoms as specified by the suffix. Use the numeral, if given, to define the first carbon atom in the chain or ring at which the multiple bond starts. If no numeral is included in a compound's name, then the double or triple bond in an acyclic molecule starts at the end of the chain. In a cyclic molecule, the position of the multiple bond defines C1 (and C2) of the ring.

a.   **3-Hexene**

 hex   6 carbon atoms
 ene   one double bond that starts at C3

$$\underset{6}{CH_3}\underset{5}{CH_2}\underset{4}{CH}=\underset{3}{CH}\underset{2}{CH_2}\underset{1}{CH_3}$$

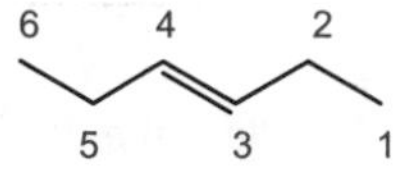

b.   **4-Octyne**

 oct   8 carbon atoms
 yne   one triple bond that starts at C4

$$\underset{8}{CH_3}\underset{7}{CH_2}\underset{6}{CH_2}\underset{5}{C}\equiv\underset{4}{C}\underset{3}{CH_2}\underset{2}{CH_2}\underset{1}{CH_3}$$

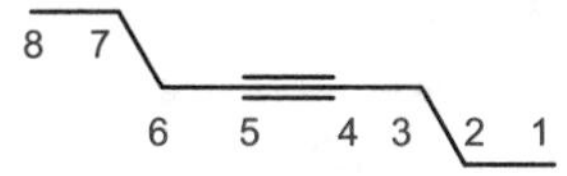

c.   **1-Butene-3-yne**

 but   4 carbon atoms
 ene   one double bond that starts at C1
 yne   one triple bond that bond that starts at C3

$$\underset{4}{HC}\equiv\underset{3}{C}-\underset{2}{CH}=\underset{1}{CH_2}$$

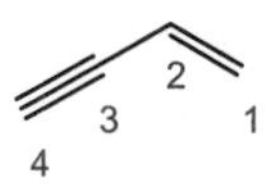

d.   **Cyclobutene**

 cyclobut   four-membered ring of carbon atoms
 ene   one double bond that defines C1

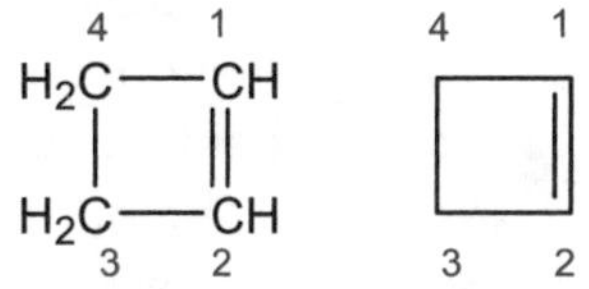

1.7.   The principal functional group in a molecule is denoted by the suffix of the compound's name. The identities and structures of common functional groups are summarized in Table 1.1 of the text. The structures shown below represent general structures of the functional groups, not the structures of the whole molecules.

1.7. (continued)

Compound:	a. **2,2-Dimethyl-3-hexanone**	b. **3-Methoxybenzaldehyde**	c. **2-Methyl-2-butanol**
Suffix:	–one	–aldehyde	–ol
Functional group:	ketone	aldehyde	alcohol

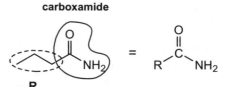

1.8. To simplify a given structural formula, identify the principal functional group in each molecule (circled below in color), and then replace the remainder of the molecule with the letter "R." If the compound is a derivative of benzene, use "Ar" instead of "R" for the benzene portion.

a.   **aldehyde**

b.   **carboxamide**

c.   **alkyne**

d.   **phenol**

1.9. The prefix associated with the substituents must match the quantity of numerals in front of the name, except in cases when there is no ambiguity.

a. 2,3,4-Hydroxyhexanal should be 2,3,4-<u>tri</u>hydroxyhexanal because there are three numerals in front of the name.

b. 2,2,4,4-Tetrachloropentane is correct because there are four numerals in front of the name and the prefix "tetra" precedes the substituent name "chloro".

c. Triiodomethane is correct because there is only one carbon atom to which the three iodine atoms can be attached, so the name is unambiguous without numerals.

1.10. Break each name into its constituent parts, and then combine the pieces to form the structure.

a.   **4-Fluorobutanal**

but	4 carbon atoms
an	no double or triple bonds
al	aldehyde functional group; its carbon atom defines C1
4-fluoro	fluorine atom at C4

b.   **3-Mercapto-2-pentanol**

pent	5 carbon atoms
an	no double or triple bonds
ol	OH group at C2
3-mercapto	SH group at C3

1.10.   (continued)

c.   **Trichloroethanenitrile**

eth	2 carbon atoms
ane	no double or triple bonds
nitrile	CN group; its carbon atom defines C1
trichloro	three chlorine atoms on C2

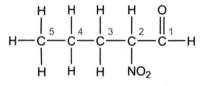

1.11.   Break each name into its constituent parts, and then combine the pieces to form the structure.

a.   **3-Oxopentanoic acid**

pent	5 carbon atoms
an	no double or triple bonds
oic acid	COOH group; its carbon atom defines C1
3-oxo	double-bonded oxygen atom at C3

b.   **2-Nitropentanal**

pent	5 carbon atoms
an	no double or triple bonds
al	aldehyde functional group;
	its carbon atom defines C1
2-nitro	NO$_2$ group at C2

c.   **4-Hydroxy-2-hexyne**

hex	6 carbon atoms
yne	triple bond starting at C2
4-hydroxy	OH group at C4

1.12.   Break each name into its constituent parts, and then combine the pieces to form the structure.

a.   **2-Bromobenzoic acid**

benzoic acid	benzene ring with –COOH group attached;
	the attachment point defines C1 of the ring
2-bromo	bromine atom attached at C2 of the ring

b.   **3-Fluoro-2-cyclohexenone**

cyclohex	six-membered ring of carbon atoms
en	one double bond, starts at C2
one	ketone functional group; its carbon atom defines C1
3-fluoro	F atom attached at C3

c.   **3-Chloro-4-nitrophenol**

phenol	benzene ring with the OH group attached;
	the attachment point defines C1 of the ring
3-chloro	chlorine atom attached at C3 of the ring
4-nitro	nitro group attached at C4 of the ring

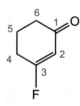

1.13.   First, draw each structure in its expanded form, showing all of the hydrogen atoms (follow the procedure given in the solution to Exercise 1.3). Then, assign the atom type to each carbon atom having four single bonds: the carbon atoms attached to only one other carbon atom are primary (1°); to two other carbon atoms are secondary (2°); to three other carbon atoms are tertiary (3°); and to four other carbon atoms are quaternary (4°).

a.

b.

1.14.   Break each name into its constituent parts, and then combine the pieces to form the structure.

a.   **Isobutylbenzene**

benzene	six-membered ring with alternating single and double bonds
isobutyl	a four-carbon fragment with a branch, attached via one of its 1° carbon atoms to the ring

b. **3-*tert*-Butylhexanol**

hex	6 carbon atoms
an	no double or triple bonds
ol	OH group; because no number is given, this group is at C1, by convention
3-*tert*-butyl	a four-carbon, branched fragment attached via its 3° carbon atom to C3 of the principal chain

c. **3-Ethylcyclopentanone**

cyclopent	a ring of 5 carbon atoms
an	no double or triple bonds
one	ketone; this group defines C1 of the ring
3-ethyl	a 2-carbon substituent, attached to the ring at C3

1.15.   Break the name into its constituent parts, and then combine the pieces to form the structure.

2-(1,1-Dimethylpropyl)hexanoic acid

hex	6 carbon atoms
an	no double or triple bonds
oic acid	carboxylic acid, its carbon atom defines C1 of the chain
2-(X)	substituent X; attached at C2

X = substituent

1.15.   (continued)

X = substituent: **1,1-Dimethylpropyl**	
propyl	a 3 carbon-atom substituent, by convention attached through its C1
1,1-dimethyl	two methyl groups attached at C1 of the substituent

$$CH_3-CH_2 \cdot CH_2-CH_2-\overset{\overset{\displaystyle O}{\|}}{C}-OH$$

with at the CH position:

$$H_3C-\overset{1}{C}-CH_3$$
$$\overset{2}{C}H_2$$
$$\overset{3}{|}$$
$$CH_3$$

1.16.   Break the name into its constituent parts, and then combine the pieces to form the structure.

3-Allyl-5-chlorobenzoic acid
    benzoic acid   benzene ring with –COOH group attached; the point of attachment of the COOH group defines C1 of the ring
    3-allyl   allyl group (CH_2=CH–CH_2–) attached at C3
    5-chloro   chlorine atom attached at C5

1.17.   Follow the procedures outlined in Examples 1.12-1.16 in the text.

a. This compound has the carboxylic acid functional group in a carbon chain, so the name ends in "oic acid." The longest carbon chain that includes both the carboxylic acid carbon atom and the double bond has four carbon atoms, so the root is *but–*. There is a carbon–carbon double bond, so the multiple bond index is *–en–*. The name so far is but/en/oic acid = butenoic acid.

    The principal functional group is one that must be at the end of the chain (Table 1.1), so numbering begins at the right end of the structure (as drawn). This numbering order means that the double bond begins at C3: 3-butenoic acid.

    The positions of attachment for the methyl groups are subsequently established (C2 and C3). The name of this molecule is **2,3-dimethyl-3-butenoic acid**.

b. This compound is a cyclic ketone with five carbon atoms in the ring, which also has a double bond: cyclopent/en/one = cyclopentenone.

    The principal functional group defines C1 by its position. This numbering order means that the double bond begins at C2: 2- cyclopentenone.

    A *tertiary*-butyl group is attached at C2. The name of this molecule is **2-*tert*-butyl-2-cyclopentenone.**

c. This compound is an alcohol with seven carbon atoms in the chain, and there are no double or triple bonds. The name so far is hept/an/ol = heptanol.

    The principal functional group (alcohol) can be at any position in the chain. Numbering is started at the end that gives the OH attachment position the lowest number. Numbering from the left end places the OH group at C3: 3-heptanol.

    This numbering order places the methyl group at C5. The name of this molecule is **5-methyl-3-heptanol**.

1.18.   To draw all of the possible isomers of a hydrocarbon, you must generate all of the possible carbon skeletons that can exist. Isomers may differ in the number of carbon atoms in the longest chain or by the placement of substituents.

For this exercise, first draw a straight chain with the requisite number of carbon atoms. Then reproduce that structure with one less carbon atom, but connect the omitted carbon atom at different sites along the chain (except at the terminal positions). Then draw a chain with two fewer carbon atoms, and attach the two omitted carbon atoms at unique positions along the chain. Consider putting the omitted carbon atoms together, too (see part b. of this exercise, where the two carbon atoms are combined to form an ethyl group instead of two methyl groups). Repeat this process, drawing the chain each time with one less carbon atom in the row. Number the chain from each end to make certain you have not just drawn a molecule in a different orientation.

a.   Draw the five isomers of C_6H_{14}.

Following the procedure outlined above:

6C chain

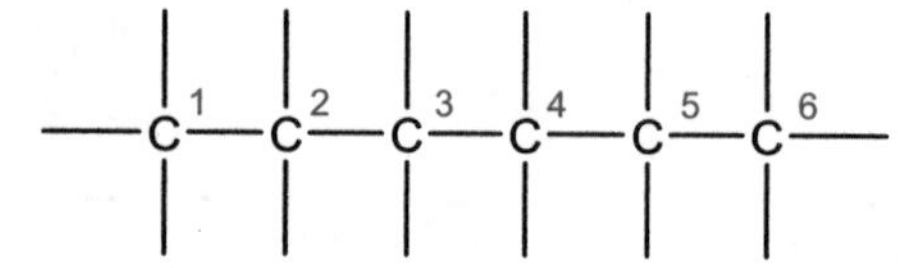

5C chain

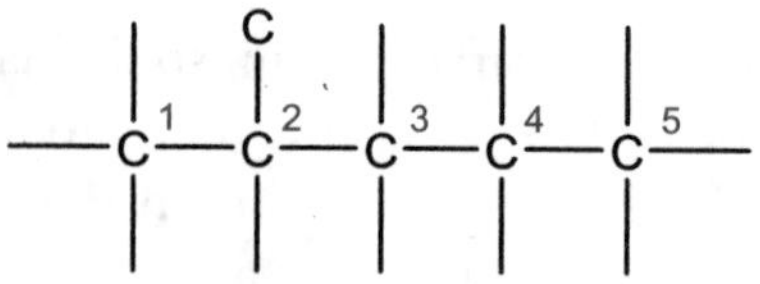

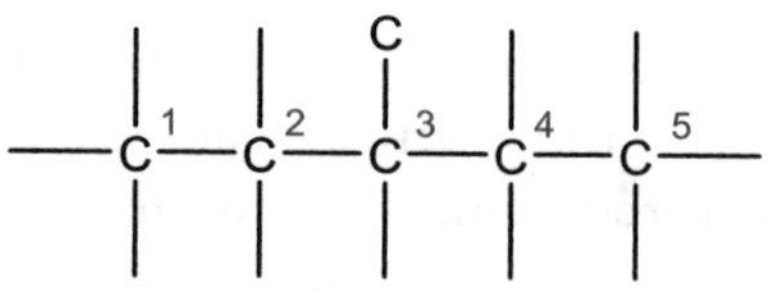

4C chain

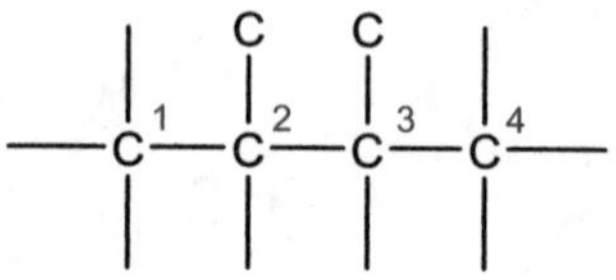

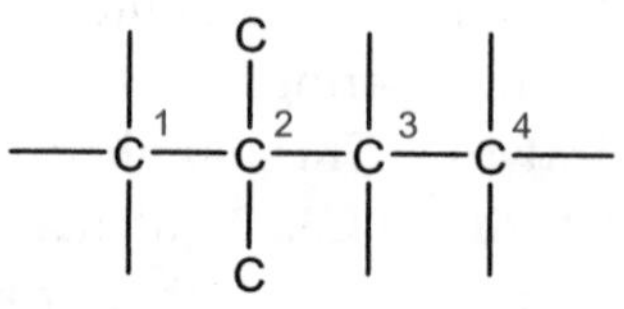

Finish by adding hydrogen atoms so that each carbon atom forms four bonds.

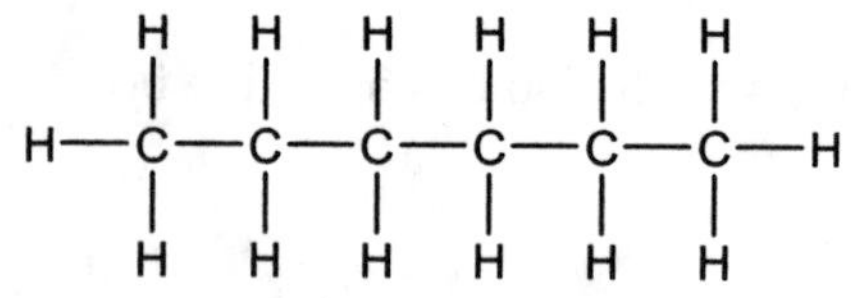

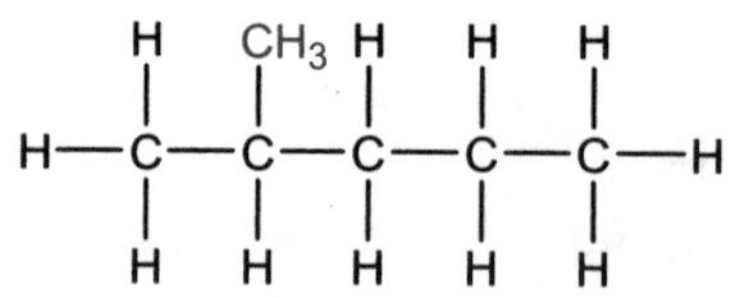

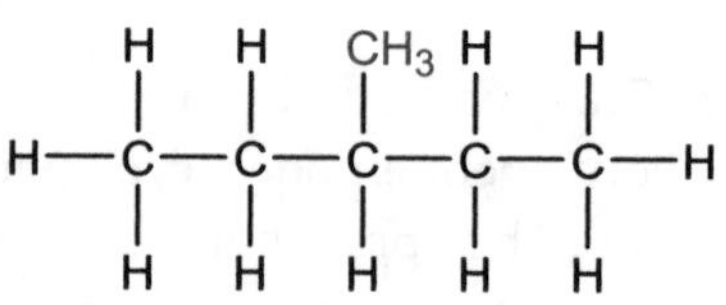

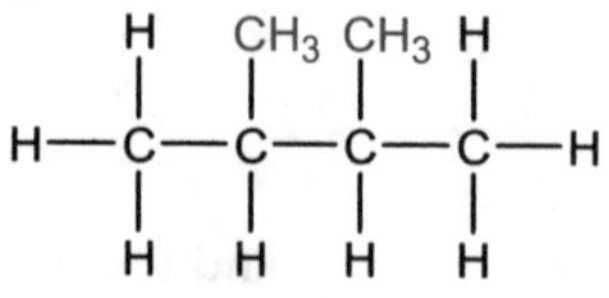

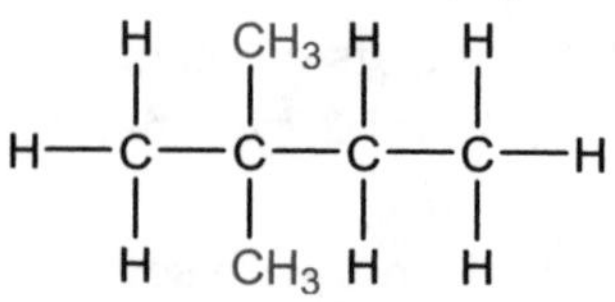

- - - - - - - - - - - - - - - -

b.   Draw the nine isomers of C_7H_{16}

7C chain

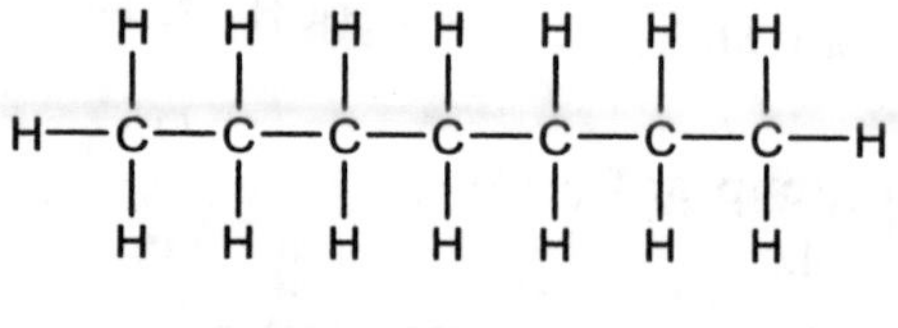

1.18.   (continued)

6C chain

$$\begin{array}{ccccccc} & & CH_3 & & & & \\ & H & \mid & H & H & H & H \\ \mid & \mid & \mid & \mid & \mid & \mid \\ H-C-C-C-C-C-C-H \\ \mid & \mid & \mid & \mid & \mid & \mid \\ H & H & H & H & H & H \end{array}$$

Methyl group attached at C2

$$\begin{array}{ccccccc} & & & CH_3 & & & \\ H & H & \mid & H & H & H \\ \mid & \mid & \mid & \mid & \mid & \mid \\ H-C-C-C-C-C-C-H \\ \mid & \mid & \mid & \mid & \mid & \mid \\ H & H & H & H & H & H \end{array}$$

Methyl group attached at C3

5C chain

$$\begin{array}{ccccc} & CH_3 & CH_3 & & \\ H & \mid & \mid & H & H \\ \mid & \mid & \mid & \mid & \mid \\ H-C-C-C-C-C-H \\ \mid & \mid & \mid & \mid & \mid \\ H & H & H & H & H \end{array}$$

Methyl groups attached at C2, C3

$$\begin{array}{ccccc} & CH_3 & & CH_3 & \\ H & \mid & H & \mid & H \\ \mid & \mid & \mid & \mid & \mid \\ H-C-C-C-C-C-H \\ \mid & \mid & \mid & \mid & \mid \\ H & H & H & H & H \end{array}$$

Methyl groups attached at C2, C4

$$\begin{array}{ccccc} & CH_3 & & & \\ H & \mid & H & H & H \\ \mid & \mid & \mid & \mid & \mid \\ H-C-C-C-C-C-H \\ \mid & \mid & \mid & \mid & \mid \\ H & CH_3 & H & H & H \end{array}$$

Methyl groups attached at C2 and C2,

$$\begin{array}{ccccc} & & CH_3 & & \\ H & H & \mid & H & H \\ \mid & \mid & \mid & \mid & \mid \\ H-C-C-C-C-C-H \\ \mid & \mid & \mid & \mid & \mid \\ H & H & CH_3 & H & H \end{array}$$

Methyl groups attached at C3 and C3

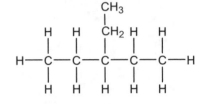

Ethyl group attached at C3

4C chain

$$\begin{array}{cccc} & CH_3 & CH_3 & \\ H & \mid & \mid & H \\ \mid & \mid & \mid & \mid \\ H-C-C-C-C-H \\ \mid & \mid & \mid & \mid \\ H & CH_3 & H & H \end{array}$$

Methyl groups attached at C2, C2, and C3

1.19.   To draw the condensed formulas for each of the following, express the hydrocarbon units with the groups CH, CH_2, or CH_3.

a.

$CH_3CH_2CH_2CH_2CH_2CH_3$

$$\begin{array}{l} CH_3 \\ \mid \\ CH_3-CH-CH_2CH_2CH_3 \end{array} \qquad \begin{array}{l} CH_3 \\ \mid \\ CH_3CH_2-CH-CH_2CH_3 \end{array} \qquad \begin{array}{l} CH_3 \quad CH_3 \\ \mid \qquad \mid \\ CH_3-CH-CH-CH_3 \end{array} \qquad \begin{array}{l} CH_3 \\ \mid \\ CH_3-C-CH_2CH_3 \\ \mid \\ CH_3 \end{array}$$

- - - - - - - - - - - - - -

b.

$CH_3CH_2CH_2CH_2CH_2CH_2CH_3$

$$\begin{array}{l} CH_3 \\ \mid \\ CH_3-CH-CH_2CH_2CH_2CH_3 \end{array} \quad \begin{array}{l} CH_3 \\ \mid \\ CH_3CH_2-CH-CH_2CH_2CH_3 \end{array} \quad \begin{array}{l} CH_3 \quad CH_3 \\ \mid \qquad \mid \\ H_3C-CH-CH-CH_2CH_3 \end{array} \quad \begin{array}{l} CH_3 \qquad CH_3 \\ \mid \qquad\quad \mid \\ H_3C-CH-CH_2-CH-CH_3 \end{array}$$

$$\begin{array}{l} CH_2CH_3 \\ \mid \\ CH_3CH_2-CH-CH_2CH_3 \end{array} \quad \begin{array}{l} CH_3 \\ \mid \\ H_3C-C-CH_2CH_2CH_3 \\ \mid \\ CH_3 \end{array} \quad \begin{array}{l} CH_3 \\ \mid \\ CH_3CH_2-C-CH_2CH_3 \\ \mid \\ CH_3 \end{array} \quad \begin{array}{l} CH_3 \quad CH_3 \\ \mid \qquad \mid \\ H_3C-C-CH-CH_3 \\ \mid \\ CH_3 \end{array}$$

1.20.   Follow the procedure outlined in the solution to Exercise 1.3.

Carvone

Progesterone

Pinene

Thromboxane A₂

1.21.   Follow the procedures outlined in the solution to Exercise 1.8.

a. R—OH	b. R—O—R'	c. R—COOH	d.
alcohol	ether	carboxylic acid	ketone

e. R—SH	f. R—NH₂	g. R—CHO	h. R—C≡CH
thiol	amine	aldehyde	alkyne

1.22.   An exercise such as this reinforces your knowledge about the identities of the functional groups and compound types. Table 1.1 in the text lists the structures of the common functional groups. An aromatic compound is a derivative of benzene. An alicyclic compound has a ring but does not have alternating single and double bonds.   Two specific compounds are illustrated for each portion of this exercise, but many other possible answers could be given.

a.   **A ketone with the formula C₅H₁₀O**
       The possible structures are limited to those with a saturated aliphatic carbon skeleton (no multiple bonds) and a ketone group.

b.   **A chloro ketone with four carbon atoms**
       The possible structures are limited to those with a ketone group and a chlorine atom substituent. It cannot be a benzene derivative because that would require six carbon atoms; the molecule may be cyclic, however.

1.22.   (continued)

c. **An aromatic amine**
    The NH₂ group must be attached to
    a benzene ring, but no limitations exist
    on the number or types of substituents.

d. **An aldehyde with six carbon atoms**
    The total number of carbon atoms
    must be six, but the molecule can
    be aliphatic or alicyclic..

e. **A hydroxy aldehyde**
    The molecule must have the CHO
    and OH groups, but it can have any
    number of carbon atoms and may
    be aliphatic, alicyclic, or aromatic.

f. **An alicyclic carboxylic acid**
    The molecule must have the COOH
    group and a non-aromatic ring, but
    there are no other structural limitations.
    The COOH group should be attached
    directly to a carbon atom in the ring.

1.23.   Follow the procedure given in the solution to Exercise 1.13.

a.

b.

c.

d.

e.

f.

1.24. Follow the procedures given in the solutions to Exercises 1.3 and 1.1.

Novocain

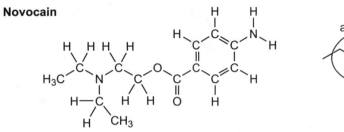

Captopril

Prozac

1.25. Follow the procedure given in the solution to Exercise 1.13 and in the examples in Section 1.4c.

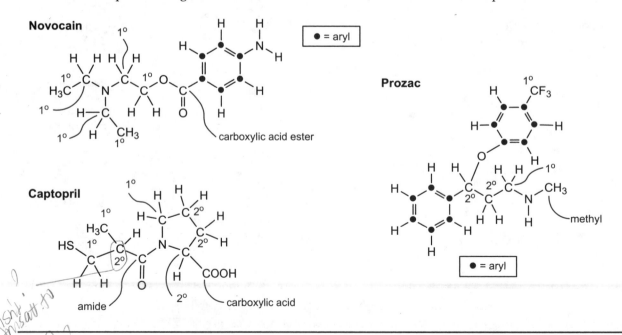

1.26. Some possible sources of errors in a name include the following:

- The substituent can be located at several positions, so a numeral is required to indicate the point of attachment.
- The longest carbon chain containing the principal functional group was not chosen as the root word.
- The quantity of numerals does not match the substituent prefix (di-, tri-, etc.).

a. **Methylheptane:** the methyl group can be placed at several positions along the 7C chain, therefore a numeral is needed at the beginning of the name.

b. **3-Propylhexane:** the longest chain is seven carbon atoms.
This compound should be named **4-ethylheptane.**

c. **2,2-Dimethyl-3-ethylbutane:** the longest chain is five carbon atoms.
This compound should be named **2,2,3-trimethylpentane.**

d. **2-Dimethylpentane:** there should be two numerals in the prefix.
The name should read **2,2-dimethylpentane.**

e. **2-Isopropyl-1-propanol:** the longest chain that also contains the OH group has four carbon atoms.
This compound should be named **2,3-dimethyl-1-butanol.**

f. **Dichloroheptane:** the two chlorine atoms can be placed at several positions along the 7C chain, therefore numerals need to appear at the beginning of the name.

1.27. Follow the procedure given in the solution to Exercise 1.18, incorporating the triple bond at various positions.

6C chain

$$H-C\equiv C-CH_2CH_2CH_2CH_3 \qquad CH_3-C\equiv C-CH_2CH_2CH_3 \qquad CH_3CH_2-C\equiv C-CH_2CH_3$$

1-hexyne   **2-hexyne**   **3-hexyne**

5C chain

$$H-C\equiv C-\underset{\underset{CH_3}{|}}{C}HCH_2CH_3 \qquad H-C\equiv C-CH_2-\underset{\underset{CH_3}{|}}{C}HCH_3 \qquad CH_3-C\equiv C-\underset{\underset{CH_3}{|}}{C}HCH_3$$

3-methyl-1-pentyne   **4-methyl-1-pentyne**   **4-methyl-2-pentyne**

4C chain

$$H-C\equiv C-\underset{\underset{CH_3}{|}}{\overset{\overset{CH_3}{|}}{C}}-CH_3 \qquad \text{3,3-dimethyl-1-butyne}$$

1.28. The structures of isomeric alcohol molecules can be generated from the structures of the corresponding alkanes by inserting an oxygen atom into each unique carbon-hydrogen bond. Therefore, the first step in this exercise is to identify the possible skeletal isomers of C_5H_{12}. This can be done by following the procedures outlined in the solution to Exercise 1.18. There are three such alkanes.

1.28. (continued)

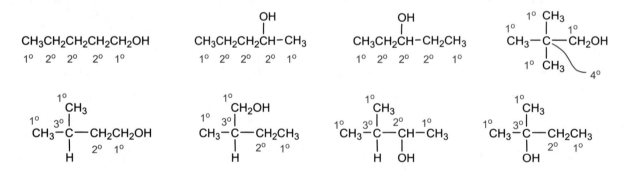

CH₃CH₂CH₂CH₂CH₃

pentane

$$CH_3-\overset{\overset{CH_3}{|}}{\underset{\underset{H}{|}}{C}}-CH_2CH_3$$

2-methylbutane

$$CH_3-\overset{\overset{CH_3}{|}}{\underset{\underset{CH_3}{|}}{C}}-CH_3$$

2,2-dimethylpropane

Inserting an oxygen atom into each unique C–H bond of each alkane generates the possible alcohols.

5   4   3   2   1
CH₃CH₂CH₂CH₂CH₂OH

1-pentanol

$$CH_3CH_2CH_2\overset{\overset{OH}{|}}{C}H-CH_3$$

2-pentanol

$$CH_3CH_2\overset{\overset{OH}{|}}{C}H-CH_2CH_3$$

3-pentanol

$$CH_3-\overset{\overset{CH_3}{3|}}{\underset{\underset{H}{|}}{C}}-\overset{2}{C}H_2\overset{1}{C}H_2OH$$

3-methylbutanol

$$CH_3-\overset{\overset{1}{C}H_2OH}{\underset{\underset{H}{|}}{\overset{2|}{C}}}-CH_2CH_3$$

2-methylbutanol

$$CH_3-\overset{\overset{CH_3}{3|}}{\underset{\underset{H}{|}}{C}}-\overset{2}{\underset{\underset{OH}{|}}{C}}H-\overset{1}{C}H_3$$

3-methyl-2-butanol

$$\overset{1}{C}H_3-\overset{\overset{CH_3}{2|}}{\underset{\underset{OH}{|}}{C}}-CH_2CH_3$$

2-methyl-2-butanol

$$CH_3-\overset{\overset{CH_3}{|}}{\underset{\underset{CH_3}{|}}{C}}-CH_2OH$$

2,2-dimethylpropanol

Classify each carbon atom according to the number of other carbon atoms that are attached. Follow the procedure given in the solution to Exercise 1.13.

CH₃CH₂CH₂CH₂CH₂OH
1° 2° 2° 2° 1°

$$CH_3CH_2CH_2\overset{\overset{OH}{|}}{C}H-CH_3$$
1° 2° 2° 2° 1°

$$CH_3CH_2\overset{\overset{OH}{|}}{C}H-CH_2CH_3$$
1° 2° 2° 2° 1°

1° CH₃
1°   1°
$$CH_3-\overset{\overset{1°\ CH_3}{|}}{\underset{\underset{1°\ CH_3}{|}}{C}}-CH_2OH$$
1° CH₃   4°

$$CH_3-\overset{\overset{1°}{\overset{CH_3}{3°|}}}{\underset{\underset{H}{|}}{C}}-CH_2CH_2OH$$
1°   2° 1°

$$CH_3-\overset{\overset{1°}{\overset{CH_2OH}{3°|}}}{\underset{\underset{H}{|}}{C}}-CH_2CH_3$$
1°   2° 1°

$$CH_3-\overset{\overset{1°}{\overset{CH_3}{3°|}}}{\underset{\underset{H}{|}}{C}}-\overset{2°}{\underset{\underset{OH}{|}}{C}}H-\overset{1°}{C}H_3$$
1°

$$CH_3-\overset{\overset{1°}{\overset{CH_3}{3°|}}}{\underset{\underset{OH}{|}}{C}}-CH_2CH_3$$
1°   2° 1°

1.29. Break each name into its constituent parts and then combine the pieces to form the structure.

a. **2-Phenylethanol**

eth	2 carbon atoms
an	no double or triple bonds
ol	OH group (alcohol) at C1
2-phenyl	phenyl group at C2

b. **1,3-Dibromo-2-pentanol**

pent	5 carbon atoms
an	no double or triple bonds
ol	OH group (alcohol) at C2
1,3-dibromo	two bromine atoms, one at C1 and one at C3

1.29. (continued)

c. **3-Chloropropanol**

prop	3 carbon atoms
an	no double or triple bonds
ol	OH group (alcohol) at C1
3-chloro	chlorine atom at C3

d. **2-Methyl-3-buten-2-ol**

but	4 carbon atoms
en	double bond starting at C3
ol	OH group (alcohol) at C2
2-methyl	CH_3 group at C2

e. **2,2,2-Trifluoroethanol**

eth	2 carbon atoms
an	no double or triple bonds
ol	OH group (alcohol) at C1
trifluoro	three fluorine atoms at C2

f. **2-Amino-2-methylbutanol**

but	4 carbon atoms
an	no double or triple bonds
ol	OH group (alcohol) at C1
2-amino	NH_2 group at C2
2-methyl	CH_3 group at C2

g. **2,3-Butadienol**

but	4 carbon atoms
dien	two double bonds, one starting at C2 and one at C3
ol	OH group (alcohol) at C1

h. **4-Hexynol**

hex	6 carbon atoms
yn	triple bond, starting at C4
ol	OH group at C1

1.30. Break each name into its constituent parts, and then combine the pieces to form the structure.

a. **2-Aminobenzoic acid**

benzoic acid	the carboxylic acid derivative of benzene; the point of attachment of the COOH group defines C1
2-amino	NH_2 group at C2

b. **2,2-Difluorobutanoic acid**

but	4 carbon atoms
an	no double or triple bonds
oic acid	carboxylic acid functional group; the COOH carbon atom is C1 of the chain
2,2-difluoro	two fluorine atoms at C2

1.30.   (continued)

c.   **2,3-Dibromopropanoic acid**

prop	3 carbon atoms
an	no double or triple bonds
oic acid	carboxylic acid functional group, the COOH carbon atom is C1 of the chain
2,3-dibromo	two bromine atoms, one at C2 and one at C3

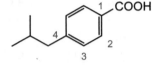

d.   **4-Isobutylbenzoic acid**

benzoic acid	the carboxylic acid derivative of benzene; the point of attachment of the COOH group defines C1
4-isobutyl	CH₂[CH(CH₃)₂] group at C4

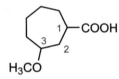

e.   **3-Methoxycycloheptanecarboxylic acid**

cyclohept	7 carbon atoms in a ring
an	no double or triple bonds
carboxylic acid	carboxylic acid functional group; the point of attachment of the COOH group defines C1
3-methoxy	OCH₃ group at C3

f.   **3-Mercapto-4-hexenoic acid**

hex	6 carbon atoms
en	double bond; starts at C4
oic acid	carboxylic acid functional group; the point of attachment of the COOH group defines C1
3-mercapto	SH group at C3

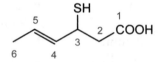

g.   **5-Hydroxy-3-heptenoic acid**

hept	7 carbon atoms
en	double bond; starts at C3
oic acid	carboxylic acid functional group; the COOH carbon atom is C1 of the chain
5-hydroxy	OH group at C5

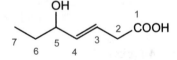

h.   **2,5-Dimethylbenzoic acid**

benzoic acid	the carboxylic acid derivative of benzene; the point of attachment of the COOH group defines C1
2,5-dimethyl	two CH₃ groups, one at C2 and one at C5

1.31.   To assess whether two compounds are identical, confirm that the longest carbon chain (or ring system) is the same size. Next, make certain that the substituents are attached at the same positions on the chain (or ring). Number the chains from each end when checking the substitution patterns. To decide if compounds are isomers, make certain that the molecular formulas are identical. If the formulas are the same, but the compounds are not identical, then the two substances are isomers.

1.31. (continued)

a.   **A** and **C** are identical because each has a chain four carbon atoms long.   Structure **B** is a constitutional isomer of the other two (it is a skeletal isomer).

b.   All of these compounds are the same.   Each has a chain of four carbon atoms with a chlorine atom attached at C2.

c.   **B** and **C** are identical.   Each has a methyl and propyl group attached to an oxygen atom. Structure **A** is a constitutional isomer of the other two (it is a positional isomer).

d.   **A, B,** and **C** are identical.   Each has a chain of four carbon atoms with an OH group attached at C2.   Compound **D** is a constitutional isomer of the others (it is a skeletal isomer) because its longest chain has only three carbon atoms.

1.32.   Follow the procedure given in the solution to Exercise 1.17.

a.   This compound is a ketone with five carbon atoms in the chain, and there are no double or triple bonds: pent/an/one = pentanone.

   The principal functional group (ketone) can be at any position in the chain except C1.   Numbering is started at the end that gives the C=O group the position with the lowest number.   Numbering from the right end places the OH group at C2: 2-pentanone.

   This numbering order places the bromine atom at C3 and the methyl group at C4.   The name of this molecule is **3-bromo-4-methyl-2-pentanone**.

1.32.   (continued)

b.   This compound is a cyclic alcohol with five carbon atoms in the ring, which also has a double bond: cyclopent/en/ol = cyclopentenol.

    The principal functional group (alcohol) defines its position of attachment as C1.   This numbering order means that the double bond begins at C2.   The name of this molecule is **2-cyclopentenol**.

c.   This compound is an aldehyde with five carbon atoms in the chain, and there is a triple bond: pent/yn/al = pentynal.

    The principal functional group (aldehyde) has to be at the end of the chain, and its carbon atom is C1.   Numbering from the right end places the triple bond at C3:   3-pentynal. This numbering order places the phenyl group at C5.   The name of this molecule is **5-phenyl-3-pentynal**.

d.   This compound is a carboxylic acid with five carbon atoms in the chain, and there is a double bond: pent/en/oic acid = pentenoic acid.

    The principal functional group (carboxylic acid) has to be at the end of the chain, and its carbon atom is C1.   Numbering from the right end places the double bond at C4:   4-pentenoic acid.   The two methyl groups are at C3, and the name of this molecule is **3,3-dimethyl-4-pentenoic acid**.

e.   This compound is an alcohol with five carbon atoms in the chain, and there is a double bond: pent/en/ol = pentenol.

    The principal functional group (alcohol) can be at any position in the chain.   Numbering is started at the end that gives the OH attachment position the lowest number.   Numbering from the right end places the OH group at C1, which puts the double bond at C2.   The name of this molecule is **2-pentenol** (by convention, the OH group is at C1 if no number is given).

f.   This compound is a cyclic ketone with four carbon atoms in the ring, which has no double or triple bond: cyclobut/an/one = cyclobutanone.

    The principal functional group defines C1 by its position.   A phenyl group is attached at C2.   The name of this molecule is **2-phenylcyclobutanone**.

1.33.   Break each name into its constituent parts, and then combine the pieces to form the structure.

a.   *N,N*-**Dimethylaniline**

    aniline   amine derivative of benzene; the point of attachment
                        of the NH_2 group defines C1

    N,N-dimethyl   two CH_3 groups are attached to the nitrogen atom

1.33. (continued)

b. **1,2-Diaminocyclohexane**

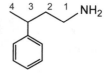

cyclohex	6 carbon atoms in a ring
ane	no double or triple bonds
1,2-diamino	two NH$_2$ groups, one at C1 and one at C2

c. **1-Amino-3-phenylbutane**

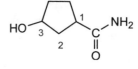

but	4 carbon atoms
ane	no double or triple bonds
1-amino	NH$_2$ group at C1
3-phenyl	phenyl ring at C3

d. **3-Hydroxycyclopentanecarboxamide**

cyclopent	5 carbon atoms in a ring
ane	no double or triple bonds
carboxamide	CONH$_2$ group attached to the ring; its point of attachment defines C1 of the ring
3-hydroxy	OH group at C3

e. *N*-**Methylbutanamide**

but	4 carbon atoms
an	no double or triple bonds
amide	CONH$_2$ group at the end of the chain; its C atom defines C1
N-methyl	a CH$_3$ group replaces a H atom on the N atom

f. **2,4-Dimethylaniline**

aniline	amine derivative of benzene; the attachment point of the NH$_2$ group defines C1
2,4-dimethyl	two CH$_3$ groups, one at C2 and one at C4

1.34. Break each name into its constituent parts, and then combine the pieces to form the structure.

a. **3-Methyl-2-butanone**

but	4 carbon atoms
an	no double or triple bonds
one	ketone functional group at C2
3-methyl	methyl group at C3

b. **1-Chloro-3-hexene-2-one**

hex	6 carbon atoms
ene	double bond starting at C3
one	ketone: the carbonyl group is at C2
1-chloro	chlorine atom at C1

c. **3-Methoxypentanal**

pent	5 carbon atoms
an	no double or triple bonds
al	aldehyde; its carbon atom defines C1
3-methoxy	OCH$_3$ group at C3

1.34.   (continued)

d.   **3-Isopropylcyclohexanecarbaldehyde**

cyclohex	6 carbon atoms in a ring
an	no double or triple bonds
carbaldehyde	aldehyde functional group; its point of attachment defines C1 of the ring
3-isopropyl	CH(CH₃)₂ group at C3

e.   **4-Bromobenzaldehyde**

benzaldehyde	the aldehyde derivative of benzene; the attachment point of the CHO group defines C1 of the ring
4-bromo	bromine atom attached at C4

f.   **3-Ethoxy-2-hexanone**

hex	6 carbon atoms
an	no double or triple bonds
one	ketone; the carbonyl group is at C2
3-ethoxy	OCH₂CH₃ group at C3

g.   **2-Cyclohexenone**

cyclohex	6 carbon atoms in a ring
en	double bond starting at C2
one	carboxyl group defines C1

h.   **3-*tert*-Butylcyclobutanone**

cyclobut	4 carbon atoms in a ring
an	no double or triple bonds
one	carbonyl group defines C1
3-*tert*-butyl	C(CH₃)₃ group at C3

1.35.   Follow the procedure given in the solution to Exercise 1.17.

a.   This compound is a cyclic alcohol with six carbon atoms in the ring, which also has a double bond: cyclohex/en/ol = cyclohexenol.

The principal functional group (alcohol) defines its position of attachment as C1. This numbering order means that the double bond begins at C2: 2-cyclohexenol. This numbering order places the chlorine atom at C6. The name of this molecule is **6-chloro-2-cyclohexenol**.

b.   This compound is a ketone with seven carbon atoms in the chain, which also has a double bond: hept/en/one = heptenone.

The principal functional group (ketone) can be at any position in the chain. Numbering is started at the end that gives the carbonyl group the position with the lowest number. Numbering from the left end places the carbonyl group at C3 and means that the double bond begins at C4: 4-heptene-3-one. A methyl group is attached at C6. The name of this molecule is **6-methyl-4-heptene-3-one**.

1.35.   (continued)

c.   This compound is the carboxylic acid derivative of benzene, which has the root = benzoic acid.

   The position of attachment of the –COOH group defines C1. Numbering around the ring places the nitro group at C2 and the bromine atom at C4. The names of the substituents are arranged in alphabetical order. The name of this molecule is **4-bromo-2-nitrobenzoic acid**.

d.   This compound is an aldehyde with seven carbon atoms in the chain, and there are no double or triple bonds. The name so far is hept/an/al = heptanal.

   The principal functional group (aldehyde) has to be at the end of the chain, and its carbon atom is C1. Numbering from the right end places the two fluorine atoms at C4. The name of this molecule is **4,4-difluoroheptanal**.

1.36.   Break each name into its constituent parts, and then combine the pieces to form the structure.

a.   **1-Chloro-2-hexyne**

hex	6 carbon atoms
yne	triple bond starts at C2
1-chloro	chlorine atom at C1

b.   **1,3-Dicyanobenzene**

benzene	6-membered ring arene
1,3-dicyano	two CN groups, one at C1 and one at C3

c.   **6-Bromohexanoic acid**

hex	6 carbon atoms
an	no double or triple bonds
oic acid	carboxylic acid functional group; its carbon atom defines C1
6-bromo	bromine atom at C6

d.   **2-Nitrobenzaldehyde**

benzaldehyde	the aldehyde derivative of benzene; the attachment point of the CHO group defines C1 of the ring
2-nitro	NO_2 group at C2

e.   **4-Nitrotoluene**

toluene	the methyl derivative of benzene; the attachment point of the methyl group defines C1
4-nitro	NO_2 group at C4

f.   **2-Allyl-6-chlorophenol**

phenol	the OH derivative of benzene; the attachment point of the OH group defines C1
6-chloro	chlorine atom attached to the ring at C6
2-allyl	allyl group attached to the ring at C2

1.36.   (continued)

g.   **4-Pentynol**

pent	5 carbon atoms
yne	triple bond starting at C4
ol	alcohol functional group: OH at C1

h.   **3,5-Heptadienal**

hept	7 carbon atoms
diene	two double bonds; one at C3 and one at C5
al	aldehyde functional group, defines C1

i.   **1-Decanethiol**

dec	10 carbon atoms
an	no double or triple bonds
thiol	SH functional group at C1

j.   **1-Nitropropane**

prop	3 carbon atoms
ane	no double or triple bonds
1-nitro	NO₂ group at C1

k.   **2, 3, 4-Hexanetriol**

hex	6 carbon atoms
an	no double or triple bonds
triol	three OH groups, one each at C2, C3, and C4

l.   **2-Butyl-3-chlorobenzonitrile**

benzonitrile	the nitrile derivative of benzene; the attachment point of the CN group defines C1 of the ring
2-butyl	CH₂CH₂CH₂CH₃ group at C2
3-chloro	chlorine atom at C3

1.37.   Follow the procedure given in the solution to Exercise 1.17.

a.   This compound is the carboxylic acid derivative of benzene, which has the root = benzoic acid.

The position of attachment of the –COOH group defines C1. Numbering around the ring places the OH group at C2 and the allyl group at C6. The names of the substituents are arranged in alphabetical order.   The name of this molecule is **6-allyl-2-hydroxybenzoic acid**.

b.   This compound is the methoxy derivative of benzene, which has the root = anisole.

The position of attachment of the –OCH₃ group defines C1. Numbering of the ring places the F atom at C3. The name of this molecule is **3-fluoroanisole**.

1.37.   (continued)

c.   This compound is the amine derivative of benzene, which has the root = aniline.

The position of attachment of the –NH₂ group defines C1. Numbering around the ring to give the substituents the lowest possible numbers places the methyl group at C2 and the isopropyl group at C4.   The names of the substituents are arranged in alphabetical order.   The name of this molecule is **4-isopropyl-2-methylaniline**.

d.   This compound is the aldehyde derivative of benzene, which has the root = benzaldehyde.

The position of attachment of the –CHO group defines C1. Numbering around the ring to give the substituents the lowest possible numbers places Br atoms at C3 and C4. The name of this molecule is **3,4-dibromo-benzaldehyde**

1.38. Break each name into its constituent parts, and then combine the pieces to form the structure.

a.   **4-(1,1-Dimethylethyl)-4-octanol**

oct	8 carbon atoms
an	no double or triple bonds
ol	alcohol; OH group attached at C4
4-(X)	substituent X; attached at C4

X = substituent: **1,1-Dimethylethyl**	
ethyl	2 carbon atoms; by convention attached through its C1
1,1-dimethyl	two CH₃ groups attached at C1 of the substituent chain

- - - - - - - - - - - - - - -

b.   **3-Chloro-2-(1-hydroxyethyl)-6-nitrophenol**

phenol	the OH derivative of benzene; the attachment point of the OH group defines C1
6-nitro	NO₂ group at C6
3-chloro	chlorine atom at C3
2-(X)	substituent X; attached at C2

X = substituent: **1-hydroxyethyl**	
ethyl	2 carbon atoms; by convention attached through its C1
1-hydroxy	OH group attached at C1 of the substituent chain

1.38.   (continued)

c.   **3-(2-Fluoro-2-propenyl)-5-hepten-2-one**

hept	7 carbon atoms
en	double bond, starts at C5
one	ketone; the carbonyl group is at C2 of the chain
3-(X)	substituent X; attached at C3

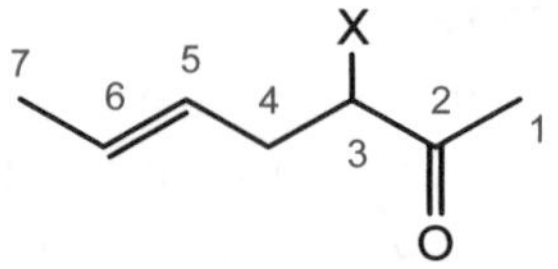

X = substituent: **2-Fluoro-2-propenyl**

prop	3 carbon atoms; by convention attached through its C1
en	double bond, starts at C2 of this substituent chain
yl	a suffix that indicates this group is a substituent
2-fluoroa	fluorine atom at C2 of the substituent chain

- - - - - - - - - - - - - -

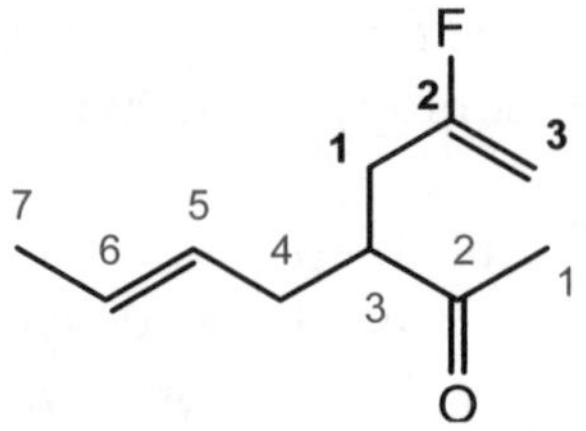

d.   **4-(1-Methylethyl)-5-methyl-3-hexenal**

hex	6 carbon atoms
en	double bond, starts at C3
al	aldehyde, its carbon atom is C1 of the chain
5-methyl	CH₃ group attached at C5
4-(X)	substituent X; attached at C4

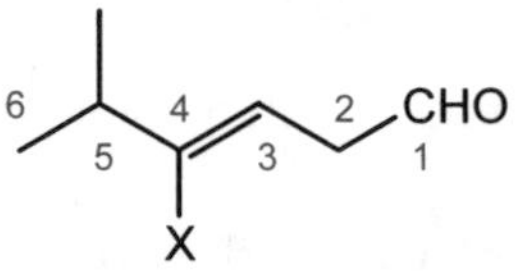

X = substituent: **1-Methylethyl**

ethyl	2 carbon atoms; by convention attached through its C1
1-methyl	CH₃ group attached at C1 of the substituent chain

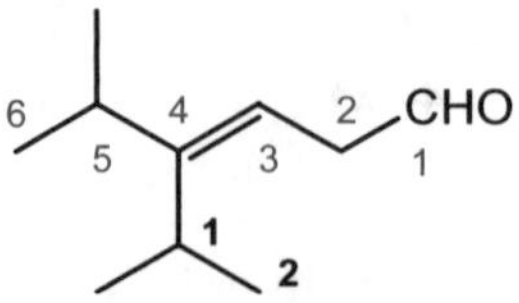

BONDING IN ORGANIC MOLECULES

2.1. To draw the Lewis structure of a given molecule, follow the procedure given in Section 2.1a of the text. Briefly, this procedure is as follows:

1) Draw the expanded structural formula.
2) Calculate the total number of valence electrons.
3) Subtract the number of electrons in the bonds already shown (2 electrons per bond) to calculate how many electrons are to be added to the structural formula.
4) Distribute the remaining electrons in pairs to give eight electrons to each non-hydrogen atom.

a.

Tally the total number of valence electrons and subtract the number of bonding electrons.

C	3 x 4e	= 12e
H	7 x 1e	= 7e
Cl	1 x 7e	= 7e
TOTAL		26e
− 10 bonds x 2e	=	−20e

6e to distribute (3 pairs)

- -

b.

Tally the total number of valence electrons and subtract the number of bonding electrons.

C	4 x 4e	= 16e
H	9 x 1e	= 9e
F	1 x 7e	= 7e
O	1 x 6e	6e
TOTAL		38e
− 14 bonds x 2e	=	−28e

10e to distribute (5 pairs)

- -

c.

Tally the total number of valence electrons and subtract the number of bonding electrons.

C	1 x 4e	= 4e
O	2 x 6e	= 12e
TOTAL		16e
− 4 bonds x 2e	=	− 8e

8e to distribute (4 pairs)

2.1.   (continued)

d.

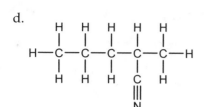

Tally the total number of valence electrons
and subtract the number of bonding electrons.

C	6 x 4e	= 24e
H	11 x 1e	= 11e
N	1 x 5e	= 5e
TOTAL		40e
− 19 bonds x 2e		= −38e

2e to distribute (1 pair)

e.

Tally the total number of valence electrons
and subtract the number of bonding electrons.

C	4 x 4e	= 16e
H	6 x 1e	= 6e
O	1 x 6e	= 6e
TOTAL		28e
− 12 bonds x 2e		= −24e

4e to distribute (2 pairs)

2.2.   Draw the Lewis structure for each compound by following the procedure outlined in the solution
to Exercise 2.1. To calculate the formal charge on each atom, apply the following formula:

Formal charge = # valence electrons − (# bonds + # nonbonding electrons)

For the following compounds, add the unshared electrons to the given structure to generate a Lewis
structure, and then calculate the formal charge on each nonhydrogen atom.

a.

C	3 x 4e	= 12e
H	7 x 1e	= 7e
N	1 x 5e	= 5e
O	2 x 6e	= 12e
TOTAL		36e
− 13 bonds x 2e		= −26e

10e to distribute (5 pairs)

Nitrogen atom: formal charge = 5 - (4 + 0) = +1
Single-bonded oxygen atom: formal charge = 6 - (1 + 6) = −1
Double-bonded oxygen atom: formal charge = 6 - (2 + 4) = 0

---

b.

C	1 x 4e	= 4e
H	2 x 1e	= 2e
N	2 x 5e	= 10e
TOTAL		16e
− 6 bonds x 2e		= −12e

4e to distribute (2 pairs)

Terminal nitrogen atom: formal charge = 5 - (2 + 4) = −1
Other nitrogen atom: formal charge = 5 - (4 + 0) = +1

2.2.   (continued)

c.

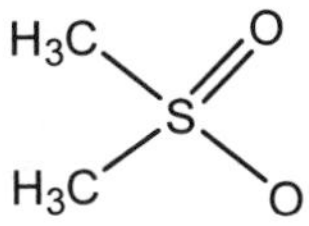

C	2 x 4e	=	8e
H	6 x 1e	=	6e
O	2 x 6e	=	12e
S	1 x 6e		6e
TOTAL			32e
– 11 bonds x 2e		=	–22e

10e to distribute (5 pairs)

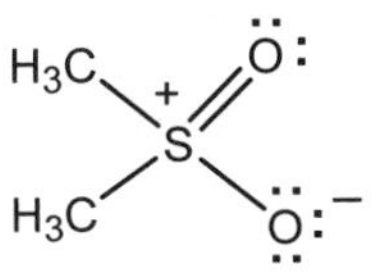

Sulfur atom: formal charge = 6 - (5 + 0) = +1
Oxygen atom: formal charge = 6 - (1 + 6) = –1.
Double-bonded oxygen atom: formal charge = 6 - (2 + 4) = 0

2.3.   To assign the polarity of a bond, look at the atoms attached at the ends of the bond: The element with the greater electronegativity value will be δ– and the other atom will be δ+. The assumption is generally made that carbon-carbon and carbon-hydrogen bonds are not polarized to a significant extent.

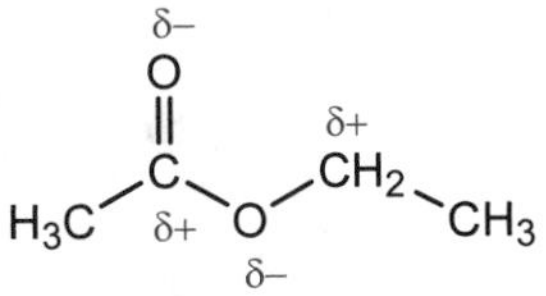

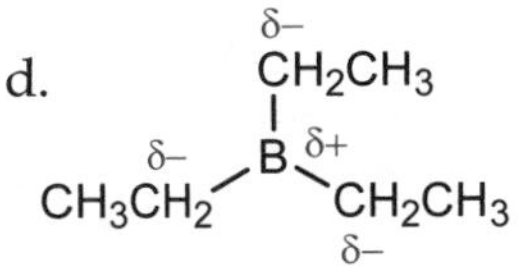

2.4.   Follow the procedure outlined in Example 2.5. Briefly, this procedure is as follows:

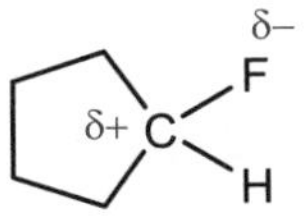

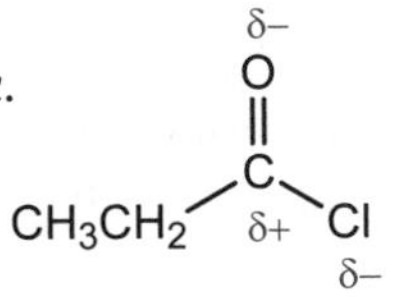

1) Draw a Lewis structure for the molecule.

2) As many times as necessary, reproduce the atom positions of the first structure and move the electrons to generate the other structures. Make certain that you use all of the electrons and calculate the formal charges correctly.

3) Evaluate the forms according to the criteria outlined in Section 2.3c of the text.

a. **Phosphate ion**
    1) Lewis structure (with formal charges):

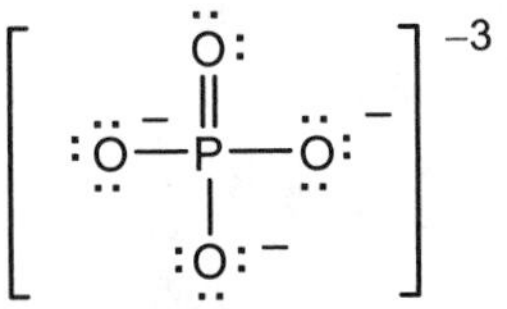

    2) Reproduced structures with redistributed electrons:

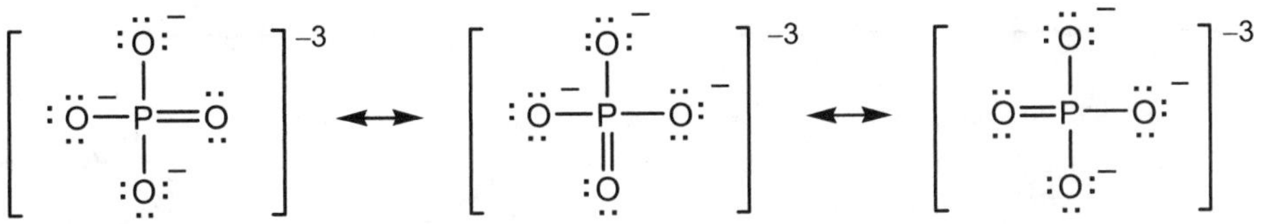

    3) The phosphate ion has resonance structures in which a double bond exists between the P atom and one of the four oxygen atoms. Each single-bonded oxygen atom carries a –1 charge to give an overall charge of –3. The two following structures are less important:

2.4.   (continued)

$$\left[\begin{array}{c} :\overset{..}{\underset{..}{O}}:^- \\ | \\ :\overset{..}{\underset{..}{O}}{}^- \!\!-\! \overset{+}{P} \!-\! \overset{..}{\underset{..}{O}}:^- \\ | \\ :\overset{..}{\underset{..}{O}}:^- \end{array}\right]^{-3}$$ $$\left[\begin{array}{c} \overset{..}{\underset{..}{O}}: \\ \| \\ :\overset{..}{\underset{..}{O}}{}^- \!\!-\! P \!-\! \overset{..}{\underset{..}{O}}:^- \\ \| \\ :\overset{..}{O} \end{array}\right]^{-3}$$

Other less important structures (not shown) would have even fewer bonds and more formal charges or a negative formal charge on the phosphorus atom, which is a less electronegative element than oxygen.

- -

b. **Methyl azide**

1) Lewis structure (with formal charges): $H_3C-\overset{..}{N}=\overset{+}{N}=\overset{..}{\underset{..}{N}}:^-$

2) Reproduced structure with redistributed electrons: $H_3C-\overset{..}{\underset{..}{N}}{}^- -\overset{+}{N}\equiv N:$

3) Methyl azide has two excellent resonance structures (shown above). Any other structure puts only six electrons on one of the nitrogen atoms.

$H_3C-\overset{..}{\underset{..}{N}}{}^- -\overset{2+}{N}=N:^-$ $H_3C-\overset{..}{\underset{..}{N}}{}^- -N\equiv N:^+$ $H_3C-\overset{+}{\underset{..}{N}} -\overset{..}{N}=\overset{..}{\underset{..}{N}}:^-$

N has only six electrons in these structures

2.5.   Draw the Lewis structure for water by following the procedure outlined in Exercise 2.1. Draw the valence bond representation by following the example of methane described in the text.

a. The Lewis structure for water has an octet of electrons around the oxygen atom and a bond to each of the hydrogen atoms.

$$H-\overset{..}{\underset{|}{O}}:$$
$$H$$

b. The valence bond representation for water that employs hybrid orbitals has four sp^3 orbitals arranged in a tetrahedral geometry. Overlap with the hydrogen 1s orbitals creates the O–H bonds. The idealized ∠H–O–H would be 109.5°.

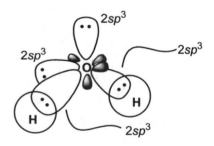

2.6.   To draw the valence bond representation for propane, first generate the Lewis structure to make certain that the electrons are located appropriately. Then decide on the hybridization of the carbon atoms according to whether they have four single bonds (sp^3), two single bonds and a double bond (sp^2), one single bond and a triple bond (sp), or two double bonds (sp). (Hybrid orbitals in the figures on the following pages are shown in black; p orbitals are shown in color.)

2.6. (continued)

To draw the orbital picture for each carbon atom, create overlap between the hybrid orbitals of adjacent carbon atoms to form the carbon–carbon bonds. If multiple bonds are present, show the overlap between adjacent *p* orbitals. Overlap between a hybrid orbital of carbon with the hydrogen 1*s* orbital generates a C–H bond. This molecule has no pi bonds.

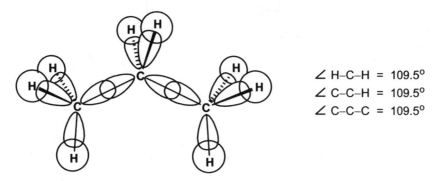

$\angle$ H–C–H = 109.5°
$\angle$ C–C–H = 109.5°
$\angle$ C–C–C = 109.5°

2.7. Follow the procedure outlined in the solution to Exercise 2.6.

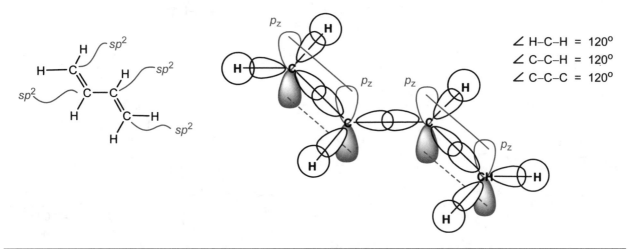

$\angle$ H–C–H = 120°
$\angle$ C–C–H = 120°
$\angle$ C–C–C = 120°

2.8. Follow the procedure outlined in the solution to Exercise 2.6.

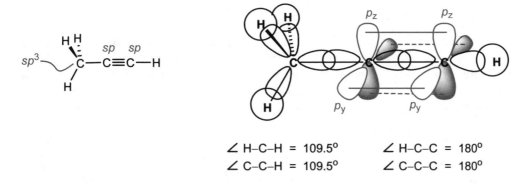

$\angle$ H–C–H = 109.5° $\angle$ H–C–C = 180°
$\angle$ C–C–H = 109.5° $\angle$ C–C–C = 180°

2.9. Follow the procedure outlined in the solution to Exercise 2.6. When a heteroatom is present, assign its hybridization to be the same as the carbon atom(s) to which it is attached. If a heteroatom is attached to carbon atoms with different types of hybridization, assign the heteroatom's hybridization as the one with less *p*-orbital character (that is, *sp²* rather than *sp³*) and place an unshared electron pair in the *p* orbital.

2.9.   (continued)

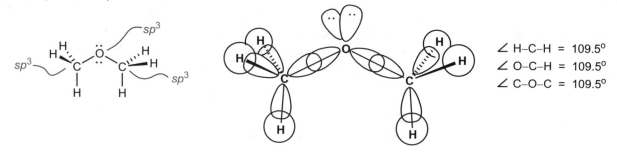

∠ H–C–H = 109.5°
∠ O–C–H = 109.5°
∠ C–O–C = 109.5°

2.10.   Follow the procedure outlined in the solution to Exercise 2.9.

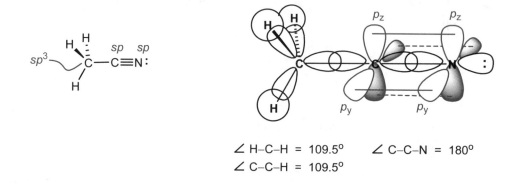

∠ H–C–H = 109.5°   ∠ C–C–N = 180°
∠ C–C–H = 109.5°

2.11.   Follow the procedure outlined in the solution to Exercise 2.9.

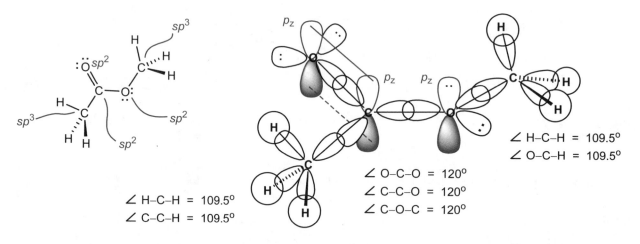

∠ H–C–H = 109.5°
∠ O–C–H = 109.5°

∠ O–C–O = 120°
∠ C–C–O = 120°
∠ C–O–C = 120°

∠ H–C–H = 109.5°
∠ C–C–H = 109.5°

2.12.   In the acetate ion, each oxygen atom has *sp²* hybridization. Two equivalent resonance forms can be drawn for the acetate ion, which explains why the two carbon–oxygen bond lengths are equal. A composite Lewis formula for the acetate ion is equivalent to the valence bond representation, which shows overlap between the *p* orbital of each oxygen atom with the *p* orbital of the carbon atom.

Resonance structures   Composite   Valence bond representation
                                        Lewis structure

2.13.   Hydrogen bonds are formed when a hydrogen atom attached to a heteroatom (usually O, N, or S) is proximal to an unshared pair of electrons on a second heteroatom, usually O or N.

a.   In ethanol, the proton donor is the alcohol OH group, and the acceptor is the oxygen atom of another molecule of ethanol with its unshared pair of electrons.

b.   In pure water, the proton donor is the water OH group, and the acceptor is the oxygen atom of another molecule of water with its unshared pair of electrons.

c.   In a mixture of water and ethanol, the proton donor is the OH group of either water or the alcohol molecule, and the acceptor is the oxygen atom of either water or the alcohol molecule.   Four combinations are therefore possible (ethanol/ethanol; water/water; ethanol/water; water/ethanol).

d.   Dimethyl ether cannot form hydrogen-bonds with itself because there is no proton donor group.   All of the hydrogen atoms are attached to carbon.

e.   In a mixture of water and dimethyl ether, the proton donor must be the OH group of water, but the acceptor can be the oxygen atom of either water or the ether molecule.   Two combinations are possible (dimethyl ether/water and water/water).

2.14.   Follow the procedure outlined in Exercise 2.1.

a.

Tally the total number of valence electrons and subtract the number of bonding electrons.

C	6 x 4e	= 24e
H	12 x 1e	= 12e
O	1 x 6e	= 6e
TOTAL		42e
– 19 bonds x 2e		= –38e

4e to distribute (2 pairs)

2.14.   (continued)

b.

Tally the total number of valence electrons
and subtract the number of bonding electrons.

C	3 x 4e	=	12e
H	4 x 1e	=	4e
Br	1 x 7e	=	7e
N	1 x 5e	=	5e
TOTAL			28e
– 10 bonds x 2e		=	–20e

8e to distribute (4 pairs)

c.

Tally the total number of valence electrons
and subtract the number of bonding electrons.

C	6 x 4e	= 24e
H	10 x 1e	= 10e
O	3 x 6e	= 18e
TOTAL		52e
– 20 bonds x 2e	=	–40e

12e to distribute (6 pairs)

d.

Tally the total number of valence electrons
and subtract the number of bonding electrons.

C	8 x 4e	= 32e
H	8 x 1e	= 8e
O	2 x 6e	= 12e
TOTAL		52e
– 22 bonds x 2e	=	–44e

8e to distribute (4 pairs)

2.15.   To assign the hybridization type for a given carbon atom, look at the number of bonds it forms with the neighboring elements. If the carbon atom forms four single bonds, then it has sp^3 hybridization. If the carbon atom has a double bond and two single bonds, it has sp^2 hybridization.   If the carbon atom has a triple and a single bond or if it has two double bonds, it has sp hybridization.

A heteroatom is normally assigned the same hybridization type as the carbon atom to which it is attached. If a heteroatom is attached to carbon atoms with different types of hybridization, assign the heteroatom's hybridization the same as the carbon atom with less p-orbital character (that is, sp^2 rather than sp^3) and place an unshared electron pair in the p orbital. Hydrogen atoms have only a $1s$ orbital and are not hybridized.

a.

black: sp^3 red: sp^2

b.

black: sp^3 red: sp

2.15.   (continued)

c.

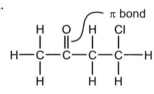

black: *sp³*   red: *sp²*

d.

black: *sp³*   red: *sp²*

2.16.   A single bond between any two atoms is a sigma bond. If two atoms are connected by a double or triple bond, then one bond is a sigma (σ) bond, and any other bonds are pi (π) bonds.

a.

π bond

π bond

π bond

b.

π bond

π bond

π bond

c.

π bond

π bond

π bond

2.17.   Bond lengths correlate with the types of bonds: a single bond between any two atom types is longer than a double bond between the same atom types, which in turn is longer than a triple bond. If the same type of bond is being evaluated (for example, single bonds), then bonds to smaller atoms are shorter. Atomic sizes decrease toward the upper right corner of the periodic table. In the following figures, the shorter bond within a given molecule is indicated by the arrow. For the comparisons between different molecules (c. and d.), only the molecule with the shorter specified bond is shown below.

a.

$H_3C-O-N=O$

b.

$H_3C-C-O-CH_3$

c.

$H-C\equiv N$

d.

CH_3CH_2-Br

2.18.   Follow the procedure outlined in the solution to Exercise 2.2.

a.

Tally the total number of valence electrons and subtract the number of bonding electrons.

C	6 x 4e	= 24e
H	5 x 1e	= 5e
N	1 x 5e	= 5e
O	2 x 6e	12e
TOTAL		46e
− 18 bonds x 2e	=	−36e

10e to distribute (5 pairs)

Nitrogen atom: formal charge = 5 - (4 + 0) = +1
Single-bonded oxygen atom: formal charge = 6 - (1 + 6) = −1
Double-bonded oxygen atom: formal charge = 6 - (2 + 4) = 0

2.18.   (continued)

b.

Tally the total number of valence electrons
and subtract the number of bonding electrons.

C	3 x 4e	= 12e
H	8 x 1e	= 8e
O	1 x 6e	= 6e
S	1 x 6e	6e
TOTAL		32e
– 13 bonds x 2e	=	–26e

6e to distribute (3 pairs)

Sulfur atom: formal charge = 6 - (4 + 2) = 0
Oxygen atom: formal charge = 6 - (2 + 4) = 0

- .

c.

Tally the total number of valence electrons
and subtract the number of bonding electrons.

C	4 x 4e	= 16e
H	7 x 1e	= 7e
O	2 x 6e	= 12e
⊖	1 x 1e	1e
TOTAL		36e
– 13 bonds x 2e	=	–26e

10e to distribute (5 pairs)

Single-bonded oxygen atom: formal charge = 6 - (1 + 6) = –1
Double-bonded oxygen atom: formal charge = 6 - (2 + 4) = 0

2.19.   Follow the procedure outlined in the solutions to Exercises 2.2 and 2.4. If you consider the resonance forms in which each oxygen atom has an octet of electrons and each sulfur atom has ten electrons, you will find that three equivalent resonance forms can be drawn for each species. Sulfur trioxide should have shorter sulfur–oxygen bonds than the sulfite ion because each sulfur–oxygen bond in sulfur trioxide has double-bond character in two resonance forms , whereas each sulfur–oxygen bond in the sulfite ion has double-bond character in only one resonance form. The sulfur atom in sulfur trioxide has a positive formal charge in each resonance form, and the oxygen atoms in the sulfite ion have a greater overall density of negative charge.

Sulfite ion

Sulfur trioxide

2.20. Follow the procedure outlined in the solution to Exercise 2.15. In the structures shown below, non-hydrogen atoms shown in color have sp^2 hybridization; the remaining non-hydrogen atoms have sp^3 hybridization. There are no atoms with sp hybridization in these molecules. The H atoms are not hybridized.

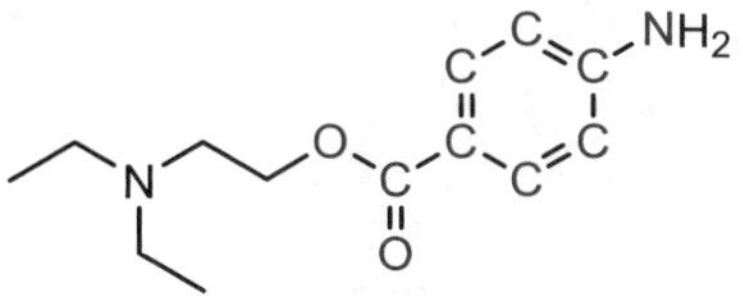

Novocain	**Captopril**	**Prozac**
Local anesthetic	Antihypertensive	Antidepressant

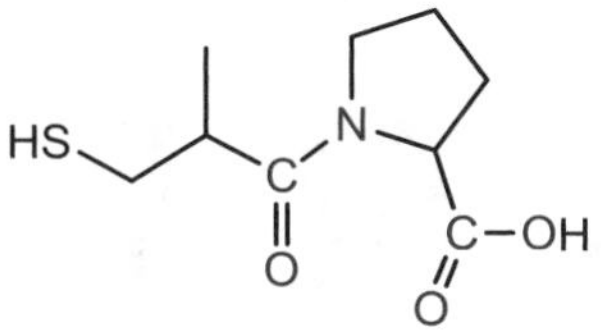

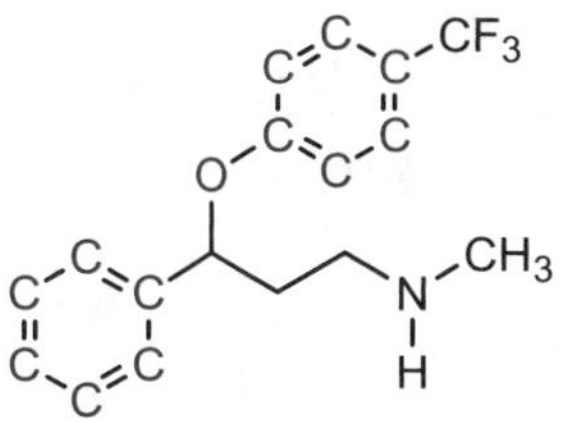

2.21. When drawing resonance structures, nuclei positions must not change, and the number of electrons and atoms must be the same.

a. These are isomers, not resonance structures: a hydrogen atom has changed positions.

b. These are resonance forms; only the electrons have moved.

c. These are isomers, not resonance structures: a hydrogen atom has changed positions.

d. These are resonance forms; only the electrons have moved.

e. These are not resonance forms: the number of atoms in each structure is different.

f. These are resonance forms; only the electrons have moved.

2.22. The best resonance structures are those in which each non-hydrogen atom has an octet of electrons and no formal charge. The next best structures may have formal charges, but the charges are small (+1 or –1), and a negative charge resides on the more electronegative atom.

a. Structure **I** is the most important because each atom has an octet of electrons. Structure **II** is also a reasonable contributor because the negative formal charge is on oxygen, a highly electronegative element. Structure **III** makes an insignificant contribution: the oxygen atom lacks an octet and it carries a positive charge even though it is the most electronegative element in the structure.

b. Structure **I** is more important because each atom has an octet of electrons and there are no charges on the atoms. Structure **II** is also reasonable because the atoms have octets and the negative formal charge is on oxygen, a highly electronegative element.

c. Structure **I** is the most important because each atom has an octet of electrons, and no atom carries a formal charge. Structure **II** is reasonable because each atom has an octet of electrons, but it is less important than **I** because formal charges are present (notice that the oxygen atom carries a positive formal charge: this situation is acceptable if every atom has an octet). Structure **III** has a carbon atom with only six electrons, so it is the least important of these resonance forms.

d. Structures **I** and **II** are equally good and contribute significantly to the resonance hybrid. Structure **III** is much less important because one carbon atom lacks an octet of electrons.

2.23. Follow the procedure outlined in the solution to Exercise 2.2.

a. The Lewis structure for BF$_3$:

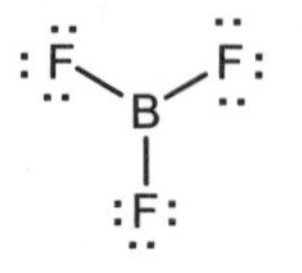

b. The Lewis structure for boron trifluoride etherate: It is not surprising that the adduct between boron trifluoride and diethyl ether is stable because each non-hydrogen atom has an octet of electrons.

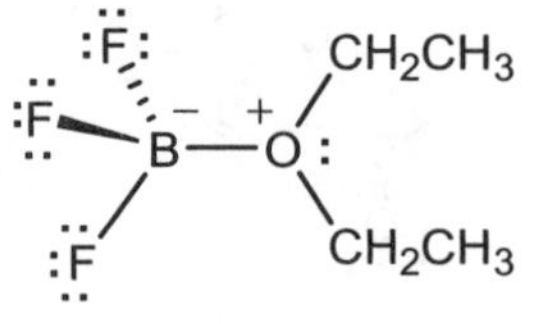

2.24. Follow the procedure outlined in the solution to Exercise 2.4.

a. The anion of 2,4-pentanedione has an overall charge of –1, so each resonance form must have a –1 charge. Three good structures can be drawn in which each atom has an octet of electrons.

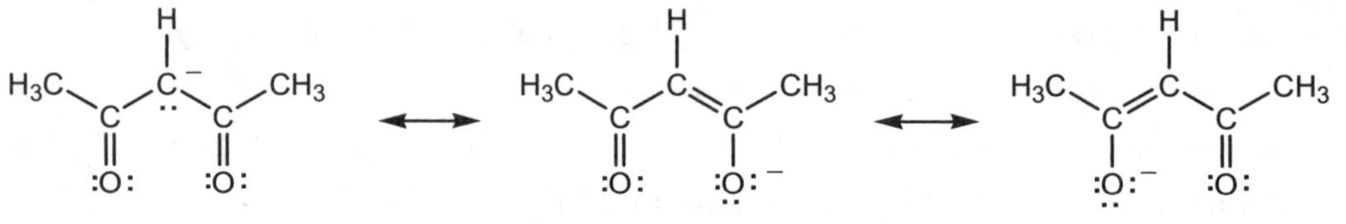

b. For methyl nitrite, two good structures can be drawn in which each non-hydrogen atom has an octet of electrons.

c. Seven good resonance forms can be drawn for the cycloheptatrienyl cation. Each structure has a carbon atom with only six electrons and a positive charge. It is not possible to draw a structure in which each non-hydrogen atom has an octet of electrons.

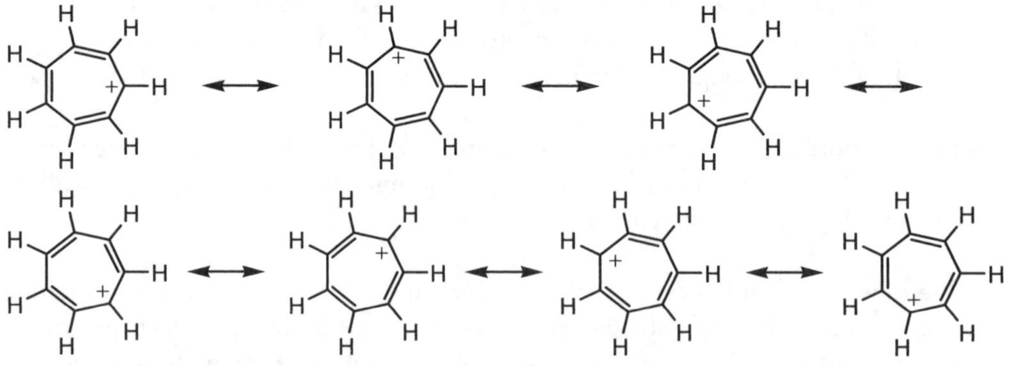

2.25. Triplet methylene has two electrons with the same spin (indicated in the following figure [part (a.)] with arrows pointing in the same direction), and those electrons will repel each other. If each single electron can be placed in a separate orbital, then the repulsive forces will be minimized. Therefore, triplet methylene is likely to have a carbon atom with sp^3 hybridization (tetrahedral).

Singlet methylene has two electrons with opposite spins (indicated in the following figure [part (b.)] with arrows pointing in opposite directions), so they can occupy the same orbital. According to the VSEPR model, three electron pairs (the two pairs that form the C–H bonds plus the unshared pair) will adopt a trigonal arrangement, so the hybridization is likely to be sp^2.

2.25.   (continued)

a.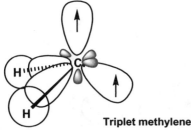

Triplet methylene

b.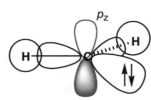

Singlet methylene

2.26.   Follow the procedure and examples outlined in the solution to Exercises 2.9-2.11.

a.   Acrylonitrile: the valence bond representation has a double bond between two carbon atoms so each of those atoms has sp^2 hybridization. The carbon and nitrogen atoms that are connected via a triple bond each have sp hybridization.

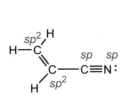 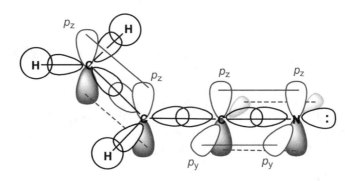

b.   Acetamide: the valence bond representation has a double bond between carbon and oxygen and each atom has sp^2 hybridization. The nitrogen atom is also assigned sp^2 hybridization, and its unshared pair of electrons can partially overlap with the carbonyl carbon atom.

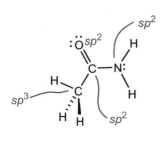 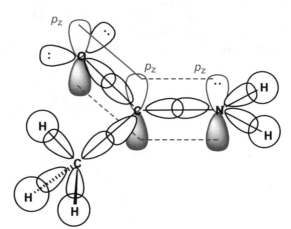

c.   Methyl vinyl ether: the valence bond representation has a double bond between the two carbon atoms, both of which have sp^2 hybridization. The oxygen atom is also assigned sp^2 hybridization, and its p orbital can overlap with the adjacent carbon atom.

2.26. (continued)

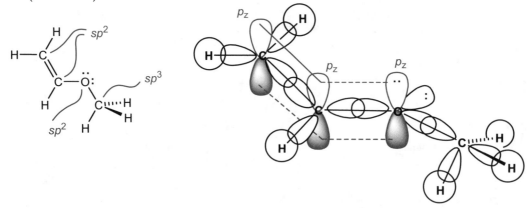

- -.

d. 2-Methylpropene: the valence bond representation has a double bond between two of the carbon atoms that have *sp²* hybridization. The carbon atoms of the two methyl groups have *sp³* hybridization.

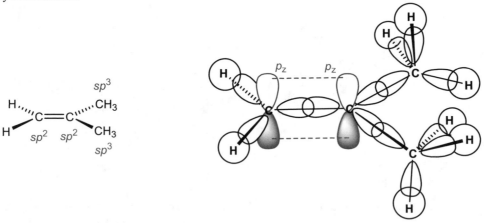

2.27. Bond angles are determined by the hybridization of the middle atom among the three defining the angle. If the central atom has *sp* hybridization, the angle is 180°. If the central atom has *sp²* hybridization, the angle is 120°. If the central atom has *sp³* hybridization, the angle is 109.5°

2.28.   A compound that forms hydrogen bonds among molecules of itself will have strong attractive forces that have to be overcome to reach the boiling point. More heat, therefore, has to be applied to such substances to vaporize them, and their boiling points will be high.

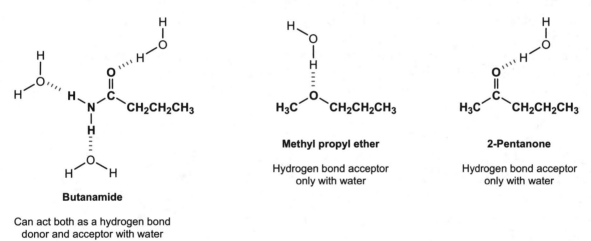

Butanamide forms hydrogen bonds between neighboring molecules, so the boiling point is high (bp = 220 °C)

H₃C–O–CH₂CH₂CH₃   bp = 72 °C

H₃C–C(=O)–CH₂CH₂CH₃   bp = 105 °C

These two molecules cannot form hydrogen bonds because they lack O–H or N–H bonds.

2.29.   A compound that forms hydrogen bonds with molecules of water will be more soluble than those that do not. Even though the ketone and ether molecules can act as hydrogen bond acceptors toward water, butanamide can be both a hydrogen bond donor and acceptor, so it likely will be more soluble.

Butanamide

Can act both as a hydrogen bond donor and acceptor with water

Methyl propyl ether

Hydrogen bond acceptor only with water

2-Pentanone

Hydrogen bond acceptor only with water

2.30.   Hydrogen bonds form when an N–H or O–H group comes within about 3Å of a heteroatom that has an unshared electron pair. In the compounds shown, the amide N–H groups form hydrogen bonds with the electron pairs on the carbonyl group oxygen atoms of the other compound.

2.31.   A hydrogen bond *donor* is a compound that has at least one O–H, N–H, or S–H bond. A hydrogen bond *acceptor* has a heteroatom—usually O or N—that also has an unshared pair of electrons (therefore, R_4N^+ compounds are not in this category). Any compound that has a hydrogen bond donor group is often a hydrogen bond acceptor as well because it has a heteroatom with an unshared pair of electrons.

The compounds shown below that can act as hydrogen bond donors are circled. These will be hydrogen bond donors toward water and toward other like molecules. Those that are *only* hydrogen bond acceptors are enclosed in a box. These will form a hydrogen bond with water in which the O–H bond of water is the donor. They will not, however, be hydrogen bond donors toward like molecules. The alkene (part e.) is neither a hydrogen bond donor nor acceptor.

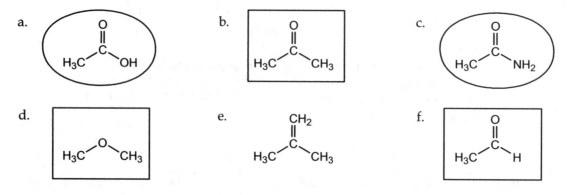

2.32.   Hydrogen bonds form when an N–H or O–H group comes within about 3Å of a heteroatom that has an unshared electron pair. In the reverse turn of a protein, the amide N–H groups form hydrogen bonds with the unshared electron pairs on the carbonyl group of another amide group 3 or 4 amino acids away as shown below in (a.).

It is also possible that the NH group bonded to the carbon atom with R_3 attached can form a hydrogen bond with the carbonyl oxygen atom of the amino acid with R_1 (below, b.)

THE CONFORMATIONS OF ORGANIC MOLECULES

3.1. To construct the sawhorse representation of a molecule, draw a horizontal line that corresponds to the central carbon-carbon bond. Next, attach the two carbon atoms that will become the methyl groups— in the *syn*-periplanar conformation, the methyl group carbon atoms will be eclipsed with each other. Finally, attach hydrogen atoms to each carbon atom.

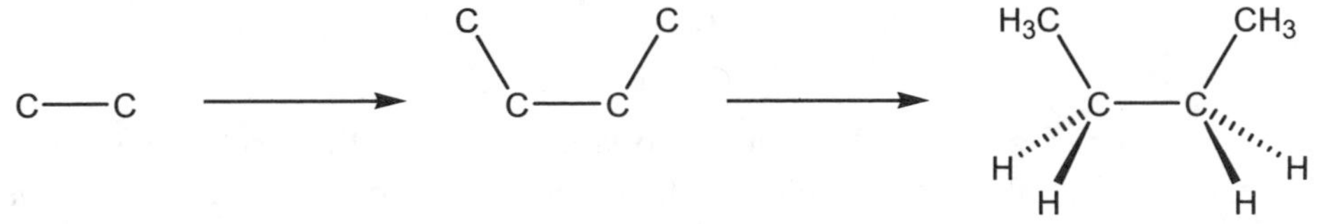

The hydrogen atoms on the two central carbon atoms are eclipsed, but the hydrogen atoms of the methyl groups are staggered with respect to those of the methylene groups. These staggered relationships between a carbon-hydrogen bond and the central carbon-carbon bond are emphasized using colored lines and atoms in the figures below.

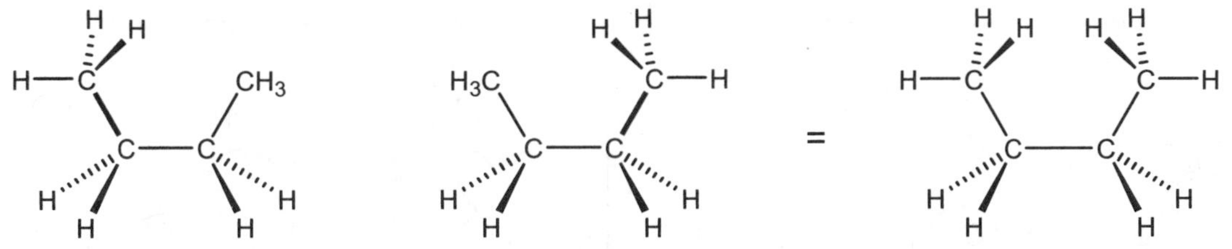

3.2. The most stable conformation of an alkane exists when the carbon-carbon bonds are staggered with respect to the neighboring bonds. If groups bigger than hydrogen atoms are on adjacent carbon atoms, they will tend to be as far as possible from each other. To draw the sawhorse projection, follow the steps shown in the solution to Exercise 3.1. To convert that representation to a Newman projection, sight along the bond that you drew originally, and add the substituents as they appear in the sawhorse projection.

a. **Propane (C1-C2 bond).** This compound is not unlike butane (Exercise 3.1) except that it has one less methyl group. The staggered conformation is the most stable, so the methyl group (C3) should not be eclipsed with any of the hydrogen atoms on C1.

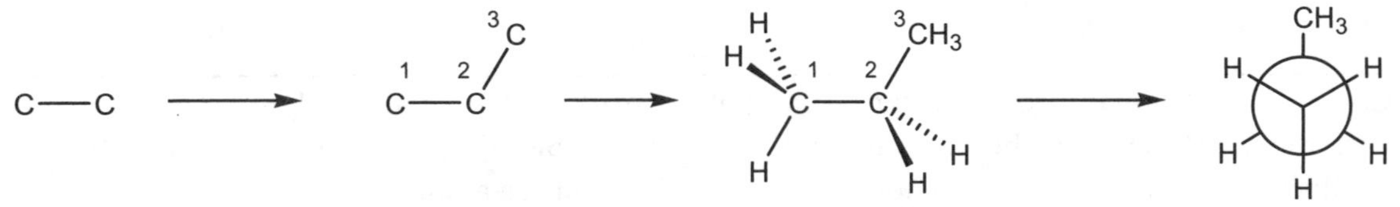

3.2.   (continued)

b.   **2-Methylpropane (C1-C2 bond).** This compound looks like propane [part (a.) of this exercise] except that one hydrogen atom attached to C2 is replaced by a methyl group.

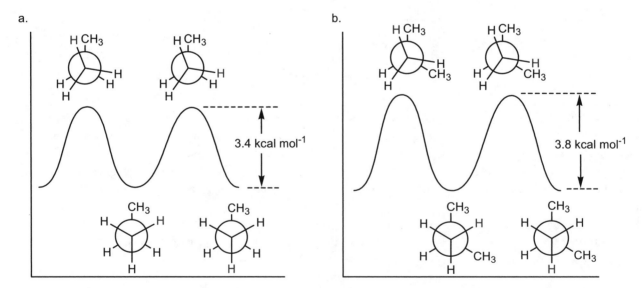

c.   **2,2-Dichlorobutane (C2-C3 bond).** This compound looks like butane except that the two hydrogen atoms at C2 are replaced by chlorine atoms. A chlorine atom is smaller than a methyl group, so the two methyl groups (C1 and C4) will be *anti* to each other in the most stable conformation.

Energy diagrams for propane (a.) and 2-methylpropane (b.) reflect the limiting conformations that can exist. For each compound, only two are unique: staggered and eclipsed. Therefore, a plot of energy versus the dihedral angle looks like the diagram that was presented in the text for the conformations of ethane. The only differences are the magnitudes of the energy values for the limiting conformations. Each H•••H eclipsing interaction is about 1 kcal mol^{-1}, and each H•••CH$_3$ eclipsing interaction is about 1.4 kcal mol^{-1}. Therefore, the difference in energy between the eclipsed and staggered conformations for propane has a value of 1 + 1 + 1.4 = 3.4 kcal mol^{-1}, and that for 2-methylpropane has a value of 1 + 1.4 + 1.4 = 3.8 kcal mol^{-1}. (The hydrogen atoms in color on the following diagrams are included only to show the relative orientations of substituents as rotation occurs about the central carbon–carbon bond.)

3.3.   Construct a Newman projection by sighting along each of the carbon–carbon bonds in turn, and then place the substituents in their appropriate orientations. Sighting along the C1-C2 and the C1-C6 bonds is illustrated in Figure 3.9 of the text. The views that you see along the other bonds are illustrated in the following schemes.

3.3.   (continued)

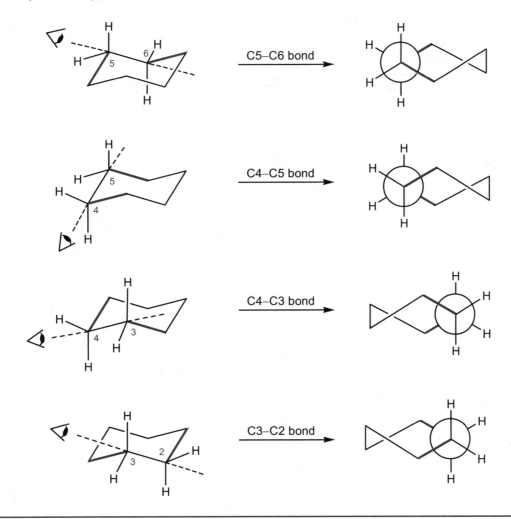

3.4.   The Newman projection along the C1-C2 bond in the boat conformation of cyclohexane is generated by sighting along the C1–C2 bond and placing the substituents in their appropriate orientations. The C2–C3 bond has a pseudo*gauche* orientation (dihedral angle between 0° and 60°) with respect to C1–C6 (give yourself full credit if you said "*gauche*".)

3.5.   Within its cyclic framework, cyclopentane has two *gauche* and two pseudo*gauche* interactions (in a pseudo*gauche* conformation, the dihedral angle is between 0° and 60°). When an axial methyl group is included, two additional *gauche* interactions are created (shown in the following scheme with colored lines). When the methyl group is equatorial, its relationships to the carbon-carbon bonds within the ring are *anti*. Therefore, the conformation having an equatorial methyl group is expected to be more stable.

3.5.   (continued)

Cyclopentane

| gauche | gauche | pseudogauche | pseudogauche |

Methylcyclopentane (axial)   **Methylcyclopentane (equatorial)**

| gauche | gauche | anti | anti |

3.6.   Use Equations 3.4 and 3.5 in the text to calculate K_{eq} and the percentage of species B.

a. $\Delta G° = +5.0$ kcal mol^{-1} $K_{eq} = 2.14 \times 10^{-4}$ %B = 0.02%

b. $\Delta G° = -5.0$ kcal mol^{-1} $K_{eq} = 4.67 \times 10^{3}$ %B = 99.98%

3.7.   For the substituted cyclohexane derivatives given in Table 3.2, use Equations 3.4 and 3.5 in the text to calculate K_{eq} and the percentage of the equatorial isomer ("B").

Substituent	$\Delta G°$	K_{eq}	% equatorial isomer
$-H$	0.0	1.0	50
$-F$, $-CN$	-0.2	1.4	59
$-Cl$, $-Br$, $-C\equiv CH$	-0.5	2.3	70
$-OCH_3$	-0.6	2.8	74
$-OH$	-1.0	5.5	85
$-COOH$	-1.4	10.8	92
$-CH_3$, $-CH=CH_2$	-1.7	18.1	95
$-CH_2CH_3$	-1.8	21.5	96
$-CH(CH_3)_2$	-2.1	35.9	97
$-C_6H_5$	-2.9	140.0	99
$-C(CH_3)_3$	-5.4	9.9×10^4	100

3.8.   To draw the structures for the chair conformations of the 1,4-dimethylcyclohexane isomers, first draw the top view of the structure, using filled and dashed wedges to depict the stereochemical relationships (cis and trans). Then, convert these structural formulas to their chair forms by placing the methyl groups in the proper positions. Draw the form that occurs after a ring flip and decide which of the two is more stable. Remember that a ring flip converts axial to equatorial positions and vice versa.

cis-**1,4-Dimethylcyclohexane**

conformers of equal stability

3.8.   (continued)

trans-1,4-Dimethylcyclohexane

3.9.   Follow the procedures outlined in Example 3.5 and in Exercise 3.8.

a. *trans*-1-Bromo-3-fluorocyclohexane

b. *cis*-2-Ethylcyclohexanecarboxylic acid

c. *trans*-4-Chlorocyclohexanol

d. *cis*-1-*tert*-Butyl-3-methylcyclohexane

e. *cis*-1-Isopropyl-4-phenylcyclohexane

f. *trans*-1,2-Dimethoxycyclohexane

3.10. To generate the chair conformers of 2-methyldecalin, first draw the ring system for the two isomers, *trans-* and *cis*-decalin. Next attach a methyl group at C2 in either the axial or equatorial position of each isomer. For *cis*-decalin, perform a ring flip and decide which conformer is more stable. In the structures of the cis isomers drawn below, the more stable conformation is the one with fewer alkyl substituents in the axial positions. These axial bonds to carbon atoms are indicated with colored lines.

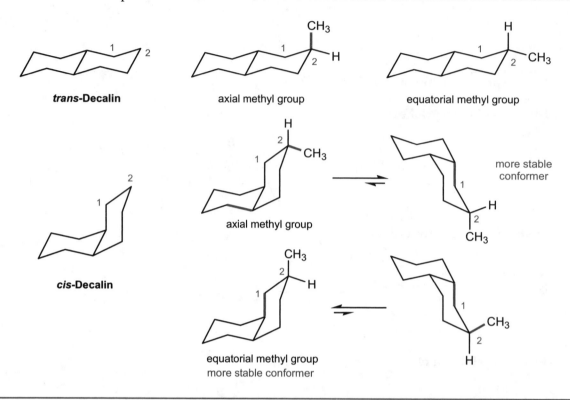

trans-Decalin   axial methyl group   equatorial methyl group

cis-Decalin

axial methyl group

more stable conformer

equatorial methyl group
more stable conformer

3.11. The boat conformation of cycloheptane is generated by flipping one end so that it points in the same direction (up or down) as that in which the other end points. There are six eclipsed interactions that can be identified in the boat conformation (shown in two separate structures for clarity).

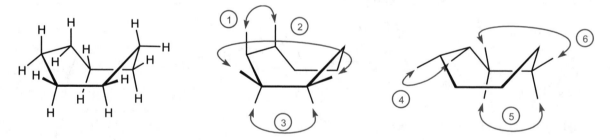

Besides the eclipsed interactions, there are two *gauche* and four pseudo*gauche* interactions, as shown with the colored lines in the following structures.

gauche (dihedral angle = 60°)

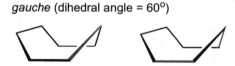

pseudo*gauche* (dihedral angle between 0° and 60°)

3.12. A spiro compound is a bicyclic molecule in which the two rings share a common atom, and the numbers within the brackets indicate how many atoms are in the links that form each ring. Bridged bicyclic compounds have two carbon atoms in common, and the numbers in brackets indicate the numbers of atoms that link these two carbon atoms, which are the bridgehead positions. When one of the numbers within the brackets is zero, a bridged bicyclic compound has fused rings. The total number of carbon atoms in all bicyclic compounds is given by the root name.

a. **Spiro[3.5]nonane**

    nonane: 9 carbon atoms
    spiro: the shared atom is bridged by chains with
                   3 and 5 carbon atoms

b. **Bicyclo[3.2.1]octane**

    octane: 8 carbon atoms
    bicyclo: the bridgehead carbon atoms are linked
                   by chains with 1, 2, and 3 carbon atoms

c. **Bicyclo[3.3.0]octane**

    octane: 8 carbon atoms
    bicyclo: the bridgehead carbon atoms are linked by
                   two chains with 3 carbon atoms each

3.13. Follow the procedure outlined in the solution to Exercise 3.1.

a. **1-Chloropropane (C1–C2 bond).** This compound is like propane (Exercise 3.1a) except that a chlorine atom replaces one of the terminal methyl groups. The *anti* conformation places the C3 methyl group and chlorine atom as far from each other as possible.

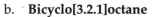

There are two eclipsed conformations: one has the two largest groups (Cl and CH_3) pointing in the same direction. The other has the chlorine atom in front of a hydrogen atom in the Newman projection.

The *gauche* conformation places the methyl group and the chlorine atom 60° apart as you look along the C1–C2 bond in the Newman projection.

3.13.   (continued)

b.   **2-Iodobutane (C2–C3 bond).**   The *anti* conformation is that in which the methyl groups attached to C2 and C3 are as far from each other as possible.   The iodine atom is *gauche* to the C4 methyl group.

methyl groups *anti*

There are three eclipsed conformations: one in which a methyl group is in front of the other methyl group, one in which the iodine atom is in front of the C4 methyl group, and one in which the hydrogen atom on C2 is in front of the C4 methyl group.

methyl groups eclipsed

iodine atom and methyl group eclipsed

hydrogen atom and methyl group eclipsed

The conformation in which the methyl groups are *gauche* puts the iodine atom either *gauche* or *anti* to the methyl group. The conformation with the iodine atom *anti* to the C4 methyl group is more stable than the other because it puts the two largest groups farthest apart.

methyl groups *gauche*
I and CH$_3$ *gauche*

methyl groups *gauche*
I and CH$_3$ *anti*

3.14.   To calculate $\Delta G°$ for an equilibrium process that interconverts **A** and **B**, first determine the value of K_{eq}, which is equal to the concentration of **B** (*gauche* conformation) over the concentration of **A** (*anti* conformation).

$K_{eq} = [\mathbf{B}]/[\mathbf{A}] = [gauche]/[anti] = 11/89 = 0.1236$ $\Delta G° = -2.303RT\log K = -2.303(1.986)(298)\log(0.1236)$

$$= 1238 \text{ cal mol}^{-1}$$

$$= 1.24 \text{ kcal mol}^{-1}$$

Compare this value (1.24 kcal mol^{-1}) with those calculated in Exercise 3.7 to see if these results are reasonable (Yes: the –COOH group has a slightly larger $\Delta G°$ value of 1.4 and the ratio is 8:92).

3.15. Follow the procedure outlined in the solution to Exercise 3.1.

a. **Propanal (C2–C3 bond).** This compound is like propane (Exercise 3.2) except that an aldehyde group takes the place of the methyl group. The staggered conformation is the more stable and the eclipsed conformation is the least stable.

most stable conformer

least stable conformer

b. **1-Pentyne (C3–C4 bond).** This compound has the same conformations as butane (Figure 3.3) except that a methyl group has been replaced by a –C≡CH group. The *anti* conformation is the most stable and the *syn*-periplanar conformation is the least stable.

most stable conformer

least stable conformer

3.16. Break each name into its constituent parts, and then combine the pieces to form the structure.

a. *cis*-**1-Bromo-2-methylcyclopentane**

cyclopentane	five-membered ring of carbon atoms with no double or triple bonds
1-bromo	a bromine atom is attached to and defines C1
2-methyl	a CH₃ group is attached at C2
cis	the bromine atom and the methyl group are on the same side of the ring plane

b. **2,2-Difluorocyclohexanone**

cyclohex	six-membered ring of carbon atoms
an	no double or triple bonds
one	ketone functional group; its carbon atom defines C1
2,2-difluoro	two F atoms are attached at C2

3.16. (continued)

c. **2-Cyclohexenol**

cyclohex	six-membered ring of carbon atoms
en	one double bond, starts at C2
ol	alcohol functional group; an OH group is attached to and defines C1

d. *cis*-**3-Chlorocyclobutanol**

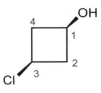

cyclobut	four-membered ring of carbon atoms
an	no double or triple bonds
ol	alcohol functional group; an OH group is attached to and defines C1
3-chloro	a Cl atom is attached at C3
cis	the chlorine atom and the OH group are on the same side of the ring plane

e. *trans*-4-*tert*-**Butylcyclohexanecarboxylic acid**

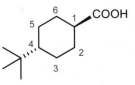

cyclohexane	six-membered ring of carbon atoms with no double or triple bonds
carboxylic acid	carboxylic acid functional group; a COOH group is attached to and defines C1
4-*tert*-butyl	a –C(CH₃)₃ group is attached at C4
trans	the *tert*-butyl and COOH groups are on opposite sides of the ring plane

---

3.17. To evaluate the stereochemical and isomeric relationships that exist for a pair of compounds, consider the position of attachment for each substituent, determine the stereochemical relationship of the substituents (cis or trans), and then evaluate the conformational relationships (axial, equatorial, staggered, *gauche*, eclipsed, *anti*, *syn*-periplanar). If the point of attachment of substituents is different, then the compounds are structural isomers. If the stereochemical relationships of substituents differ, then the compounds are geometric isomers. If the conformational relationships are different, then the substances are conformers.

a. The attachment points of the substituents are the same (1,2), and their orientations are cis in compounds **A** and **B** and trans in **C**. Therefore, **A** and **C** and **B** and **C** represent sets of geometric isomers. **A** and **B** are identical.

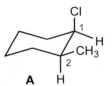 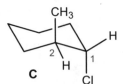

A	**B**	**C**
1-chloro (axial)	1-chloro (axial)	1-chloro (axial)
2-methyl (equatorial)	2-methyl (equatorial)	2-methyl (axial)
cis	cis	trans

b. The attachment points of the substituents are the same in each compound (1,2) and there is no stereochemistry to consider. Only the conformation of **C** is different. Therefore, **A** and **B** are identical and the pairs **A** & **C** as well as **B** & **C** represent sets of conformers.

3.17.   (continued)

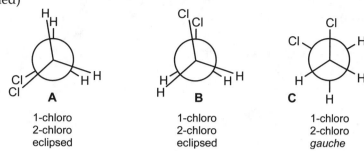

A
1-chloro
2-chloro
eclipsed

B
1-chloro
2-chloro
eclipsed

C
1-chloro
2-chloro
gauche

c.   The attachment points for the substituents are the same in compounds **A** and **C** (C1 and C3), but compound **B** has the substituents attached at C1 and C4. Compounds **A** and **C** differ in the stereochemical relationships of their substituents (**A** is trans, and **C** is cis). Therefore, **A** and **B** are structural isomers and **A** and **C** are geometric isomers. If the compounds are isomers, then we do not have to consider conformations.

A
1-hydroxy (axial)
3-methyl (equatorial)
trans

B
1-hydroxy (equatorial)
4-methyl (axial)
cis

C
1-hydroxy (equatorial)
3-methyl (equatorial)
cis

3.18.   Follow the procedure outlined in Example 3.5.

a. *cis*-**1-Bromo-3-ethylcyclohexane**

more stable conformer

b. *cis*-**4-Hydroxycyclohexanecarboxylic acid**

more stable conformer

c. *trans*-**2-Isopropylcyclohexanol**

more stable conformer

d. *trans*-**3-Chloro-1-bromocyclohexane**

more stable conformer

3.19. The nitrile group is linear, so it has minimal steric requirements. The 1,3-diaxial interactions are therefore slight when a CN group is in the axial position. (Remember that "1,3-diaxial interactions" is a relative term referring to the steric repulsion between the substituents that are three carbon atoms apart.) A methyl group occupies a larger volume, which leads to greater repulsive interactions with the axial hydrogen atoms at C3 and C5.

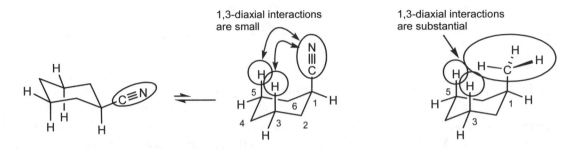

3.20. For a cyclohexane ring with three substituents, follow the same procedure that was outlined in the solution to Exercise 3.18.

a.

b.

c.

d.

3.21. Follow the procedure outlined in the solution to Exercise 3.12. To carry out the conformational analysis of a bicyclic compound, consider the typical conformations that exist for each of the different rings. A six-membered ring will be in the chair form and atoms in the other ring will be in its equatorial positions, if possible. A five-membered ring will be in the open-envelope form, and a four-membered ring has a creased-square shape. Bridged bicyclic compounds with three bridges will most likely be locked into conformations based on the overall shape of the molecule. Numbering of the carbon skeleton is only included in the structures having substituent groups (to show their placements).

a.   **Spiro[3.5]nonane**

nonane	9 carbon atoms
spiro	the shared atom is bridged by chains with 3 and 5 carbon atoms

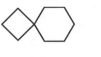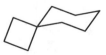

3.21.   (continued)

b.   ***trans*-Bicyclo[3.3.0]octane**

octane   8 carbon atoms

bicyclo   the bridgehead carbon atoms are linked
                        by two chains with 3 carbon atoms each

trans   the hydrogen atoms at the ring junction will
                        be on opposite sides of the ring plane

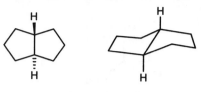

c.   **Spiro[2.4]heptane**

heptane   7 carbon atoms

spiro   the shared atom is bridged by chains with
                        2 and 4 carbon atoms

d.   ***trans*-2,2-Dibromobicyclo[4.4.0]decane**

decane   10 carbon atoms

bicyclo[4.4.0]   the bridgehead carbon atoms are linked
                        by two chains with 4 carbon atoms each

trans   the hydrogen atoms at the ring junction will
                        be on opposite sides of the ring plane

2,2-dibromo   two bromine atoms are attached at C2
                        (see Section 3.4a for the numbering scheme)

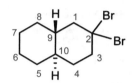

e.   **2-Methylspiro[3.3]heptane**

heptane   7 carbon atoms

spiro[3.3]   the shared atom is bridged by chains with
                        3 carbon atoms each

2-methyl   a methyl group is attached at C2 (numbering
                        starts adjacent to the spiro carbon atom,
                        which cannot bear a substituent)

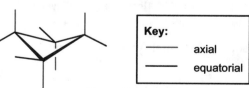

f.   **Bicyclo[4.3.1]decane**

decane   10 carbon atoms

bicyclo[4.3.1]   the bridgehead carbon atoms are linked by
                        three chains with 1, 3, and 4 carbon atoms

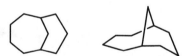

3.22.   The process for drawing the isomers and conformers of a four-membered ring is not much
different than that used for cyclohexane derivatives. Attach the substituents at the appropriate carbon
atoms so that each combination of *cis* and *trans* geometries
is formed. Then perform a ring flip. Conformers with a
greater number of equatorial substituents are expected to
be more stable (see the structure at the right to see how to
locate the axial and equatorial substituents).

> Key:
> _____ axial
> _____ equatorial

The structures of the dimethylcyclobutane isomers
are shown following:

1,1-Dimethylcyclobutane

no other
conformations

3.22. (continued)

cis-1,2-Dimethylcyclobutane

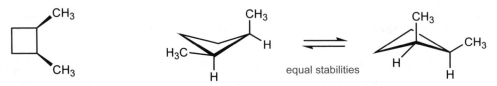

equal stabilities

trans-1,2-Dimethylcyclobutane

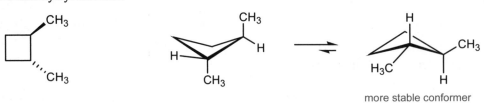

more stable conformer

cis-1,3-Dimethylcyclobutane

more stable conformer

trans-1,3-Dimethylcyclobutane

equal stabilities

3.23. For a bridged bicyclic hydrocarbon, the root name reflects the total number of carbon atoms in the parent structure (the first step is to erase the methyl and isopropyl substituents from the structures presented in this exercise). Next, add the prefix bicyclo-. Within the brackets, list in decreasing numerical values the numbers of carbon atoms needed to link the two bridgehead carbon atoms (indicated by the colored dots in the structures shown below).

(Key: ● = bridgehead carbon atom)

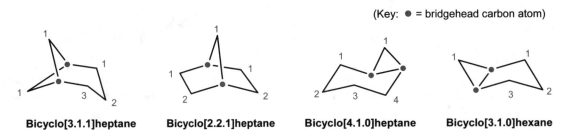

Bicyclo[3.1.1]heptane   **Bicyclo[2.2.1]heptane**   **Bicyclo[4.1.0]heptane**   **Bicyclo[3.1.0]hexane**

Isoprene is 2-methyl-1,3-butadiene.

2-Methyl-1,3-butadiene
    buta   4 carbon atoms
    diene   two double bonds, starting at C1 and C3
    2-methyl   a CH_3 group is attached at C2

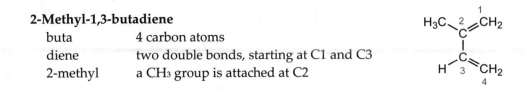

3.24. Follow the process illustrated in the solution to Exercise 3.20.

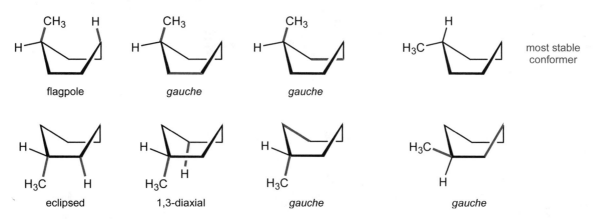

more stable conformer

3.25. In the boat conformation of cyclohexane, there are two types of carbon atoms: the four in the plane and the two out of the plane. Four unique conformations can be created by placing a substituent in the axial or equatorial position of each of the two unique carbon atom environments (the possible positions are numbered in the structure shown at the right).

The unfavorable interactions that exist in each conformer are indicated below each structure and the problematic bonds are shown in color. Only one conformer lacks unfavorable interactions between the methyl group and bonds in the ring, and this conformer is the most stable (top right, below).

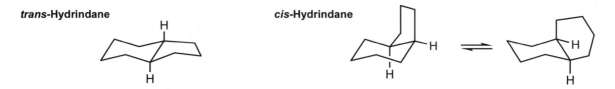

flagpole   *gauche*   *gauche*   most stable conformer

eclipsed   1,3-diaxial   *gauche*   *gauche*

3.26. Like decalin, the *trans* isomer of hydrindane has only a single conformation. The *cis* isomer can undergo a ring flip, and the two conformations have equal energies because the five-membered ring is attached in both axial and equatorial positions to the six-membered ring.

trans-Hydrindane   **cis-Hydrindane**

3.27. *trans*-Decalin has no *gauche* interaction besides the ones already present within each six-membered rings. *cis*-Decalin, in contrast, has three *gauche* interactions between bonds that lie in *separate* rings. Two sets of 1,3-diaxial interactions also exist between the carbon-carbon bonds in one ring and two of the hydrogen atoms attached to the other ring, as shown below. All of these steric interactions that exist in the *cis* isomer make it less stable than *trans*-decalin.

gauche interactions between bonds in adjacent rings   1,3-diaxial interactions between the rings

3.28.   The most stable conformation of a cyclohexane derivative is the chair form with the larger groups in equatorial positions. Because a methyl group is larger than a carboxylic acid group, as indicated by the greater value of $\Delta G°$ given in Table 3.2, the methyl groups occupy the equatorial positions in Kemp's triacid.

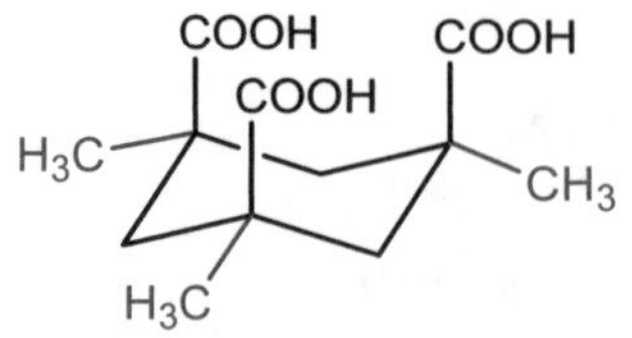

3.29.   The most stable conformation of a cyclohexane derivative is the chair form in which groups larger than hydrogen atoms occupy the equatorial positions. The most stable isomer of inositol is expected to have all six OH groups in the equatorial positions (shown in color in the following structure).

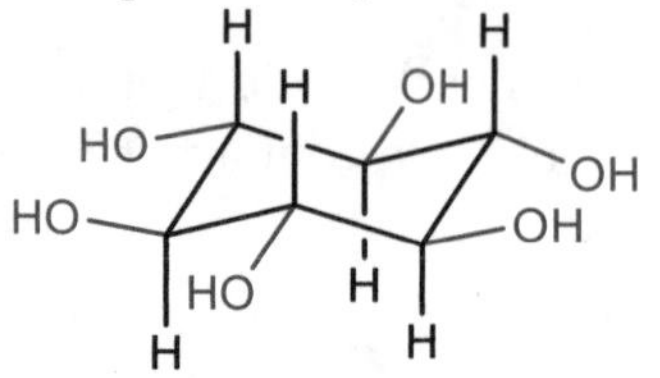

THE STEREOCHEMISTRY OF ORGANIC MOLECULES

4.1. To interpret a compound's name when stereochemical descriptors are present, first draw the structural formula using the procedures outlined in Chapter 1. For alkenes, the prefix *cis* or *trans* indicates that the double bond is disubstituted: *cis* means that the hydrogen atoms are on the same side of the line connecting the double-bonded carbon atoms, and *trans* means they are on opposite sides of the C=C bond.

a. *cis*-**3-Octene**

    oct   eight carbon atoms

    ene   carbon-carbon double bond starting at C3

    cis   the hydrogen atoms are on the same side
                     of the C=C bond

b. *trans*-**1,1-Dichloro-2-pentene**

    pent   five carbon atoms

    ene   carbon-carbon double bond starting at C2

    1,1-dichloro   two Cl atoms at C1

    trans   the hydrogen atoms are on opposite sides
                     of the C=C bond

c. *trans*-**2-Hexenal**

    hex   six carbon atoms

    en   carbon-carbon double bond starting at C2

    al   aldehyde; its carbon atom defines C1

    trans   the hydrogen atoms are on opposite sides
                     of the C=C bond

4.2. The stereochemical classification of an alkene double bond is made as follows: for the pair of substituents at each end of the double bond, one is assigned a higher priority ranking based on the atomic number of the atom (or atoms) that constitute the substituent. In the answers given below, the group that has the higher priority or the portion that gives a substituent its higher priority is shown in color. Functional groups that have multiple bonds are first converted to their single bond equivalents (Figure 4.2).

    When the higher priority groups at the different ends of the double bond are on the same side of the line connecting the double-bonded carbon atoms, the prefix (Z) is used; when the higher priority groups at the ends of the double bond are on opposite sides of the C=C bond, then the stereochemistry of the double bond is designated (E). When identical groups are attached to the carbon atom at one or both ends of the double bond, then the alkene does not exist in isomeric forms.

    Follow the procedures outlined in Examples 4.1 and 4.2 in the text. (AN = atomic number).

4.2.   (continued)

a.

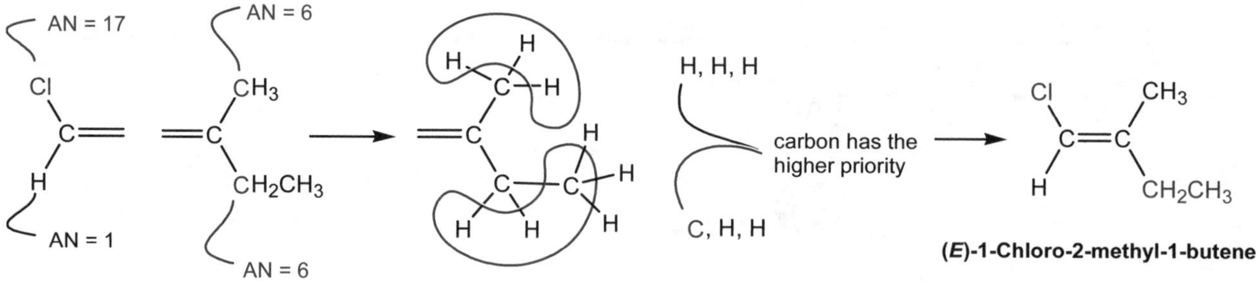

(E)-1-Chloro-2-methyl-1-butene

- -

b.

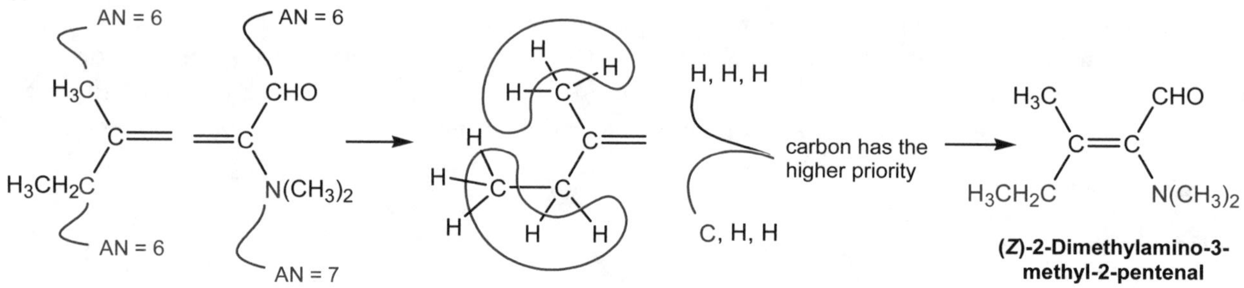

**(Z)-2-Dimethylamino-3-
methyl-2-pentenal**

- -

c.

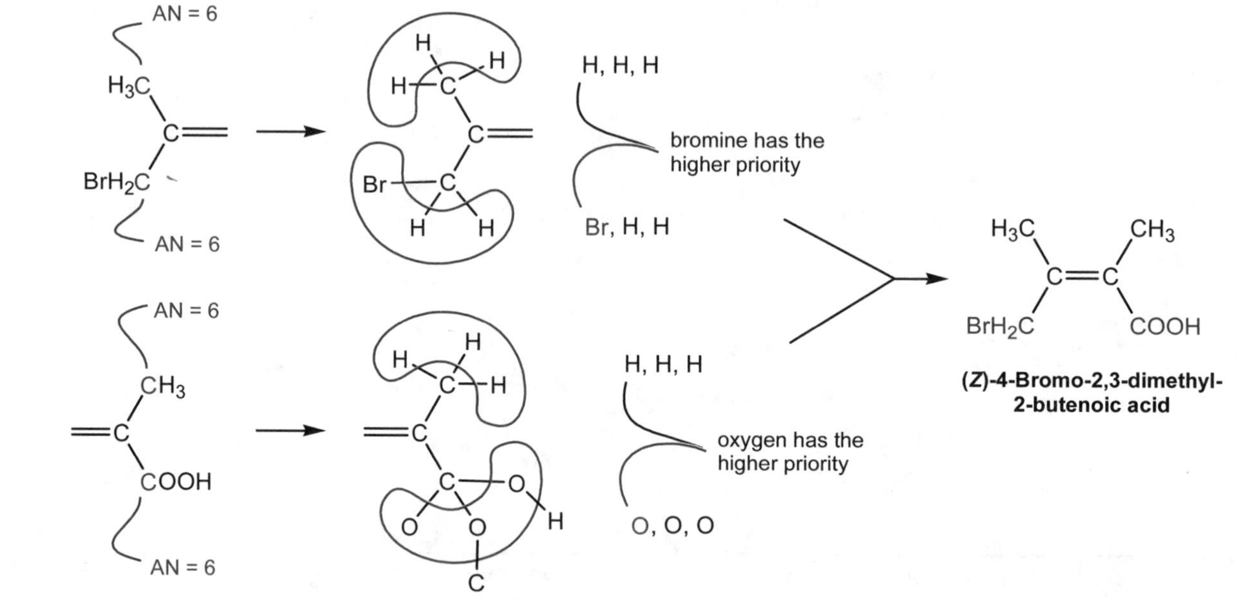

**(Z)-4-Bromo-2,3-dimethyl-
2-butenoic acid**

4.3.   To interpret a compound's name when stereochemical descriptors are present, first draw the structural formula using the procedures outlined in Chapter 1. A prefix of (E) or (Z) denotes that the Cahn-Ingold-Prelog system is being used to specify the double-bond stereochemistry according to which sides have the higher priority groups attached, as shown in the solution to Exercise 4.2.

a.   **(Z)-3-Heptene**

hept	seven carbon atoms	
ene	carbon-carbon double bond starting at C3	
(Z)	the higher priority groups at the ends of the π bond are on the same side of the C=C bond	

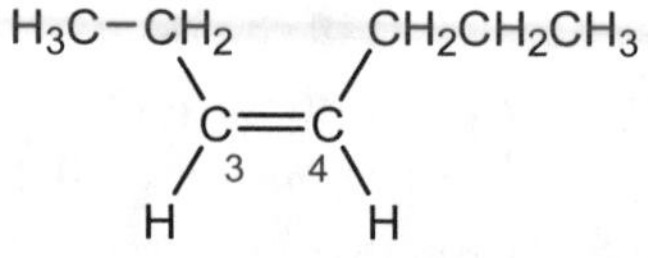

4.3.   (continued)

b.   **(Z)-3-Methoxy-2-octenal**

oct	eight carbon atoms
en	carbon-carbon double bond starting at C2
al	aldehyde; its carbon atom defines C1
3-methoxy	the OCH$_3$ group is attached at C3
(Z)	the higher priority groups at the ends of the π bond are on the same side of the C=C bond

c.   **(E)-1,3-Dichloro-2-methyl-2-hexene**

hex	six carbon atoms
ene	carbon-carbon double bond starting at C2
2-methyl	the CH$_3$ group is attached at C2
1,3-dichloro	two chlorine atoms, one at C1 and one at C3
(E)	the higher priority groups at the ends of the π bond are on opposite sides of the C=C bond

d.   **(E)-3,4-Dibromo-3-heptene**

hept	seven carbon atoms
ene	carbon-carbon double bond starting at C3
3,4-dibromo	two bromine atoms, one at C3 and one at C4
(E)	the higher priority groups at the ends of the π bond are on opposite sides of the C=C bond

4.4.   To evaluate whether compounds are identical or stereoisomers, make a model of the first compound, and then see if it superimposes on a model of the second. If they are superimposable, then the molecules are identical.   Only the molecules in (c.) are stereoisomers.

a.

b.

c.

4.5.   To draw the enantiomer of a compound, first make certain that it is chiral (not superimposable on its mirror image). Then change the configuration of every stereogenic center by switching any two of the substituents attached to each chiral center.

4.5. (continued)

a.

b.

c.

4.6. To convert the structure of a D compound to its L isomer, interchange any two groups on each asymmetric carbon atom. In glyceraldehyde there is only one stereogenic center, so switching two groups (H and OH, below) produces the enantiomeric structure.

D-Glyceraldehyde   **L-Glyceraldehyde**

4.7. The stereochemical classification of a stereogenic center is made by first assigning a priority ranking to each substituent based on its atomic number. (Functional groups with multiple bonds are first converted to their single bond equivalents according to Figure 4.2.) Then follow the procedures outlined in Examples 4.3 and 4.4.

a.

b.

4.8. To draw the structure of a chiral molecule, first draw its structural formula (ignoring stereochemistry) using the procedures outlined in Chapter 1. Then use the Cahn-Ingold-Prelog system to designate a priority for each substituent attached to any carbon atom with four different groups. If the carbon atom has an attached hydrogen atom, include it in the structure so that it points away from you, then assign the configuration, R or S. If that configuration is incorrect, simply exchange any two groups, which inverts the configuration.

4.8.   (continued)

a.   **(R)-2-Bromopentanoic acid**

pent	five carbon atoms
an	no double or triple bonds
oic acid	carboxylic acid; the carbon atom of the COOH group defines C1
2-bromo	bromine atom at C2

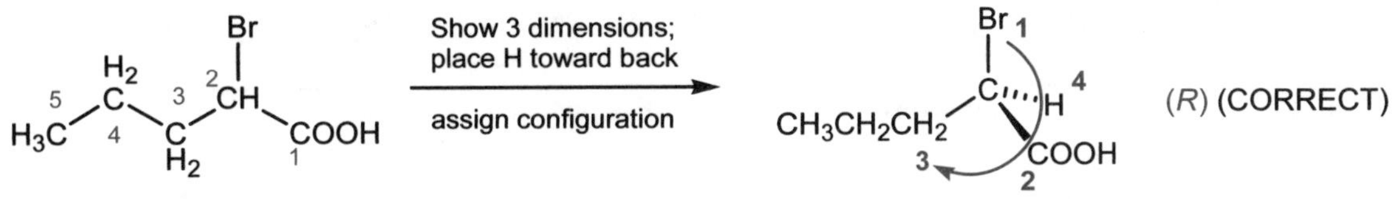

b.   **(S)-3-Hexanol**

hex	six carbon atoms
an	no double or triple bonds
ol	alcohol; an OH group is attached at C3

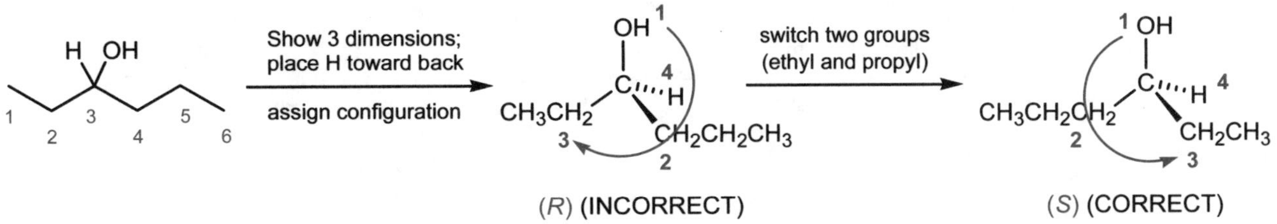

c.   **(R)-3-Chloro-2-butanone**

but	four carbon atoms
an	no double or triple bonds
2- one	ketone; the carbonyl group is at C2
3-chloro	a Cl atom is attached at C3

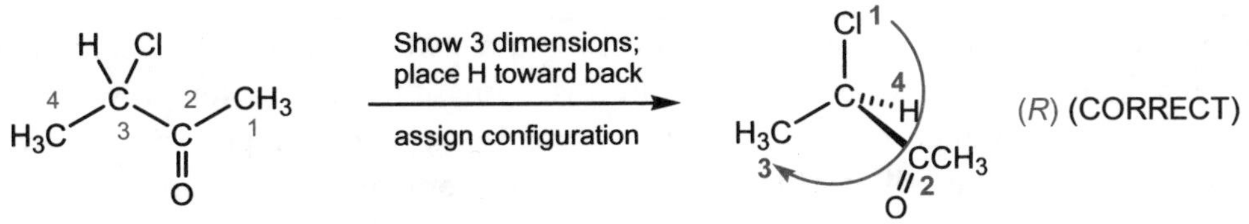

4.9.   To assign the absolute configuration of a stereogenic carbon atom in a ring, break the ring at the bond or atom directly opposite the stereogenic center, and then assign priorities to each substituent of the chiral carbon atom. Follow the procedures shown in Example 4.6.

a.

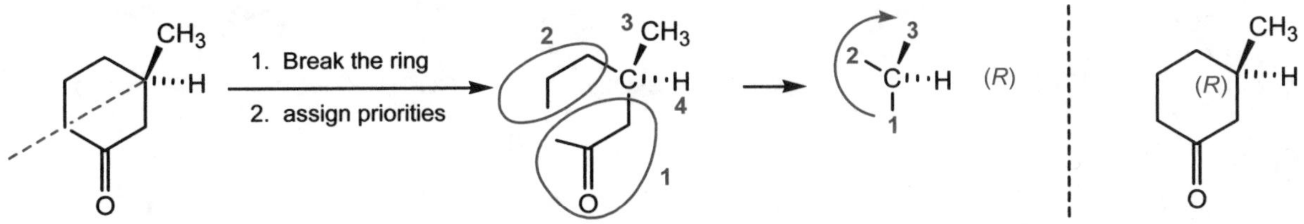

4.9.   (continued)

b.

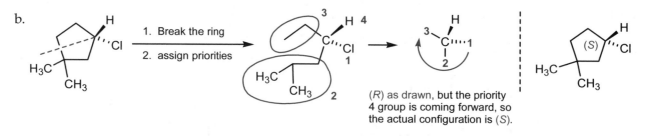

4.10.   Follow the procedure outlined in the solution to Exercise 4.7.

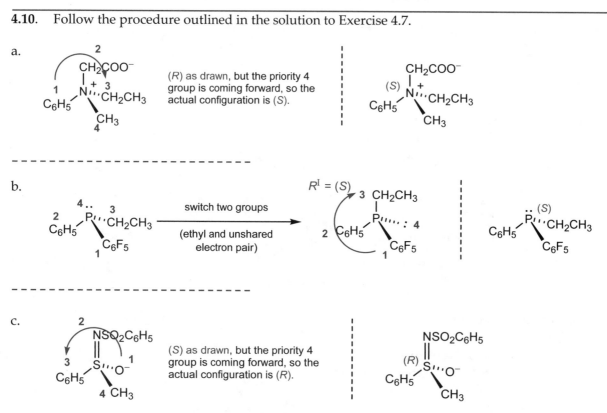

a.

(R) as drawn, but the priority 4 group is coming forward, so the actual configuration is (S).

- -

b.

switch two groups

(ethyl and unshared electron pair)

- -

c.

(S) as drawn, but the priority 4 group is coming forward, so the actual configuration is (R).

4.11.   The total number of stereoisomers that exist for a compound without symmetry is 2^n, where n is the total number of stereogenic centers. The compound in this Exercise has two potentially stereogenic centers, so a maximum of 4 stereoisomers can exist. All four stereoisomers are unique because there is no symmetry. The first step is to draw all of the isomers. Then, compare the configurations of the stereogenic centers in any two isomers [you can make the assignments of (R) and (S) if necessary]. Pairs of compounds in which the configurations of all of the chiral centers are inverted relative to the other are enantiomers. If fewer than all of the configurations have changed, then the compounds are diastereomers.

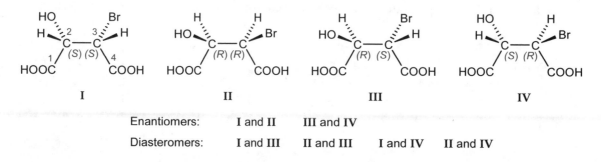

Enantiomers: I and II   III and IV

Diasteromers: I and III   II and III   I and IV   II and IV

4.12.   To decide whether a compound can have *meso* forms, you first interpret the name and draw all of the possible stereoisomers. If the molecule has an internal mirror plane (as shown by a dashed colored line in the structures below), it is a *meso* compound. Compounds (a.) and (b.) exist only in three stereoisomeric forms because one isomer is a *meso* compound.

a.   **2,4-Dichloropentane**

pent	five carbon atoms
ane	no double or triple bonds
2,4-dichloro	two Cl atoms, one at C2 and one at C4

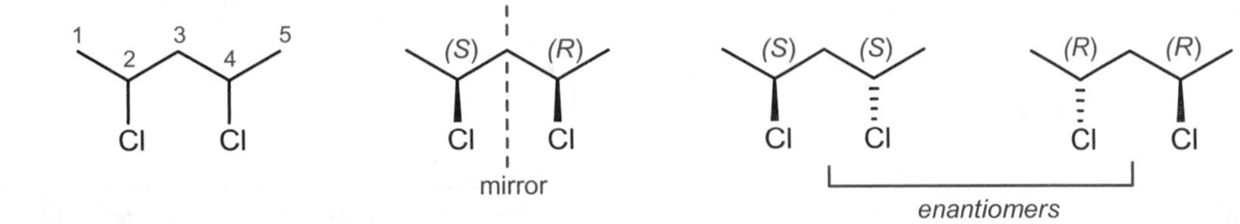

b.   **1,3-Dimethylcyclohexane**

cyclohexane	six-membered ring with no double or triple bonds
1,3-dimethyl	CH₃ groups, one at C1 and one at C3

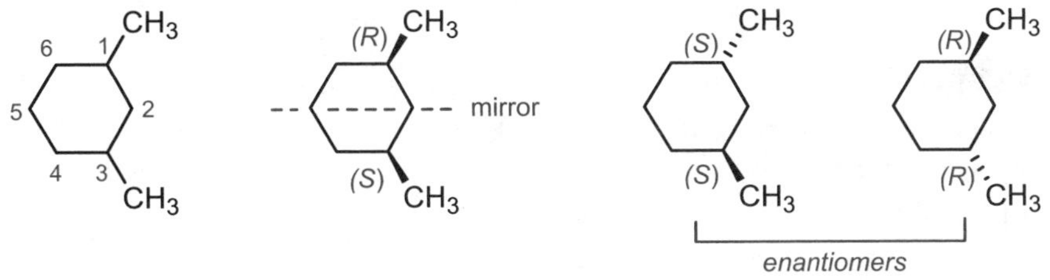

c.   **2,3-Dibromopentane**

pent	five carbon atoms
ane	no double or triple bonds
2,3-dibromo	two Br atoms, one at C2 and one at C3

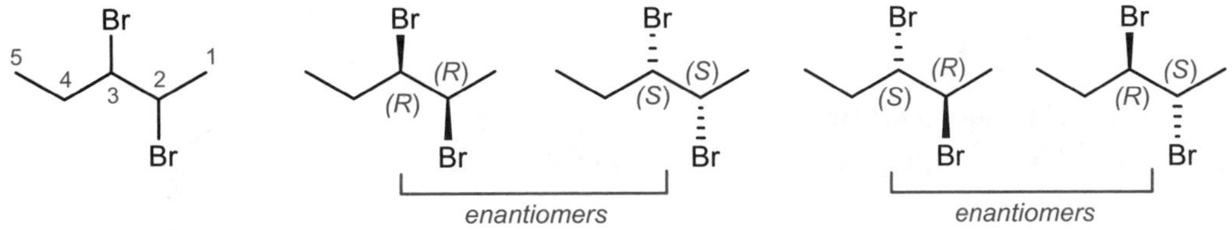

4.13.   To assign the absolute configurations of atoms in a Fischer projection, first convert the Fischer form to a perspective representation by drawing filled wedges to indicate that the groups attached via the horizontal bonds are in front of the plane of the page and dashed wedges to show the substituents attached via the vertical bonds are behind the plane of the page. Then assign the configurations according to the procedure outlined in the solution to Exercise 4.7.

a.

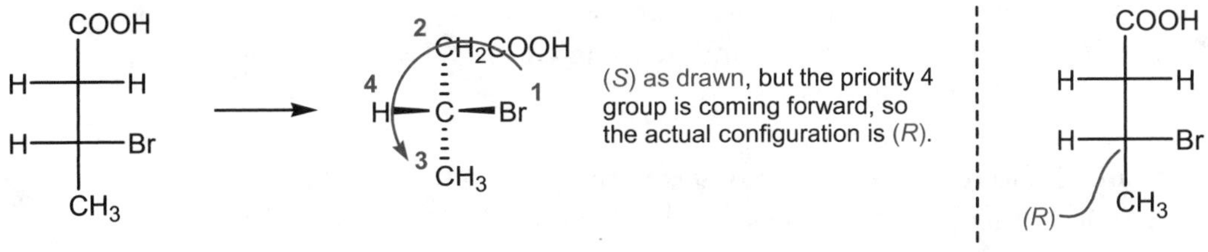

(S) as drawn, but the priority 4 group is coming forward, so the actual configuration is (R).

4.13.   (continued)

b.

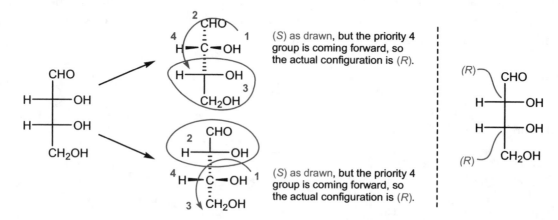

4.14   For each name, first generate a structural formula according to the procedures outlined in Chapter 1, but ignore the stereochemical descriptor if given. Then consider how the stereochemistry should be indicated based on the structural characteristics. If the prefix that is given is different from the one needed, then modify the name appropriately. If no prefix is present, draw the structures with all of the possible prefixes.

a.   ***trans*-2-Chloro-3-methyl-2-pentene**

pent	five carbon atoms
ene	double bond, starting at C2
2-chloro	Cl atom, attached at C2
3-methyl	CH₃ group, attached at C3
trans	(ignore to begin)

The stereochemistry of a trisubstituted double bond must be designated using the prefixes (*E*) or (*Z*), not *cis* or *trans*.

b.   **3,4-Dimethylcyclohexene**

cyclohex	six-membered ring of carbon atoms
ene	double bond, starting at and defining C1
3,4-dimethyl	two CH₃ groups, one at C3 and one at C4

The substituents attached to a ring can be either above or below the ring plane. If they are on the same side, the name should have the prefix "*cis*-" and if they are on opposite sides, the name should have the prefix "*trans*-." [You could also specify the stereochemistry of the chiral carbon atoms in this molecule, as in (c.) directly below.]

c.   **3-Methylcyclopentene**

cyclopent	five-membered ring of carbon atoms
ene	double bond, starting at and defining C1
3-methyl	a CH₃ group, attached at C3

A carbon atom with four different substituents is chiral, so the stereochemical descriptor (*S*) or (*R*) should appear as a prefix to the name.

4.14. (continued)

d. 3-Hexene

hex	six carbon atoms
ene	double bond, starting at C3

The stereochemistry of a disubstituted double bond must be designated using a prefix: either *cis*, *trans*, (*E*), or (*Z*).

trans
(*E*)

cis
(*Z*)

4.15. To interpret a compound's name when stereochemical descriptors are given, first draw the structural formula using the procedures outlined in Chapter 1. Then consider which geometric isomers can be generated by switching substituents at each end of the double bond(s). Finally, use the Cahn-Ingold-Prelog system to designate the stereochemistry of any double bond for which a stereochemical descriptor is valid. (When identical groups are attached to the carbon atom at one or both ends of the double bond, then the alkene does not exist in isomeric forms.)

a. 1,3-pentadiene

penta	five carbon atoms
diene	two double bond, one starting at C1 and one at C3: the double bond at C1 does not have isomeric forms because C1 has two H atoms attached.

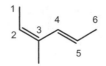

(*E*)-1,3-Pentadiene   (*Z*)-1,3-Pentadiene

b. 3-methyl-2,4-hexadiene

hexa	six carbon atoms
diene	two double bond, one starting at C2 and one at C4: both exist in isomeric forms
3-methyl	CH_3 group attached at C3

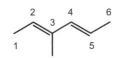

(*2E, 4E*)-3-Methyl-2,4-hexadiene   (*2E, 4Z*)-   (*2Z, 4E*)-   (*2Z, 4Z*)-

c. 2,3-dimethyl-2-butene

but	four carbon atoms
ene	double bond, starting at C2; this double bond does not have isomeric forms
2,3-dimethyl	two CH_3 groups, one at C2 and one at C3

d. 3-ethyl-4-octene

oct	eight carbon atoms
ene	double bond, starting at C4; this double bond can exist in isomeric forms
3-ethyl	an ethyl group, attached at C3

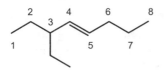

(*E*)-3-Ethyl-4-octene

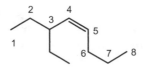

(*Z*)-3-Ethyl-4-octene

4.16.   Follow the procedure given in the solution to Exercise 4.3.

a.   **(Z)-3-Bromo-2-hexene**

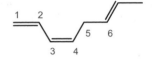

hex	six carbon atoms
ene	carbon-carbon double bond starting at C2
3-bromo	a Br atom is attached at C3
(Z)	the higher priority groups at the ends of the
	π bond are on the same side of the C=C bond

b.   **(3Z,6E)-1,3,6-Octatriene**

octa	eight carbon atoms
triene	three carbon-carbon double bonds,
	starting at C1, C3, and C6
(3Z)	the higher priority groups at the ends of the π bond
	that starts at C3 are on the same side of the C=C bond
(6E)	the higher priority groups at the ends of the π bond
	that starts at C6 are on opposite sides of the C=C bond

c.   **(Z)-1-Methylcyclononene**

cyclonon	nine-membered ring of carbon atoms
ene	carbon-carbon double bond starting at C1
1-methyl	a CH₃ group is attached at C1
(Z)	the higher priority groups at the ends of the
	π bond are on the same side of the C=C bond

d.   **(E)-2-Methoxy-2-pentene**

pent	five carbon atoms
ene	carbon-carbon double bond starting at C2
2-methoxy	an –OCH₃ group is attached at C2
(E)	the higher priority groups at the ends of the
	π bond are on opposite sides of the C=C bond

4.17.   Follow the procedure outlined in the solution to Exercise 4.2.

a. H_3C, CH_3, H, Br **(E)**

b. H, CH_3, Br, CH_3 **same** neither

c. C_6H_5, CH_3, H_3C, C_6H_5 **(E)**

d. H_3C, $COOH$, H, CH_2OCH_3 **(Z)**

4.18.   To evaluate whether two compounds are conformers, isomers, or identical, follow the scheme shown in Figure 4.12. If necessary, make a model of the first compound, and then see if it superimposes on a model of the second. If they are superimposable, then the molecules are identical. If you first have to rotate around sigma bonds to make the two substances superimposable, then they are conformers. If the connectivities of the bonds differ, then the compounds are isomers. Assigning (R), (S), (E), or (Z) will help you to evaluate whether the stereochemistry is different for the two compounds being compared.

4.18. (continued)

a.

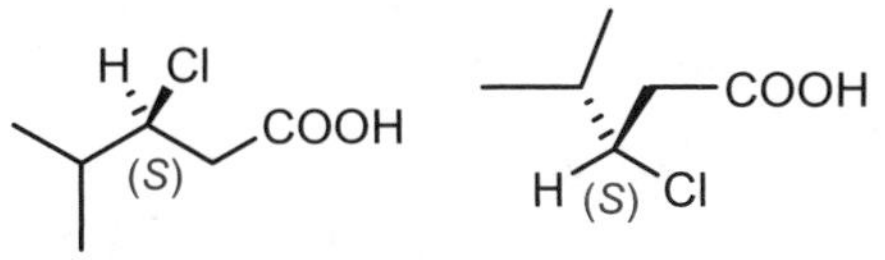

Identical

b.

Diastereomers

c.

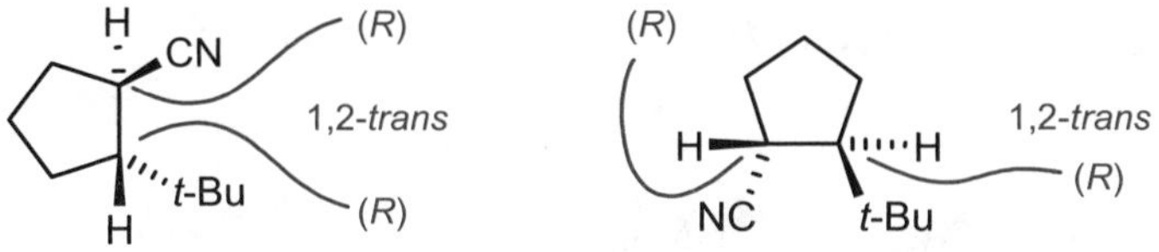

Identical

d.

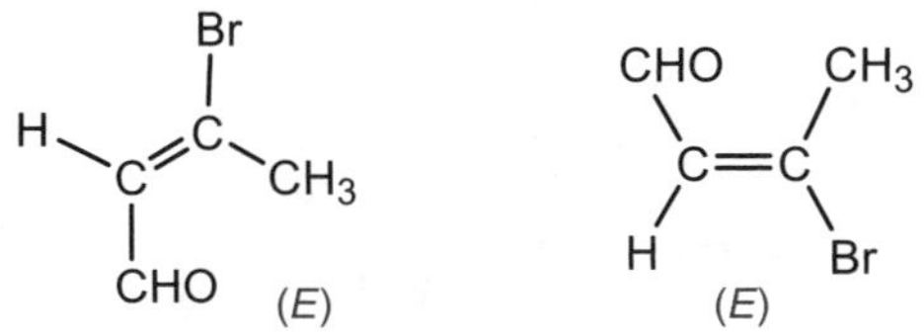

Identical

e.

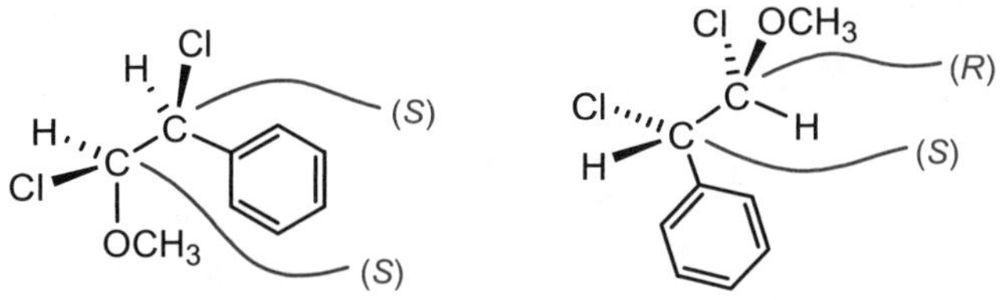

Diastereomers

f.

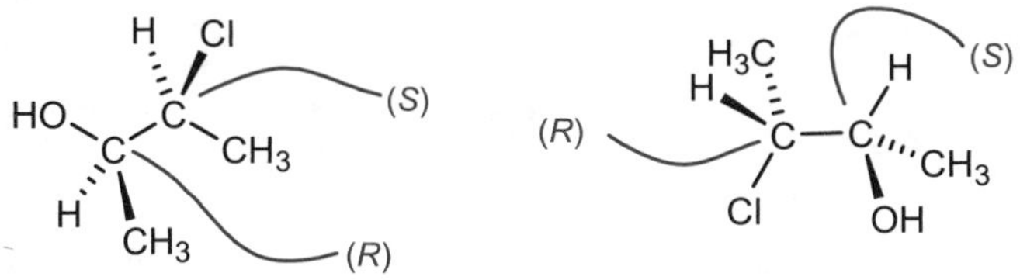

Enantiomers

4.19. The priority of a group in the Cahn-Ingold-Prelog system is related to the atomic number of the atom through which the group is attached. If that atom is the same for two different groups, atomic numbers of the elements in the next shell are compared. A higher atomic number for any one of those atoms gives the entire group a higher priority. Functional groups that have multiple bonds are first converted to their single bond equivalents (Figure 4.2) before comparing the shells of atoms beyond the initial point of attachment.

a. $-Cl$ > $-OCH_3$ > $-NH_2$ > $-H$

b. $-OCH_3$ > $-COOH$ > $-CN$ > $-CH_2CH_3$

c. $-Br$ > $-OH$ > $-CHO$ > $-CH_2CH_3$

d. $-COOCH_3$ > $-CN$ > $-C{\equiv}CH$ > $-CH_3$

4.20. Follow the procedure outlined in the solution to Exercise 4.15.

a. **2-Heptene**

hept	seven carbon atoms
ene	double bond, starting at C2

The stereochemistry of a disubstituted double bond can exist as (*E*) and (*Z*) (or cis and trans) isomers.

trans-2-Heptene *or*
(*E*)-2-Heptene

cis-2-Heptene *or*
(*Z*)-2-Heptene

b. **1-Chlorocyclobutene**

cyclobut	four-membered ring of carbon atoms
ene	double bond, starting at and defining C1
1-chloro	a chlorine atom, attached at C1

The stereochemistry of a double bond in a ring with seven or fewer atoms can only be *cis* with respect to the ring carbon atoms so a descriptor is not needed.

c. **1,3-Butadiene**

buta	four carbon atoms
diene	two double bonds, one starting at C1 and one at C3

A double bond with two identical groups attached to a carbon atom at one or both ends does not exist in isomeric forms.

d. **3,4-dimethyl-3-hexene**

hex	six carbon atoms
ene	double bond, starting at C3
3,4-dimethyl	two CH₃ groups, one attached at C3 and the other at C4

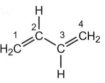

The stereochemistry of a tetrasubstituted double bond can exist as (*E*) and (*Z*) isomers.

(*E*)-3,4-Dimethyl-3-hexene   **(*Z*)-3,4-Dimethyl-3-hexene**

e. **3-phenyl-2-pentene**

pent	five carbon atoms
ene	double bond, starting at C2
3-phenyl	a phenyl group, attached at C3

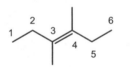

The stereochemistry of a trisubstituted double bond can exist as (*E*) and (*Z*) isomers.

(*Z*)-3-Phenyl-2-pentene   **(*E*)-3-Phenyl-2-pentene**

4.21. Follow the procedure outlined in the solution to Exercise 4.2.

a.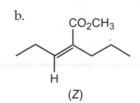

b.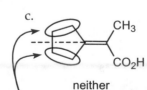

c., d.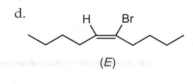

4.22. An object is chiral if it is not superimposable on its mirror image. Often, an item that cannot be used in the same way by right- and left-handed persons is chiral.

a. a golf club   chiral   b. a pair of scissors   chiral

c. a baseball glove   chiral   d. a corkscrew   chiral

e. a telephone   chiral   f. a pencil   achiral

g. a basketball   achiral   h. a hammer   achiral

i. a spiral staircase   chiral

4.23. The prefix *"trans"* is not valid for this molecule because the double bond is trisubstituted. The name should be **(E)-2-methyl-2-butenal**.

Tiglic aldehyde

(*E*)-2-Methyl-2-butenal

4.24. The prefix *"cis"* is not valid for this molecule because the double bond is trisubstituted. The name should be **(Z)-aconitate**.

(Z)-Aconitate

4.25. Follow the procedure outlined in the solution to Exercise 4.7.

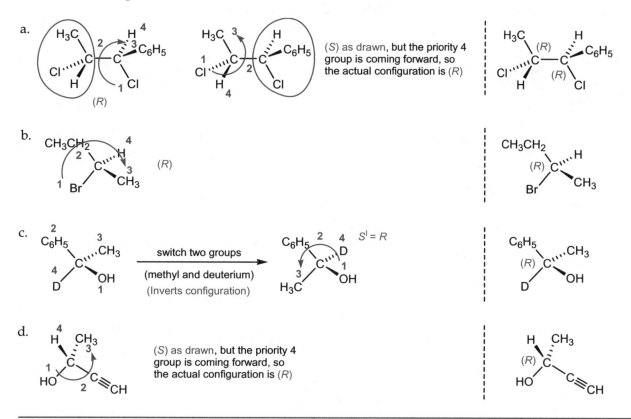

a. (*S*) as drawn, but the priority 4 group is coming forward, so the actual configuration is (*R*)

b.

c. switch two groups (methyl and deuterium) (Inverts configuration) $S^I = R$

d. (*S*) as drawn, but the priority 4 group is coming forward, so the actual configuration is (*R*)

4.26. Follow the procedure outlined in the solution to Exercise 4.8. For compounds that are referred to by the relative stereochemical relationship of the chiral centers (parts b. and d. below), make certain that the configurations meet the indicated criteria. For a *meso* compound, the chiral carbon atoms should have opposite configurations; an optically active isomer will have a preponderance of a particular configuration.

4.26.   (continued)

a.   **(3S,4S)-4-Methyl-3-hexanol**

hex	six carbon atoms
an	no double or triple bonds
ol	alcohol functional group; the OH group is attached at C3
4-methyl	a CH_3 group, attached at C4

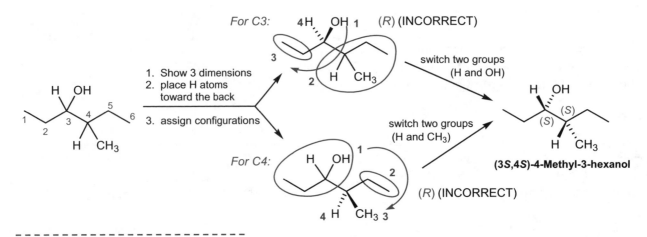

----

b.   an optically active isomer of **1,2-dimethylcyclopentane**

cyclopentane	six-membered ring of carbon atoms with no double or triple bonds
1,2-dimethyl	two CH_3 groups, attached at C1 and C2

----

c.   **(1R, 3S)-3-Methylcyclohexanol**

cyclohex	six-membered ring of carbon atoms
an	no double or triple bonds
ol	alcohol functional group; the OH group is attached at and defines C1
3-methyl	a CH_3 group, attached at C3

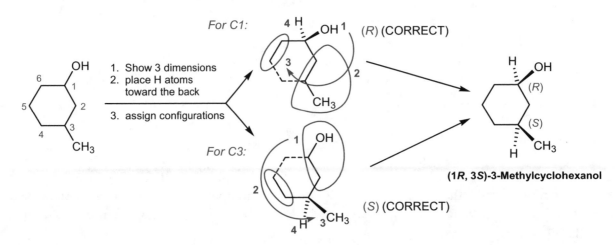

4.26. (continued)

d. the *meso* isomer of **1,3-Dichlorocyclopentane**

cyclopentane   six-membered ring of carbon atoms with no double or triple bonds
1,3-dichloro   two Cl atoms, attached at C1 and C3

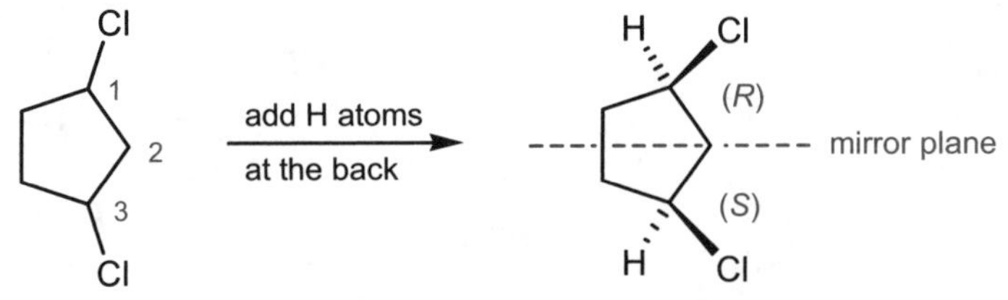

***meso*-1,3-Dichlorocyclopentane**

4.27. If a molecule has no symmetry, then the total number of stereoisomers is 2^n, where n is the total number of stereogenic centers (or double bonds that can exist as (E) and (Z) isomers). It is best to expand the structure so that the carbon atoms with four different groups can be identified readily. If the compound has a mirror plane, then one isomer will be *meso*. The enantiomeric pairs are identified below. All other relationships represent pairs of diastereomers.

a.

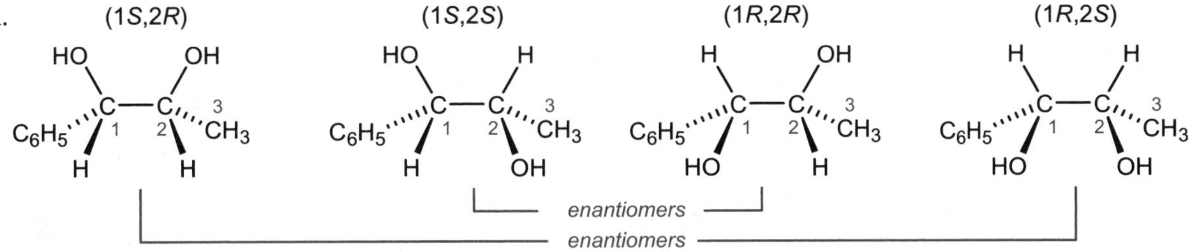

b.

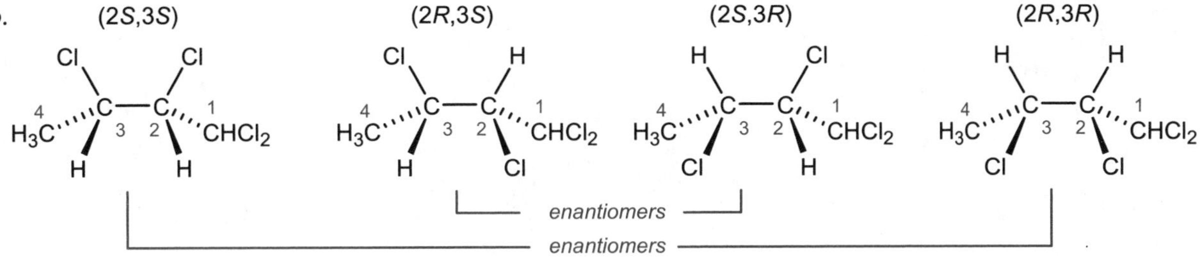

c.

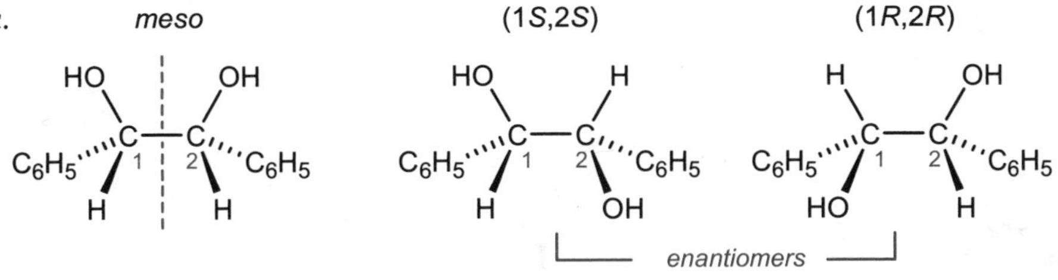

d.

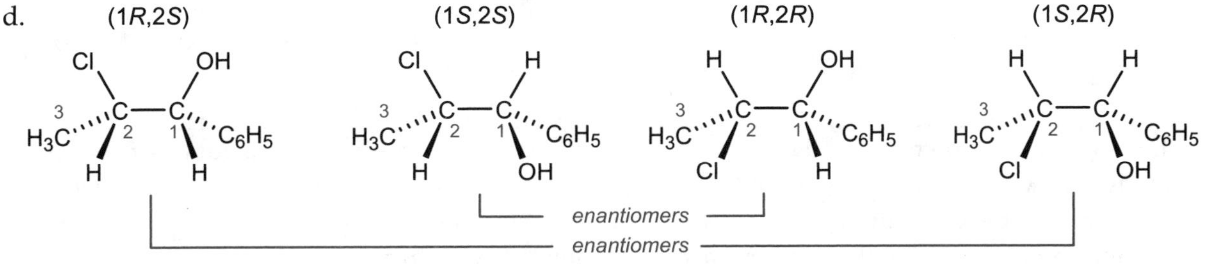

4.28.   Follow the procedure outlined in the solution to Exercise 4.9.

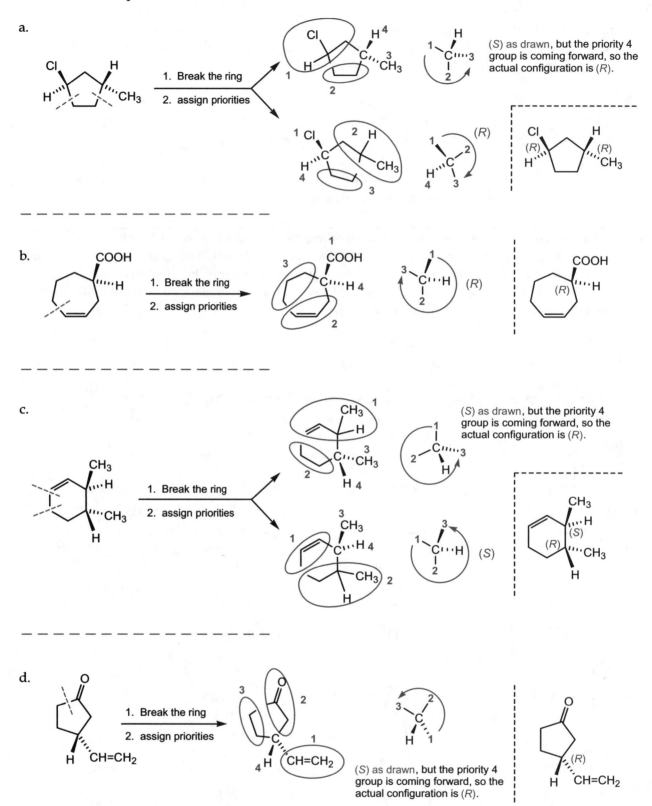

4.29.   Looking at the structures of the isomeric alcohols that were drawn as the solutions to Exercise 1.28, identify the ones with a carbon atom having four different groups. Then draw a dimensional formula and assign the configuration of the chiral carbon atom. Switching any two groups attached to the stereogenic center generates the compound with the opposite configuration.

4.29. (continued)

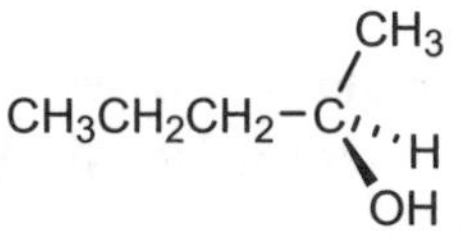

(R)-2-Pentanol 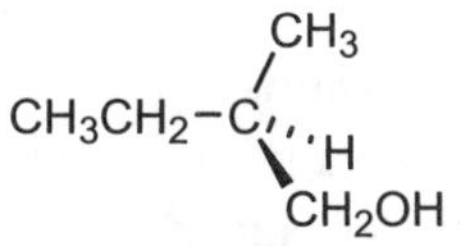   (R)-2-Methylbutanol 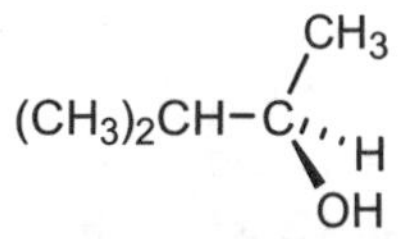   (R)-3-Methyl-2-butanol

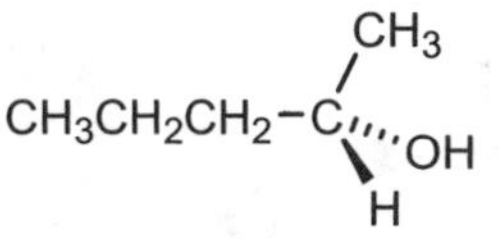

(S)-2-Pentanol 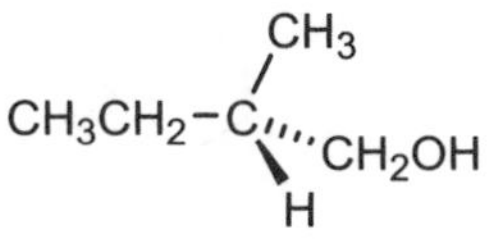   (S)-2-Methylbutanol 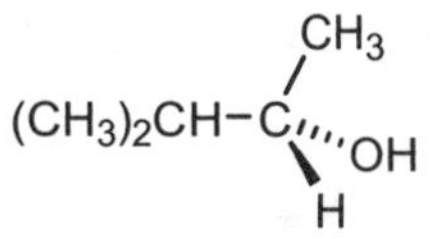   (S)-3-Methyl-2-butanol

---

4.30.   To convert a dimensional structure to a Fischer projection, orient the molecule so that the groups at the top and bottom of the asymmetric carbon atom recede behind the plane of the paper and those on the horizontal bonds are in front of the paper plane. The longest carbon chain should run from top to bottom in a Fischer projection, so if necessary, switch three groups on the chiral center (as shown in Figure 4.10) to create that orientation. Use a model if necessary to get the proper orientation.

a.

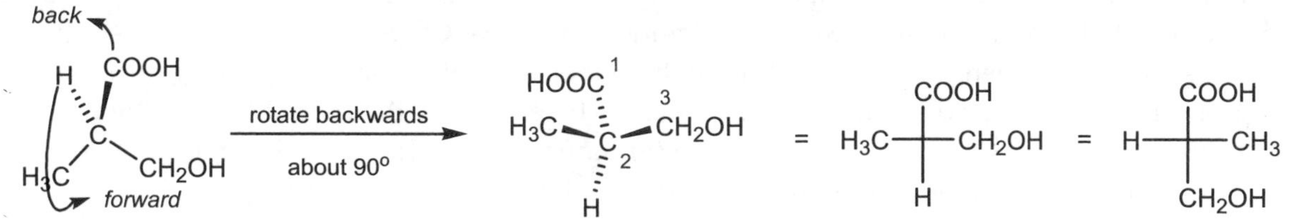

b.

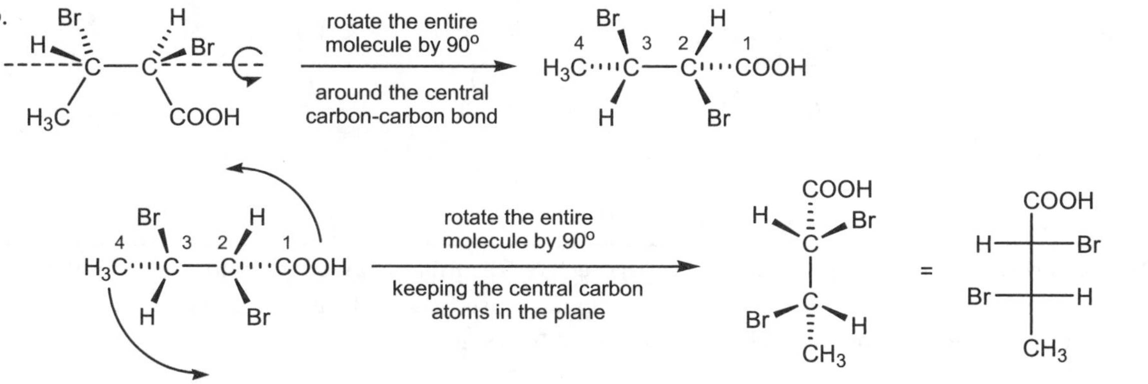

c.

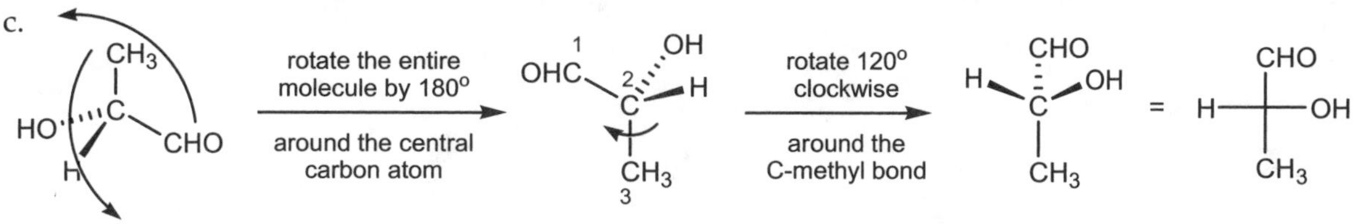

4.31.   For a molecule with n stereogenic centers, the total number of stereoisomers is 2^n. The compound under consideration has three potentially stereogenic centers, so a maximum of 8 stereoisomers might exist. The symmetry of the compound reduces that number, however, so only four stereoisomers actually exist, two of which are *meso*.

4.31. (continued)

Draw each of the eight isomers, then determine which are identical (making models may be helpful for this exercise). The *meso* compounds have a plane of symmetry that passes through the central carbon atom (shown as a dashed line in the structures below).

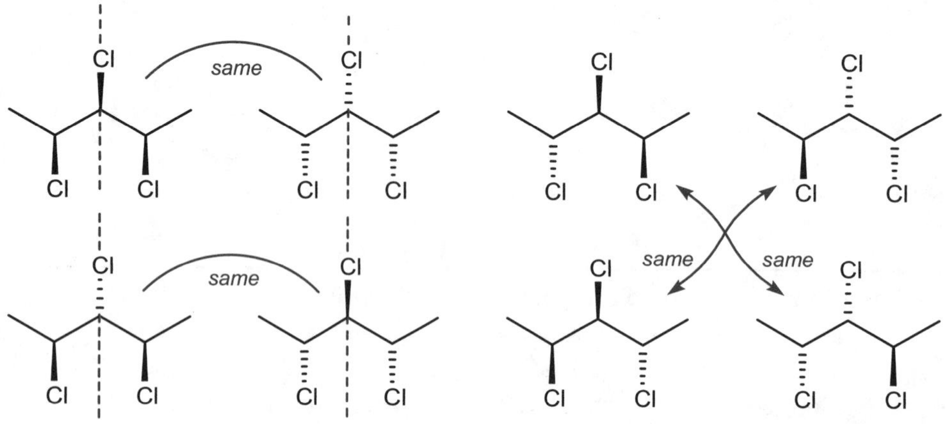

When a stereogenic center in a *meso* compound sits on a mirror plane, it is chiral if there are four different groups attached to it. For structure **I**, below, the four groups are H, Cl, (*S*)–CHClCH$_3$ and (*R*)–CHClCH$_3$. When two structurally "identical" groups differ only in their configurations, the priority rules state that an (*R*) group has a higher priority than one that is (*S*). Therefore, C3 (the central carbon atom) in **I** has the (*R*) configuration, and C3 in structure **II** has the (*S*) configuration. The central carbon atom in **III** and **IV** is not chiral, even though the other two centers are chiral.

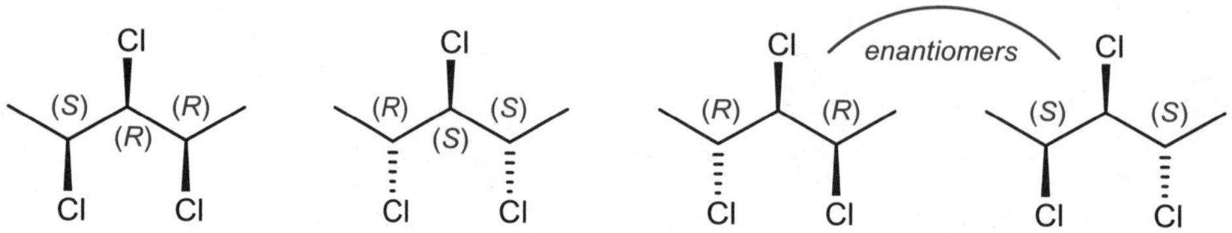

4.32. To convert from a Fischer projection to a perspective formula, use wedge bonds to show that the horizontal groups are in front of the plane of the page and used dashed lines to indicate that the two substituents at the top and bottom are behind the plane of the page.

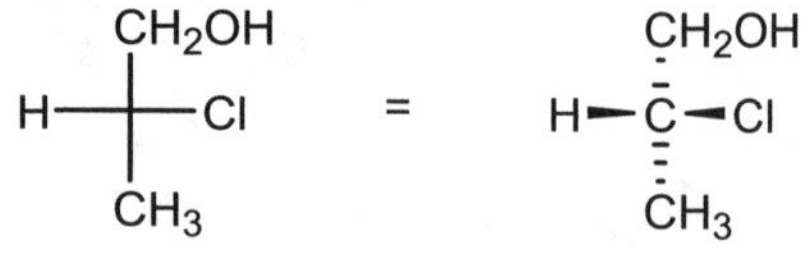

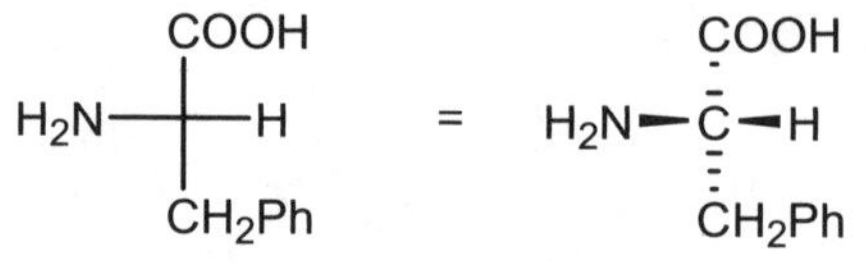

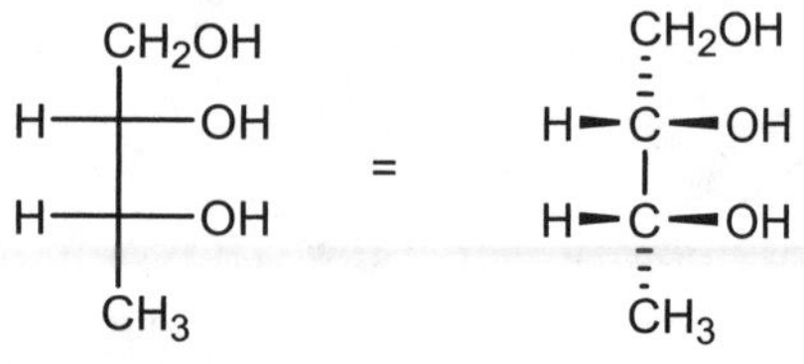

4.33.   Follow the procedures outlined in the solution to Exercise 4.31.

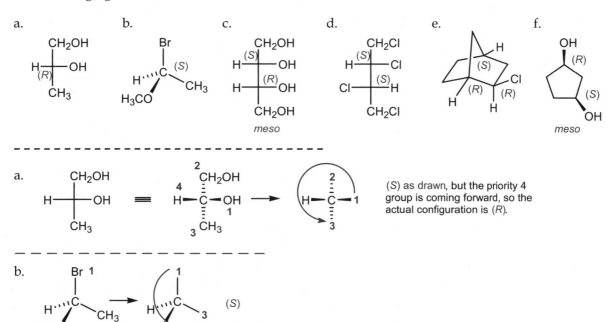

A stereogenic center in a *meso* compound that sits on the mirror plane is chiral if there are four different groups attached to it. When two structurally "identical" groups differ in their configurations, the priority rules state that an (R) group has a higher priority than one that is (S). Therefore, C1 (the carbon atom bearing the COOH group) in **I** has the (S) configuration, and the (R) configuration in structure **II**.

4.34.   Follow the procedures outlined in the solution to Exercises 4.7, 4.9, and 4.13. The final assignments are shown directly below; the procedures showing how the assignments were made are shown in the following figures.

4.34. (continued)

c.

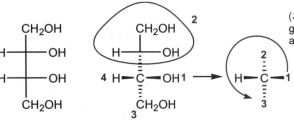

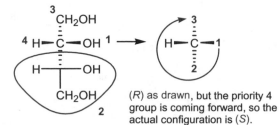

(S) as drawn, but the priority 4 group is coming forward, so the actual configuration is (R).

(R) as drawn, but the priority 4 group is coming forward, so the actual configuration is (S).

---

d.

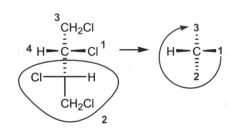

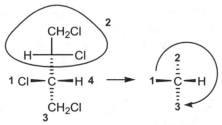

(R) as drawn, but the priority 4 group is coming forward, so the actual configuration is (S).

(R) as drawn, but the priority 4 group is coming forward, so the actual configuration is (S).

---

e.

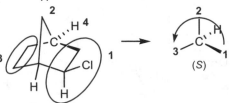

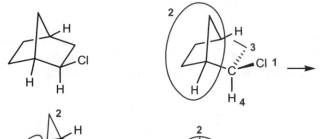

(S) as drawn, but the priority 4 group is coming forward, so the actual configuration is (R).

---

f.

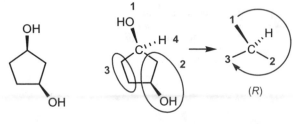

4.35. Follow the procedures outlined in the solution to Exercise 4.27.

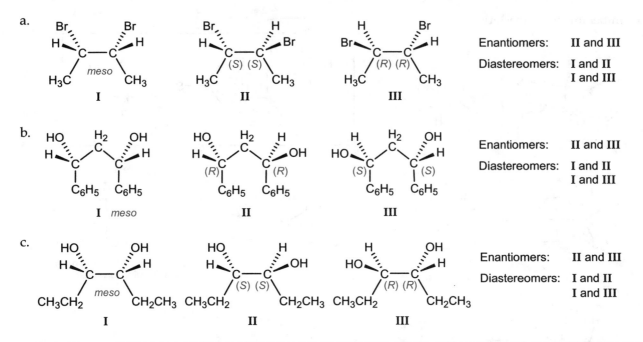

a.

			Enantiomers:	II and III
			Diastereomers:	I and II
				I and III

I (meso)   II (S)(S)   III (R)(R)

b.

			Enantiomers:	II and III
			Diastereomers:	I and II
				I and III

I meso   II (R)(R)   III (S)(S)

c.

			Enantiomers:	II and III
			Diastereomers:	I and II
				I and III

I meso   II (S)(S)   III (R)(R)

4.36. Follow the procedures outlined in the solution to Exercises 4.7 and 4.9. The assignments are shown on the structures in the boxes, and the procedures used to make those assignments follow. The asterisks in the original structures indicate which carbon atoms are chiral.

a. Brefeldin A, an antiviral agent

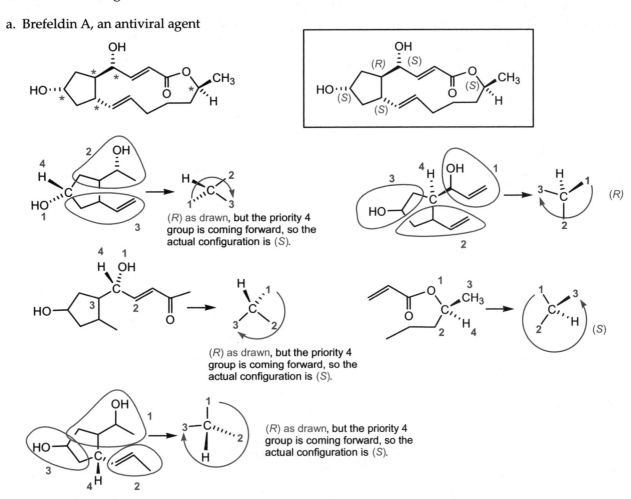

(R) as drawn, but the priority 4 group is coming forward, so the actual configuration is (S).

(R) as drawn, but the priority 4 group is coming forward, so the actual configuration is (S).

(R) as drawn, but the priority 4 group is coming forward, so the actual configuration is (S).

4.36. (continued)

b. Imipenim, an antibacterial

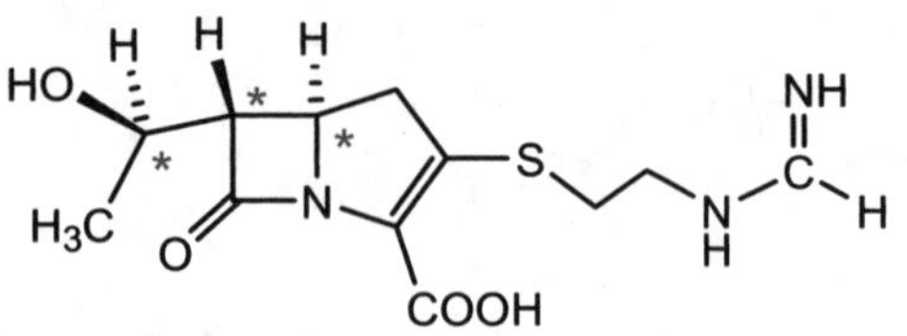

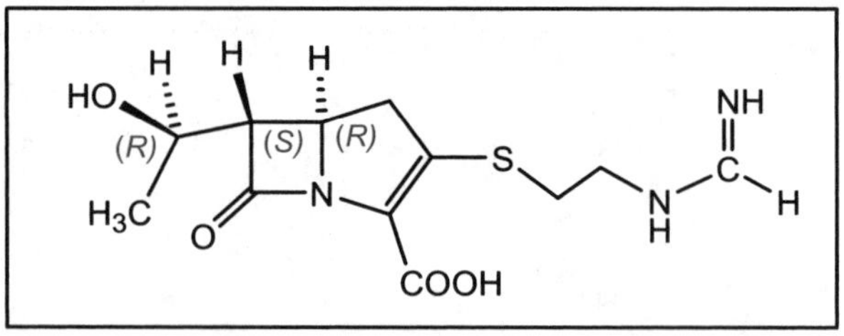

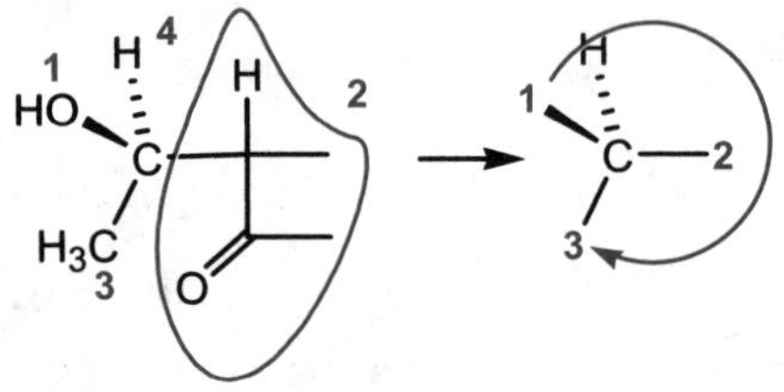

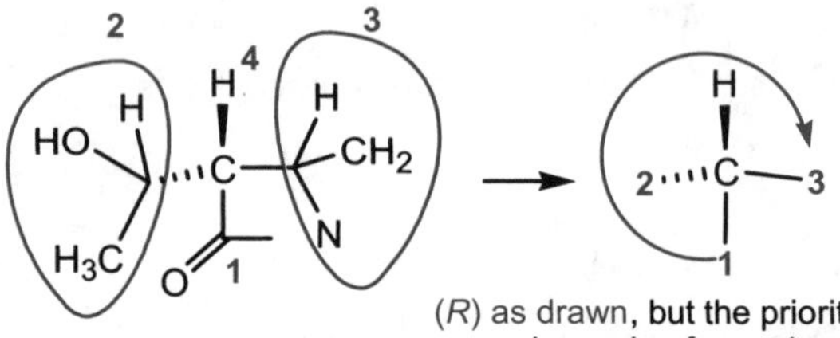

(R) as drawn, but the priority 4 group is coming forward, so the actual configuration is (S).

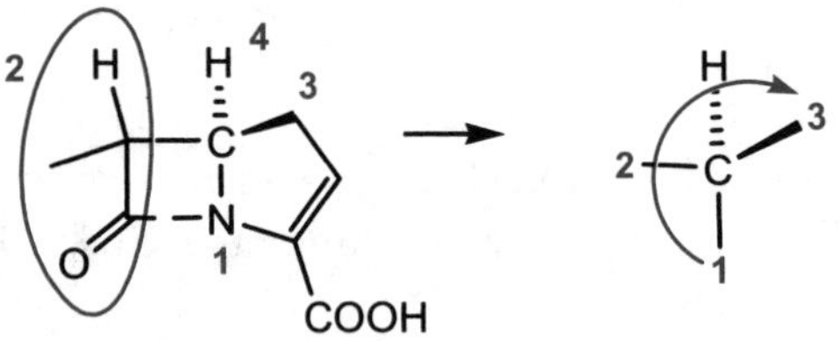

CHEMICAL REACTIONS AND MECHANISMS

5.1. A reactant containing atoms that do not appear in the product is most likely acting as a catalyst.

5.2. To classify a reaction by one of the seven given types, make use of the definitions given in the text. Draw the full structural formulas of the reactants and products to examine the functional groups involved in the transformation. Evaluate those structures to see which atoms are replaced or whether π bonds are made or broken.

a. **Addition:** methyllithium (CH$_3$Li) adds to the π bond of a ketone carbonyl group to form a product having only single bonds.

b. **Substitution:** iodide ion replaces the methanesulfonate ion (CH$_3$SO$_3^-$)

c. **Elimination:** a molecule of water is lost from the starting alcohol to yield an alkene, which means that a carbon–carbon π bond is formed. Sulfuric acid is a catalyst for this reaction.

5.3. The conjugate base of a protic acid (an acid that donates a proton) is the species in which the most acidic proton has been removed; its charge is one unit more negative than the charge on the starting acid. Often, the most acidic proton in a molecule or ion is attached to a heteroatom. The acidic proton in each of the following substances is shown in color.

a. $CH_3CH_2COOH \longrightarrow CH_3CH_2COO^-$

b. $HN_3 \longrightarrow N_3^-$

c. $(CH_3)_3CSH \longrightarrow (CH_3)_3CS^-$

d. $C_6H_5OH \longrightarrow C_6H_5O^-$

e. $NH_4^+ \longrightarrow NH_3$

5.4. Acid-base reactions occur so as to produce a weaker acid from a stronger one. Identify the acid on each side of the equation and decide which proton is the most acidic. (These are shown in color in each equation.) Then, look up the approximate pK_a values in Table 5.1. (These values are given below each acid in the following equations.) If the pK_a values are approximately the same (as in part d., below), use the table printed on the inside front cover of the textbook to find a more precise value. The equilibrium lies toward the side of the equation with the acid having the larger pK_a value.

a. CH_3CO_2H + Cl^- ⇌ $CH_3CO_2^-$ + HCl

   $pK_a \sim 5$ $pK_a \sim -10$

b. CH_3OH + CN^- ⇌ CH_3O^- + HCN

   $pK_a \sim 15$ $pK_a \sim 10$

c.   + CH_3O^- ⇌   + CH_3OH

   $pK_a \sim 10$ $pK_a \sim 15$

d. NH_4^+ + $CH_3CH_2S^-$ ⇌ NH_3 + CH_3CH_2SH

   $(pK_a \sim 10)$ $(pK_a \sim 10)$

   $pK_a \sim 9$ $pK_a \sim 11$

5.5. Identify which proton is the most acidic for each molecule in the given pair. (These protons are shown in color.) Then, look up the approximate pK_a values in Table 5.1. (These values are given below each acid.) The stronger acid is the one with the lower pK_a value and it is circled.

a.

$pK_a \sim 5$ $pK_a \sim 15$

b.

$pK_a \sim 15$ $pK_a \sim 40$

c.

$pK_a \sim 10$ $pK_a \sim 15$

5.6. Follow the procedure outlined in Example 5.4.

a. Compounds **I**, **II**, and **IV** each have a substituent at C2. The more electron-withdrawing a substituent is, the more acidic that compound will be. A methyl group is electron-donating, and fluorine is more electronegative (more electron-withdrawing) than oxygen. Therefore, the order of acidity among compounds with a substituent adjacent to the COOH group will be **I** > **II** > **IV**. Compounds **I** and **III** each have a fluorine atom (at C2 and C4, respectively). The closer an electron-withdrawing group or atom is to the acidic proton, the stronger that acid will be. Therefore, the order of acidity will be **I** > **III**. Overall, compound **I** is the most acidic.

Compound **IV** is the only one in the series with an electron-donating substituent. Therefore, it will be the least acidic.

I   II   III   IV

5.6. (continued)

b. Each compound has a substituent at C3. The more electron-withdrawing a substituent is, the more acidic that compound will be. The heteroatoms are all electron-withdrawing, and only the methyl group is electron-donating, so compound **I** is the least acidic. Among the others, the order of electronegativity values is O > Cl > Br. Therefore, the order of increasing acid strength will be **III** > **II** > **IV**. Compound **III** is the most acidic.

5.7. 4-Methoxyaniline is a stronger base than aniline as a result of the resonance form shown in the box below. When this species is protonated on its nitrogen atom, the positive charge of the ammonium ion is offset by the negative charge on the neighboring carbon atom (circled in the structure below at the right).

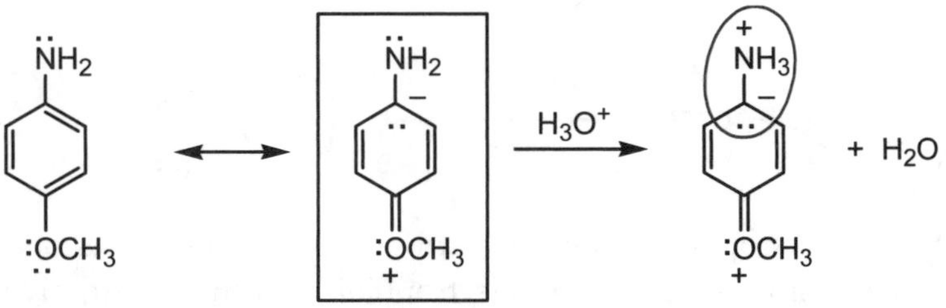

5.8. Follow the procedure outlined in Example 5.5.

a. Boron is a group 13 element that exists in violation of the octet rule: when B has only three bonds, it has six electrons, which makes it a Lewis acid. After drawing a bond between the atoms of the electron-pair donor (O) and the electron-pair acceptor (B), include formal charges on the atoms in the product.

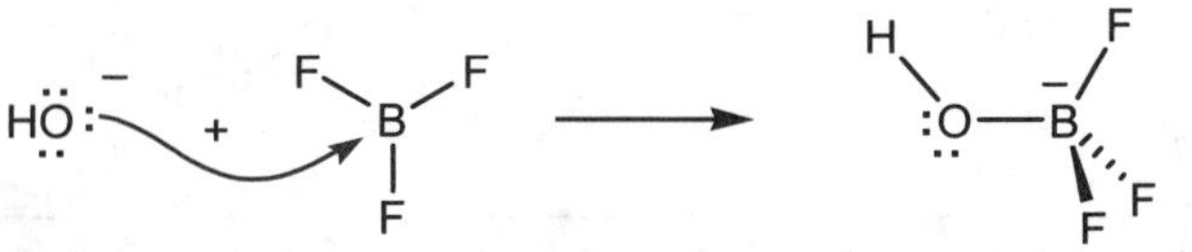

b. Aluminum is another element that forms molecules that violate the octet rule, and many of its compounds are Lewis acids. After drawing a bond between the electron pair donor atom (Cl) and the electron pair acceptor atom (Al), include formal charges on the atoms in the product.

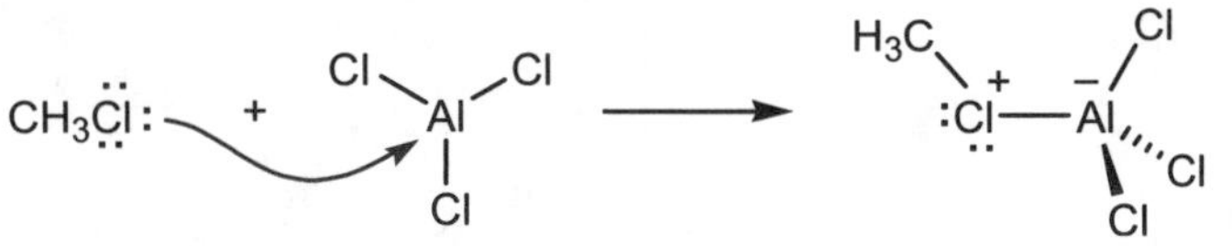

5.9. Follow the procedures outlined in Example 5.6.

a. A species in which an atom has a charge of +1 and no unshared electron pairs is an electrophile.

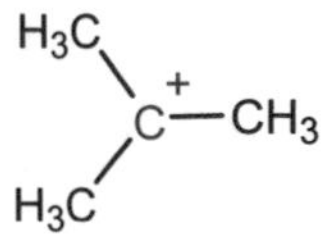

b. A species with electron pairs and/or π bonds, especially one that carries a negative charge, is a nucleophile.

5.9. (continued)

c. Benzene has three carbon-carbon double bonds. These π bonds will act as nucleophiles in most polar reactions.

d. A species that bears an overall charge of +1, even if unshared electron pairs are present, reacts mainly as an electrophile.

$$:O{=}\overset{+}{N}{=}O:$$

5.10. Follow the procedure outlined in Example 5.7.

a. A reactant with a negative charge usually functions as the nucleophile in a polar transformation. Cyanide is a good nucleophile, and its carbon atom bears a –1 charge. The carbon atom of the carbonyl group in the other reactant plays the role of the electrophile—the C atom has a partial positive charge because O is the more electronegative atom of the pair. The first step in this reaction, as in any polar reaction, is the interaction between the nucleophile and the electrophile, which is depicted using a curved arrow that starts at the electron pair of the nucleophile and points to the electrophile. The electrons in the π bond of the carbonyl group have to move so that the carbon atom maintains four bonds.

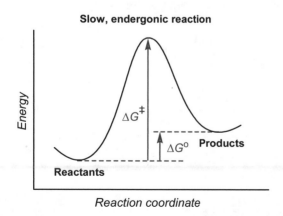

b. The proton is a common electrophile, so when a strong acid is present as one of the reactants, it usually functions in a transformation as the electrophilic reagent by supplying a proton. Any heteroatom or π bond in the other reactant will play the role of the nucleophile. The first step in this reaction, as in any polar reaction, is the interaction between the nucleophile and the electrophile, which is depicted using a curved arrow that starts at the electron pair of the nucleophile and points to the electrophile. Because H normally forms only one bond, the bond between H and Cl is broken as the new bond between the oxygen atom and hydrogen atom is made.

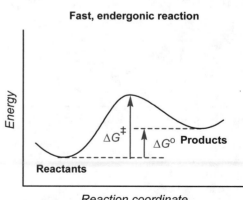

5.11. A slow reaction has a high free energy of activation whereas a fast reaction has a low free energy of activation. An endergonic process is one in which the free energy of the products is higher than the free energy of the reactants.

Slow, endergonic reaction

Energy

$\Delta G^{\ddagger}$

ΔG° **Products**

Reactants

Reaction coordinate

Fast, endergonic reaction

Energy

$\Delta G^{\ddagger}$ ΔG° **Products**

Reactants

Reaction coordinate

5.12.   To calculate values of K_{eq} from the given values of $\Delta G°$, make use of equation 5.16 ($\Delta G° = -RT\ln K_{eq}$) where R = 1.986 cal • mol^{-1} • K^{-1} and T is the temperature in Kelvins.

a. $\Delta G° = +3.00$ kcal • mol^{-1} $K_{eq} = 6.29 \times 10^{-3}$

b. $\Delta G° = -3.00$ kcal • mol^{-1} $K_{eq} = 1.59 \times 10^{2}$

c. $\Delta G° = -6.00$ kcal • mol^{-1} $K_{eq} = 2.53 \times 10^{4}$

5.13.   Similar to a reaction that proceeds via a carbocation intermediate, a reaction that occurs with involvement of a radical intermediate comprises reactants, products, and the intermediate itself. The energy diagram, therefore, looks much like that in Figure 5.4, in which an activation barrier exists between reactants and intermediate, as well as between intermediate and products. There are two transition states as well.

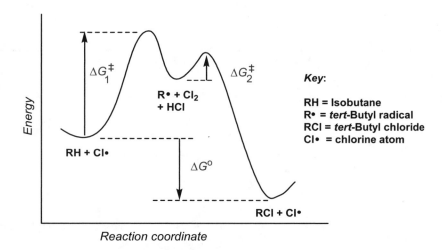

5.14.   An amino acid with the D configuration is the mirror image (enantiomer) of an amino acid with the L configuration. Therefore, to draw the indicated structural formulas, switch any two groups attached to the stereogenic carbon atom in each of the L compounds. Starting with the structures shown in the text, the H and side chain groups have been switched to generate the structures depicted below.

D-Aspartic acid   **D-Phenylalanine**

5.15.   Follow the procedure outlined in Example 5.8.

a.   b.

5.16. Follow the procedure outlined in the solution to Exercise 5.2.

a. **Elimination:** the elements of water are removed to form a π bond.

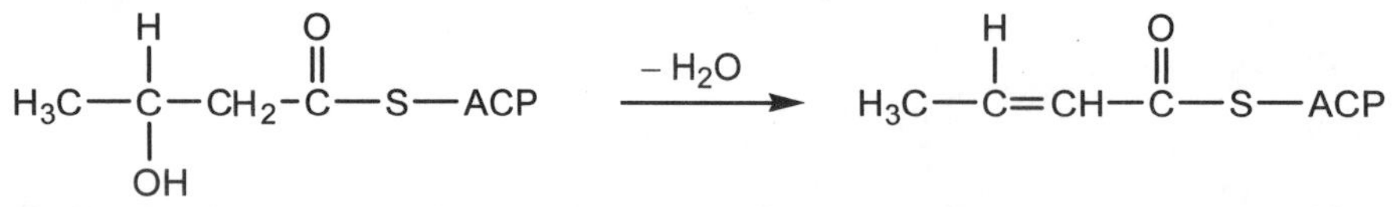

b. **Addition:** two hydrogen atoms are added to the carbonyl π bond. This transformation can also be considered as a **reduction** reaction because only hydrogen atoms are added.

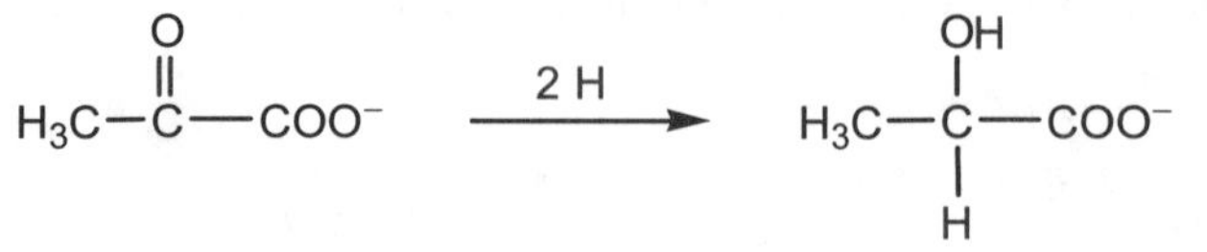

5.17. Follow the procedure outlined in the solution to Exercise 5.2.

a. **Addition:** two hydrogen atoms are added to the carbonyl π bond of the ketone. This transformation can also be considered as a **reduction** reaction because only hydrogen atoms are added.

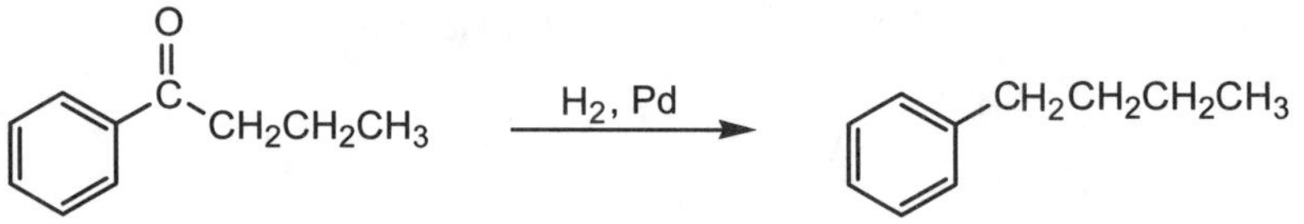

b. **Addition:** the cyano group and a hydrogen atom are added to the carbonyl π bond of the aldehyde.

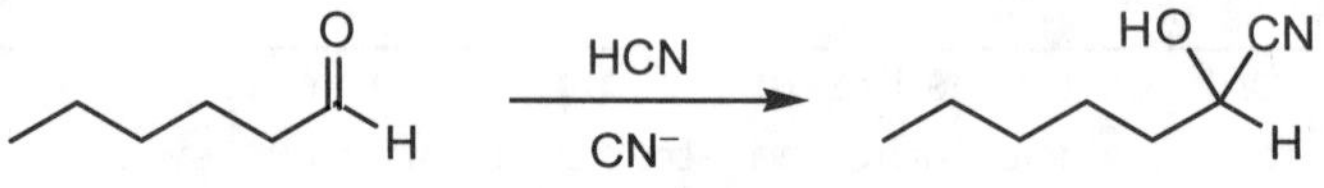

c. **Rearrangement:** the same atoms are found in both the reactant and the product, but the ways they are bonded has changed.

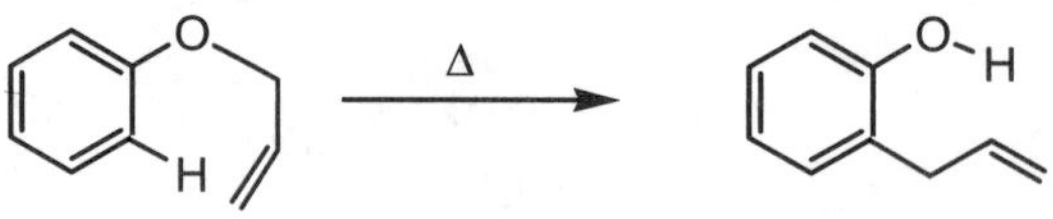

d. **Addition:** a hydrogen atom and a bromine atom are added to the carbon–carbon π bond of the alkene.

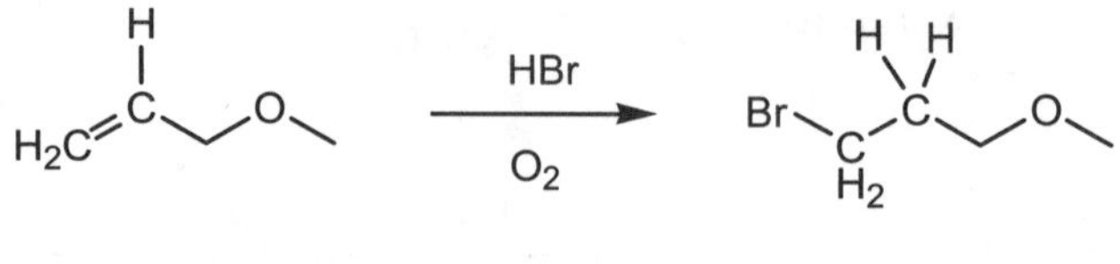

5.18. The stoichiometry of a reaction is defined by the ratio of reactant molecules. These ratios are most often 1:1, but 2:1 is not uncommon. A catalyst is identifiable because its atoms do not appear in the product.

a. The stoichiometry is 1:1 for the reaction between 1-phenyl-1-butanone and molecular hydrogen. Pd is a catalyst for this reaction.

5.18.   (continued)

b.   The stoichiometry is 1:1 for the reaction between hexanal and hydrogen cyanide.   Cyanide ion is a catalyst for this reaction.

c.   There is only one reactant, so stoichiometry has no meaning. There is no catalytic reagent for this reaction.

d.   The stoichiometry is 1:1 for the reaction between 3-methoxy-1-propene and hydrogen bromide. Molecular oxygen is a catalyst for this reaction.

5.19.   To write an equation for a reaction described by words, first interpret each of the compound names for the reactants and products. Then, place the reactant molecules at the left, draw an arrow, and write the product molecules at the right. Include the other information on the arrow, numbering separate steps if needed. The chemical equations for each part of this exercise are shown in the boxes.

    To classify each reaction, follow the procedure outlined in the solution to Exercise 5.2.   The changes that occur during the transformation are indicated by bonds and/or atoms in color in the equations with the dashed arrows.

a.   This is an **addition reaction** because a π bond is broken. It can also be classified as a **reduction reaction** because two H atoms add to the carbonyl group.

b.   This is an **addition reaction** because a π bond is broken. It can also be classified as a **reduction reaction** because two H atoms add to the carbonyl group. This transformation is a two-step procedure.

c.   This is a **substitution reaction** because one group replaces another (ArS replaces I or a propyl group replaces a proton).

5.19. (continued)

d. This is a **substitution reaction** because an oxygen atom replaces two hydrogen atoms. It can also be classified as an **oxidation reaction** because two H atoms are removed from the molecule.

5.20. The yield of a chemical reaction is the ratio, expressed as a percentage, of the number of moles of product obtained divided by the number of moles product expected. The amount of expected product is based on the number of moles of starting material and the stoichiometric ratio of reactant to product molecules.

a. The reaction begins with $(8.33 \div 90.19) = 0.0924$ mol of the thiol starting material, and the molar ratio of reactant to product (1:1) indicates that 0.0924 mol of product should be obtained. Instead, $(15.9 \div 180.3) = 0.0882$ mol are obtained. Therefore the yield is $(0.0882 \div 0.0924) \times 100 = 95.4\%$.

b. The reaction begins with $(3.38 \div 84.12) = 0.0402$ mol of the starting material, and the molar ratio of reactant to product (1:1) indicates that we expect to obtain 0.0402 mol of product. Instead, $(3.16 \div 84.12) = 0.0376$ mol are obtained. Therefore the yield is $(0.0376 \div 0.0402) \times 100 = 93.5\%$.

5.21. Follow the procedure outlined in Example 5.6. The following structural properties are important for the classification process:
- Species with electron pairs and/or π bonds, especially those that carry negative charges, are nucleophiles.
- Species in which an atom has a charge of +1 and no unshared electron pairs is an electrophile.
- Species that bear an overall charge of +1, even if unshared electron pairs are present, react mainly as electrophiles.
- Species that have an atom with only 6 electrons are electrophiles.
- Species with single electrons are radicals, and are not usually classified as nucleophiles or electrophiles because their reactions are not polar in nature.

Among the following species, the nucleophiles are (b.), (c.), (f.), (g.), and (i.). The electrophiles are (a.), (e.), and (h.) and are shown in color. The species that are neither nucleophiles nor electrophiles are (d.) and (j.) and are circled.

5.22. For bases that are amine derivatives, any substituent that increases the electron density on the nitrogen atom makes the base stronger. Electron-withdrawing influences will make the base weaker.

a. The second compound in this pair has an electron-donating methyl group closer to the nitrogen atom, which makes that molecule the stronger base.

b. The second compound in this pair has an electron-donating methyl group, which makes that molecule the stronger base. The other compound has an electron-withdrawing group (CF_3) attached to the benzene ring.

c. The second compound in this pair has an electron-donating methyl group, which makes that molecule the stronger base. The other compound has only a hydrogen atom attached to the carbon atom that bears the amino group.

d. Both molecules have a heteroatom two atoms away from the amino group. Because sulfur is less electronegative than oxygen, the first compound of the pair will have greater electron density at the nitrogen atom, so that molecule is the stronger base.

5.23. Follow the procedure outlined in Exercise 5.4. The pK_a values can be found in Table 5.1 and in Section 5.2e (see Tables 5.5 and 5.6).

a.

Acid ($pK_a \sim 15$)   Base   Base   Acid ($pK_a \sim 25$)

b.

+ CH_3CH_2COOH + $CH_3CH_2COO^-$

Base   Acid ($pK_a = 4.87$)   Acid ($pK_a = 5.3$)   Base

c.

—NH_2 + $(CH_3CH_2)_2NH_2$ —NH_3 + $(CH_3CH_2)_2NH$

Base   Acid ($pK_a = 11.0$)   Acid ($pK_a = 4.62$)   Base

5.24. An endergonic process is one in which the free energy of the products is higher than the free energy of the reactants. A one-step reaction has a single maximum in its reaction coordinate curve, which is the transition state, and the height of this maximum is equal to the free energy of activation.

5.24.   (continued)

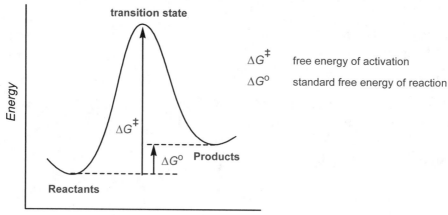

5.25.   An exergonic process is one in which the free energy of the products is lower than the free energy of the reactants. A two-step reaction has two maxima in its reaction coordinate curve, and the top of each is the transition state for each step. The minimum between the transition states corresponds to the energy of a reaction intermediate. The height of each maximum is equal to the free energy of activation for each step. For a reaction in which the first step is slower than the second, the first free energy of activation value has to be greater than the free energy of activation value for step 2. (The transition state energy for step 2 does *not* have to be lower than the transition state energy of the first step, although it is illustrated that way in the following figure.)

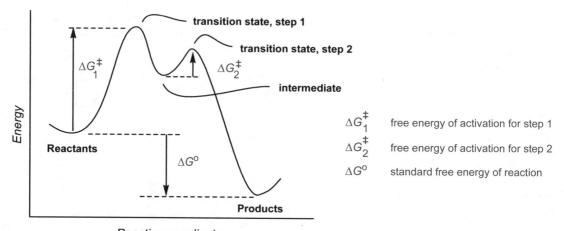

5.26.   Acid-base reactions occur so as to produce a weaker acid from a stronger one. Determine the approximate pK_a value of each acid from Table 5.1 or from the table on the inside front cover of the text. (These values are given below each acid in the following equations.) The product in each case is water, which has a pK_a value of about 15. The equilibrium lies toward the side of the equation having the acid with the larger pK_a value.

a. $H_3C-COOH$ + NaOH $\rightleftharpoons$ H_3C-COO^- Na^+ + H_2O **right**
            pK_a ~ 5   pK_a ~ 15

b.

    OH   + NaOH $\rightleftharpoons$ O^- Na^+ + H_2O **right**

    Cl   Cl
pK_a ~ 10   pK_a ~ 15

5.26.   (continued)

c.

SH + NaOH $\rightleftharpoons$ S⁻ Na⁺ + H_2O **right**

$pK_a \sim 11$ $pK_a \sim 15$

d.

+ NaOH $\rightleftharpoons$ + H_2O **left**

$pK_a \sim 16$ N⁻ Na⁺ $pK_a \sim 15$

e.

OH + NaOH $\rightleftharpoons$ O⁻ Na⁺ + H_2O **left**

$pK_a \sim 17$ $pK_a \sim 15$

5.27.   Follow the procedure outlined in the solution to Example 5.4

a.   All of the compounds except **III** have at least one substituent on the carbon atom adjacent to the COOH group. Compound **II** has two electron-donating groups at that position, making it the least acidic substance, and compound **IV** has the most electron-withdrawing substituent, which makes it the most acidic.

b.   All of the compounds have a heteroatom close to the COOH group. Compound **I** has the more electronegative substituent, which is also closest to the COOH group, so it is the most acidic molecule. Compound **II** has an oxygen atom, which is the less electronegative than F, and it is farthest from the COOH group, so compound **II** is the least acidic.

5.28.   Acid-base reactions occur so as to produce a weaker acid from a stronger one. Hydrochloric acid is an aqueous solution of HCl. Its pK_a value is about –1.7.

HCl + H_2O $\rightleftharpoons$ H_3O^+ + Cl⁻

$pK_a \sim -7$ $pK_a = -1.7$

Determine the approximate pK_a value of each acid from Table 5.1 or from the table on the inside front cover of the text. (These values are given below each acid in the following equations.) The equilibrium lies toward the side of the equation having the acid with the larger pK_a value.

5.28. (continued)

a. $CH_3CH_2NH_2$ + H_3O^+ ⇌ $CH_3CH_2\overset{+}{N}H_3$ + H_2O **right**

pK_a = -1.7   pK_a ~ 9

b.

OH + H_3O^+ ⇌ $\overset{+}{O}H_2$ + H_2O **~ equal**

pK_a = -1.7   pK_a ~ -2

c.

$SO_3^-\ Na^+$ + H_3O^+ ⇌ SO_3H + H_2O **left**

pK_a = -1.7   pK_a ~ -6.5

d.

+ H_3O^+ ⇌ + H_2O **left**

pK_a = -1.7   pK_a ~ -6

5.29.   Follow the procedure outlined in the solution to Exercise 5.28.

a. $CH_3CH_2NH_2$ + HOAc ⇌ $CH_3CH_2\overset{+}{N}H_3$ + OAc^- **right**

pK_a ~ 5   pK_a ~ 9

b.

OH + HOAc ⇌ $\overset{+}{O}H_2$ + OAc^- **left**

pK_a ~ 5   pK_a ~ -2

c.

$SO_3^-\ Na^+$ + HOAc ⇌ SO_3H + NaOAc   **left**

pK_a ~ 5   pK_a ~ -6.5

d.

+ HOAc ⇌ + OAc^- **left**

pK_a ~ 5   pK_a ~ -6

5.30.   First, interpret the names according to the procedures outlined in Chapter 1.

a.   b.   c.   d.

CF_3-CH_3

Next, assess the acidity values of the different types of protons in each molecule. For the ketone (a.), the most acidic protons are attached adjacent to the carbonyl group. For the amide (c.), the protons attached to the N atom are the acidic ones. For (d.), the alcohol OH group bears the acidic proton. Compound (b.) is the least acidic because it is a hydrocarbon derivative.   Amide (c.) is probably slightly more acidic than the alcohol molecule (both have pK_a values ~ 15) because the aromatic ring is electron withdrawing.

5.30. (continued)

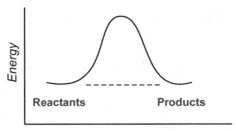

$pK_a \sim 20$ CF_3-CH_3 $pK_a \sim 15$ $pK_a \sim 15$

             $pK_a \sim 40\text{-}50$

5.31. Using the equation $\Delta G° = -RT\ln K_{eq}$, we calculate that when $K_{eq} = 1$, $\Delta G° = 0$. This means that the reactants and products have the same energy values as shown in the reaction coordinate diagram below.

Energy

Reactants   **Products**

Reaction coordinate

5.32. The side chains of glutamic and aspartic acids each have a carboxylic acid group. The conjugate base of a carboxylic acid is the carboxylate ion. The structures of these amino acid derivatives are shown below.

D-Aspartic acid   **D-Glutamic acid**

5.33. Follow the procedure outlined in the solution to Example 5.7.

a. A reactant with a positive charge usually functions as an electrophile in a polar transformation. A π bond of the benzene ring plays the role of the nucleophile. The first step in this reaction, as in any polar reaction, is the interaction between the nucleophile and the electrophile, which is depicted using a curved arrow that starts at the electron pair of the nucleophile (the π bond) and points to the electrophile, the nitrogen atom.

nucleophile (π bond)

NO_2^+

electrophile (N atom)

b. The proton is a common electrophile, so when a strong acid (often indicated as H⁺) is present as one of the reactants, it is the electrophile. Any heteroatom or π bond in the other reactant can play the role of the nucleophile. The first step in this reaction, as in any polar reaction, is the interaction between the nucleophile and the electrophile, which is depicted using a curved arrow that starts at the electron pair of the nucleophile—the oxygen atom of the alcohol molecule—and points to the electrophile, H⁺.

H^+

electrophile

nucleophile (O atom)

5.33. (continued)

c. A reactant with a negative charge usually functions as the nucleophile in a polar transformation. Methoxide ion is a good nucleophile, and its oxygen atom bears a −1 charge. The carbon atom of the carbonyl group in the other reactant plays the role of the electrophile—the carbon atom has a partial positive charge because oxygen is the more electronegative atom of the pair. The first step in this process is the interaction between the nucleophile and the electrophile, which is depicted using a curved arrow that starts at the electron pair of the nucleophile and points to the electrophile, the carbon atom. The electrons in the π bond of the carbonyl group move so that the carbon atom maintains four bonds.

d. The proton is a common electrophile, so when a strong acid (often indicated as H⁺) is present as one of the reactants, H⁺ is the electrophile. Any heteroatom or π bond in the other reactant can play the role of the nucleophile. The first step in this reaction, as in any polar reaction, is the interaction between the nucleophile and the electrophile, which is depicted using a curved arrow that starts at the electron pair of the nucleophile—the π bond of the alkene—and points to the electrophile, H⁺.

5.34. Lysine has the amino group in its side chain, so its conjugate acid form is the ammonium ion, which will have a pK_a value equal to about 9, according to the data in the table on the inside cover of the text. The other amino group will also be protonated. Histidine has an imidazole ring, which has a nitrogen atom capable of being protonated by acid.

5.35. Follow the procedures outlined in the solution to Exercise 5.21. The side chains of Lys, His, and Cys all have heteroatoms with unshared pairs of electrons. These groups are nucleophilic.

5.35. (continued)

The side chains of Asp, Glu, and Tyr all have OH groups that can donate a proton, which is an excellent electrophile. Therefore we classify these side chains as electrophilic.

Asp   **Glu**   **Tyr**

The side chains of Asp, Glu, and Tyr can be converted to their conjugate base forms, which would be expected to be nucleophilic.

Asp   **Glu**   **Tyr**

The other amino acids in Table 5.8 have side chains that are neither nucleophilic nor electrophilic.

Val   **Phe**

5.36. Follow the procedure outlined in the solution to Exercises 5.2 and 5.16.

a. **Elimination:** two hydrogen atoms are removed to form a π bond. This transformation can also be classified as an **oxidation reaction** (removal of H atoms).

b. **Substitution:** a hydrogen atom is replaced by an OH group. This transformation can also be considered as an **oxidation reaction** because a heteroatom (oxygen) is incorporated into the product.

c. **Addition:** a hydrogen atom and an OH group are added to the alkene π bond.

SUBSTITUTION REACTIONS OF ALKYL HALIDES

6.1. As described in Chapter 5, a nucleophile is the species in a reaction that carries a negative charge or has at least one unshared pair of electrons. To identify the nucleophile in the reactions shown in Figure 6.1, look for the portion of the product that has changed in relation to the structure of the starting material. If a protic acid is a reactant, the nucleophile is likely its conjugate base.

a. CH_3S^-

b. Br^- (the conjugate base of the acid HBr)

c. $(CH_3)_3N$: The nitrogen atom of an amine is a nucleophile because it has an unshared electron pair.

d. I^-

6.2. Nucleophilicity increases with basicity within a period (horizontal row of the Periodic Table) or for a given atom. Within a group (a vertical column of the Periodic Table), nucleophilicity increases as atoms become larger (S > O; I > Br > Cl, etc.).

a. **SH^- or OH^-**
   Sulfur is larger than oxygen and they are in the same group (16), so SH^- is the better nucleophile.

b. **$P(CH_3)_3$ or $N(CH_3)_3$**
   Phosphorus is larger than nitrogen and they are in the same group (15), so $P(CH_3)_3$ is the better nucleophile.

c. **NH_3 or H_2O**
   Both N and O are second period elements and because ammonia is a stronger base than water, NH_3 is the better nucleophile.

6.3. To classify an organohalide according to its type, first characterize the type of carbon atom to which the halogen atom is bonded.

   To predict the type of substitution reaction that each will participate in, make use of the data in the following summary:

   S_N1:   all types of 3°, all types of 2°, 1°-benzylic, 1°-allylic

   S_N2:   Methyl, all types of 1°, all types of 2°

   Neither: alkenyl, alkynyl, aryl

a. $CH_3CH_2CH_2Br$	b.	c.	d. $CH_3-CH-CH_3$	e.
S_N2	S_N2	neither	S_N1 or S_N2	S_N1

6.4. The organohalide in this transformation is tertiary alkyl halide, so substitution will occur by the SN1 mechanism. The first step is dissociation of chloride ion to form a carbocation. Then the nucleophile, methanol, reacts with the electrophilic carbon atom (step 2). Finally, the solvent acts as a base to deprotonate the oxonium ion intermediate, forming the ether molecule product (step 3).

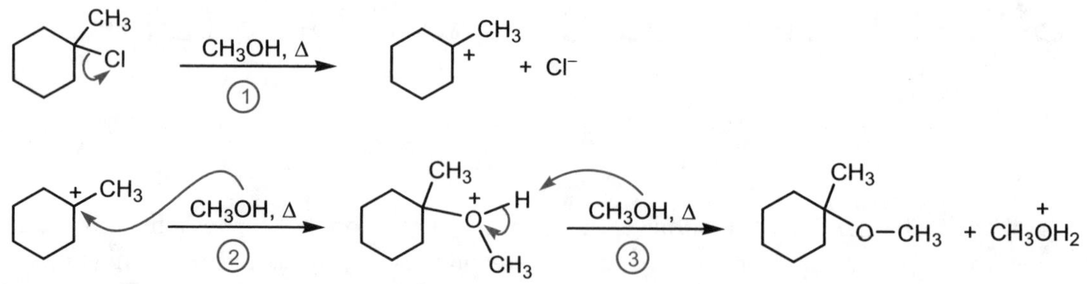

6.5. The so-called "no-bond" resonance forms that are used to represent hyperconjugation place a positive charge on one of the protons (shown in color) originally attached to a carbon atom *adjacent* to the one that bears the positive charge. This delocalization also produces a carbon-carbon double bond. Only the carbon-hydrogen bond that is aligned with the *p* orbital on the cationic carbon atom is able to participate, which explains why each methyl group contributes only one resonance form (the choice of the hydrogen atoms involved in these no-bond resonance forms was made arbitrarily).

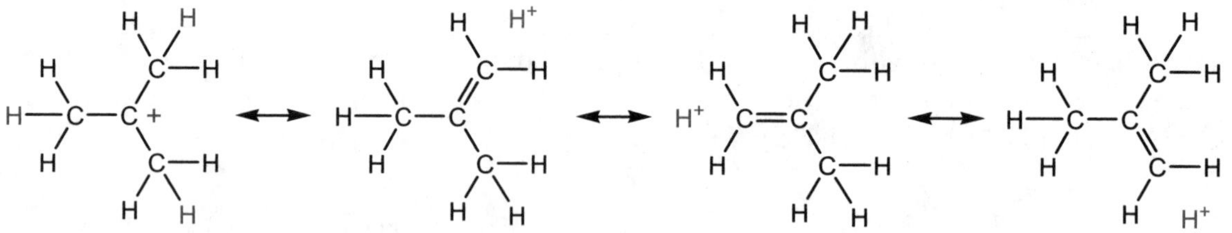

6.6. Assign a type to each carbocation, and then rank them according to the recognized order of carbocation stabilities (most to least stable):

benzylic ~ allylic > 3° > 2° >> 1° > methyl > phenyl, alkenyl

(Within the benzylic and allylic carbocation categories, the stability order is 3° > 2° > 1°.)

3° benzylic ~ 3° allylic > 3° > 2° >> aryl

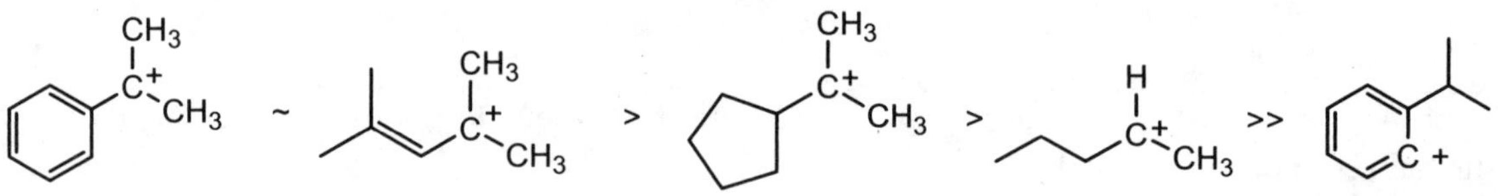

6.7. The rates of SN1 reactions mirror the order of carbocation stability: benzylic ~ allylic > 3° > 2°. The following do not react by the SN1 pathway: 1°, methyl, phenyl, and alkenyl. Therefore if you classify the types of organohalides, you can predict the relative rates of their reactions according to their types (no reaction = slowest rate).

a. 2-Bromoheptane 3-Bromo-3-methylhexane 2-Bromo-2-heptene

2° 3°: Fastest Alkenyl: N.R. = Slowest

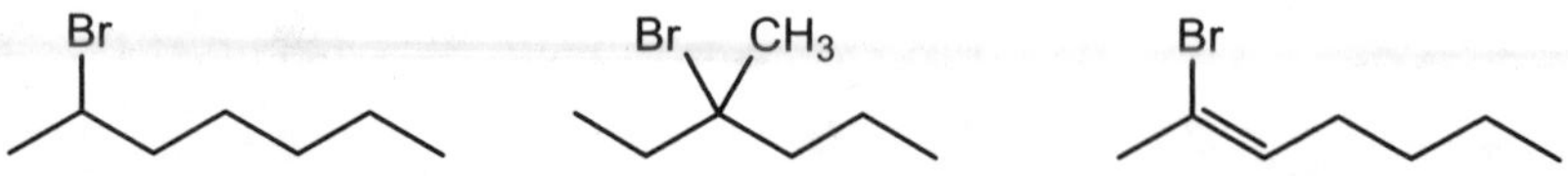

6.7. (continued)

b.

Aryl: N.R. = Slowest   2°   3°: Fastest

6.8.   As stated, the reaction given in this exercise occurs by the S_N1 pathway, which is recognizable because the solvent is aqueous methanol, a reagent combination that comprises only weak base nucleophiles. A carbocation is formed in the first step of the S_N1 reaction. For direct substitution, the nucleophile (water) reacts with the first-formed carbocation (step 2) and is followed by an acid-base reaction (step 3). This product is **2-phenylcyclohexanol**.

To predict likely rearrangement products, look at the structure of the initially generated carbocation and see if it is possible to generate a more stable carbocation simply by moving a hydrogen atom or alkyl group from an adjacent atom(s). If a more stable carbocation *can* be generated (step R below), the subsequent substitution products will result from reaction of the nucleophile with this second-formed carbocation (steps 2′ and 3′). This product is **1-phenylcyclohexanol**.

2° carbocaton   3° and benzylic   + H_3O^+
                             carbocaton

6.9.   This reaction occurs by the S_N1 pathway, recognizable because the solvent that is used is aqueous–methanol, a reagent combination that comprises only weak base nucleophiles. In addition, the substrate is a secondary benzylic alkyl bromide, which will form a stable carbocation upon dissociation of the leaving group. When a carbocation forms, the stereochemical course of a reaction is racemization at that center because the carbocation is achiral. The nucleophile (water) can react at each face of the carbocation (top and bottom as drawn), so the products are enantiomers.

3° and benzylic
carbocaton
achiral

6.10. These reactions proceed via the S$_N$2 pathway, inferred because an aprotic solvent and a good nucleophile [hydroxide ion in part (a.) and iodide ion in part (b.)] are present in each. The stereochemical course of an S$_N$2 reaction is inversion of configuration at the carbon atom bearing the leaving group. Part (b.) is a rare example of a chiral primary substrate; deuterium (D) is an isotope of hydrogen (H), so the carbon atom that bears the leaving group has four different groups and is chiral.

a.

b.

6.11. The reaction between a primary amine and a primary alkyl halide normally gives both mono- and disubstitution products, in which the nitrogen atom of the amine replaces the leaving group (in this case a bromide ion). The initial substitution product (reaction 1) is the hydrobromide salt of the dialkylamine. It reacts with the original amine in an equilibrium process (reaction 2). This dialkylamine reacts with a second molecule of the alkyl bromide (reaction 3), so the second product has two alkyl groups from the original alkyl halide attached to the nitrogen atom of the original amine component. This second product is also formed as its hydrobromide salt. Upon treatment with aqueous base, these hydrobromide salts (**A** and **B**) are converted to the neutral amine products.

6.12. The reaction between an amine and a dihaloalkane proceeds via successive S$_N$2-type reactions. Normally, a base is added to remove the HX that is formed. In this transformation, the excess amine that is added from the start can react in steps (2) and (4) to get rid of the HBr that is produced.

6.13. Follow the procedures outlined in Example 6.4: classify the type of substrate and nucleophile that are present, and then use the data in Table 6.4 to predict which mechanism will operate.

6.13. (continued)

a. The substrate in this reaction is a primary, benzylic alkyl bromide. The nucleophile is neutral and a weak base, and the solvent is protic. The data in Table 6.4 indicate that this reaction should occur via the S_N1 mechanism. The starting materials are achiral, so there is no reaction stereochemistry to evaluate.

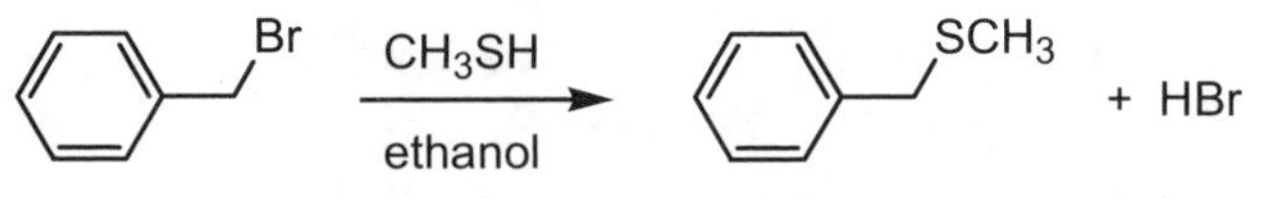

b. The substrate in this reaction is a secondary alkyl chloride. The nucleophile is good and a moderate base, and the solvent is aprotic. The data in Table 6.4 indicate that this reaction should occur via the S_N2 mechanism. The starting material is chiral at the carbon atom bearing the leaving group, so the reaction should occur by inversion of configuration at the chiral center.

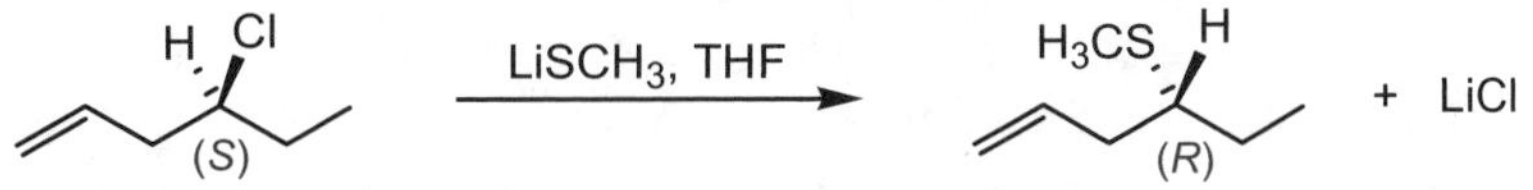

6.14. A substitution reaction is viable when three conditions are met: (1) a nucleophile is present, (2) the substrate molecule has at least one sp^3-hybridized carbon atom, which (3) is attached to a good leaving group. Each of the following reactions meets these criteria. If the leaving group is attached to a 1° or 2° carbon atom, then the mechanism is likely S_N2, and the configuration of a chiral center to which the leaving group is attached will be inverted.

a. The nucleophile is the cyanide ion, which replaces the bromide ion with inversion of configuration at the 2° carbon atom.

b. The nucleophile is the iodide ion, which replaces the chloride ion with inversion of configuration at the 2° carbon atom.

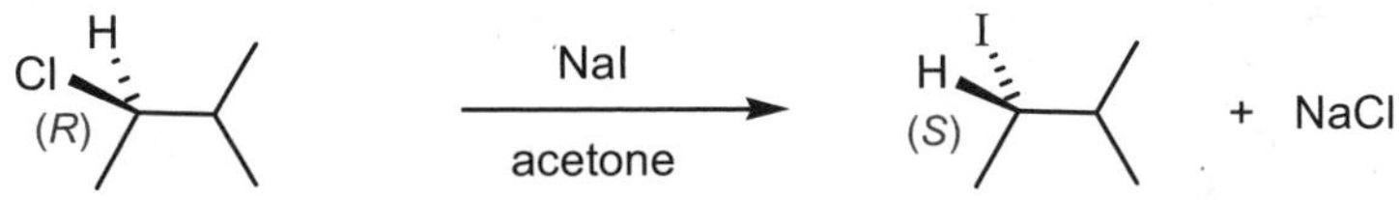

c. The nucleophile is dimethylamine, an uncharged nucleophile. It replaces the bromide ion, but the product is ionic because both reactants are neutral. Amines are reasonable bases so the proton on the nitrogen atom is retained. The substrate molecule is achiral, so stereochemistry is not an issue.

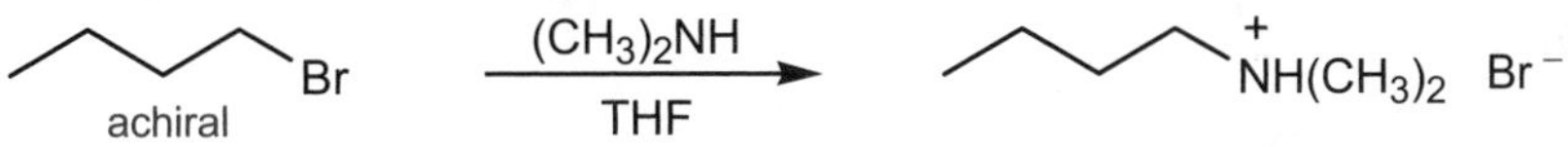

d. The nucleophile is the methanethiolate ion, which replaces the bromide ion with inversion of configuration at the 2° carbon atom. Even though the carbon atom bearing the leaving group is not chiral, its stereochemistry is defined by the presence of another substituent attached to the ring, and a geometric isomer is formed.

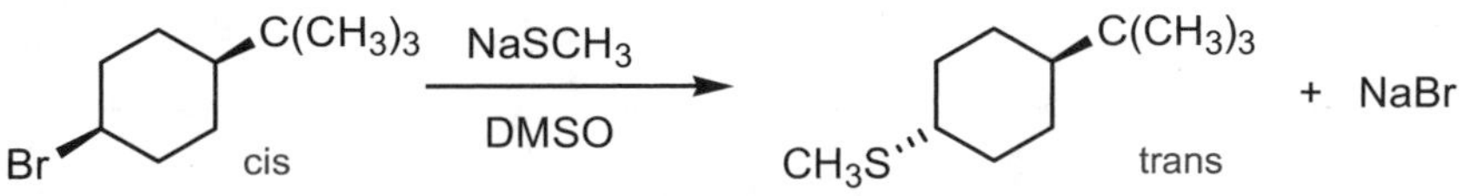

6.15. The rate of a substitution reaction depends mainly on the strength of the nucleophile (better nucleophiles react faster) and the structure of organic substrate undergoing substitution. For an S$_N$1 reaction (usually a solvolysis process), the rate decreases in the order 3° > 2° >> 1°. For an S$_N$2 reaction, the rate decreases in the order methyl > 1° > 2° >> 3°. The faster reaction in each pair is enclosed in the colored box.

a. The substrate in each reaction is the same, so we have to consider the relative strengths of the nucleophiles. For a given atom type with the same charge, nucleophilicity parallels base strength. Hydroxide ion is a stronger base than acetate ion (in both, the oxygen atom is the nucleophilic atom), so the first reaction goes faster. Methyl substrates react via the S$_N$2 pathway.

$$CH_3I \ + \ OH^- \ \xrightarrow{\text{DMSO}} \ CH_3OH \ + \ I^-$$

$$CH_3I \ + \ OAc^- \ \xrightarrow{\text{DMSO}} \ CH_3OAc \ + \ I^-$$

b. The nucleophile in each reaction is the same, so we have to consider the substrate structures. The nucleophile is the solvent, which is water, so each transformation is a solvolysis reaction and probably occurs via the S$_N$1 pathway. Therefore, the order of reactivity is 3° > 2°, so the first reaction occurs faster.

$$(CH_3)_3CBr \ + \ H_2O \ \longrightarrow \ (CH_3)_3COH \ + \ HBr$$

$$(CH_3)_2CHBr \ + \ H_2O \ \longrightarrow \ (CH_3)_2CHOH \ + \ HBr$$

c. The nucleophile in each reaction is the same, so we have to consider the substrate structures. The nucleophile is a good base and the solvent is aprotic, so the reaction occurs via the S$_N$2 pathway. Therefore, the order of reactivity is 1° > 2°, and the second reaction occurs faster.

d. The nucleophile in each reaction is the same, so we have to consider the substrate structures. The nucleophile is a charged species, and both substrates are 2° bromoalkanes, so these transformations probably occur by the S$_N$2 pathway. In an S$_N$2 reaction, the carbon atom in the transition state is five-coordinated, and the carbon–carbon bond angles within the substrate approach 120°. The cyclobutyl compound has more angle strain to begin with, so to go from 109° in the reactant to 120° in the transition state is more difficult for the substrate with the four-membered ring. Therefore, bromocyclohexane reacts faster.

6.16. In solvolysis reactions, a carbocation is formed as an intermediate, so rearrangement reactions are possible. Consider, therefore, whether an atom or group can readily migrate within the intermediate to form a more stable carbocation. If so, then rationalize the formation of the rearranged product by composing a suitable mechanism. First, write out the overall reaction:

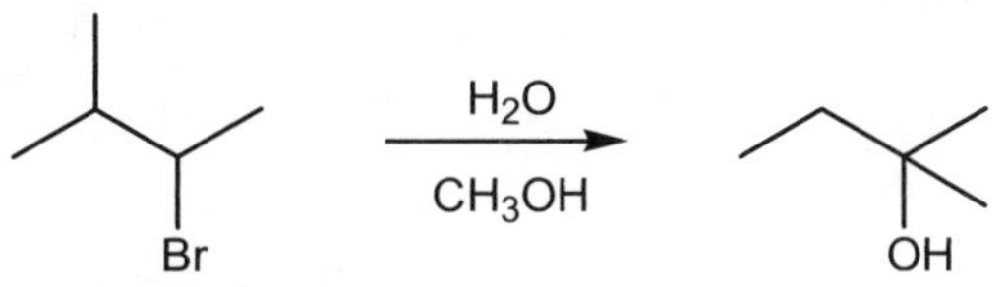

Then, write the steps of the mechanism. The bromide ion dissociates (step 1), and then a hydrogen atom migrates to convert the 2° carbocation to a 3° carbocation (step 2). Water intercepts this 3° carbocation (step 3), and an acid-base reaction yields the alcohol product (step 4).

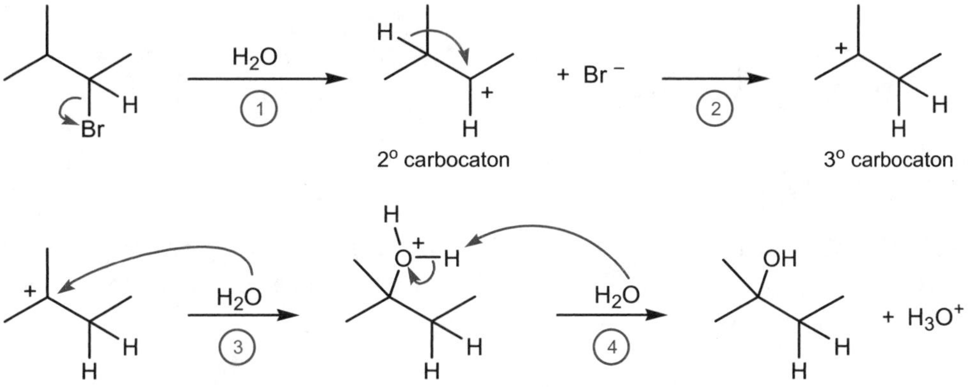

6.17. The reaction described in this exercise is straightforward, comprising substitution of a bromide ion by the iodide ion with concomitant racemization.

The fact that the reaction rate depends upon the concentrations of both substrate and nucleophile tells us that the mechanism is S$_N$2. The normal stereochemical result of an S$_N$2 reaction is inversion of configuration which is in contrast with the experimental results. Look at the mechanism.

The *initial* S$_N$2 reaction occurs as expected:

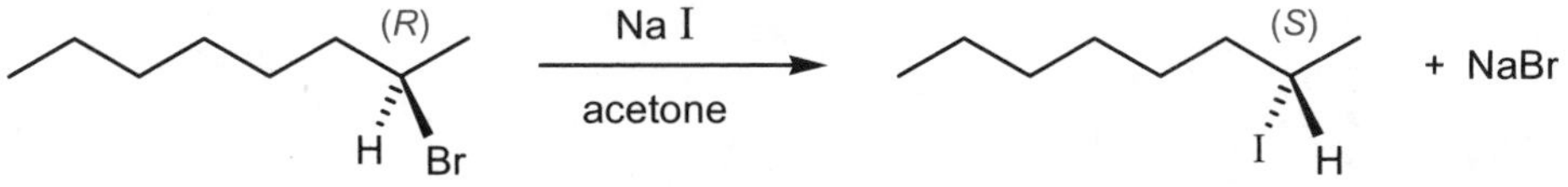

This product has a good leaving group (iodide ion) attached to an aliphatic carbon atom, so it can also undergo the S$_N$2 reaction. The product of this second substitution reaction has the same bonding structure as the starting material, but it has the inverted stereochemistry. A racemic product is formed, therefore, after some time has passed.

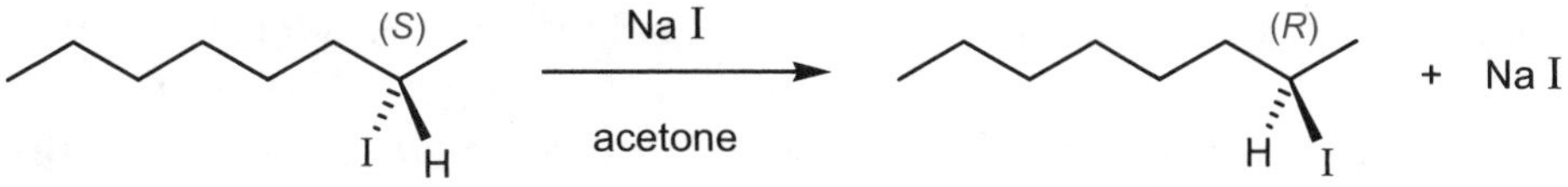

Racemization of a carbon atom with a good leaving group will occur any time the nucleophile and the leaving group are the same. This process normally only occurs when the nucleophile is a halide ion, especially the iodide ion.

6.18. Draw Lewis structures (and resonance forms) for the given ions according to the procedures outlined in the solution to Exercise 2.4. Because the electrons are delocalized among several atoms, the nucleophilicity of any one atom is low, which makes the ion itself only a weak nucleophile.

6.19. When an alkyl chloride is treated with a soluble silver salt, the silver(I) ion reacts with the halide atom (chlorine in this example) to form an adduct (step 1). The insoluble nature of AgCl leads to its precipitation with formation of the 2° carbocation (step 2).

This carbocation is intercepted by a molecule of methanol (step 3), and an acid-base reaction yields the methoxy compound (step 4).

6.20. To predict the structure of the product of a given substitution reaction, you must first decide whether the reaction will occur. Specifically, look to see whether a good leaving group (a chloride, bromide, or iodide ion) is attached to an sp^3-hybridized carbon atom, and then whether a good nucleophile is present. A chiral center attached to a leaving group undergoes inversion of configuration in the S$_N$2 mechanism. The stereochemical outcome of an S$_N$1 reaction is racemization.

a. The substrate is a 3° allylic alkyl chloride, the nucleophile is water, a weak base, and the solvent is protic. These features are associated with the S$_N$1 mechanism. An OH group will replace the chlorine atom. The substrate is achiral and no chiral center is generated during the course of the reaction, so the product is achiral as well.

6.20.   (continued)

b.   The substrate is a 2° benzylic alkyl chloride, the nucleophile is the iodide ion, a weak base, and the solvent is protic. These features are associated with the S$_N$1 mechanism. The iodine atom will replace the chlorine atom. The substrate is chiral, so a racemic mixture of product molecules will be formed.

c.   The substrate is an alkenyl chloride, which means the leaving group is attached to an sp^2-hybridized carbon atom. Therefore, no reaction will occur.

d.   The substrate has two 1° alkyl bromide groups, and the nucleophile is an amine molecule, a moderate base. These features are associated with the S$_N$2 mechanism. The amine will react twice (see Exercise 6.12 for a similar reaction and its mechanism). The substrate is achiral and no chiral center is generated during the course of the reaction, so the product is achiral as well.

e.   The substrate is a 1° alkyl bromide, the nucleophile is the azide ion, a weak to moderate base but an excellent nucleophile, and the solvent is aprotic. These features are associated with the S$_N$2 mechanism. The azide group will replace the bromine atom. The substrate is achiral and no chiral center is generated during the course of the reaction, so the product is achiral as well.

f.   The substrate is a 3° benzylic alkyl chloride, the nucleophile is a thiol molecule, a weak base, and the solvent is protic. These features are associated with the S$_N$1 mechanism. The CH$_3$S group will replace the chlorine atom. The substrate is achiral and no chiral center is generated during the course of the reaction, so the product is achiral as well.

g.   The substrate is an aryl iodide, which means the leaving group is attached to an sp^2-hybridized carbon atom. Therefore, no reaction will occur.

6.20.   (continued)

h.   The substrate is a 2° alkyl chloride, the nucleophile is the $CH_3CH_2S^-$ ion, a moderate base and an excellent nucleophile, and the solvent is aprotic.   These features are associated with the S_N2 mechanism.   The CH_3CH_2S group will replace the chlorine atom.   The substrate is chiral, so the product will have the opposite configuration (inversion).

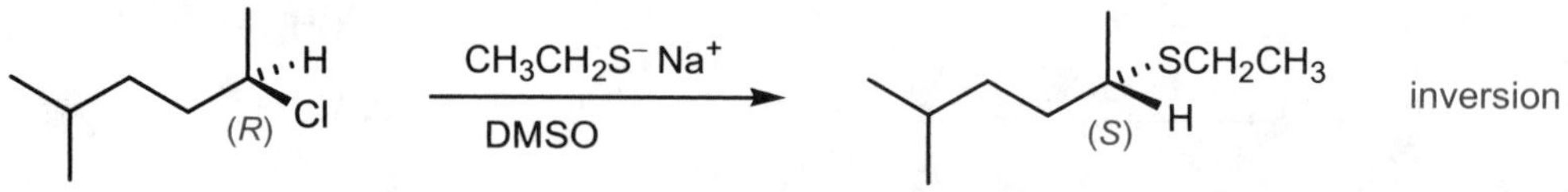

i.   The substrate is a 2° alkyl bromide, the nucleophile is the I^- ion, a weak base and an excellent nucleophile, and the solvent is aprotic.   These features are associated with the S_N2 mechanism.   The substrate is achiral and no chiral center is generated during the course of the reaction, so the product is achiral as well.

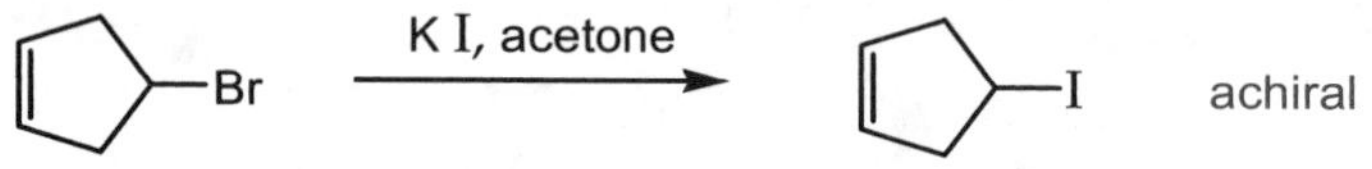

j.   The substrate is a 1° alkyl bromide, the nucleophile is water, a weak base, and the solvent is protic.   The nucleophile and solvent are features normally associated with the S_N1 mechanism, but a primary alkyl halide cannot undergo carbocation formation.   Therefore, no reaction occurs

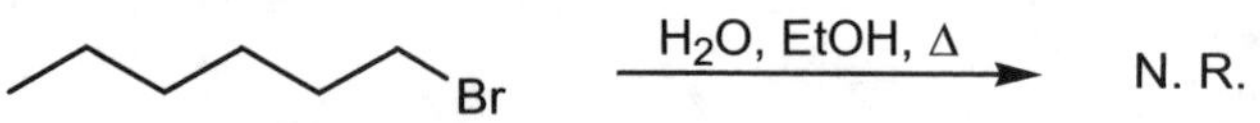

6.21.   The equation given in the exercise shows that the bond to be made is the one between the carbon atom adjacent to the carboxylic acid group and the oxygen atom of the phenolic compound.   This bond is indicated below by the wavy, colored line.

Thus, the reactants on the left side of the equation are the ones needed to make the aryloxy carboxylic acids illustrated in the exercise, and you need only convert the general groups "R" and "Ar" to their specific structures as shown below.

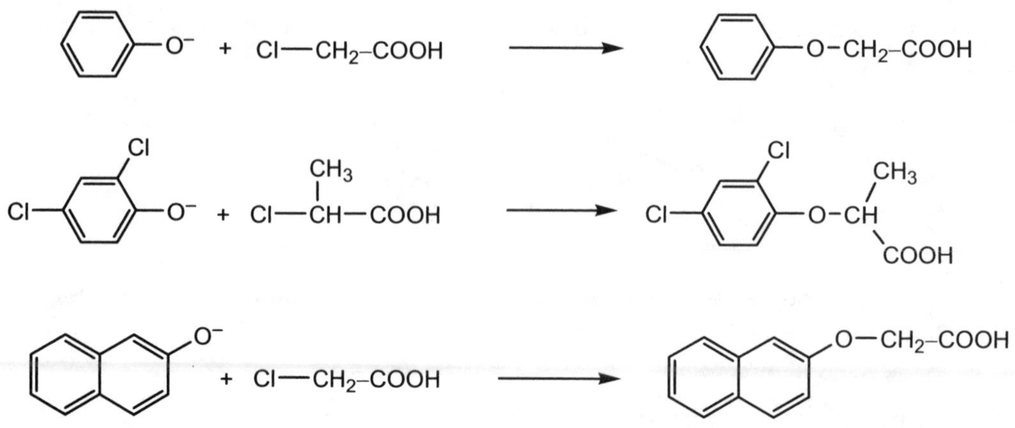

6.22.   For this type of exercise, you first decide what type of reaction is occurring—they are all examples of substitution reactions (one group replaces another). Therefore, use the structural type of the substrate to decide which type of nucleophile and solvent to use according to the mechanism that is needed.

a.   The substrate is a 1° alkyl bromide, so S_N2 conditions are required. The nucleophile should be a form of C_6H_5S (the replacing group) that is a moderate base and an excellent nucleophile, and the solvent should be aprotic.

b.   The substrate is a 2° alkyl chloride, and inversion of the configuration has occurred, so S_N2 conditions are required. The nucleophile should be a form of N_3 (the replacing group) that is a weak to moderate base and an excellent nucleophile, and the solvent should be aprotic.

c.   The substrate is a 3° benzylic alkyl bromide, so S_N1 conditions are required. The nucleophile should be a form of $CH_3CH_2CH_2S$ (the replacing group) that is a weak base (preferably neutral), and the solvent should be protic.

d.   The substrate is a 3° alkyl iodide, so S_N1 conditions are required. The nucleophile should be a form of CH_3CH_2O (the replacing group) that is a weak base (preferably neutral), and the solvent should be protic.

e.   The substrate is a 1° benzylic alkyl bromide, so either S_N1 or S_N2 conditions could be used. For the S_N2 pathway, the nucleophile should be a form of $(CH_3)_2CHO$ (the replacing group) that is a moderate to strong base, and the solvent should be aprotic.   For the S_N1 pathway, the nucleophile should be a form of $(CH_3)_2CHO$ (the replacing group) that is a weak base (preferably neutral), and the solvent should be protic.

6.23. In order for a nucleophilic substitution reaction to occur, the substrate molecule should have a good leaving group (weak base) attached at an sp^3-hybridized carbon atom, and a suitable nucleophile should be present.

a. The chloride ion is a good leaving group, and the iodide ion is an excellent nucleophile. This reaction will occur readily.

$$CH_3CH_2Cl \ + \ I^- \longrightarrow CH_3CH_2I \ + \ Cl^-$$

b. The bromide ion is a good nucleophile, but the hydroxide ion is a poor leaving group because it is a strong base. This reaction will not occur.

$$CH_3CH_2OH \ + \ Br^- \longrightarrow CH_3CH_2Br \ + \ OH^- \qquad \textbf{N.R.}$$

c. The bromide ion is a good leaving group, and the cyanide ion is a good nucleophile. This reaction will occur readily.

$$CH_3CH_2Br \ + \ CN^- \longrightarrow CH_3CH_2CN \ + \ Br^-$$

d. The methanethiolate ion is an excellent nucleophile, but the cyanide ion is a moderate base and will be a poor leaving group. This reaction will not occur.

$$CH_3CH_2CN \ + \ CH_3S^- \longrightarrow CH_3CH_2SCH_3 \ + \ CN^- \quad \textbf{N.R.}$$

6.24. In the reaction of a primary allylic halide, the S_N2 pathway is straightforward: the nucleophile will react backside at the carbon atom that bears the leaving group, displacing the bromide ion.

The S_N2' process works in a similar way except that the π electrons of the carbon-carbon double bond are involved, and it is those electrons that actually react at the backside at the carbon atom bearing the leaving group. The nucleophile reacts at the end of the double bond farthest from the position bearing the leaving group.

6.25. In the S_N2' pathway illustrated in the solution to Exercise 6.24, the π electrons of the carbon–carbon double bond are involved as an "auxiliary" nucleophile. In the S_N1 pathway, a carbocation is formed, and resonance delocalization creates two reactive centers:

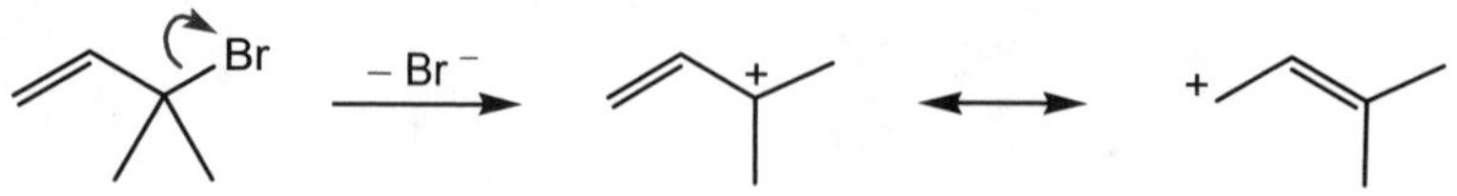

The nucleophile can react with either resonance form of this carbocation. In this example, water is the nucleophile, so the products are isomeric alcohols.

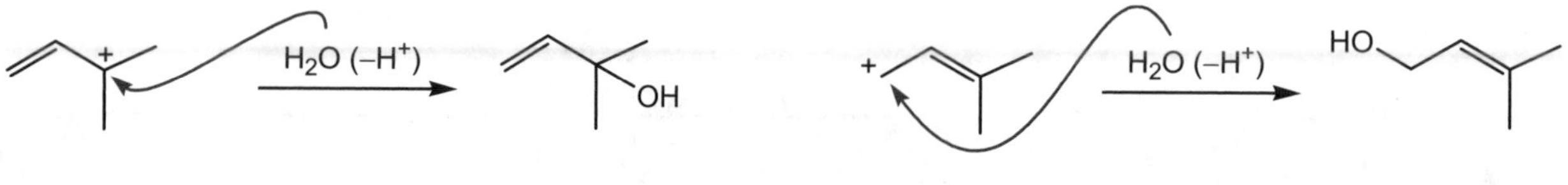

SUBSTITUTION REACTIONS OF ALCOHOLS AND RELATED COMPOUNDS

7.1. The substitution reactions of primary alcohols must occur via the S_N2 pathway because a primary carbocation cannot be formed under these conditions. The OH group is protonated (step 1), and then the bromide ion nucleophile displaces the good leaving group, H_2O (step 2).

7.2. When you see a reaction in which one group has replaced another and the carbon skeleton is unchanged, then the reaction type is *substitution*. If the starting molecule is chiral and the product has the opposite configuration at the carbon atom at which the replacement process has occurred, then the mechanism by which substitution occurs is S_N2. When the group being replaced is an OH group, then it first has to be converted to a good leaving group by a process that occurs with *retention* of configuration. Making an alkyl sulfonate ester is the best way to convert OH to a good leaving group while retaining the stereochemistry of the carbon atom to which the OH group is attached. Then, a good nucleophile is used in the second step to carry out the substitution process. Because the mechanism is S_N2, the nucleophile should be a weak or moderate base and the solvent should be aprotic.

7.3. PBr₃ reacts with an alcohol group to replace the OH proton. This step creates a good leaving group and generates bromide ion as a nucleophile. If the process is concerted, as shown in the text, then the mechanism is S_N2, and the stereocenter undergoes inversion of configuration in the second step.

7.4.   When an alcohol reacts with thionyl chloride, the OH group is replaced by a chlorine atom. The reaction can be done in the presence of pyridine to remove the HCl that is formed, or it can be heated as shown in part (b.) to drive gaseous HCl from the reaction mixture.

a.

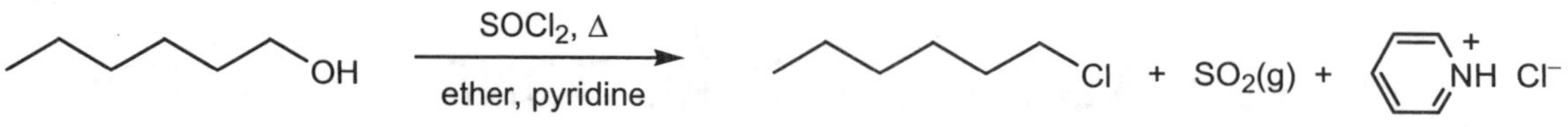

b.

7.5.   The Mitsunobu reaction occurs via the S_N2 mechanism with primary and secondary alcohols as the substrates. The reactants are triphenylphosphine, diethyl diazodicarboxylate, and an acidic reagent HX, in which the conjugate base X (shown in color in the following equations) is the nucleophile that replaces the OH group of the starting alcohol. In part (a.), $X = SC_6H_5$ and in part (b.), $X = N_3$. If the starting alcohol is chiral at the carbon atom that bears the OH group, then the configuration of that carbon atom is inverted.

a.

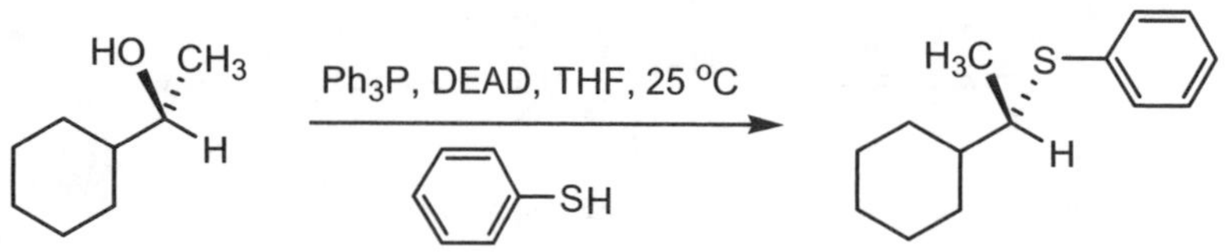

b.

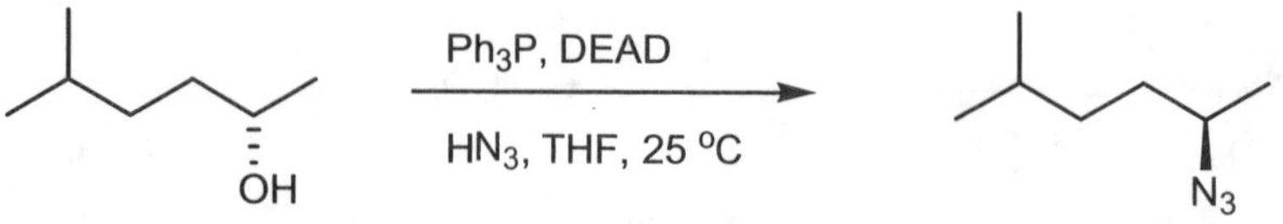

7.6.   The substrate molecule in this exercise is a secondary, benzylic alkyl bromide and the nucleophile is a weak base, so the S_N1 mechanism will operate in this transformation. An S_N1 reaction occurs via dissociation of the leaving group to form a carbocation (step 1) followed by reaction of the nucleophile with the cation. When the nucleophile is an alcohol molecule, loss of a proton (step 3) completes the mechanism.

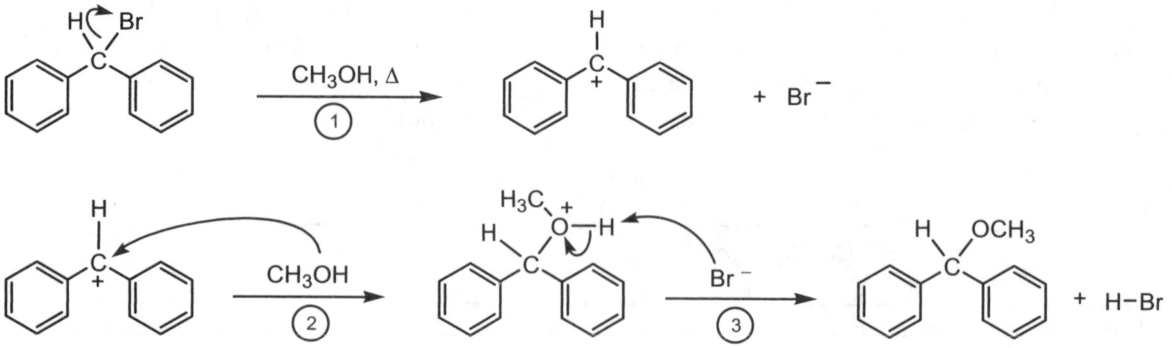

7.7.   The intramolecular reaction between an alkoxide (or phenoxide) ion and an alkyl halide generates a cyclic ether as the product. In each of the following reactions, the base (hydride ion) reacts with the alcohol OH group to generate the anionic nucleophile, and then the substitution step occurs. It is important to count the number of atoms correctly so that you draw the correct ring size for the product.

7.7.   (continued)

a.

7-membered ring

b.

5-membered ring

7.8.   In the intramolecular reaction to form an epoxide, the bond between the carbon and oxygen atom of the original OH group is not broken, so the carbon atom retains its configuration. The bond to the leaving group (bromide ion in these equations) will be broken, so the configuration of the carbon atom bonded to the halogen atom will change. The product molecules from the mechanisms are redrawn to match the orientation of the molecules shown in the text, confirming the stereochemical results.

7.9.   In the reactions between alcohols or ethers and the strong acids HI or HBr, the oxygen atom becomes protonated (steps 1 and 3) to generate a good leaving group (water in the case of an alcohol and an alcohol molecule in the case of an ether molecule). The halide ion is the nucleophile that displaces the leaving group if the S_N2 mechanism operates, as in this example (steps 2 and 4). (If the oxygen atom is attached to a 2° or 3° carbon atom, then the S_N1 mechanism occurs, which would involve formation of a carbocation intermediate.)

7.10.   Iodotrimethylsilane reacts with the oxygen atom of an ether molecule to produce a good leaving group. The silicon–iodine bond is polarized in such a way that the silicon atom is electrophilic. The ether oxygen atom is the nucleophile and reacts with the trimethylsilyl group, displacing the iodide ion (step 1). In the second step of this sequence, the iodide ion nucleophile reacts at the aliphatic carbon atom adjacent to the oxygen atom to displace the leaving group. A reaction does not occur at the sp^2-hybridized carbon atom of the benzene ring, so only one carbon–oxygen bond of the original ether linkage is cleaved.

7.10.   (continued)

7.11.   When the azide ion (or any good nucleophile) reacts with a symmetrical epoxide ring, reaction can occur at either carbon atom. As shown in the equations below, the two products are enantiomers, which means that a racemic mixture is formed. Ring opening is accompanied by proton transfer from water (the solvent) to generate the alcohol OH group.

7.12.   In the first step of this reaction, the oxygen atom of the epoxide ring is protonated. This cationic intermediate is in equilibrium with a partially-opened form that places a positive charge on the more highly-substituted carbon atom.

The chloride ion can react in the next step either at the less highly-substituted carbon atom, which is less hindered (step 2a) or at the carbon atom with the positive charge (step 2b). The products are structural isomers.

7.13.   Lithium aluminum hydride reacts by transferring a hydride ion (H with the two electrons in its bond to aluminum) to the less-hindered carbon atom of the epoxide ring. The resulting alkoxide ion is protonated (step 2) by the acid added during workup. Because the product has two methyl groups attached to the carbon atom that was originally chiral, the product molecule is achiral.

7.13.   (continued)

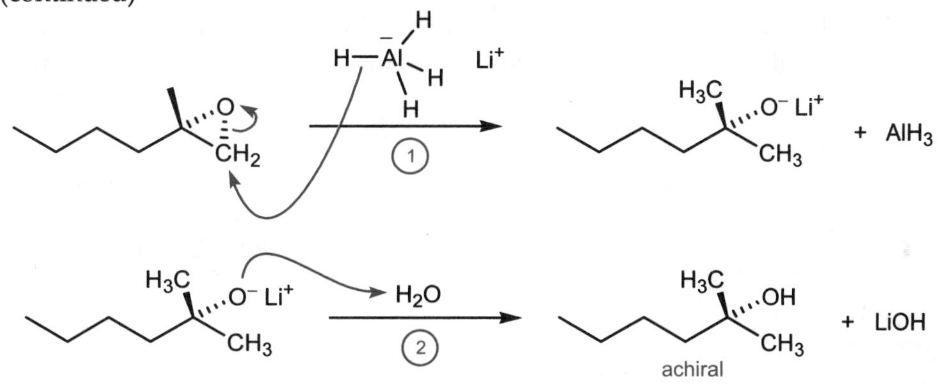

7.14.   In the Mitsunobu reaction, the alcohol, the nucleophile (in its acid form—a thiol group in this case), triphenylphosphine, and diethyl diazodicarboxylate (DEAD) all react to convert the alcohol OH group into a good leaving group and to generate the nucleophile in its conjugate base form.   In the second step, the thiolate ion reacts to displace the leaving group, triphenylphosphine oxide.   This step creates the six-membered ring product.   If the carbon atom bearing the leaving group is chiral, it will undergo inversion in the Mitsunobu reaction.

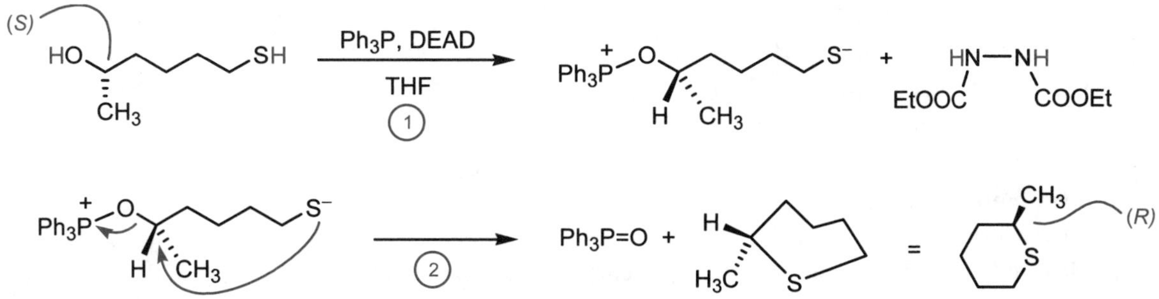

7.15.   Both of the given reactions are examples of the S$_N$2 reaction.

a.   The triethylamine first reacts with the thiol to product the triethylammonium salt of the thiolate ion. This nucleophile displaces the tosylate leaving group in the substrate molecule.   Because the substrate molecule is chiral, inversion of configuration occurs.

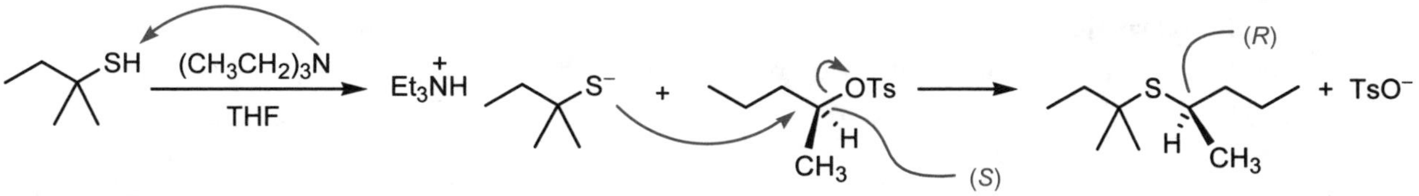

b.   The thiolate ion is already present, so it displaces the chloride ion leaving group.   The reactant molecules are achiral, as is the product molecule.

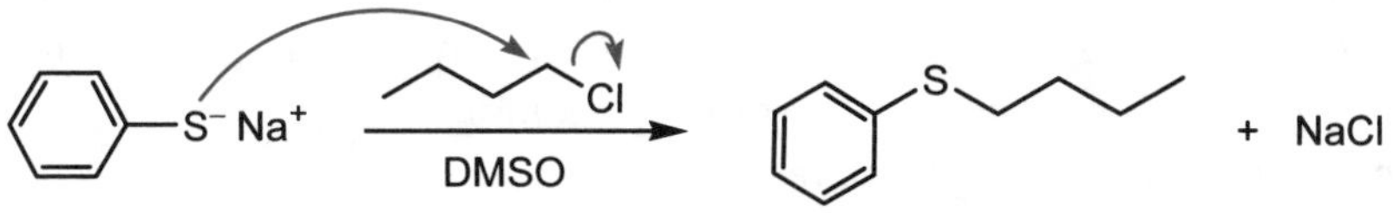

7.16. This reaction occurs as shown in the text for the methylation of phosphatidylethanolamine. In this exercise, the nucleophile is the amino group of norepinephrine. This amino group reacts at the methyl carbon atom of *S*-adenosylmethionine, displacing *S*-adenosylhomocysteine as the leaving group. The mechanism is S$_N$2.

Norepinephrine   Epinephrine (Adrenaline)

7.17. Follow the procedures outlined in the solution to Exercise 1.29.

a. **2-Methyl-2-hexanethiol**

hex	6 carbon atoms
an	no double or triple bonds
thiol	SH group (thiol) at C2
2-methyl	CH$_3$ group at C2

b. **Cyclobutyl methanesulfonate**

methanesulfonate	–OSO$_2$CH$_3$ group
cyclobutyl	this is the group attached to the – OSO$_2$CH$_3$ group through the oxygen atom

c. *trans-*2-Phenyl-1-cyclohexanethiol**

cyclohexane	a saturated ring with 6 carbon atoms
thiol	SH group (thiol) at C1
2-phenyl	phenyl group at C2

d. **Isobutyl phenyl sulfide**

sulfide	a sulfur atom with two organic groups attached
isobutyl	(CH$_3$)$_2$CHCH$_2$- group
phenyl	C$_6$H$_5$- group

7.18. Follow the procedures outlined in the solution to Exercise 1.17.

a. This compound has the alcohol functional group in a carbon chain, so the name ends in "–ol." The longest carbon chain has five carbon atoms, so the root is *pent–*. There are no carbon–carbon double or triple bonds, so the multiple bond index is *–an–*: pent/an/ol = pentanol.

The position of attachment for the thiol group (as a substituent, –SH = mercapto) is C4, and the configuration of this chiral carbon atom is (*S*). The name of this molecule is **(*S*)-4-mercaptopentanol**.

7.18. (continued)

b. This compound is a saturated cyclic sulfide (five-membered ring), so the root word is thiacyclopentane (see Section 7.3b).

Numbering begins with the sulfur atom and goes in the direction that gives the lowest possible number to the substituent. An isopropyl group is attached at C3, and the name of this molecule is **3-isopropylthiacyclopentane.**

- -

c. This compound is a derivative of aniline; the position of attachment of the amino group defines C1. Numbering goes in the direction that gives the lowest possible numbers to the substituents.

A mercapto group (–SH) is attached at C2 and a chlorine atom is attached at C4. The substituents are added to the root word in alphabetical order. The name of this molecule is **4-chloro-2-mercapto-aniline**.

- -

d. This compound is a thiol with six carbon atoms and a double bond: hex/en/thiol = hexenethiol.

The chain is numbered to give the principal function group the lowest possible number, so numbering begins at the right end of the chain. The double bond therefore begins at C4 and a methyl group is attached at C2. The geometry of the double bond is trans or (E), so the name of this molecule is **_trans_-2-methyl-4-hexene-2-thiol** or **(E)-2-methyl-4-hexene-2-thiol**.

7.19. The Williamson ether synthesis makes use of the reaction between an alkoxide (or phenoxide) ion and an alkyl halide. For S_N2 reactions, the alkyl halide must be methyl, 1°, or 2°. It cannot be 3° or aryl. To decide which combination of reactants to use, determine the types of carbon atoms attached to the oxygen atom. If at least one of those carbon atoms is methyl, 1°, or 2°, then the corresponding alkyl halide (bromide in the molecules below) can be used as one reactant, and the alkoxide (or phenoxide) ion that corresponds to the remainder of the molecule is the other reactant. If there is a choice as to which alkoxide ion to use, then the combination with the less substituted alkyl halide is preferred.

The alkoxide (phenoxide) ion in each reaction is formed by treating the corresponding alcohol or phenol with NaH in DMF.

Alcohol or phenol required

a.

b.

c.

7.20.   The sulfide ion is a potent nucleophile, and it reacts with an alkyl halide to replace the leaving group (bromide ion in this reaction). The resulting thiolate ion is also a potent nucleophile, so it reacts with another equivalent of alkyl halide to form the dialkyl sulfide product.

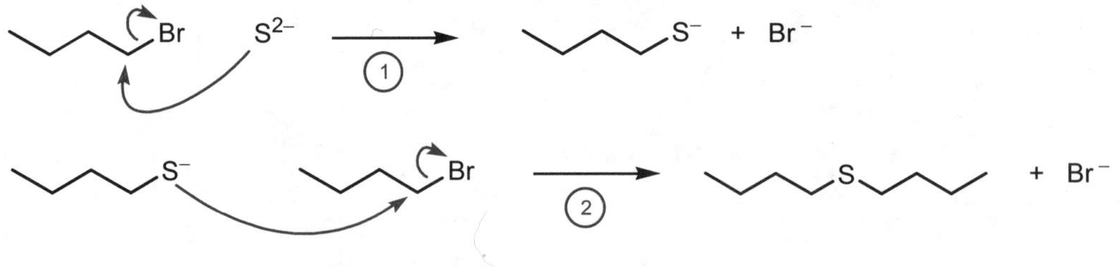

7.21.   The course of this transformation is analogous to the reaction between an amine and an alkyl halide. The first product, a thiol, is formed by replacement of the bromide ion by HS^-.

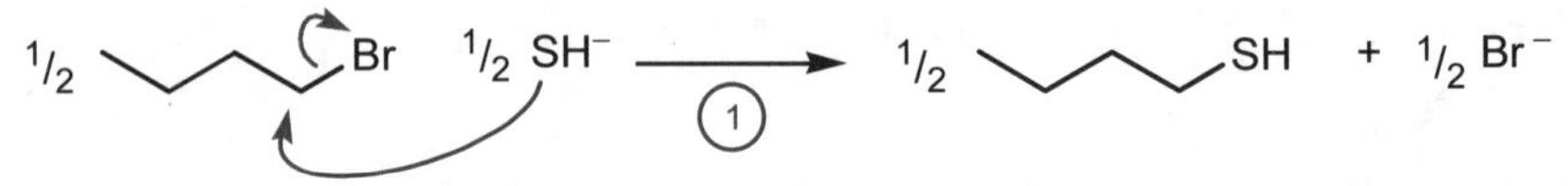

This thiol product can undergo an acid-base reaction with the original nucleophile to form an alkylthiolate ion, which itself is a good nucleophile. This acid-base reaction occurs because the pK_a of a thiol is about 10, which is nearly the same as the pK_a value for H_2S.

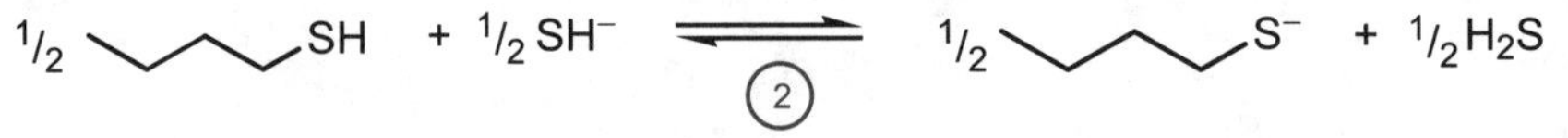

The alkylthiolate ion then reacts with the original alkyl halide to form the dialkylsulfide.

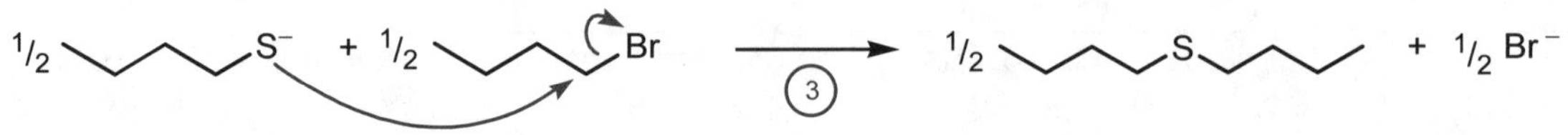

Overall:

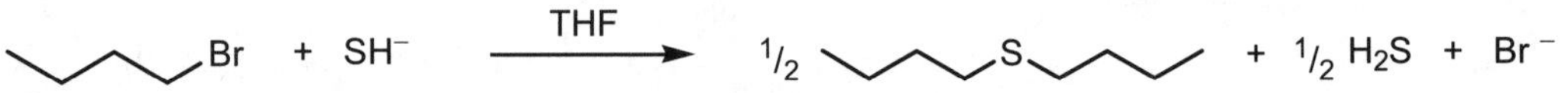

7.22.   Potassium thioacetate, K^+ $^-SCOCH_3$, reacts as a nucleophile with an alkyl halide (bromide in the following equation) to form a thioester, $R-SCOCH_3$, which is readily hydrolyzed with dilute, aqueous KOH solution. When treated with aqueous acid in a second step, the thiol, RSH, and acetic acid are formed.

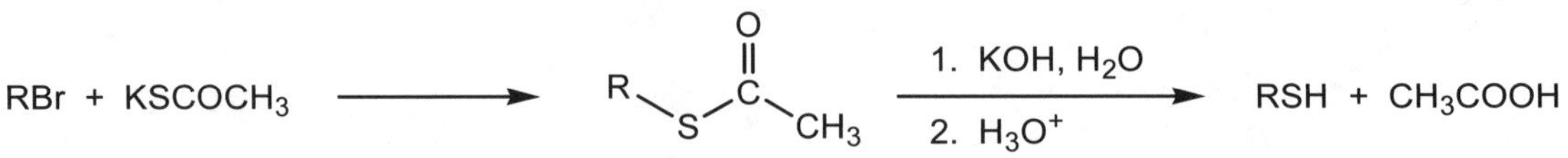

The Mitsunobu reaction can be used to form the thioester, too. Hydrolysis and aqueous acid workup yield the thiol and acetic acid.

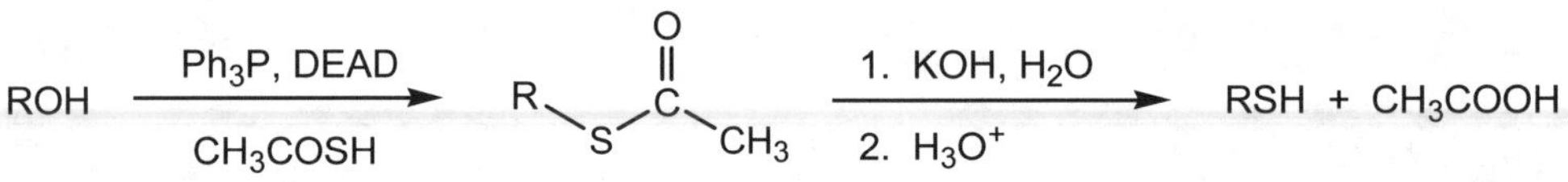

7.23.   For a substitution reaction to be viable, two features are required: a good leaving group attached to an sp^3-hybridized carbon atom and a nucleophile. An S$_N$2 process occurs with 1° and many 2° substrates. A chiral center in a secondary alkyl halide undergoes inversion of configuration. An S$_N$1 reaction occurs most readily with 3° substrates (and 2° substrates in solvolysis processes).   The stereochemical outcome of an S$_N$1 reaction is racemization.

a.   The OH group is converted to a good leaving group (mesylate ion) by the first reaction in this sequence, and formation of this alkyl sulfonate ester proceeds with retention of configuration. Azide ion replaces the mesylate group in the second step. This is an S$_N$2 reaction (strong nucleophile, aprotic solvent), so inversion of configuration occurs.

b.   Ethers are cleaved only under conditions in which a good leaving group can be formed. HI and HBr are two common reagents that promote cleavage of carbon-oxygen bonds in ethers. No stereogenic carbon atoms are present in this transformation, so stereochemistry is not an issue.

c.   The alcohol OH group is converted to a good leaving group with strong acid. Bromide ion (the conjugate base of HBr) is a good nucleophile, so it replaces the leaving group, a water molecule. No stereogenic carbon atoms are present in product, so it is achiral.

d.   Epoxide rings are opened by good nucleophiles at the less highly-substituted carbon atom. The chiral carbon atom in the substrate does not change its configuration because none of its bonds are broken during the transformation.

e.   In the Mitsunobu reaction, an alcohol molecule, a nucleophile (in its acid form—a carboxylic acid in this case), triphenylphosphine, and diethyl diazodicarboxylate (DEAD) all react to convert the alcohol OH group to a good leaving group with formation of the nucleophile in its conjugate base form.   In the second step, the carboxylate ion (nucleophile) replaces the leaving group. The carbon atom attached to the leaving group is chiral so it undergoes inversion with respect to its configuration.

f.   Bromide ion is attached to a carbon atom that has sp^2 hybridization.   Aryl and vinyl halides do not undergo S$_N$1 and S$_N$2 reactions. The answer is "N.R."

7.23.   (continued)

g.   The alcohol OH group is replaced by chlorine when the alcohol is heated with thionyl chloride. This alcohol is achiral, so stereochemistry is not an issue.

h.   This is an example of an intramolecular Williamson ether synthesis (see the solution given for Exercise 7.7). No stereogenic carbon atoms are present product, so it is achiral.

7.24.   In the Mitsunobu reaction, an alcohol molecule, a nucleophile (in its acid form), triphenylphosphine, and diethyl diazodicarboxylate (DEAD) all react to convert the alcohol OH group into a good leaving group with formation of the nucleophile in its conjugate base form.   In the second step, the nucleophile replaces the leaving group. If the carbon atom attached to the leaving group is chiral, its configuration undergoes inversion.

a.   The nucleophile is an alkylthiolate ion. Inversion of configuration occurs.

b.   The nucleophile is a carboxylate ion. Inversion of configuration occurs.

7.25.   When an alcohol is treated with a strong acid such as phosphoric acid, protonation of the oxygen atom will occur in the first step. If the alcohol is 3° or benzylic, then a molecule of water dissociates in the second step to generate a carbocation. For this molecule, two benzylic alcohol groups are present. The one that reacts faster is the one that can form the more stable carbocation.

Once the carbocation forms, the other alcohol OH group functions as a nucleophile to intercept this positive center (step 3) and deprotonation (step 4) completes the sequence.

7.26.   In the substitution reactions that occur in biochemical systems, the phosphate or diphosphate ion often functions as a leaving group. The fact that the diphosphate ion (also called "pyrophosphate ion") is replaced by the nitrogen atom of the amine tells us that the nitrogen atom acts as the nucleophile in this transformation.   Therefore, we write the mechanism as follows.   In step 1, the pyrophosphate ion dissociates to form a resonance-stabilized carbocation.

LG = $(PO_2-O-PO_3)^{3-}$ resonance-stabilized allylic carbocation

After substitution occurs (step 2), an acid-base reaction occurs to form the neutral amino group (step 3).

Configurations of the stereocenters in Domoic acid:

7.27.   In order for a nucleophilic substitution reaction to be facile, the substrate must have a good leaving group, and a good nucleophile must be present.   The structure of the substrate and/or stereochemistry of the transformation will often suggest which mechanism (S_N1 or S_N2) has to operate.

a.   This transformation begins with a primary alcohol, so S_N2 conditions are required. The nucleophile is the carboxylate ion (look to see what group replaces the OH group), so a Mitsunobu reaction should be employed.

Ph_3P, DEAD, PhCOOH

b.   This transformation starts with a tertiary alcohol, so S_N1 conditions are required. The nucleophile must be a weak base (probably neutral with respect to its charge), so a thiol molecule will be used to form the sulfide product. First, however, the alcohol OH group has to be converted to a good leaving group. Use of HX (X = Cl in this case) will convert the alcohol to the corresponding alkyl chloride. Then, a thiol is used to form the product.

1. HCl   2. CH_3CH_2SH, EtOH

7.27.   (continued)

c.   This transformation starts with phenol. The OH group of the phenol is attached to an *sp²*-hybridized carbon atom, so it does not participate in either S$_N$1 or S$_N$2 reactions. Therefore, the phenol must be the nucleophile and it will react with an alkyl halide. This is an example of the Williamson ether synthesis. The phenol is first converted to the phenolate ion, and then it reacts with 2-bromopropane to yield the product.

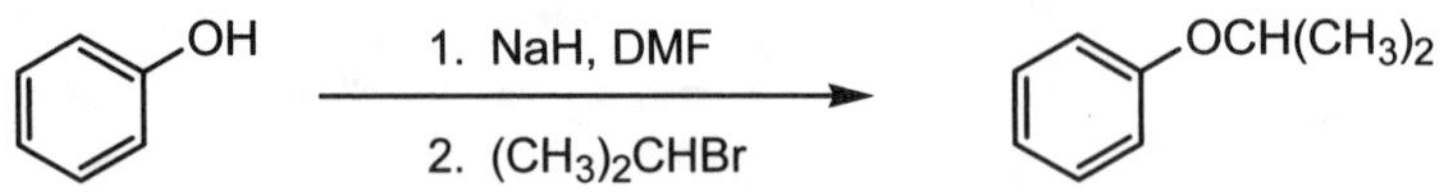

d.   This transformation begins with a secondary alcohol, so either S$_N$1 or S$_N$2 conditions can be used. However, the stereochemistry of the carbon atom bearing the OH group is inverted, so S$_N$2 conditions are required. A Mitsunobu reaction is employed using HN₃ as the acidic component.

7.28.   The reactions of epoxides with acids begin with protonation of the oxygen atom (step 1). If the nucleophile is weak or unreactive, then the ring can first open to form a carbocation. Formation of this carbocation (step 2) creates a more reactive center at the more highly-substituted carbon atom, and the nucleophile can subsequently react there. If the nucleophile is a good one, then it is likely to react with the epoxide ring before (or at the same time) the ring is opened; this pathway occurs at the less highly-substituted carbon atom (step 3), because that one is less hindered.

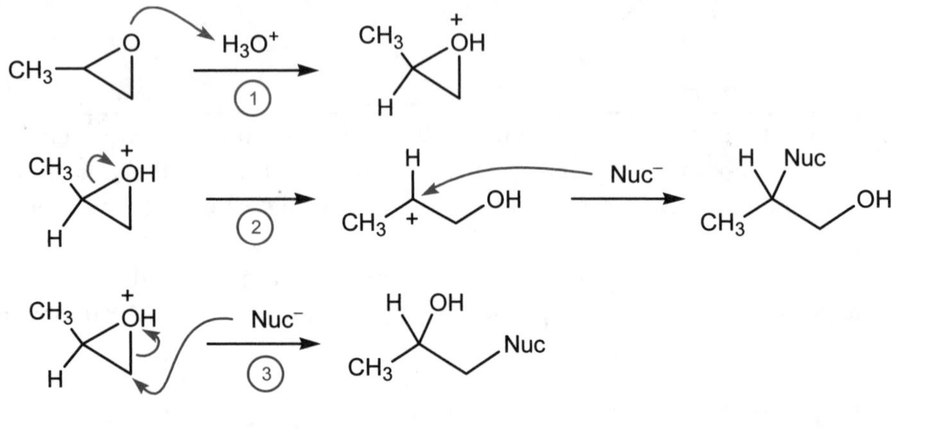

In the given reactions, methanol is a weak nucleophile and chloride ion is a good one. Thus the reaction with methanol occurs predominantly via pathway 2 (above) and the one with chloride ion occurs more likely via pathway 3 (above).

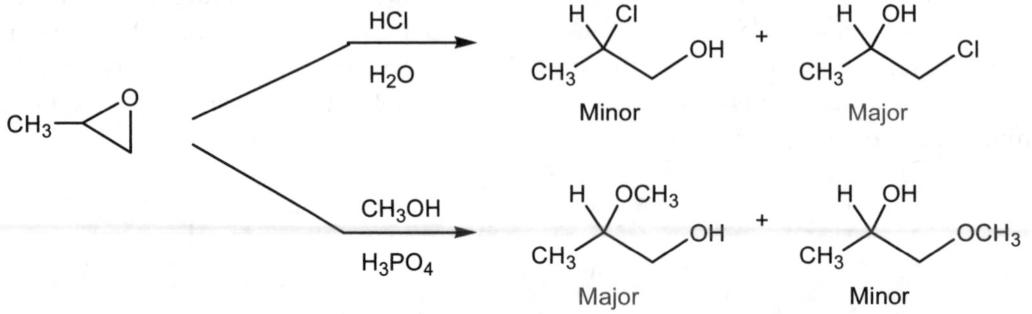

7.29.   In order for a substitution reaction to be viable, two features are required: a good leaving group attached to an sp^3-hybridized carbon atom and a nucleophile. An S$_N$2 process occurs with 1° and many 2° substrates. A chiral center in a secondary alkyl halide undergoes inversion of configuration. An S$_N$1 reaction occurs most readily with 3° substrates (and 2° substrates in solvolysis processes).   The stereochemical outcome of an S$_N$1 reaction is racemization.

a.   The SH group is first converted by its reaction with NaH to the corresponding thiolate ion. The alkyl chloride is the substrate with the good leaving group. Therefore, the thiolate ion replaces the chlorine atom to give a sulfide product. No stereogenic carbon atoms are present in this transformation, so stereochemistry is not an issue and the product is achiral.

b.   Alcohols react with HBr, and the OH group is replaced by a Br atom. No stereogenic carbon atoms are present in this transformation, so stereochemistry is not an issue.

c.   An ether molecule reacts with iodotrimethylsilane and undergoes cleavage. Because a methyl group reacts the fastest among organic substrates with nucleophiles, methyl iodide is one product and the trimethylsilyl group remains attached to the oxygen atom.

d.   Epoxides are opened at their less highly-substituted carbon atoms by reactions with good nucleophiles. Both carbon atoms attached to the oxygen atom in the starting epoxide of this exercise are chiral; the more highly-substituted carbon atom in the substrate does not change its configuration because none of its bonds are broken during the course of the transformation. The other center—the one at which the nucleophile reacts—undergoes inversion of configuration.

e.   Formation of an alkyl sulfonate ester from an alcohol proceeds with retention of configuration at the carbon atom attached to the OH group because only the O–H bond of the substrate molecule is broken. The tosyl group replaces the proton of the OH group.

7.29.　(continued)

f.　A primary alkyl sulfonate ester reacts with a good nucleophile via the S_N2 mechanism. Iodide ion replaces the mesylate group in this transformation.

g.　Ether molecules are cleaved only under conditions in which a good leaving group can be formed. HBr is commonly used to cleavage the carbon-oxygen bonds in ether molecules. No stereogenic carbon atoms are present in this transformation, so stereochemistry is not an issue. Only the bond between the oxygen atom and the methyl group is cleaved. The other bond to oxygen comes from a carbon atom with sp^2 hybridization. These bonds are not broken in either S_N1 or S_N2 reactions.

h.　In the Mitsunobu reaction, an alcohol, a nucleophile (in its acid form—a thiol in this case), triphenylphosphine, and diethyl diazodicarboxylate (DEAD) all react to convert the alcohol OH group to a good leaving group and to generate the nucleophile in its conjugate base form.　In the second step, the thiolate ion (nucleophile) replaces the leaving group. No stereogenic carbon atoms are present in this transformation, so stereochemistry is not an issue.

7.30.　This transformation is the same as the one used for the cleavage reactions of ethers (see the solution given for Exercise 7.10). Iodotrimethylsilane reacts with the oxygen atom of the alcohol to produce a good leaving group (step 1). In the second step, the iodide ion nucleophile reacts at the aliphatic carbon atom adjacent to the positive-charged oxygen atom to displace the leaving group.

ELIMINATION REACTIONS OF ALKYL HALIDES, ALCOHOLS, AND RELATED COMPOUNDS

8.1. When an alcohol molecule is heated with a strong acid such as H_2SO_4 or H_3PO_4, dehydration (loss of water) takes place to produce the most stable alkene that can be formed. Any one of the hydrogen atoms attached to the carbon atoms adjacent to the one bearing the OH group can be removed to form the double bond. Alkene stability follows these general trends:

- more highly substituted > less highly substituted
- trans > cis and (E) > (Z)
- endocyclic (inside the ring) > exocyclic (outside the ring)

In rings with fewer than nine atoms, a double bond can exist as only one isomer (the bonds that are in the ring are cis). In an acyclic compound, both (E) and (Z) isomers are likely with the (E) isomer normally predominant.

a.

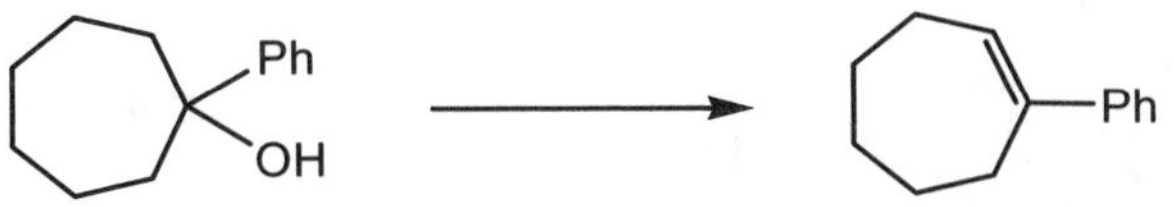

b.

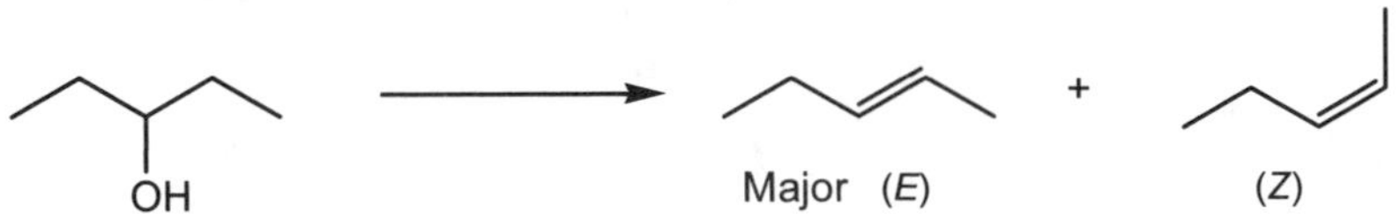

Major (E)   (Z)

8.2. As noted in the solution to Exercise 8.1, a molecule of water is eliminated when an alcohol is heated with strong mineral acid. Any of the hydrogen atoms attached to the carbon atoms adjacent to the one bearing the OH group can be removed to form the double bond. The major product in this exercise is the one with the most highly-substituted double bond, compound **A**, below.

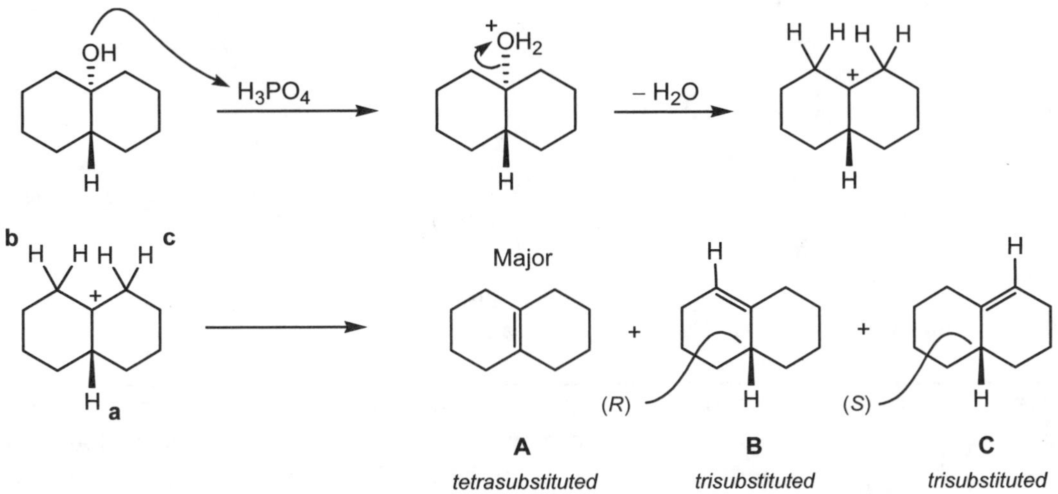

A
tetrasubstituted

B
trisubstituted

C
trisubstituted

8.3. The amount of heat evolved during hydrogenation of an alkene double bond is inversely proportional to the stability of the double bond undergoing addition. Stability follows the following order: tetrasubstituted > trisubstituted > disubstituted > monosubstituted. Also:

- trans > cis
- (E) > (Z)
- endocyclic (inside the ring) > exocyclic (outside the ring)

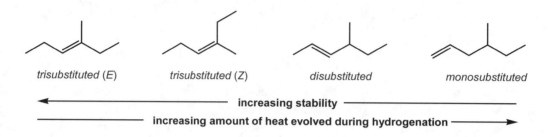

<table>
<tr><td>trisubstituted (E)</td><td>trisubstituted (Z)</td><td>disubstituted</td><td>monosubstituted</td></tr>
</table>

$\longleftarrow$ increasing stability $\longrightarrow$

$\longleftarrow$ increasing amount of heat evolved during hydrogenation $\longrightarrow$

8.4. For compounds that have an exocyclic double bond, the attached alkenyl group is treated as any substituent. Its name is made by appending the suffix "ylidene" to the name of the corresponding alkyl group root (eth ⇒ ethylidene; prop ⇒ propylidene, and so on). When the exocyclic double bond has only one carbon atom outside the ring, the substituent name is "methylene." An exocyclic double bond is sometimes required to be classified as (E) or (Z), too

a. This compound has seven carbon atoms in the ring, and there are no double or triple bonds *within* the ring. The root word is cycloheptane.

   The exocyclic double bond is connected to a single carbon atom, so the substituent is "methylene," and its point of attachment defines C1 of the ring. A chlorine atom is attached at C2, which has the (S) configuration. The name of this molecule is **(S)-2-chloro-1-methylenecycloheptane**.

b. This compound has five carbon atoms in the ring, and there are no double or triple bonds *within* the ring. The root word is cyclopentane.

   The exocyclic double bond is connected to a two-carbon fragment, so the substituent is "ethylidene." The name of this molecule is **ethylidenecyclopentane**.

c. This compound has six carbon atoms in the alicyclic ring, and there are no double or triple bonds *within* the ring. The root word is cyclohexane.

   The principal functional group (alcohol) defines C1 of the ring. The exocyclic double bond is connected to a carbon atom that bears the phenyl ring, so the substituent is "benzylidene." The double bond has the (E) geometry, so the name of this molecule is **(E)-3-benzylidenecyclohexanol**.

8.5. The dehydrohalogenation reaction of the 2° alkyl halide shown in this exercise proceeds via the E2 pathway: The substrate is an alkyl halide that has a proton attached adjacent to the carbon atom bearing the leaving group, and the reaction conditions include strong base and heat. An E2 reaction requires that the proton to be removed is *anti* to the leaving group.

8.5. (continued)

Start with the given conformation, and then rotate the carbon atom on the right side of the molecule through the various angles. A staggered conformation is required so that the substituents are *anti* to each other.

The conformation having the proton and leaving group *anti* (shown in the box directly above) is the one that undergoes the E2 reaction. In this conformation, the proton is removed by base, which causes the movement of electrons that displace the leaving group. The product is the (Z) alkene.

8.6. The reaction in this exercise proceeds by an E2 pathway: The substrate is an alkyl halide that has protons attached to the carbon atoms adjacent to the one with a leaving group, and the reaction conditions include strong base and heat. The E2 reaction of a bromocyclohexane derivative requires that the proton and leaving group be trans to one another as well as diaxial. Therefore, the given substrate must first undergo a ring flip so that the reactive groups assume the proper orientation (shown in color, below right).

After the correct orientation is achieved, the axial proton is removed by base, which causes the movement of electrons that displace the leaving group.

8.7. As shown in the solution to Exercise 8.6, you first have to look at the possible conformations that orient the H and Br atoms in the axial positions trans to each other. For the *cis*-1-bromo-4-*tert*-butylcyclohexane molecule, the most stable conformation already has the bromine atom in the axial position. Recall that the *tert*-butyl group is always equatorial (Section 3.3c). Elimination is therefore facile.

8.7.   (continued)

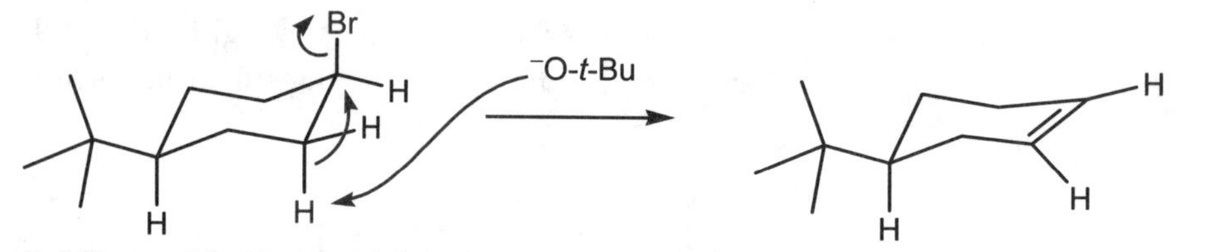

cis-1-Bromo-4-*tert*-butylcyclohexane

For the trans isomer, a ring flip has to occur first in order to place the bromine atom in the axial position. The *tert* butyl group prevents a ring flip from happening, however, so a less stable boat conformation has to form so that the reaction can proceed. Energy is required to create this boat conformation, so the overall transformation—the E2 reaction—proceeds more slowly than it does for the cis isomer.

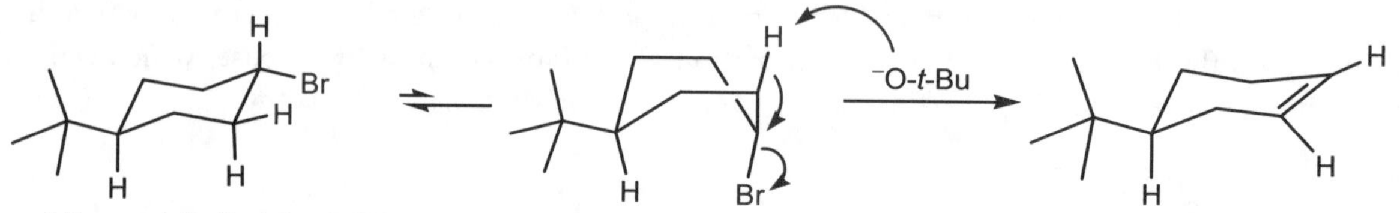

trans-1-Bromo-4-*tert*-butylcyclohexane

8.8.   The course of dehydration can be different under acidic and basic conditions, as illustrated by the reactions in this exercise. If an alcohol molecule is first converted to its alkyl sulfonate ester derivative, a base can be used for the elimination process, which provides more control of stereochemistry (and sometimes regiochemistry) because the proton and leaving group must be *anti*. This fact accounts for the possible formation of the less stable alkene isomer under E2 conditions.

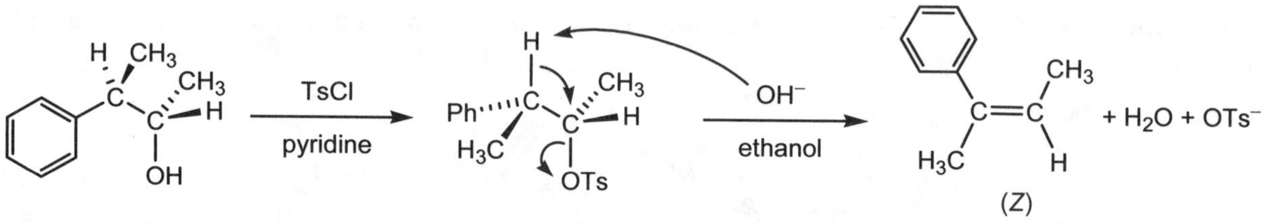

When dehydration is carried out under acid conditions, a carbocation is formed, so besides elimination, rearrangements of the carbon framework can occur. Shown below is the direct elimination reaction (no rearrangement).

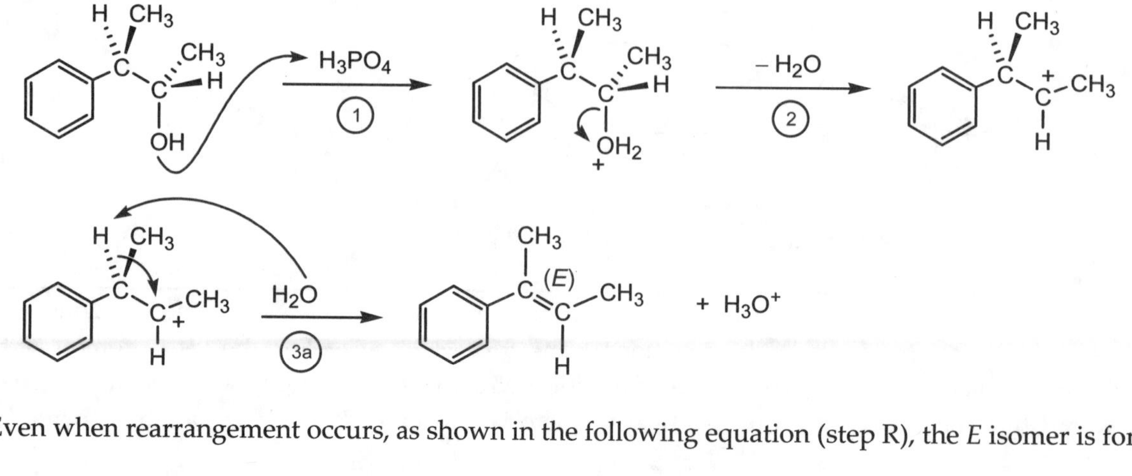

Even when rearrangement occurs, as shown in the following equation (step R), the *E* isomer is formed.

8.8.   (continued)

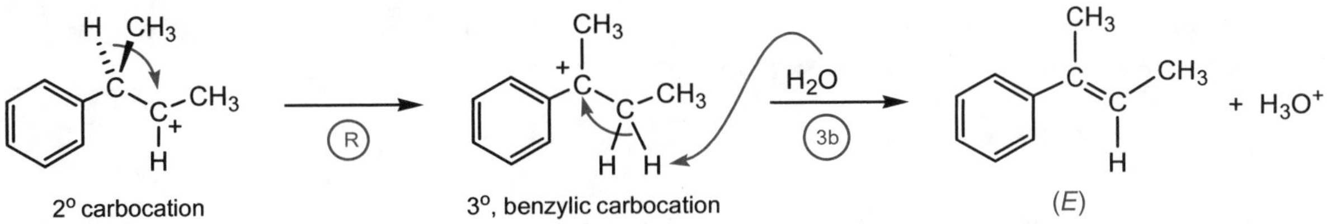

8.9.   When a vicinal dihaloalkane is treated with strong base, an alkyne forms if there are also two hydrogen atoms attached to the same carbon atoms. The base in this exercise also deprotonates the carboxylic acid groups (shown with dashed arrows in step 1a). Sulfuric acid is added in the second step to protonate the carboxylate groups, which yields the neutral diacid product.

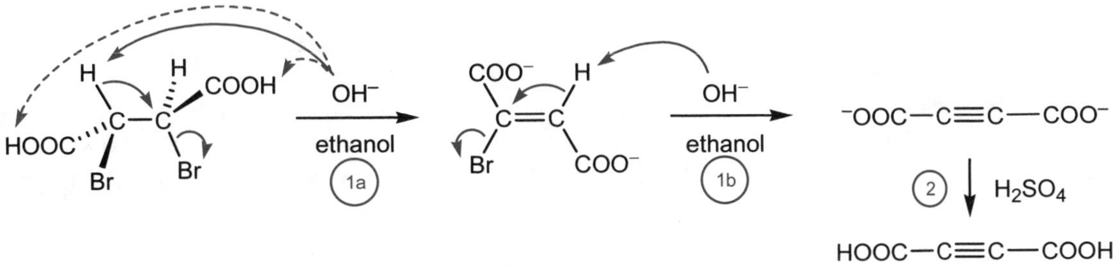

8.10.   The conjugate base of the acid reagent is bromide ion, which can react with the respective carbocation centers or with a proton attached adjacent to the respective carbocations, which leads to formation of a π bond in each case. As in Exercise 8.8, either carbocation leads to formation of the same alkene product when elimination occurs.

Substitution products

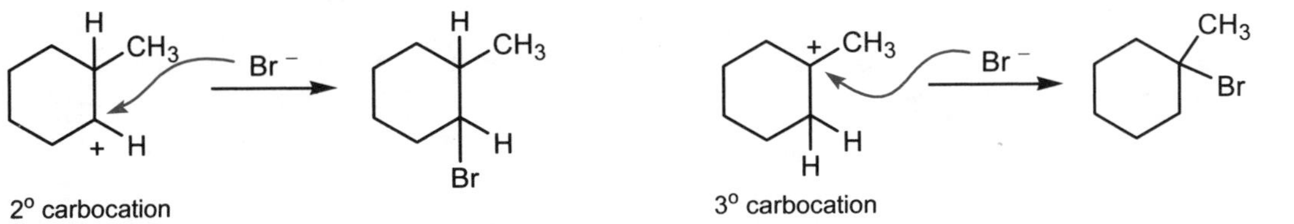

Elimination products

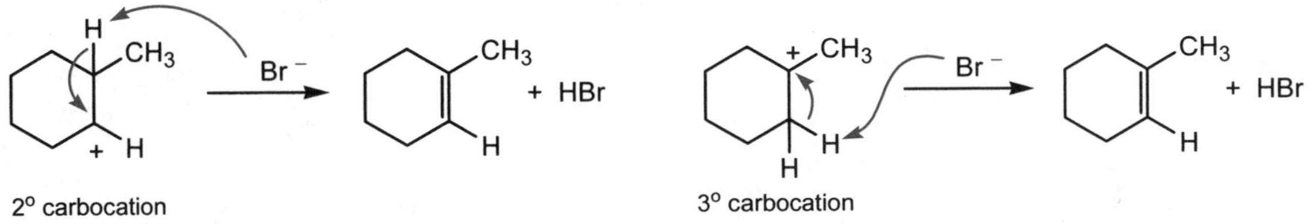

8.11.   Follow the procedures illustrated in Examples 8.3–8.5.

a.   The starting compound is a 3° alkyl sulfonate ester and the reagent is a strong base. The elimination pathway will therefore predominate. The substitution product will not be observed to an appreciable extent in this reaction.

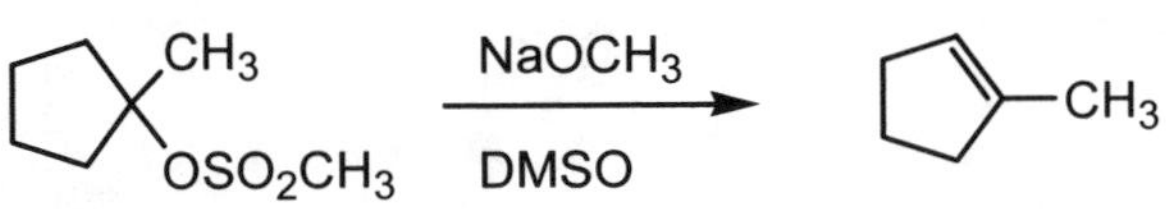

8.11.   (continued)

b.   The starting compound is a 1° alcohol, and the reagent is a strong mineral acid with a good nucleophile as its conjugate base. The substitution product will predominate.

c.   The starting compound is a 3° alkyl halide, and the reagent is a weak base in a protic solvent. These conditions are ideal for the S$_{N}$1 pathway. Some elimination product will be formed because the S$_{N}$1 and E1 mechanisms are linked. If elimination were desired, a strong base would have been used as the reagent.

8.12.   All of the elimination reactions shown in Table 8.3 require the presence of a basic site in the enzyme that catalyzes the reaction. If the substrate is an alcohol, then an acid site is needed to provide a proton that makes the OH group a better leaving group. Each of last two entries has a good leaving group already present, although the phosphate group may pick up a proton to make it a better one.

8.13.   Follow the procedures outlined in the solution to Exercise 1.18.

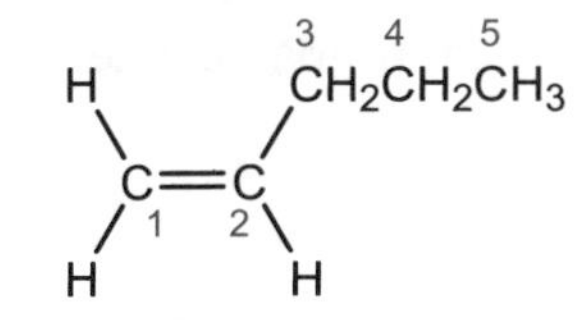

1-Pentene

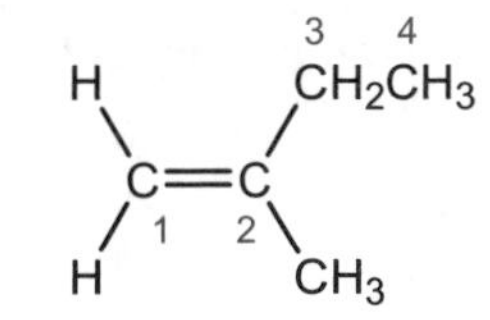

2-Methyl-1-butene

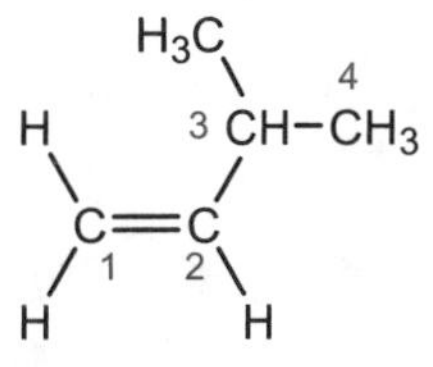

3-Methyl-1-butene

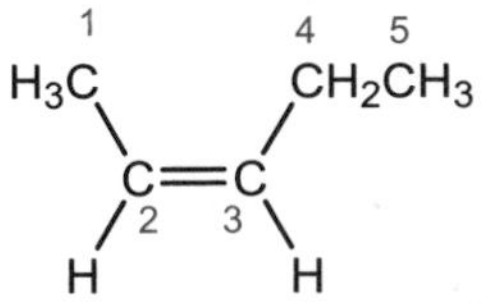

cis-2-Pentene

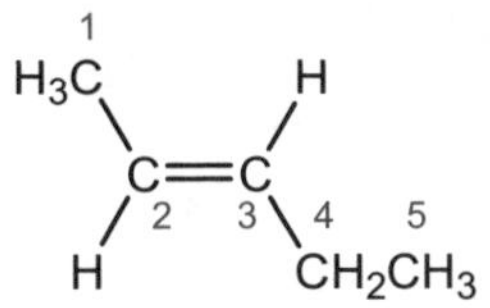

trans-2-Pentene

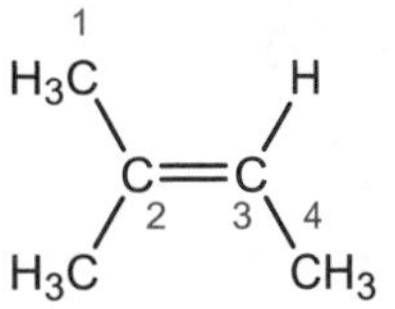

2-Methyl-2-butene

8.14.   Follow the procedures given in the solution to Exercise 8.1. If one isomer will predominate, its structure is shown below in color.

a.   The trans isomer will predominate.

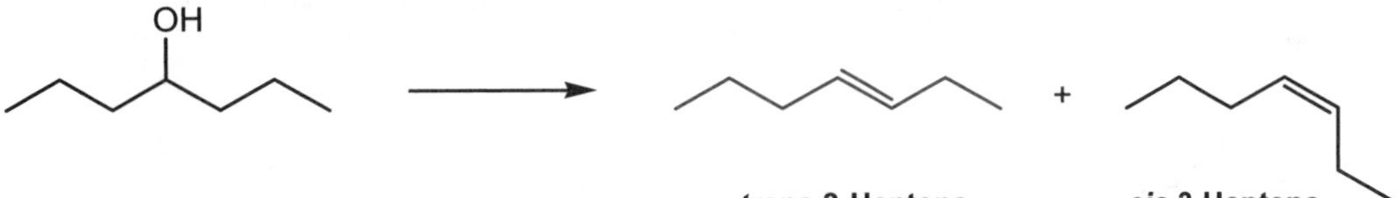

trans-3-Heptene   **cis-3-Heptene**

b.   The trisubstituted alkene will predominate.

2-Methyl-2-pentene   **2-Methylpentene**

c.   Only one alkene can form in the six-membered ring. The name of the product derives from the parent compound naphthalene.

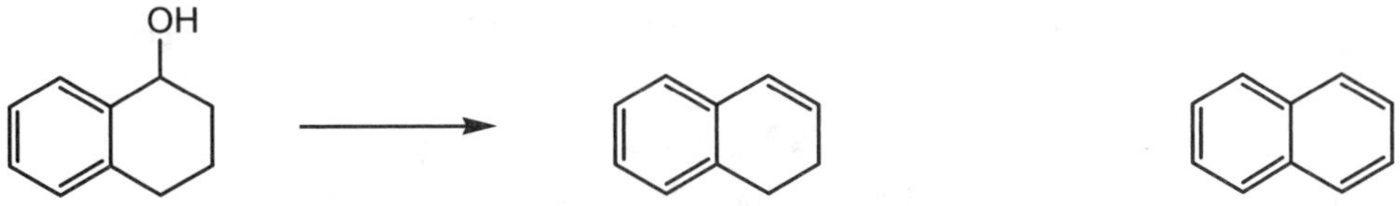

1,2-Dihydronaphthalene   **Naphthalene**

8.15.   Dehydrohalogenation reactions proceed via the E2 pathway: The substrate is an alkyl halide with protons attached to the carbon atom(s) adjacent to the one bearing the leaving group, and the reaction conditions include strong base and heat. An E2 reaction requires that the proton to be removed is *anti* to the leaving group. Therefore, the conformations of the starting compound need to be drawn in some cases to check the stereochemical relationships. If protons are attached to more than one carbon atom adjacent to the one with the leaving group, then more than one product is likely to form. Draw each possible product with a double bond and then choose the one that is most stable (the criteria are listed in the solution to Exercise 8.1).

8.15.   (continued)

a.   Elimination of HX from a molecule with the halogen atom at C1 yields the terminal alkene.

1-Hexene

b.   Elimination of HX from a cycloalkyl halide having the halogen atom attached to a 3° carbon atom gives the endocyclic cycloalkane preferentially.

Methylenecyclohexane   **1-Methylcyclohexene**
major

c.   Elimination of HX from a cycloalkyl halide with the halogen atom attached to a 2° carbon atom gives an endocyclic cycloalkene. The chair conformations of the starting compound should be drawn to make certain that a proton is in the axial position on the adjacent carbon atom(s). Formation of the more highly-substituted alkene will predominate if more than one isomer can form.

1-Methylcyclohexene   **3-Methylcyclohexene**
major

d.   Elimination of HX from an unsymmetrical alkyl halide having the halogen atom attached to a 2° carbon atom can give multiple products that are structural isomers. The major product has the most highly-substituted double bond and the (*E*) or trans geometry.

1-Butene   ***trans*-2-Butene**   ***cis*-2-Butene**
major

e.   Elimination of HX from an alkyl halide having the halogen atom attached to a 2° carbon atom can give multiple products. The major one has the most highly-substituted double bond and the (*E*) or trans geometry. In this case, the iodine atom is attached to the middle carbon atom, which eliminates the possibility that structural isomers will form.

***trans*-3-Heptene**   ***cis*-3-Heptene**
major

8.16.   Questions about rates of competing reactions require that you consider the possible conformations in which a molecule can exist. An elimination reaction carried out under basic conditions requires an *anti* relationship between the hydrogen atom to be removed and the leaving group. The most stable conformation of (1*R*,2*R*)-1-bromo-1,2-diphenylpropane has the H and Br atoms *gauche* to one another.

8.16.   (continued)

To undergo elimination, this isomer undergoes rotation to a less stable conformation in which the phenyl rings are *gauche*. Because elimination occurs through a conformation other than the most stable one, the reaction will be slowed.

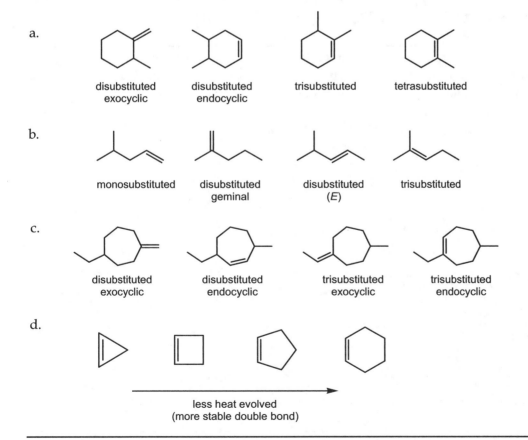

On the other hand, the most stable conformation of the (1S,2R) isomer has the bromine and hydrogen atoms in the *anti* orientation already, which is ideal for the E2 reaction. Therefore, it reacts much more rapidly than the (1R,2R) isomer does.

8.17.   Follow the procedures given in the solution to Exercise 8.3. For the compounds in part (d.), the double bond stability is related to ring size. The more strained the ring is, the less stable the double bond is because its angles deviate from the ideal angle of 120°. The six-membered ring is the most stable of the ones shown, therefore it evolves the least amount of heat during hydrogenation.

a.

| disubstituted exocyclic | disubstituted endocyclic | trisubstituted | tetrasubstituted |

b.

| monosubstituted | disubstituted geminal | disubstituted (*E*) | trisubstituted |

c.

| disubstituted exocyclic | disubstituted endocyclic | trisubstituted exocyclic | trisubstituted endocyclic |

d.

less heat evolved
(more stable double bond)

8.18. Acid-catalyzed dehydration reactions occur via formation of a carbocation intermediate after protonation of the hydroxyl group (step 1) and dissociation of a water molecule (step 2). Recall that a carbocation may rearrange, which is what occurs during this transformation [step R below]. Elimination of a proton from the carbon atom adjacent to the more stable 3° carbocation produces the endocyclic double bond, which is more stable than its exocyclic isomer.

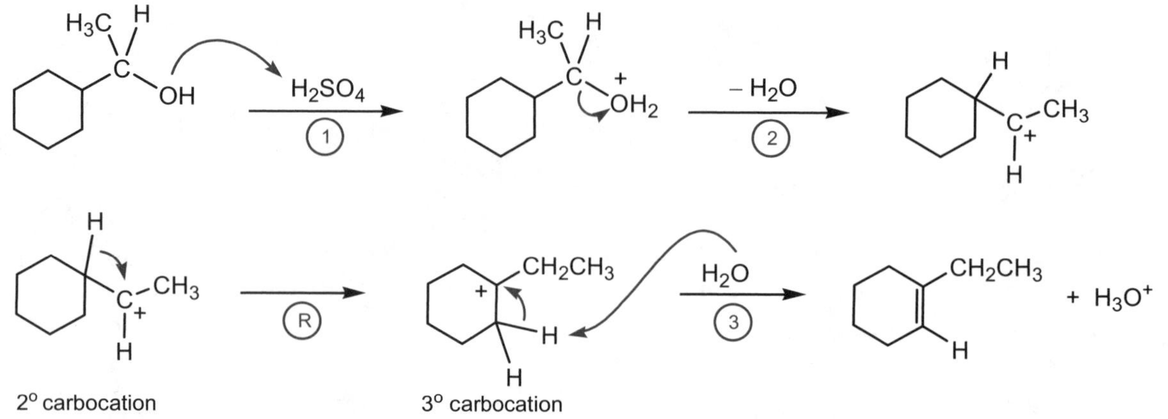

2° carbocation   3° carbocation

8.19. Elimination reactions that occur under basic conditions are not normally subject to rearrangement processes. All that is required is the presence of a good leaving group and the use of strong base. To convert an OH group to a good leaving group, prepare the alkyl sulfonate ester, and then use base to promote elimination via the E2 pathway.

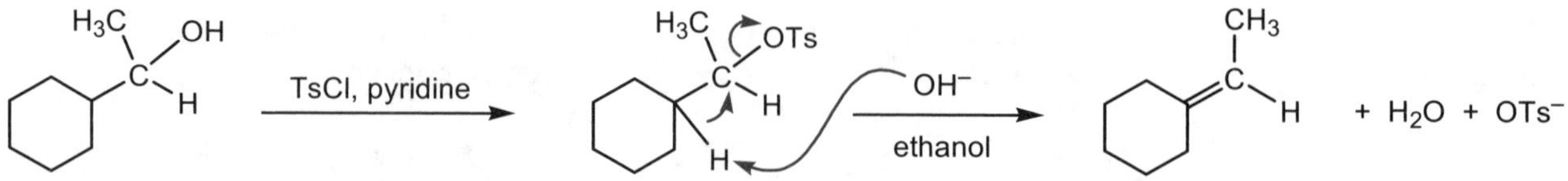

8.20. To decide what role a reagent plays in a given reaction, classify the transformation according to its fundamental type. If substitution occurs, the reagent is considered to be a nucleophile. A reagent is acting as a base if the reaction involves elimination or a proton transfer process.

a. The following is an E2 reaction, so methoxide ion functions as a base.

b. The following is a S_N2 reaction, so methoxide ion is a nucleophile.

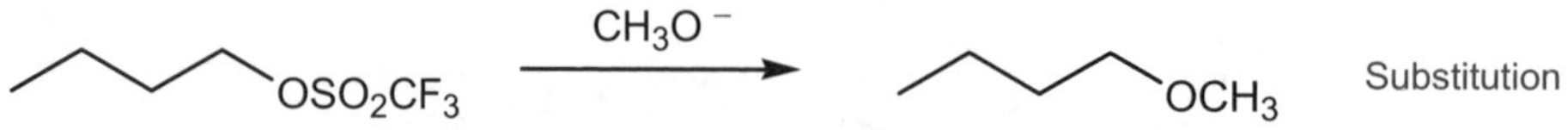

c. The following is a proton transfer process, so methoxide ion is acting as a base.

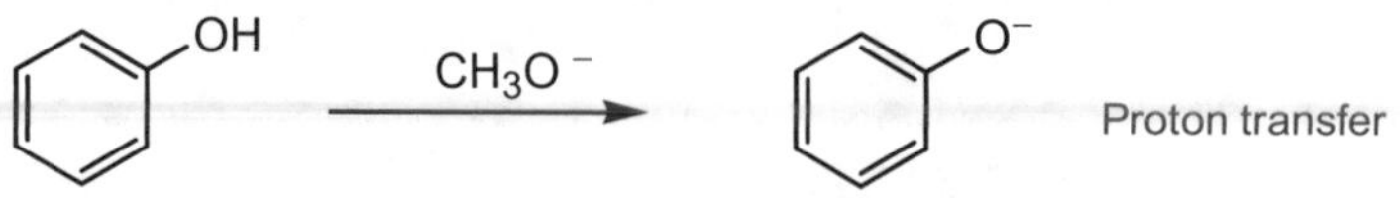

8.21.   The mechanism involved in the conversion of an acid chloride to a ketene follows the same course as any E2 process. The base triethylamine reacts with a proton attached to the carbon atom adjacent to the one bonded to chlorine. Displacement of the Cl⁻ leaving group generates the double bond.

8.22.   The mechanism for this elimination reaction is similar to the mechanism observed for dehydrohalogenation reactions. The base (methoxide ion) reacts with the hydrogen atom attached to the carbon atom adjacent to nitrogen, which bears the leaving group, acetate ion. Displacement of acetate ion generates the carbon-nitrogen triple bond.

8.23.   Substitution reactions occur when a substrate molecule has a good leaving group attached to a carbon atom with sp^3-hybridization, and a good nucleophile is also present. An S$_N$2 process occurs with 1° and many 2° substrates. A chiral center attached to the leaving group undergoes inversion of configuration. The S$_N$1 reaction occurs most readily with 3° substrates (and 2° substrates in solvolysis processes). The stereochemical outcome of an S$_N$1 reaction is racemization.

Elimination reactions occur when a substrate molecule has a good leaving group attached to a carbon atom with sp^3-hybridization, and a strong base is also present. The E2 process occurs with all types of alkyl halides and alkyl sulfonate esters. The E2 process requires that the leaving group and the proton on an adjacent carbon atom are *anti* to each other in terms of the molecule's conformation. The E1 mechanism accompanies the S$_N$1 mechanism, especially when the base is a poor nucleophile.

a.   In this reaction, a 2° bromocyclohexane derivative is treated with strong base in a protic solvent, so the E2 reaction takes place. For this mechanism, both the Br and H atoms must be *anti* and axial, so a ring flip occurs before elimination takes place.

b.   The leaving group in this substrate molecule is attached to a carbon atom with sp^2-hybridization and the nucleophile is only a moderate base, so no reaction takes place.

c.   This reaction is a substitution process that results in opening of an epoxide ring.   The nucleophile (azide ion) can react at either carbon atom because the degree of substitution at each is the same.

The configuration of the carbon atom at which the nucleophile reacts is inverted, and the configuration of the other carbon atom is retained. The starting material is meso, which is not optically active, so the product must be racemic (overall mixture not optically active).

8.23.   (continued)

d. In this reaction, a primary, benzylic alkyl chloride is treated with a strong basic nucleophile. There is no proton on the carbon adjacent to the one with the leaving group, so elimination cannot occur. Substitution therefore takes place.

e. In this reaction, a 3° alcohol is treated with strong acid having a poorly nucleophilic conjugate base. The E1 mechanism occurs to give the more stable alkene, which has an endocyclic double bond.

f. In this reaction, a 2° alcohol is treated with strong acid that has a good nucleophile as its conjugate base, so substitution most likely occurs. If this process were meant to be an elimination process, then a reagent such as sulfuric or phosphoric acid would have been employed.

g. In this reaction, a 2° alcohol is first converted to its mesylate derivative, which reacts in the same manner as a 2° alkyl halide. The use of strong base in a protic solvent leads to elimination via the E2 pathway. Because there are several protons attached to the carbon atoms adjacent to the one with the leaving group, the most stable double bond will be formed. Therefore, the disubstituted alkene with the (E) geometry is produced.

h. In this reaction a 2° alcohol is treated with phosphorus tribromide, a reagent used for substitution reactions to convert an alcohol to a bromoalkane.

i. In this reaction, a 2° alcohol is treated with sodium hydride, which generates the corresponding alkoxide ion in step 1. This species is a nucleophile and replaces the iodine atom in 1-iodopropane via the S$_N$2 pathway. This is an example of the Williamson ether synthesis.

8.23. (continued)

j.   In this reaction, a 3° alkyl bromide is treated with strong base, so the E2 reaction occurs. The Br and H atoms in the starting compound must be *anti*, but the product is achiral and symmetrical at one end, so no geometric isomers exist. Because there are several protons beta to leaving group, the most highly substituted double bond that can form is generated.

$$H_3C \underset{H_3C}{\overset{}{\underset{|}{C}}}\!\!-\!\!\underset{\underset{Br}{|}}{\overset{\overset{H\ \ CH_3}{|}}{C}}\!\!-\!\!CH_2CH_3 \quad \xrightarrow{\text{KOH, EtOH, } \Delta} \quad \underset{H_3C}{\overset{H_3C}{>}}C\!=\!C\underset{CH_2CH_3}{\overset{CH_3}{<}} \quad \text{achiral}$$

8.24.   The reaction coordinate diagram for the dehydration of *tert*-butyl alcohol reflects the mechanism of the reaction, which is shown below: (1) protonation of the OH group, (2) dissociation of a water molecule, and (3) removal of a proton to form the double bond. The reaction coordinate diagram should therefore have maxima that correspond to the activation barriers for the three steps.

$$H_3C\!-\!\underset{H_3C}{\overset{H_3C}{\underset{|}{\overset{|}{C}}}}\!-\!OH \;\; \underset{①}{\overset{H_2SO_4}{\rightleftharpoons}} \;\; H_3C\!-\!\underset{H_3C}{\overset{H_3C}{\underset{|}{\overset{|}{C}}}}\!-\!\overset{+}{O}H_2 \;\; \underset{②}{\overset{-\,H_2O}{\rightleftharpoons}} \;\; \underset{H_3C}{\overset{H_3C}{>}}\overset{+}{C}\!-\!CH_3 \;\; \underset{③}{\overset{HSO_4^-}{\rightleftharpoons}} \;\; \underset{H_3C}{\overset{H_3C}{>}}C\!=\!CH_2$$

- Step 1 is an acid-base equilibrium that lies in the direction of the protonated alcohol, which is a weaker acid than sulfuric acid. This first step has a small free energy of activation.

- Step 2 has a large free energy of activation because this step leads to formation of the high-energy carbocation intermediate; it is the slow step in the transformation.

- Step 3 has a small free energy of activation because the carbocation will react quickly to form product or to regenerate the protonated alcohol.

- The free energy of the reaction, $\Delta G°$, is the difference between the energies of the reactants and products.

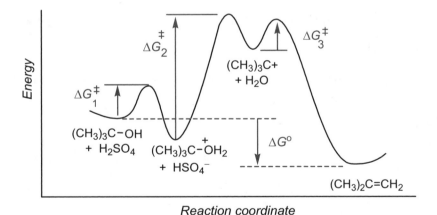

Reaction coordinate

8.25.   If an alkene is to be prepared by an E2 reaction from an organohalide, then possible starting materials are conceptualized by adding a hydrogen atom to one end of the double bond, and a bromine atom to the other end in each of the two possible orientations (the added atoms are shown in color in the following structures).

Once you have identified the possible starting materials (make certain to consider stereoisomers, especially for cyclic compounds), draw structures for the alkenes that will be formed and decide which isomer is expected to be the major one. The best starting compounds are shown in boxes.

8.25.   (continued)

a.   The best way to prepare **1-methylcyclohexene** via dehydrohalogenation starts with either 1-bromo-1-methylcyclohexane or *cis*-1-bromo-2-methylcyclohexane.

b.   The best way to prepare ***trans*-5-methoxy-2-pentene** via dehydrohalogenation starts with 4-bromo-1-methoxypentane as the starting material.

c.   The best way to prepare **(*E*)-3-methyl-3-heptene** via dehydrohalogenation starts with 4-bromo-3-methylheptane as the starting material.

8.26.   Follow the procedures outlined in the solution to Exercise 8.23.

a.   In this reaction, a 2° alcohol is first converted to its tosylate derivative, so it will react in the same manner as a 2° alkyl halide. The use of strong base in a protic solvent implies elimination via the E2 pathway. The leaving group and the proton to be removed must be *anti* to each other, and both must be in axial positions. Therefore, the disubstituted double bond is formed.

b.   In this reaction, a 2° bromoalkane is treated with strong base, so the E2 reaction takes place. The product with the most stable double bond [disubstituted and (E)] is formed.

c.   In this reaction, a 3° alcohol is treated with strong acid that has a poorly nucleophilic conjugate base, so the E1 mechanism occurs. Elimination gives the most stable alkene possible, which is tetrasubstituted.

d.   In this reaction, a 2° alcohol is treated with strong acid that has a good nucleophile as its conjugate base, so substitution occurs. There are no protons on the carbon atoms adjacent to the alcohol group, so elimination cannot occur anyway.

e.   In this reaction a 3° alkyl bromide is treated with a good nucleophile that is also a weak base. A protic solvent is used so the S_N1 pathway is indicated. The chiral center in this molecule is not attached to the leaving group, so its configuration is not affected.

f.   In this reaction, a 1° alcohol is first treated to form its mesylate derivative. Azide ion is a good nucleophile and DMF is an aprotic solvent, so the S_N2 mechanism is likely. The chiral center in this molecule is not attached to the leaving group, so its configuration is not affected.

8.26.   (continued)

g. In this reaction, a 3° alcohol is treated with strong acid that has a poorly nucleophilic conjugate base, so the E1 mechanism occurs. Rearrangement can also take place (see Exercise 8.18 for a similar example). Two isomers are formed. An exocyclic double bond is less stable than an endocyclic one, but in this case, the exocyclic double bond is tetrasubstituted, so it may form to an appreciable extent.

h. In this reaction a vicinal dichloroalkane is treated with very strong base; a proton is also attached to each carbon atom, so two E2 reactions occur to produce the corresponding alkyne.

i. In this reaction, a 2° chlorocyclohexane derivative is treated with strong base, so the E2 reaction takes place. The Cl and H are already *anti* and diaxial, so elimination occurs to form the trisubstituted double bond.

j. In this reaction, a 2° alkyl bromide is treated with an excellent nucleophile in an aprotic solvent, so the S_N2 mechanism is likely. The carbon atom bearing the leaving group is chiral, so inversion of its configuration occurs. The stereochemistry of the other chiral carbon atom is not affected, so its configuration is retained.

8.27.   In an exercise such as this, the first step is to summarize the given information by constructing a flow chart.

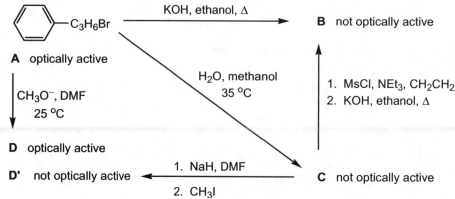

8.27.   (continued)

Consider what structures are possible for any compound in the scheme for which you have structural information (if there are too many possibilities, you may have to ignore this step).

In this exercise, we can draw five structures for compound **A**. 1-Bromo-3-phenylpropane, **I**, and 2-bromo-2-phenylpropane, **IV**, cannot be optically active so we can eliminate them from further consideration.

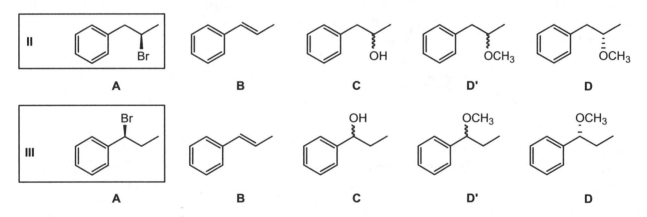

| I | II | III | IV | V |

Next, evaluate the reactions according to the specified conditions. The conversion of **A** to **B** is an E2 process; **A** to **C** is an S$_N$1 process; **C** to **B** is an E2 reaction; and **A** to **D** is an S$_N$2 reaction. Compound **V** is a primary alkyl bromide, which will not react via the S$_N$1 pathway needed to convert **A** to **C**. Therefore, we remove it from further consideration.

We now draw the structures of the compounds that will be formed via the given reactions from the two remaining candidates, **II** and **III**. We see that the structures of these compounds match the properties for **A** through **D** given in the exercise. Compound **III** is probably a better choice because the solvolysis reaction (**A** to **C**) occurs under relatively mild conditions, and a benzylic substrate is more likely to react at 35 °C than a simple secondary alkyl halide. (The squiggly bond in the structures shown below denotes that the stereochemistry is unspecified and the compound is racemic, hence not optically active.)

| **II** | A | B | C | D' | D |

| **III** | A | B | C | D' | D |

8.28.   We can choose reagents for the given reactions by classifying the type of mechanism that is involved. The mechanism in turn is deduced by classifying the reaction type and considering the stereochemistry of the transformation.

a. This is an elimination reaction that starts with an alcohol, and the alkene product shown in the exercise is the most stable of three possibilities. Strong acid can be used under E1 conditions. The alcohol can also be converted to its alkyl sulfonate ester derivative and subjected to E2 conditions (strong base). Either choice is satisfactory.

H$_3$PO$_4$, Δ

1. MsCl, NEt$_3$, CH$_2$CH$_2$
2. KOH, ethanol, Δ

8.28.   (continued)

b. This is an elimination reaction that forms an alkyne from a vicinal dihaloalkane. A very strong base is required.

NaNH$_2$, THF, Δ

c. This is an elimination reaction that starts with a 3° alkyl halide so E2 conditions (strong base) will work.

KOH, ethanol, Δ

d. This is a substitution reaction that starts with a 2° alcohol and proceeds with inversion of configuration at the carbon atom bearing the OH group. Therefore, the OH group first has to be converted to a good leaving group with retention of configuration (formation of its tosylate or mesylate derivative). Then, S$_N$2 conditions are applied (strong nucleophile, aprotic solvent).

1. TsCl, pyridine

2. NaSCH$_3$, DMF

e. This is an elimination reaction that starts with an alcohol molecule and produces an alkene product that is the less stable of two possibilities. Therefore E2 conditions are needed to control the stereochemistry. The alcohol is first converted to its alkyl sulfonate ester derivative, which is then subjected to treatment with strong base.

1. MsCl, NEt$_3$, CH$_2$CH$_2$

2. KOH, ethanol, Δ

f. This is an elimination reaction that starts with an alcohol molecule and produces an alkene product that is the more stable of two possibilities. Strong acid is required because E2 conditions would yield the product shown in part (e.), directly above.

H$_3$PO$_4$, Δ

ADDITION REACTIONS OF ALKENES AND ALKYNES

9.1. The steps involved in hydration of an alkene and dehydration of an alcohol are the reverse of each other.

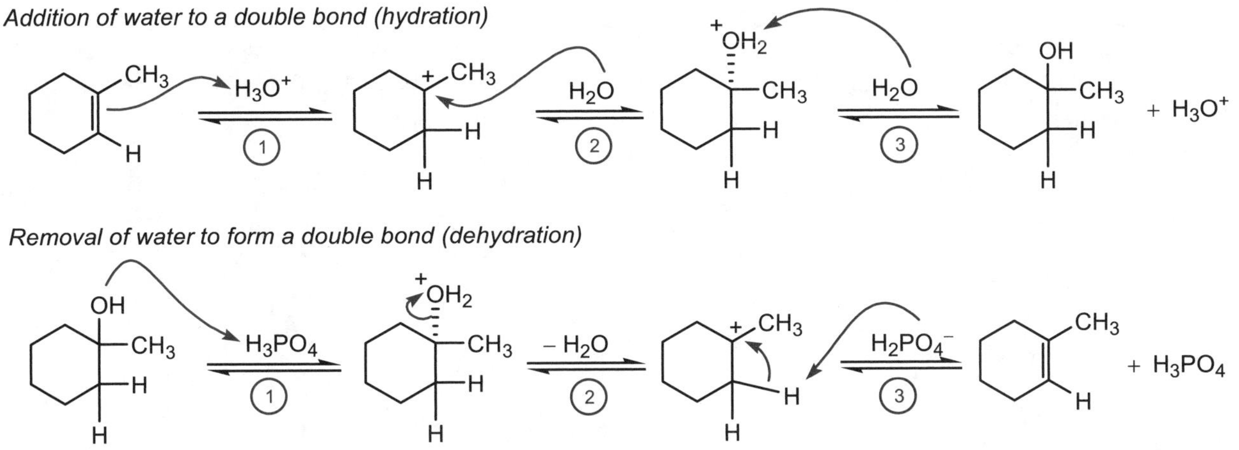

Addition of water to a double bond (hydration)

Removal of water to form a double bond (dehydration)

Differences include:

- protonation of the double bond versus the OH group in step (1);
- reaction with a water molecule versus loss of a water molecule in step (2);
- deprotonation of the OH_2^+ group versus deprotonation adjacent to the carbocation in step (3).

9.2. Hydration of 1,1-diphenylethene occurs by reaction of a proton at the end of the double bond with more hydrogen atoms and attachment of the OH group to the more highly substituted end of the double bond.

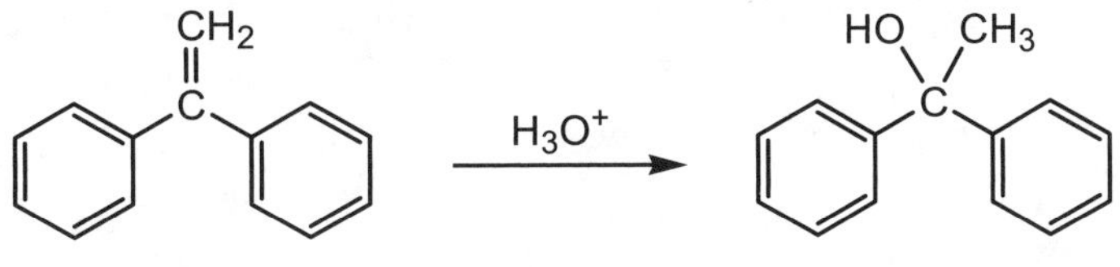

9.3. As shown in the following schemes, the reaction between chlorine and *cis*-2-butene starts with reaction of the double bond with the electrophile Cl⁺. The halogen cation creates a bridge between the carbon atoms to form a three-membered ring. The cationic intermediate is subsequently trapped by a reaction with the chloride ion, which can attach at either end of the cyclic chloronium ion, so racemic products are formed. The reaction of *trans*-2-butene occurs in the same way except that the product is the meso isomer.

Because the starting compounds are achiral, an equal mixture of chiral centers with mirror configurations has to be created. Thus, the product(s) must either consist of a racemic mixture or a meso compound.

9.3.   (continued)

cis-2-Butene

trans-2-Butene

9.4.   The reaction between bromine and 2-phenylpropene in water starts with a reaction between the double bond and the electrophile, Br⁺. The halogen cation creates a bridge between the carbon atoms to form a three-membered ring, and the bridge can form on either face of the double bond (top or bottom). The cation intermediate is subsequently trapped by a reaction with water (followed by loss of a proton). Reaction with water occurs at the more highly substituted end on the face opposite the one with the halogen atom bridge, so racemic products are formed.

9.5.   For oxymercuration, the alkene double bond reacts with the electrophilic mercury(II) ion. In the second step, a molecule of water intercepts the mercury-bridged intermediate at the more highly-substituted carbon atom. Loss of a proton (not shown as a separate step) yields the neutral OH group.

For the acid-catalyzed hydration reaction, the alkene double bond reacts with a proton to form a carbocation intermediate. The hydrogen atom on the adjacent carbon atom migrates so that the initial 2° carbocation rearranges to form the more stable 3° carbocation. A molecule of water intercepts the carbocation and a proton is lost (not shown as a separate step) to form the alcohol product.

Oxymercuration/demercuration

Acid-catalyzed hydration

9.6. The two-step process of solvomercuration/demercuration results in Markovnikov addition of a molecule of alcohol to the alkene double bond, which means that the OR group is attached to the more highly substituted end of the double bond. In these reactions, the products are achiral.

a.

b.

9.7. The addition reaction of a second equivalent of HBr to an alkyne is the same as addition of HBr to a bromoalkene: the double bond is protonated, which can generate two possible carbocations. In the next step, the more stable carbocation is trapped by bromide ion to form the geminal dibromoalkane.

The 2° carbocation is stabilized by hyperconjugation with the R group, and it is also stabilized slightly by resonance delocalization with the electron pairs on the bromine atom.

9.8. Hydrogen chloride reacts with a triple bond via steps involving Markovnikov addition. Because excess HCl is used, the addition process occurs twice. The proton becomes attached to the carbon atom with more protons, which yields the product having a terminal methyl group.

9.9. The reaction of an alkyne molecule with platinum(II) chloride and water yields a vinyl alcohol that subsequently tautomerizes to form a ketone. In this case, the alkyne is symmetrical so regiochemistry of the addition process is not an issue.

9.10. The hydrogen sulfate ion can remove a proton from a carbon atom adjacent to the carbocation center. The major isomer results from formation of the more stable double bond (trisubstituted versus disubstituted).

9.11. Dissociation of the leaving group (pyrophosphate ion) from geranyl pyrophosphate produces a resonance stabilized allylic carbocation (step 1). Next, the π bond of isopentenyl pyrophosphate reacts with the carbocation to form a new carbon–carbon bond (step 2). Addition occurs in the Markovnikov fashion to produce a tertiary carbocation. This intermediate carbocation is then deprotonated by a basic residue within the enzyme active site (step 3), producing farnesyl pyrophosphate.

9.12. The overall transformation shown in this exercise is a ring-forming reaction in which a new carbon–carbon bond is formed. Addition of a carbocation to a π bond is one way to make a carbon–carbon bond, so the first thought should be how to make a carbocation from the starting compound. Protonation of an OH group followed by dissociation of a water molecule is one such method, and the reaction conditions (strong acid) are amenable to such a process. Therefore, the first two steps lead to formation of a carbocation.

9.12. (continued)

Once formed, the neighboring π bond intercepts this carbocation, leading to formation of a second carbocation (step 3). The orientation of this addition step creates a 3° carbocation, which is subsequently intercepted by a molecule of water (step 4).

In the last step, deprotonation yields the alcohol product.

9.13. Isomerization of a double bond takes place by breaking the π bond. One way to "break" a π bond is to create an allylic carbocation so that the π electrons are delocalized. Dissociation of the pyrophosphate group creates an allylic carbocation (step 1). After rotation about the σ bond (step 2), the pyrophosphate ion can recombine with the carbocation by reaction at the terminal carbon atom to form the isomerized substrate molecule.

Geranyl pyrophosphate
(E) isomer

Geranyl pyrophosphate
(Z) isomer

9.14. In these enzyme-catalyzed processes, a base is supplied by the active site of the enzyme. For the substitution process, the carbocation reacts with a water molecule. The base deprotonates the resulting cation intermediate to form the alcohol product (step 2).

In the elimination route, the base deprotonates the carbocation intermediate itself to generate the alkene product.

9.14.   (continued)

Substitution

(base in the enzyme active site)

+ H_3O^+

Elimination

(base in the enzyme active site)

+ H_3O^+

9.15.   The statement of this exercise provides all of the information needed to draw the structure of the product: The boron atom attaches to C3 and two equivalents of the alkene react with the borane molecule. Even without being told that the boron atom is attached to C3, you can deduce this result because the boron atom reacts at the less highly substituted carbon atom of an alkene double bond, and the H atom attaches to the more highly substituted end.

| 2-Methyl-2-butene | (sia)₂BH |

9.16.   In hydroboration reactions, the boron atom becomes attached to the less highly substituted end of a carbon–carbon double bond. The hydrogen and boron atoms of the borane molecule appear cis to each other in the product, if the stereochemistry can be defined. The stoichiometry depends on steric hindrance around the ends of the double bond.

a.   This molecule is a terminal alkene, so borane can react with three equivalents of it. The boron atom is attached in the product to the CH_2 end of the double bond.

b.   This alkene is disubstituted, so borane can react with three equivalents of it. There are no electronic effects to differentiate the ends of the double bond, but the end with the isopropyl group is more hindered. The boron atom therefore becomes attached to the end of the double bond with only one methyl group.

9.16. (continued)

c. This alkene is trisubstituted, so only two equivalents are likely to react with borane (the substitution of the double bond resembles that of 2-methyl-2-butene shown in Exercise 9.15). The boron becomes attached in the product to the less highly substituted end of the double bond.

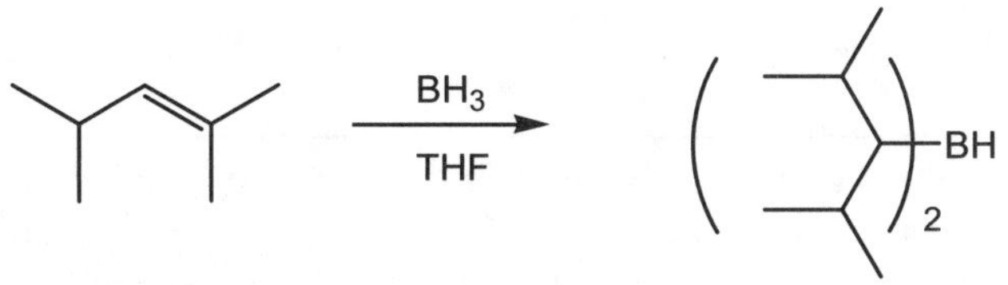

9.17. Borane itself reacts selectively with terminal alkenes to form primary alkylboranes. If the borane reagent is even more selective in its reactions with alkenes than BH_3 itself, then the product having B attached to the 1° carbon atom will be formed in even higher proportions. To draw the structures of the products, we attach the boron-containing fragment (that is, the reagent structure minus H) to the terminal carbon atom.

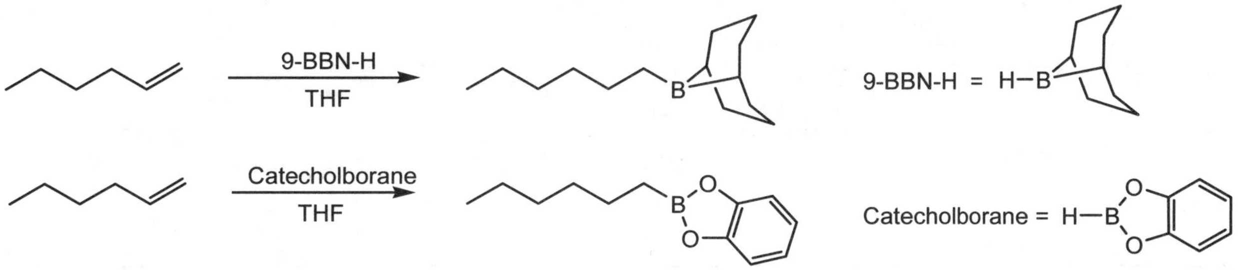

9.18. An alkyne undergoes hydroboration with one equivalent of a reagent having one active B–H bond (such as the one in 9-BBN–H) to produce an alkene that has the H and B atoms cis to each other. The starting alkyne in this example is symmetrical, so regiochemistry of the addition step is not an issue.

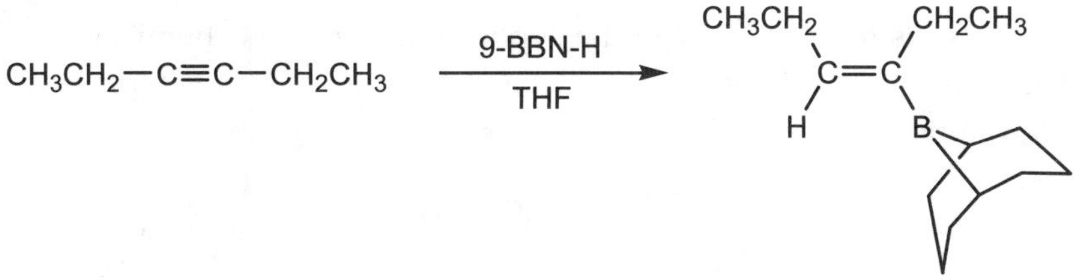

9.19. The conversion of an alkylborane to the corresponding alkoxyborane follows a two-step mechanism: One equivalent of the hydroperoxide ion binds to the boron atom, and then an alkyl group migrates to the oxygen atom of the OOH group, displacing hydroxide ion as the leaving group. These same steps occur as long as there are B—R bonds. For the overall oxidation of ethoxydiethylborane, two equivalents of hydroperoxide ion react with the boron atom (one equivalent in step 1 and one equivalent in step 3 of the following scheme).

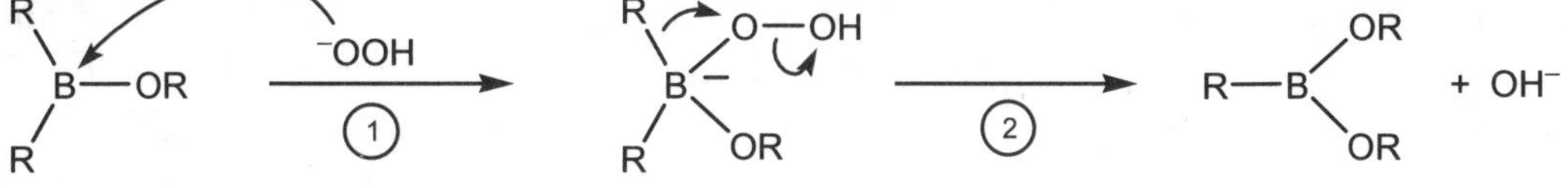

9.19. (continued)

(R = C_2H_5)   (R = C_2H_5)

9.20. In hydroboration reactions, the boron atom becomes attached to the less highly substituted end of the original carbon-carbon double bond. The oxidation/hydrolysis step replaces the boron atom with a hydroxyl group, so the overall transformation results in the addition of H and OH to the ends of the π bond. The OH group is attached in the product to the less highly substituted end of the original double bond. In the transformations given in this exercise, the product molecules cannot exist as stereoisomers. If the stereochemistry of the addition process can be determined, however, the H and OH groups add cis (on the same side) to the double bond.

a.

1. BH_3, THF
2. H_2O_2, OH^-

b.

1. BH_3, THF
2. H_2O_2, OH^-

9.21. When two H atoms are added to a π bond via an organoborane intermediate, one hydrogen atom comes from the borane reagent and the other comes from a molecule of the carboxylic acid that reacts with the organoborane in the second step. Deuterium can be incorporated into the product by using RCOOD, and it will be attached in the product to the original double bond carbon atom that has more hydrogen atoms.

The two hydrogen atoms are cis to each other in the product (if the stereochemistry can be defined) because the stereochemistry of the hydroboration step is cis, and the boron atom is replaced in the second step with retention of configuration.

a.

$CH_3CH_2CH_2C\equiv CH$

1. Catecholborane, THF
2. CH_3CH_2COOD, Δ

b.

1. BH_3, THF
2. CH_3CH_2COOH, Δ

meso

9.22. Methylene carbenoid reagents (zinc reagents that transfer a CH_2 group) react with an alkene double bond to attach the methylene group to each carbon atom of the original π bond. The stereochemistry of the alkene is retained in the reaction.

a.   The starting compound in this first equation has trans stereochemistry, so the cyclopropane product will have the alkyl groups trans to each other as well. Two chiral centers are generated from achiral reactants, so the product is racemic.

9.22. (continued)

b. The combination of chloroform and base adds the CCl_2 group to the carbon atoms of the π bond, and the stereochemistry of the starting alkene is retained. The starting alkene has the cis stereochemistry, so the cyclopropane product also has the cis stereochemistry. Two chiral centers are generated from achiral reactants, so the product is racemic.

9.23. The electrophilic addition reaction of bromine to a double bond in the presence of alcohol takes place in the same way that addition of the halogens to an alkene takes place in the presence of water. When the solvent is alcohol, a vicinal bromo ether is produced instead of the vicinal bromo alcohol formed when water is the solvent (nucleophile). Regiochemistry is not an issue in this reaction because the starting alkene molecule is symmetrical. Two chiral centers are generated from achiral reactants, so the product is racemic.

9.24. a. The reaction of 5-methyl-5-hexene-1-ol with mercury(II) acetate generates a bridged intermediate that can be trapped by the nucleophilic oxygen atom of the alcohol functional group. This intramolecular process takes place by reaction of the oxygen atom at the more highly substituted carbon atom. Ring formation (step 2) and deprotonation by acetate ion (step 3) generate the neutral product. Removal of the acetato(mercury) group with sodium borohydride yields the cyclic ether product.

b. The reaction of 5-methyl-5-hexene-1-ol with bromine generates a bromonium ion intermediate, and the cyclization can occur in the same manner as shown in part (a.). After the cyclization reaction (step 2), deprotonation is accomplished through a reaction with Br⁻ (step 3).

9.24. (continued)

racemic

9.25. To predict the products formed by addition reactions of alkenes and alkynes, answer the following questions:

- What is the nature of the reagent that is reacting: cation (H^+, Cl^+, Br^+, I^+, $HgOAc^+$, or R^+), borane (HBR_2), or carbene (R_2C:)?
- What is the stereochemistry of the addition process: *syn* (cis), *anti* (trans), or neither?
- How stereoselective is the process: nonselective, stereoselective, or stereospecific?
- What is the regiochemistry of addition: Markovnikov, anti-Markovnikov or neither?
- What is the stereochemical composition of the products: racemic, meso, achiral, diastereomeric, chiral (retention or inversion of configuration at an existing chiral center)?

a. In this reaction, a carbenoid reagent reacts with an alkene to form an achiral cyclopropane derivative. Neither regiochemistry nor stereochemistry is an issue because both the reagent and alkene are symmetrical and achiral.

b. In this reaction, bromine is the reagent that reacts with the alkene. Regiochemistry is not an issue because the reagent is symmetrical; the stereochemistry of the bromine addition reaction is *anti* and such transformations are stereospecific. Two chiral centers are generated from achiral reactants via a stereospecific process, so the product is racemic (only one enantiomer is shown).

c. In this reaction, chlorine is the reagent that, along with water, reacts with the alkene. The regiochemistry of addition is Markovnikov, which means that the electrophilic reactant (Cl^+) ends up attached to the carbon atom of the double bond that is less highly substituted to start. This process is stereospecific (trans), but the reactant is acyclic and only a single chiral center is produced, so the relative stereochemistry cannot be discerned. The product is racemic.

9.25.   (continued)

d. In this reaction, a borane reagent reacts with an alkyne in the first step. The stereochemistry of borane addition reactions is syn. In the second step, the boron atom is replaced by H, so the overall process results in addition of two hydrogen atoms, a symmetrical combination for which regiochemistry is irrelevant. The product is the (Z)-disubstituted alkene, which is achiral.

e. In this reaction, a borane reagent reacts with an alkene in the first step. The regiochemistry is "anti-Markovnikov," which means the H of the borane reagent becomes attached to the carbon atom with fewer hydrogen atoms. The stereochemistry of borane addition reactions is syn, but the alkene is symmetrical so stereochemistry is not an issue.   Oxidative hydrolysis of the organoborane (step 2) yields the alcohol. The product is achiral.

f. In this reaction, bromine is the reagent that reacts with a cis-alkene. Regiochemistry is not an issue because the reagent is symmetrical; the stereochemistry of bromine addition reactions is *anti*, and such transformations are stereospecific. Two chiral centers are generated from achiral reactants in a stereospecific process, so the product is racemic (only one enantiomer is shown).

g. In this reaction, mercury(II) acetate reacts with the alkene in the presence of water. The regiochemistry is Markovnikov, so the acetatomercury(II) group becomes attached to the carbon atom at the end of the double bond with more protons, which is the terminal position. An OH group becomes attached to the end of the alkene bond with fewer hydrogen atoms. The metal ion is replaced by H in the second step. In this transformation, a single *new* chiral center is produced, which can be (R) or (S), but the stereogenic carbon atom (R) in the starting material is unchanged, so the product is a mixture of diastereomers.

h. In this reaction, an alkyne undergoes hydration. The regiochemistry is Markovnikov, which means that the OH group becomes attached at the carbon atom that can better support a positive charge, which is the one adjacent to the benzene ring.   The resulting vinyl alcohol tautomerizes to form an achiral ketone.

9.25. (continued)

i. In this reaction, HF undergoes addition. The regiochemistry is Markovnikov, so the proton becomes attached to the carbon atom at the end of the double bond with more protons, and the fluorine atom becomes attached to the other one. The product is achiral.

j. In this reaction, mercury(II) acetate reacts with the alkene in the presence of alcohol. The regiochemistry is Markovnikov, so the acetatomercury(II) group becomes attached to the carbon atom at the end of the double bond with more protons, which is the terminal position. The OCH₃ group becomes attached to the end of the alkene bond with fewer hydrogen atoms. The metal ion is replaced by H in the second step. No new chiral centers are formed in this transformation, so the product is achiral.

9.26. The π bond in the starting compound reacts with the electrophile, a proton, to generate a carbocation intermediate that is stabilized by resonance.

The carbocation undergoes reaction with the nucleophile, methanol (step 2). Deprotonation (step 3) yields the product, which is called an acetal. This reaction will be described in detail in Chapter 19.

9.27. First, we write the steps of the mechanism for each regiochemistry outcome. The mechanism for this addition reaction follows the typical pathway: Protonation of the double bond (step 1), reaction of the carbocation intermediate with a molecule of water (step 2), and deprotonation to form the alcohol (step 3).

9.27. (continued)

The reaction coordinate diagram for these pathways will have the starting compounds and products, as well as the possible carbocation intermediates. Because the 3° carbocation is the more stable of the two possible intermediates that can form (shown in color below), the free energy of activation for its formation will be lower than the pathway leading to the "1° carbocation," which would be at an impossibly high energy.

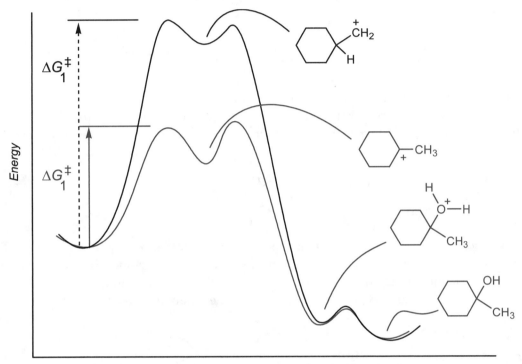

Reaction coordinate

(Look at the solution to Exercise 8.24 to see the reaction coordinate diagram for the reverse process, the E1 reaction of a tertiary alcohol. These diagrams are mirror images, as expected for processes that make use of the same steps except in reverse order.)

9.28. In the first step, the iodine atom can be either above or below the plane of the ring in the bridged intermediate.

After formation of this iodinium intermediate, the carbonyl oxygen atom of the carboxylic acid group reacts to open the three-membered ring and form a new carbon–oxygen bond. Because the carboxylic acid group is above the plane of the six-membered ring and will approach the side opposite that on which the iodine atom is bridged, the intermediate with the iodine atom below the plane will react more rapidly. In the last step, a proton is removed to generate the neutral product.

9.28. (continued)

The final stereochemistry of the product is such that the 6/5 ring junction is cis, and the iodine atom is trans to the five-membered ring. If the starting material is optically active, the product is optically active, too.

9.29. Follow the procedures outlined in the solutions to Exercise 9.25.

a. In this reaction, dichlorocarbene reacts with an alkene. Regiochemistry is not an issue because the reagent is symmetrical; the stereochemistry of carbene addition reactions is syn so the H and methyl group will be cis in the product. The reaction is stereospecific and two new chiral centers are formed, so the product is racemic.

b. In this reaction, iodine is the reagent that, along with methanol, reacts with the alkene. The regiochemistry of addition is Markovnikov, which means that the electrophilic reactant (I^+) ends up attached to the carbon atom of the double bond that is less highly substituted to start. The reactant is acyclic and only a single chiral center is produced, so the relative stereochemistry cannot be discerned.

c. In this reaction, borane reacts with an alkene in the first step. The starting alkene is meso, so the regiochemistry of addition is not an issue. The stereochemistry of borane addition reactions is syn, and the boron atom can be attached to either the top or bottom face of the double bond. Oxidative hydrolysis of the organoborane (step 2) yields the alcohol, so the OH group can also be above or below the plane. The product is therefore a mixture of four stereoisomers.

d. In this reaction, mercury(II) acetate reacts with the alkene in the presence of water. The regiochemistry is Markovnikov, so the acetatomercury(II) group becomes attached to the carbon atom at the end of the double bond with more protons, which is the terminal position. An OH group becomes attached to the end of the alkene bond with fewer hydrogen atoms. The metal ion is replaced by H in the second step. The product alcohol is achiral.

9.29.   (continued)

e. In this reaction, a borane reagent reacts with an alkyne in the first step. Each carbon atom of the triple bond has an alkyl group attached, so the regiochemistry of addition is dictated by the size of these groups and the bulkiness of the 9-BBN-H reagent (the boron atom will attach to the less hindered carbon atom). Oxidative hydrolysis yields a vinyl alcohol that undergoes tautomerism to yield the achiral ketone product.

$$(CH_3)_2CH-C{\equiv}C-CH_3 \quad \xrightarrow[\text{2. } H_2O_2,\ OH^-]{\text{1. 9-BBN-H, THF}} \quad (CH_3)_2CH-CH_2-\underset{\underset{O}{\|}}{C}-CH_3 \quad \text{achrial}$$

f. In this reaction, an allylic alcohol is treated with a strong mineral acid, which will likely generate a carbocation. The other double bond in this molecule will react with the carbocation to form a new ring. Loss of a proton after cyclization generates the double bond in the six-membered ring, and the elimination can occur in one of two ways. Each product has a new chiral carbon atom formed from an achiral carbocation intermediate, so each diene is formed as a racemic mixture.

racemic   racemic

g. In this reaction, HBr adds to the double bond. The regiochemistry is Markovnikov, so the proton becomes attached to the carbon atom at the end of the double bond with more protons, and the bromine atom becomes attached to the other one. The stereochemistry of addition is not an issue because the product is achiral.

achrial

h. In this reaction, a carbenoid reagent reacts with an alkene. Regiochemistry is not an issue because the addend (CH₂) is symmetrical; the stereochemistry of a carbene addition reaction is *syn*. The product is a cyclopropane derivative with two new chiral centers, and it is racemic because the reactants are achiral.

racemic

9.30.   To predict the product of electrophilic addition for a reagent that you have not encountered before, decide which portion of the reagent is the electrophile and which part is the nucleophile. A formula is normally written so that the electrophilic part appears first. For the reagents in this exercise, therefore, Br^+ and RS^+ are the electrophiles. They become attached to the less highly substituted end of the alkene double bond, which creates a carbocation intermediate at the more highly substituted carbon atom.

9.31. In the cyclization reaction shown in this exercise, one of the alkene double bonds is protonated by its reaction with the mineral acid. The double bonds are equivalent, so it does not matter which one reacts first. The carbocation is intercepted by the π electrons of the other double bond to create the ring in step 2 (the colored dot is used as a marker only to keep track of which carbon atom is which).

Rearrangement subsequently takes place (step R) and places the positive charge on a carbon atom in the ring. If elimination of a proton were to occur instead of the rearrangement process, an exocyclic double bond would form. After rearrangement, the compound is deprotonated to form a trisubstituted, endocyclic double bond, which is more stable than any other double bond that could form.

3° exocyclic carbocation

3° endocyclic carbocation

9.32. We can choose reagents for a given addition reaction by classifying its type (substitution, addition, etc.), and then considering the stereochemistry and regiochemistry of the transformation.

a. This transformation involves hydration of an alkene double bond in which the regiochemistry of addition is anti-Markovnikov. Therefore, hydroboration followed by oxidative hydrolysis is required because electrophilic addition reactions proceed with the Markovnikov regiochemistry.

1. BH₃, THF
2. H₂O₂, OH⁻

b. This transformation involves addition of H and Br to the alkene double bond and the regiochemistry of addition is Markovnikov. Therefore, an electrophilic process is required, and HBr will be used.

HBr

c. This transformation involves hydration of an alkyne triple bond. The alkyne is symmetrical, so the regiochemistry is not an issue. Therefore, hydration under the influence of a metal catalyst is sufficient. Hydroboration followed by oxidative hydrolysis will also work.

$H_3CH_2C-C\equiv C-CH_2CH_3$

PtCl₂, H₂O, THF
or   1. Catecholborane, THF
2. H₂O₂, OH⁻

9.32. (continued)

d. This transformation involves formation of a cyclopropane derivative from an alkene, which calls for the use of a methylene carbenoid reagent.

$$\xrightarrow{\text{Zn(Cu), CH}_2\text{I}_2}$$

or

$$(\text{CF}_3\text{COO})\text{Zn}-\text{CH}_2-\text{I}$$
$$\text{CH}_2\text{Cl}_2, \ 0\ ^\circ\text{C}$$

9.33. Hydroboration followed by oxidative hydrolysis is a procedure that can be used to make an alcohol from an alkene: the OH group will be attached to the carbon atom at the less highly substituted end of the double bond. Therefore, a double bond that has the same number of hydrogen atoms at each end will not be useful for making an alcohol unless the molecule is symmetrical. The OH group and new H atom that are added during the hydroboration sequence have to be cis to one another.

When you look at the structure of the alcohol product, remove, in turn, the OH group and a cis-hydrogen atom attached to the adjacent carbon atoms. These are the possible starting alkenes. If the substitution pattern is different at each end of the double bond and the OH group is attached to the carbon atom with more hydrogen atoms, then the alcohol can most likely be prepared by the hydroboration/oxidative hydrolysis procedure.

a. Removing H and OH from this compound creates methylenecyclohexane. The ends of the double bond are substituted with different numbers of hydrogen atoms, and the OH group is attached to the carbon atom having more hydrogen atoms, so this alkene is a suitable starting material.

b. Removing H and OH from this compound creates either 1-methylcyclohexene or 3-methylcyclohexene. The ends of the double bond in 1-methylcyclohexene are substituted with different numbers of hydrogen atoms, and the OH group is attached to the carbon atom having more hydrogen atoms, so this alkene is a suitable starting material.

c. Removing H and OH from this compound creates either ethylidenecyclohexane or 1-ethylcyclohexene; the ends of the double bond in both are substituted with different numbers of hydrogen atoms, but the OH group is attached to the carbon atom having fewer hydrogen atoms, so neither alkene is a suitable starting material.

9.33. (continued)

d.   Removing H and OH from this compound creates either 2-octene or 3-octene. The ends of the double bond in both have the same numbers of hydrogen atoms, so neither is a suitable starting material.

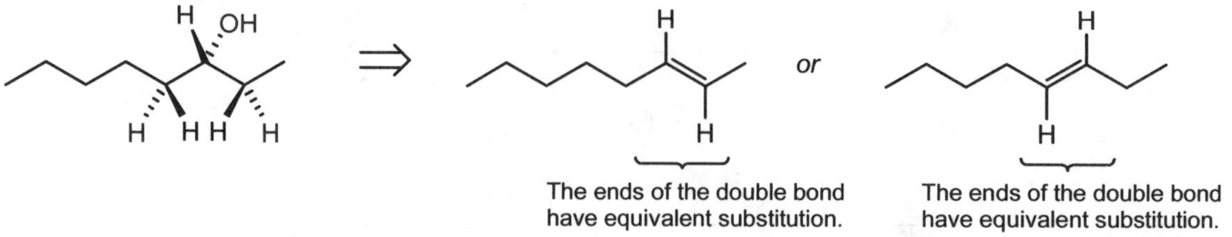

The ends of the double bond have equivalent substitution.   The ends of the double bond have equivalent substitution.

9.34. The possible starting materials were identified in Exercise 9.33. Shown below are the actual equations showing the products that will be formed from the hydroboration/oxidative hydrolysis of each alkene.

a.

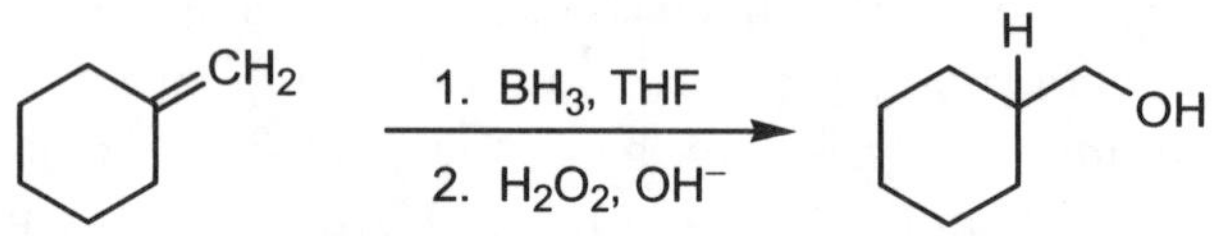

b.

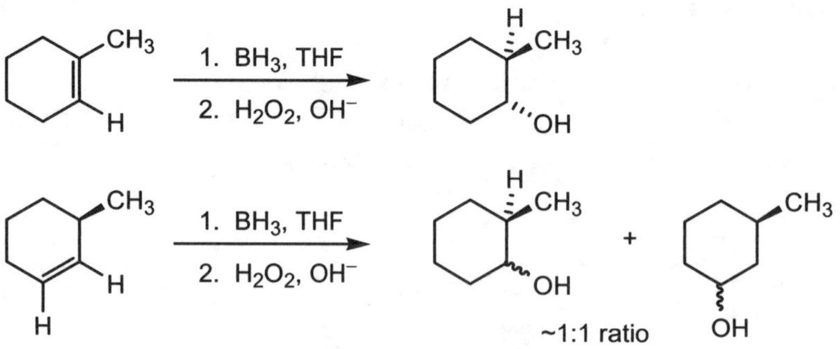

~1:1 ratio

c.

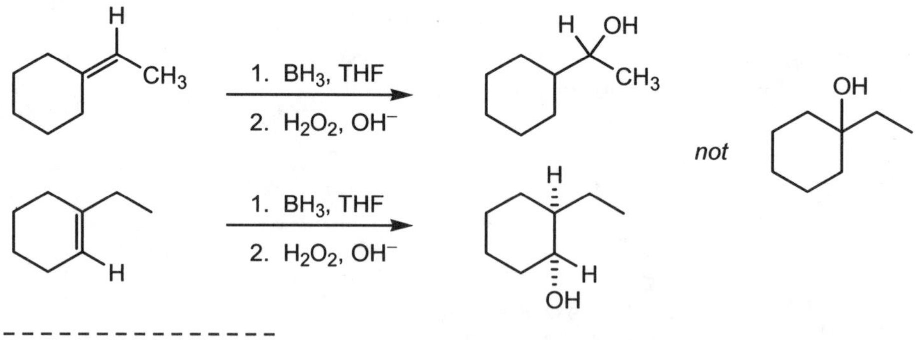

not

d.

1:1 ratio

1:1 ratio

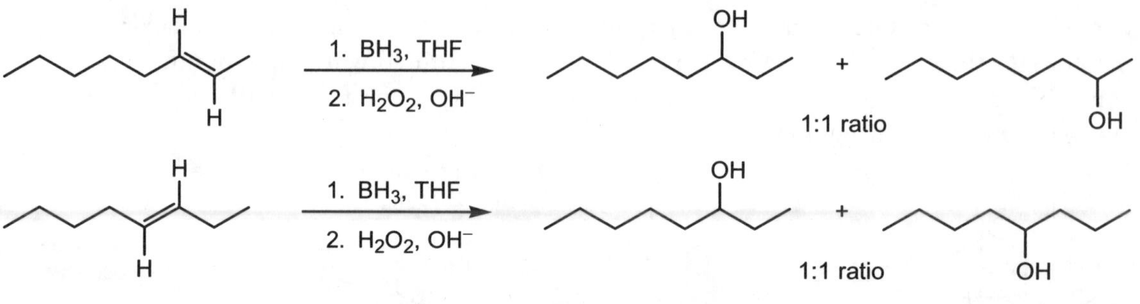

ADDITION REACTIONS OF CONJUGATED DIENES

10.1. Follow the procedures outlined in the solutions to the nomenclature exercises in Chapter 1.

a. **(2Z, 4Z)-4-Methyl-2,4-heptadiene**

hept	seven carbon atoms
2,4-diene	two double bonds, starting at C2 and C4
4-methyl	CH_3 group at C4
2Z	the higher priority groups at the ends of the
	π bond are on the same side of the C=C bond
4Z	the higher priority groups at the ends of the
	π bond are on the same side of the C=C bond

b. **2-Fluoro-1-penten-3-yne**

pent	five carbon atoms
1-en	double bond starting at C1
3-yne	triple bond starting at C3
2-fluoro	F atom at C2

c. **2-Chloro-1,3-cyclohexadiene**

cyclohexa	ring with 6 carbon atoms
1,3-diene	two double bonds, starting at C1 and C3
2-chloro	Cl atom at C2

d. **(2E, 4Z) -2,4-Hexadienal**

hexa	6 carbon atoms
2,4-diene	two double bonds, starting at C2 and C4
al	aldehyde group; its carbon atom defines C1
2E	the higher priority groups at the ends of the
	π bond are on opposite sides of the C=C bond
4Z	the higher priority groups at the ends of the
	π bond are on the same side of the C=C bond

10.2. Follow the procedures outlined in the solutions to Exercise 1.17.

a. This compound has the carboxylic acid functional group at the end of the carbon chain, so the name ends in "–oic acid." The longest carbon chain that includes both the carboxylic acid carbon atom and the double bonds has seven carbon atoms, so the root is hept. There are two carbon–carbon double bonds, so the multiple bond index is diene: hept/diene/oic acid = heptadienoic acid.

Numbering begins with the carboxylic acid carbon atom. This numbering order means that the double bonds begin at C2 and C5. A methyl group is attached at C3, and each double bonds has the (E) configuration, so the name is **(2E, 5E)-3-methyl-2,5-heptadienoic acid.**

10.2. (continued)

b. This compound has a ring of six carbon atoms with one double bond within the ring: cyclohexene. The other carbon–carbon double bond is external to the ring, so it is named as a substituent, isopropylidene (Section 8.1c). The substituent specifies the assignment of C1 for the ring, so the name is **isopropylidene-2-cyclohexene**.

c. This compound has both a double and a triple bond. Numbering can begin at either end of the chain, but the numbers would be the same (1-buten-3-yne versus 3-butene-1-yne), so we choose the first because a double bond has the higher priority. "Ene" is placed in the name before "yne" in either case. A phenyl group is attached to C1 and a chlorine atom is attached to C2. The configuration of the double bond is (Z), so the compound's name is **(Z)-2-chloro-1-phenylbutene-3-yne**.

d. This compound has a seven-membered ring and three carbon–carbon double bonds, starting at C1, C3, and C5. A *tert*-butyl group is attached at C1, so the name is **1-*tert*-butyl-1,3,5-cycloheptatriene**.

10.3. Isolated double bonds are separated by at least one sp^3-hybridized carbon atom (a. and d.). Cumulated double bonds share a carbon atom with sp hybridization (c.). The ends of conjugated double bonds are separated by a single carbon–carbon bond (shown in color in part b.).

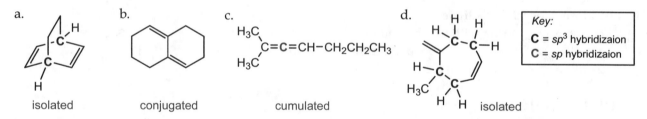

a.	b.	c.	d.	Key:
isolated	conjugated	cumulated	isolated	**C** = sp^3 hybridizaion **C** = sp hybridizaion

---

10.4. Bonding molecular orbitals are formed by the mathematical addition of the combining orbitals. The region with the highest electron density lies between the atoms. The node of each hybrid orbital at the nucleus is retained in this bonding combination.

σ_{C-C}: filled with two electrons

Antibonding molecular orbitals are formed by the subtraction of one orbital from another, which means that changes occur in the mathematical signs of the orbital being subtracted. The highest electron density lies *away* from the region between the atoms. The node of each hybrid orbital at the nucleus is retained in this combination, and a new node is created in the space between the atoms.

σ^*_{C-C}: unoccupied

10.5. The energy of a bonding molecular orbital (MO) is lower than the energy of the hybrid orbitals undergoing combination. The energy of the antibonding MO is higher than the energies of the combining hybrid orbitals. Because each hybrid orbital contributes one electron to the MOs, there are two electrons and they are used to fill the bonding MO (σ). The high energy antibonding MO (σ*) is empty.

10.6. The reaction of a conjugated diene with chlorine will produce both 1,2- and 1,4-dichloro products. If a new chiral center is produced, then both configurations will be generated at the chiral carbon atom. For the 1,4-addition process, it is necessary to consider the formation of double bond isomers, which result from rotation about the sigma bond. It is worthwhile to draw the structures of the possible carbocation intermediates. This diene is symmetrical, so it does not matter which double bond we start with.

The products will result from the reaction between the possible carbocations and the chloride ion.

10.7. Follow the procedure outlined in the solution to Exercise 10.6. The electrophile is H⁺ and the nucleophile is water (OH⁻). Both alcohol products are achiral.

10.8. Follow the procedures outlined in Example 10.3

a. Generate the possible carbocations (and their resonance forms) by protonating each double bond separately, and then decide which set of cations is more stable.

The products are formed by attaching the bromide ion to the carbocation centers in the more stable set. The kinetic product is derived from reaction with the more stable carbocation of the pair, and the thermodynamic product is the one with the more highly substituted double bond.

b. Generate the possible carbocations (and their resonance forms), and then decide which set is more stable. This diene is symmetrical, so only one set of carbocations is formed.

The products are formed by attaching the bromide ion to the two carbocation centers. The kinetic product is derived from reaction with the more stable carbocation, and the thermodynamic product is the one with the more highly substituted double bond.

c. Generate the possible carbocations (and their resonance forms), and then decide which set is more stable.

10.8. (continued)

The products are formed by attaching the bromide ion to the carbocation centers of the more stable set. The kinetic product is derived from reaction with the more stable carbocation center, and the thermodynamic product is the one with the more highly substituted double bond.

Kinetic product   **Thermodynamic product**
(trisubstituted double bond)

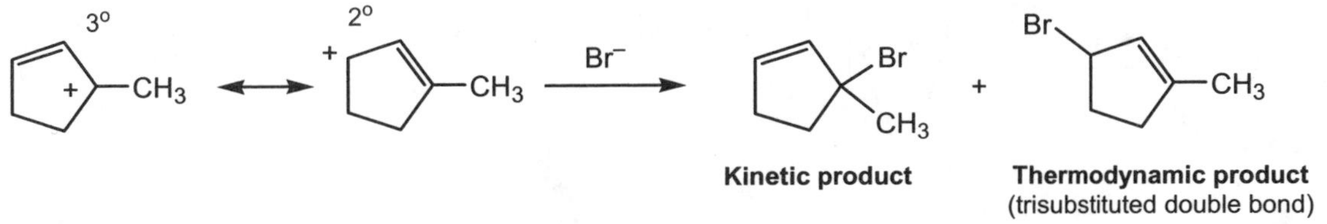

10.9. The product of the Diels-Alder reaction is a cyclohexene derivative in which the substituents have the same stereochemical relationships as the substituents in the starting diene and dienophile. We first orient the reactants next to each other, and then connect the ends of their π systems to form the six-membered ring (this procedure is shown with the dashed, colored lines below). The structure of the product has a double bond at the bond between the double bonds in the starting diene. For the reactions shown in this exercise, new chiral centers are produced from achiral starting compounds (indicated by the asterisks), so each transformation yields a racemic mixture.

a.

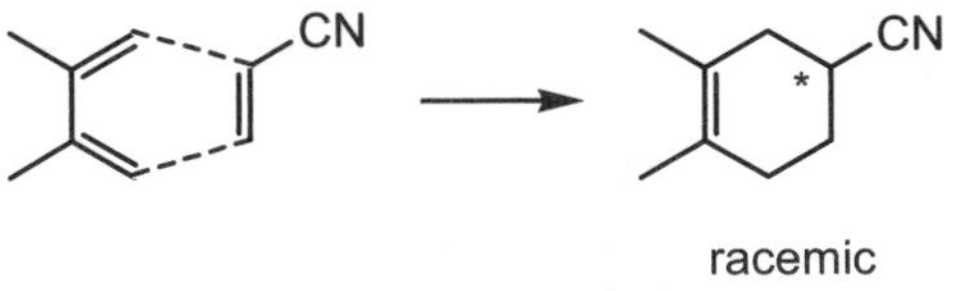

racemic

b.

racemic

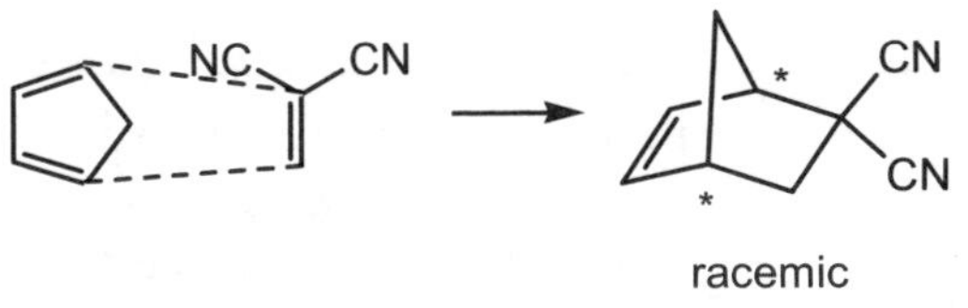

10.10. Follow the procedure outlined in Exercise 10.9. For the reactions shown in this exercise, new chiral centers are produced from achiral starting compounds (indicated by the asterisks), so each transformation yields a racemic mixture.

a. The diene in this reaction is not symmetrical, so the reactants have to be oriented so that the ester group of the dienophile is directly below (or above) the diene π system, as shown below at the right. This orientation yields the relative stereochemistry shown.

racemic

E = COOCH$_3$

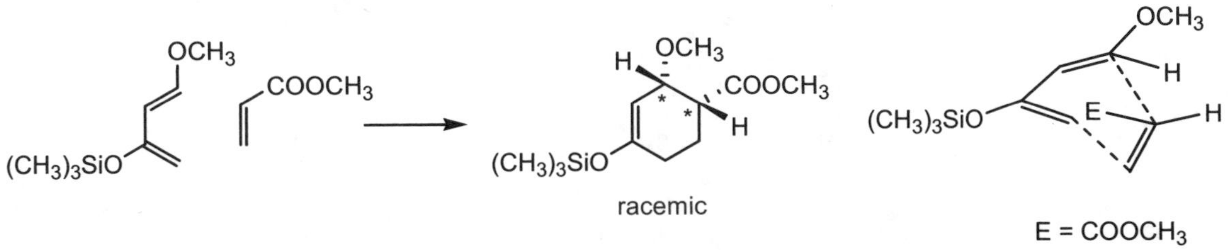

b. The diene in this reaction is symmetrical but it is cyclic, so the reactants have to be oriented so that the cyano group of the dienophile is directly below (or above) the diene π system, as shown below at the right. This orientation yields the relative stereochemistry shown.

racemic

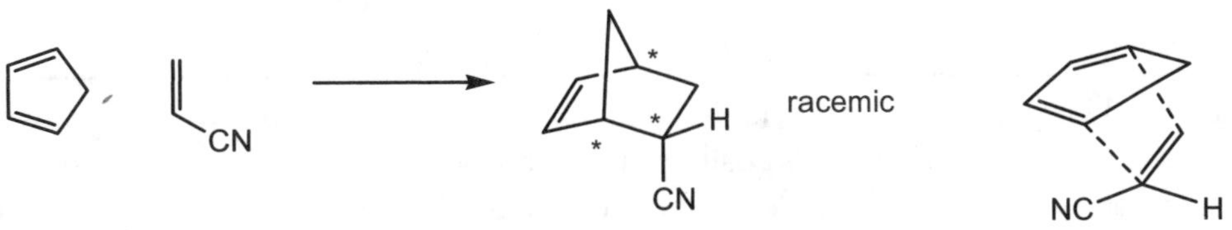

10.11. All of these reactions yield products in which the elements of H and Cl have added to one of the double bonds. When there are no substituents attached to the ring of a cyclic diene (conjugated or isolated), then the product appears as if only one of the double bonds has reacted. In other words, the 1,2- and 1,4-addition products are identical for the conjugated dienes. A new chiral center is generated from achiral reactants, so a racemic mixture is produced in each case.

a.

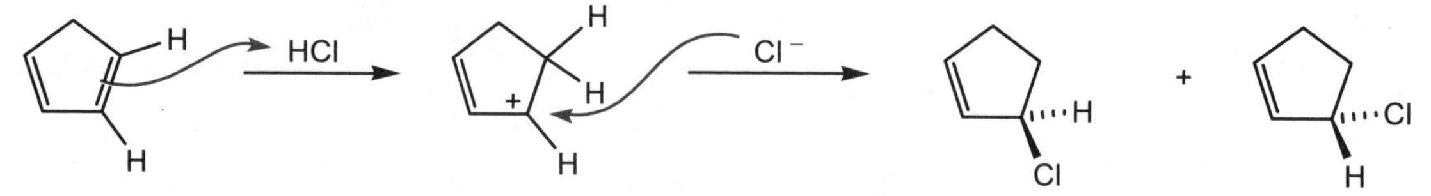

b.

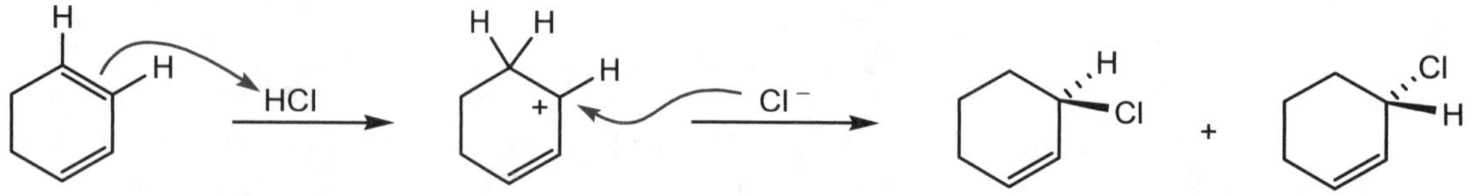

c.

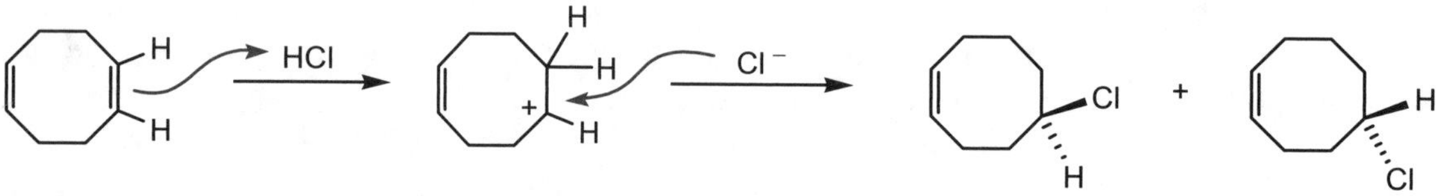

10.12. Follow the procedures outlined in Exercises 10.9 and 10.10.

a.

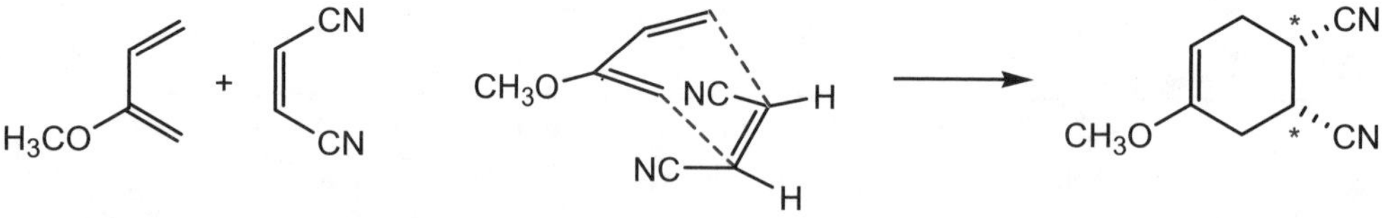

racemic

b.

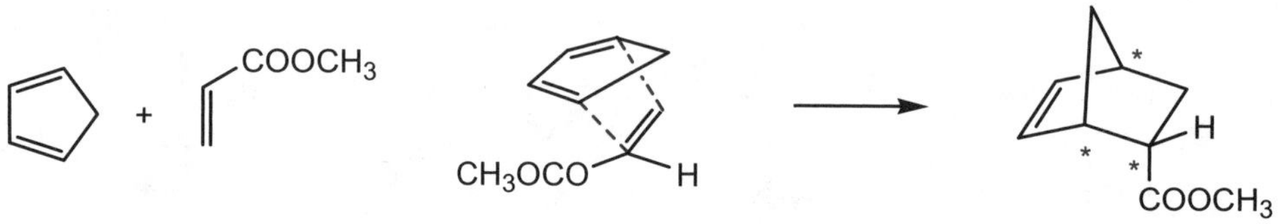

racemic

c.

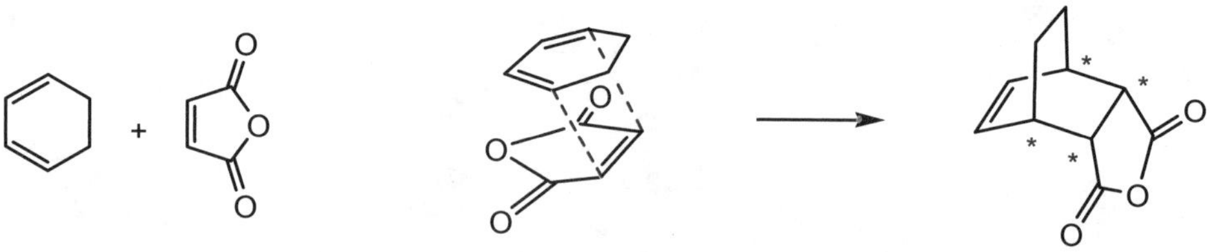

meso

d.

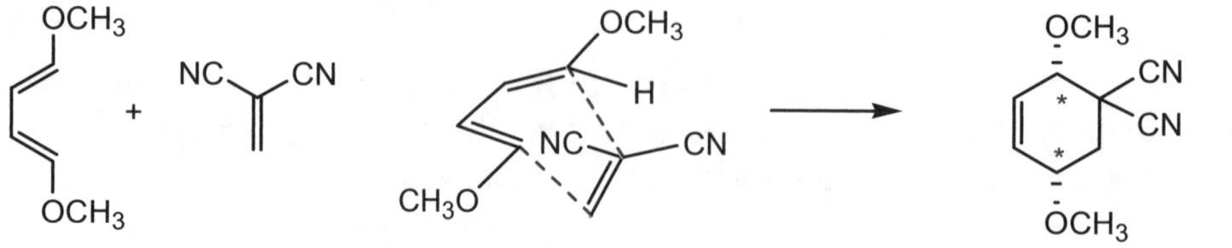

racemic

10.13. Nitroethylene and vinyltriphenylphosphonium chloride each have an atom with a positive charge attached to one of the alkene carbon atoms. A positive charge attracts electrons, which in turn makes the double bond electron deficient, the principal characteristic of a good dienophile. The boron atom of the alkenylborane has only six valence electrons, which withdraws electron density from the π bond. (The structures of these dienophiles are shown in the solution to Exercise 10.14).

10.14. To predict the expected products of these reactions, follow the procedure outlined in the solution to Exercise 10.9.

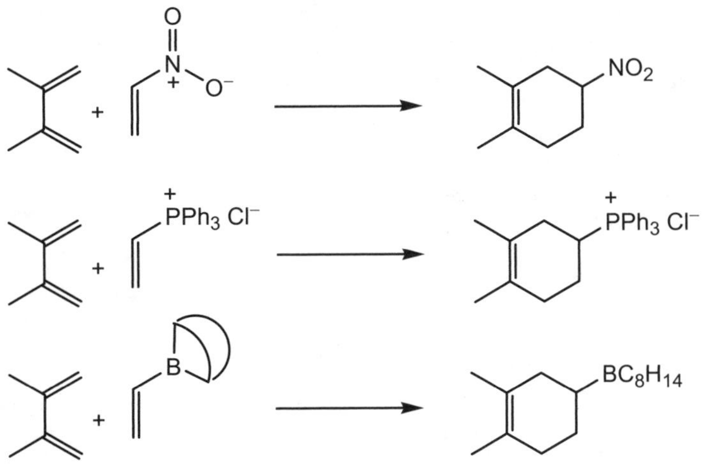

10.15. As shown in the solution to Exercise 10.14, the reaction of an alkenylborane with a diene yields an organoborane. The exo isomer will likely be the major one, and because new chiral centers are produced from achiral starting compounds, this transformation yields a racemic mixture. The second step of the given procedure converts this organoborane derivative to the corresponding alcohol, a reaction that proceeds with retention of configuration.

10.16. Given the structure of a cyclohexane derivative that has been formed via the Diels-Alder reaction, you need only break the two bonds in the six-membered ring that are connected to the carbon atoms *adjacent* to the double bond (shown as colored dots in the following structures). These carbon atoms define the termini of the diene system, and the other two carbon atoms define the double bond (or triple bond, in part c.) of the dienophile.

a.

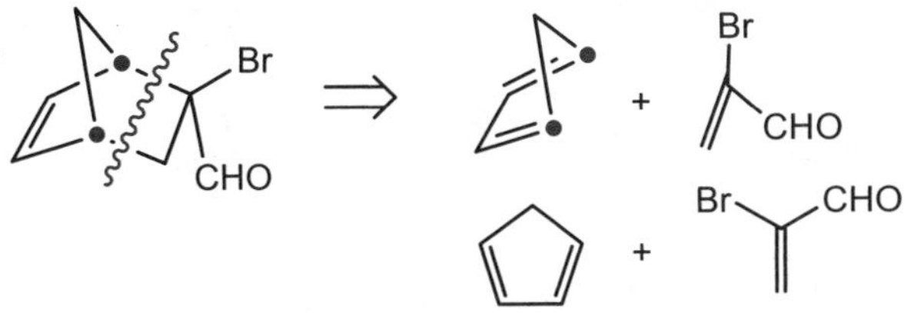

b.

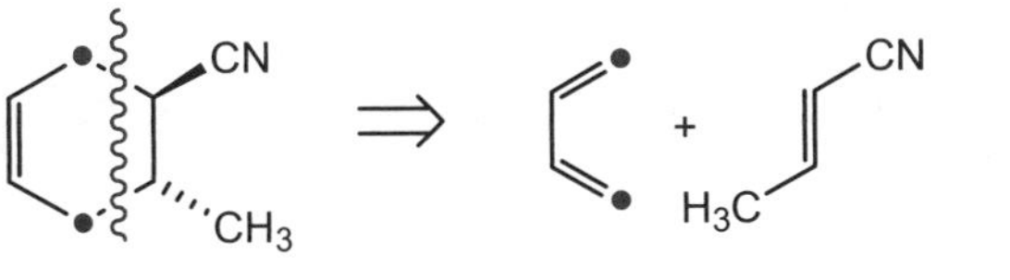

c.

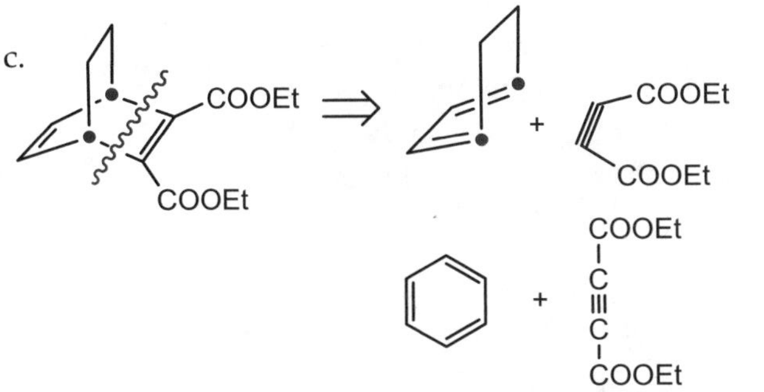

d.

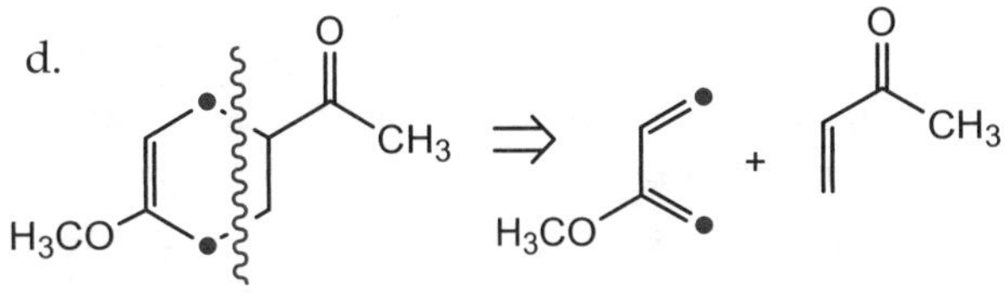

10.17. The middle carbon atom of the cumulated diene double bonds has *sp* hybridization, as do the two alkyne carbon atoms of the cyclic alkyne product.   The bond angles of a carbon atom with *sp* hybridization are 180°, so the diene product should be less strained within a 10-membered ring because only one carbon atom has this geometric constraint imposed upon it. Hence the cumulene is formed in greater quantity than the alkyne.

Ratio: 3 : 2

10.18. To solve this type of problem, first write out the synthesis scheme as a flow diagram. The last reaction, to form compound **D**, is clearly a Diels-Alder reaction, so we follow the procedure given in the solution to Exercise 10.16 to deduce the structure of the diene, which is compound **C**.

A diene can be formed by treating a dibromoalkane with strong base, which corresponds to the reaction conditions that are given. A dibromoalkane can be prepared by adding bromine to an alkene, so we identify the compounds as follows:

A	**B**	**C**	
Cyclohexene	1,2-Dibromocyclohexane	1,3-Cyclohexadiene	

10.19. Follow the procedures illustrated in the solutions to Exercise 10.8.

a.   Generate the possible carbocations (and their resonance forms), and then decide which set is more stable.

This set of carbocations is more stable.

The products are formed by attaching an OH group to the carbocations of the more stable set. The kinetic product is derived from the reaction with the more stable carbocation, and the thermodynamic product is the one with the more highly substituted double bond.

Kinetic product

Thermodynamic product (trisubstituted double bond)

10.19. (continued)

b. Generate the possible carbocations (and their resonance forms), and then decide which set is more stable. This diene is symmetrical, so only one set of carbocations is formed. The products are formed by attaching the methoxy group to the carbocation centers. The carbocations are equivalent and the double bond substitution pattern of the products is the same (disubstituted), so there is no clear choice for the identities of the thermodynamic and kinetic products. We would therefore expect to obtain equal amounts of each product no matter what the reaction conditions are.

Both intermediates are 2° carbocations.   Both products have a disubstituted double bond.

10.20. For stable alkenes and polyenes with n number of p orbitals, there are n MOs, $n/2$ of which are bonding orbitals and filled; and $n/2$ of which are antibonding orbitals and vacant. The lowest energy MO has no node and $(n-1)$ overlapping pairs. With increasing energy, an additional node is added between atoms (vertical dashed lines) and the numbers of *pairs* of overlapping orbitals decreases by one. At the highest energy MO, the number of nodes reaches its maximum, and there is no pair of overlapping orbitals. The bonding MOs have more overlapping pairs of orbitals than nodes, and the antibonding MOs have more nodes than overlapping pairs. There are six electrons to add to the orbitals, and these fill the bonding MOs.

	NODES	OVERLAPPING PAIRS
$\Psi_6{*}$	5	0
$\Psi_5{*}$	4	1
$\Psi_4{*}$	3	2
Ψ_3	2	3
Ψ_2	1	4
Ψ_1	0	5

10.21. This cycloaddition reaction takes place in the same manner that a typical Diels-Alder reaction does. The nitrogen atoms in the product have sp^2 hybridization, so they are planar and it is not necessary to consider whether the product has an endo or exo orientation (it is neither).

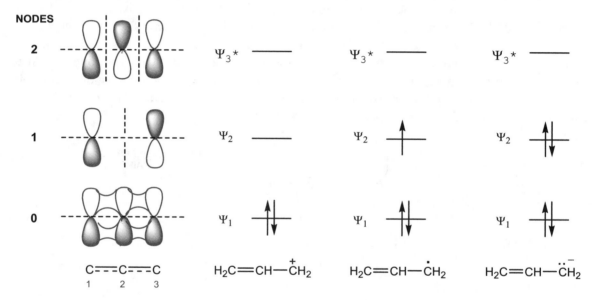

10.22. For a delocalized system with n number of p orbitals, there are n MOs. For allylic systems, there are three MOs. The lowest energy MO has no node, the next highest MO has one node, and the highest energy MO has two nodes. Notice that the node in Ψ_2 is at C2; an MO with a node at an atom position is classified as non-bonding. In filling these orbitals, we start with the lowest energy MO and add electrons. For the allyl carbocation, only Ψ_1 is filled. The allyl radical has a single electron in Ψ_2 and the allyl carbanion has both Ψ_1 and Ψ_2 filled.

The resonance forms make use of Ψ_1 and Ψ_2, which are either bonding or non-bonding orbitals. The antibonding MO is not involved.

10.23. To predict the products formed in the addition reactions of dienes, begin by assessing the types of reagents involved.

- Is the reagent electrophilic (H^+, Cl^+, Br^+) or neutral (Diels-Alder reaction)?
- If the process starts with a reaction between a diene and an electrophile, do the reaction conditions suggest kinetic or thermodynamic control?
- If a Diels-Alder reaction is taking place, what is the stereochemistry of the reactants?

For reactions that generate a diene, make use of what you learned about elimination reactions in Chapter 8.

a. In this transformation, a diene reacts with a proton in the first step. The low temperature ($< 0\ °C$) is most likely meant to indicate that kinetic control is desired. Use the procedures illustrated in the solutions to Exercise 10.8 to deduce the structures of the kinetic and thermodynamic products. Both products have a new chiral center, so both are racemic.

10.23. (continued)

racemic
Thermodynamic product   **Kinetic product**

b. In this transformation, a diene reacts with an alkene having electron-withdrawing substituents, so this is a Diels-Alder reaction. Use the procedures outlined in the solutions to Exercises 10.9 and 10.10 to deduce the structure of the product. This product has two new chiral centers as well as a mirror plane of symmetry, so it is a meso compound.

meso

c. In this transformation, a diene reacts with an alkene having an electron-withdrawing substituent, so this is a Diels-Alder reaction. Use the procedures outlined in the solutions to Exercises 10.9 and 10.10 to deduce the structure of the product, which will be endo. The product has three new chiral centers so it is obtained as a racemic mixture.

racemic

d. In this transformation, an alkyl halide reacts with a strong base, so an E2 mechanism is likely. The elimination must occur away from the bridgehead atom because a double bond to the bridgehead atom would be too strained. The product is achiral.

e. In this transformation, a diene reacts with Br^+ in the first step. The temperature (greater than room temperature) is most likely meant to indicate that thermodynamic control is desired. Use the procedures illustrated in the solution to Exercise 10.8 to deduce the structures of the kinetic and thermodynamic products. The thermodynamic product has a new chiral center, so it would be racemic. It is expected to be the major product. The kinetic products can exist as geometric isomers, and both are achiral.

racemic   achiral   achiral
**Thermodynamic   Kinetic product   Kinetic product
product**

Major

10.23. (continued)

f.   In this transformation, chlorine adds to the alkene double bond to form a vicinal dichloride. The strong base in step 2 is used to promote E2 reactions, which generate the conjugated diene. Within a ring having fewer than nine atoms, a triple bond or cumulated double bonds cannot be formed. The product is achiral.

10.24. If the mechanism for 1,4-addition were to occur by the route shown in part (a.), you would expect to isolate only the cis isomer. The actual mechanism, shown in (b.), will yield both cis and trans products because the intermediate carbocation is free to undergo rotation about the C2-C3 bond when the carbocation is localized on C2. (The 1,2-addition products are not shown.)

a.

b.

OXIDATION AND REDUCTION REACTIONS

11.1. The conversion of methanol to bromomethane occurs via the S$_N$2 mechanism: The OH group becomes protonated in the first step, which converts it to a good leaving group. The bromide ion subsequently displaces the leaving group. In the second transformation, methanethiolate ion is the nucleophile that displaces the bromide ion via the S$_N$2 pathway.

The other reaction shown is an example of the Mitsunobu reaction. The reactants crate a good leaving group plus the azide ion. The second step occurs via the S$_N$2 pathway.

11.2. Hydrogen, in the presence of a palladium catalyst, reacts with carbon-carbon π bonds, adding a hydrogen atom to each carbon atom of a double bond (or two atoms of hydrogen to each carbon atom of a triple bond). Addition of molecular hydrogen to π bonds leaves the carbon-carbon single bond intact at that site. A benzene ring does not react normally under such conditions (1 atm H$_2$, 25° C).

The stereochemistry for the addition of hydrogen to a double bond is *syn*, which means the hydrogen atoms that add will be cis to each other if the stereochemistry of the product can be defined, as in part (a.) of this exercise. If new chiral centers are produced, the product will either comprise a racemic mixture or a meso compound.

a.

b.

c.

11.3. The first step in part (a.) is a substitution reaction that produces the alkyl azide derivative. Hydrogenation converts the azido group to an amino group.

a.

In part (b.), the alcohol is first converted to a good leaving group by forming its mesylate derivative. Cyanide ion replaces the leaving group to yield the corresponding nitrile, which is subsequently reduced to produce an amine with one more carbon atom than the starting alcohol molecule.

b.

11.4. To predict the products that will be formed from ozonolysis of an alkene, "stretch" the double bond, then erase the middle of the bond and put oxygen atoms at these new ends. These are the products that will be formed when reductive workup conditions are employed. If an aldehyde is generated as a product under reductive conditions, then oxidative workup will create carboxylic acid groups (that is, the hydrogen atoms shown in color will be replaced by OH groups). Numbering the carbon atoms will help you keep track of which carbon atom is which, especially if the ozonolysis reaction is used to cleave a ring.

a.

b.

11.5. Osmium tetroxide reacts with an alkene in the presence of *tert*-butyl hydroperoxide to attach an OH group at each carbon atom of the original π bond. This addition process breaks the double bond and leaves only a single bond. The two OH groups in the product will be cis to each other; and if new chiral centers are created, then either a racemic mixture or a meso compound will be formed.

a.

(racemic mixture)

b.

(racemic mixture)

11.6. To predict the products that are formed from ozonolysis of an alkene, follow the procedures outlined in the solution to Exercise 11.4. The reagent combination of osmium tetroxide and sodium periodate converts the alkene to the same products formed from ozonolysis with reductive workup.

a. In this reaction, one alkene carbon atom in the substrate bears a hydrogen atom, but the other does not. With ozonolysis and reductive workup, the product mixture will comprise an aldehyde and a ketone molecule. The same products will be formed upon reaction with osmium tetroxide and sodium periodate. Under conditions of ozonolysis and oxidative workup, the product mixture will comprise a carboxylic acid and a ketone molecule.

O_3 followed by reductive workup
or OsO_4 + $NaIO_4$

O_3 followed by oxidative workup

b. In this reaction, both alkene carbon atoms in the substrate bear at least one hydrogen atom, so the product mixtures will consist of either two aldehyde or two carboxylic acid molecules.

O_3 followed by reductive workup
or OsO_4 + $NaIO_4$

O_3 followed by oxidative workup

c. In this reaction, each alkene carbon atoms in the substrate bears at least one hydrogen atom, so the product mixtures will consist of either two aldehyde or two carboxylic acid molecules.

O_3 followed by reductive workup
or OsO_4 + $NaIO_4$

O_3 followed by oxidative workup

11.7. The epoxidation of alkenes with use of a peracid occurs at the more highly substituted double bond if there is more than one alkene group. (A benzene ring does not undergo epoxidation reactions when treated with peracids.) The stereochemistry of the epoxide is the same as the starting alkene double bond (i.e., trans alkene = trans epoxide). If the starting material is achiral and new chiral centers are formed, then racemic products will be obtained.

11.8. The Swern oxidation procedure converts primary alcohols to aldehydes and secondary alcohols to ketones. The mechanism involves three steps. In the first step, a good leaving group is created in place of the hydroxyl group proton. In step 2, an ylide is formed by deprotonation at one of the methyl groups attached to sulfur. In the final step, elimination occurs to produce the carbonyl group and dimethyl sulfide.

11.9. Oxidation of a 1° alcohol to an aldehyde requires non-aqueous conditions and normally makes use of Collins or Corey's reagent in dichloromethane as the solvent. Jones reagent or aqueous potassium permanganate converts a 1° alcohol to a carboxylic acid. All of these reagents convert a 2° alcohol to a ketone. Tertiary alcohols are inert toward all of these reagents. Manganese(IV) oxide converts benzylic and allylic alcohols to the corresponding carbonyl compounds: A primary benzylic or allylic alcohol yields an aldehyde, and a secondary benzylic or allylic alcohol yields a ketone.

a. The reaction shown is the conversion of a secondary alcohol to a ketone: Use PCC, CrO₃/pyridine, CrO₃ in aqueous acid, or KMnO₄ in aqueous base.

b. The reaction shown is the conversion of a primary benzylic alcohol to an aldehyde: Use CrO₃/pyridine, PCC, or MnO₂.

c. The reaction shown is the conversion of a primary alcohol to an aldehyde: Use PCC or CrO₃/pyridine.

d. The reaction shown is the conversion of a primary alcohol to a carboxylic acid: Use CrO₃ in aqueous acid or KMnO₄ in aqueous base followed by workup with aqueous acid.

11.10. Alkenes and alkynes are converted to the corresponding alkanes upon reaction with molecular hydrogen and a palladium catalyst. A poisoned catalyst is used to convert an alkyne to a cis alkene.

11.11. To compare the oxidation levels of two carbon atoms, count how many heteroatoms are attached to each (expand the structures if necessary): the carbon atom with more bonds to a heteroatom is more oxidized (has a higher oxidation level). A double bond counts as two bonds to that heteroatom and a triple bond counts as three bonds. The number of heteroatom bonds is given in color for the indicated carbon atom in each molecule shown.

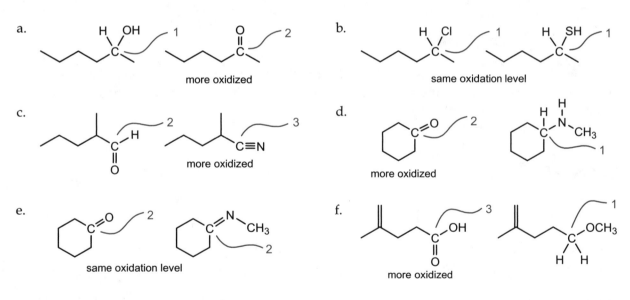

11.12. Follow the procedure outlined in the solution to Exercise 11.9. In addition, consider also the use of the Swern oxidation, which is used to convert 1° alcohols to aldehydes and 2° alcohols to ketones. The high valent iodine reagents IBX and DMP (Section 11.4d) also convert 1° alcohols to aldehydes and 2° alcohols to ketones.

a. The reaction shown is the conversion of a primary alcohol to an aldehyde: Use the Swern oxidation, CrO₃/pyridine, PCC, IBX, or DMP.

11.12. (continued)

b.   The reaction shown is the conversion of a primary alcohol to a carboxylic acid: Use CrO_3 in aqueous acid or $KMnO_4$ in aqueous base followed by acid workup.

c.   The reaction shown is the conversion of a secondary alcohol to a ketone: Use any of the oxidizing reagents mentioned in this chapter except MnO_2.

d.   The reaction shown is the conversion of a 2° benzylic alcohol to a ketone: Use any of the oxidizing reagents mentioned in this chapter.

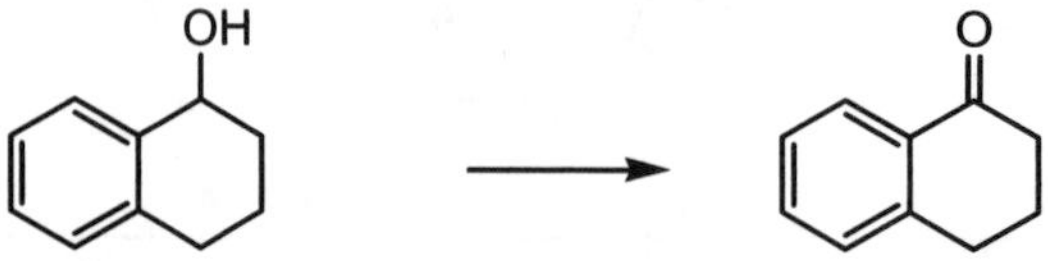

e.   The reaction shown is the conversion of a 2° benzylic alcohol to a ketone and the conversion of a primary alcohol to an aldehyde: use the Swern oxidation, CrO_3/pyridine, PCC, IBX, or DMP.

11.13. Given the structure of an epoxide, simply replace the oxygen atom with a double bond between the two carbon atoms bonded to the oxygen atom. The corresponding alkene is the one that would be used to prepare the epoxide by treatment with a peracid.

a.

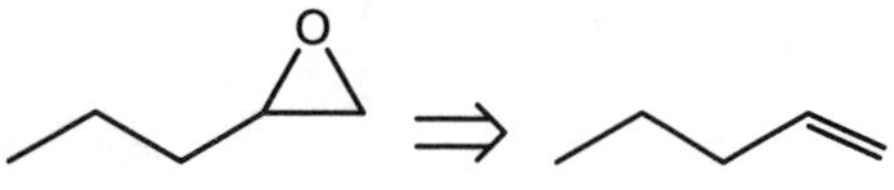

b.

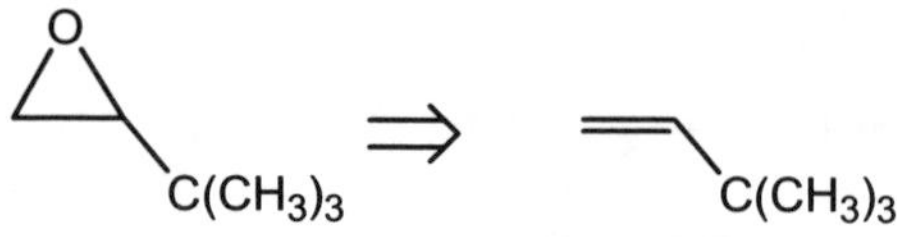

c.

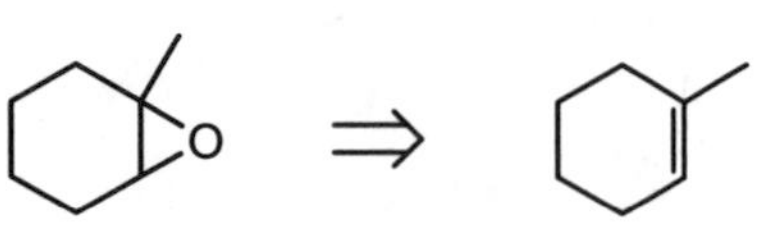

d.

11.14. The electrophilic addition reaction of bromine to a double bond in the presence of water yields a vicinal bromohydrin in which the OH group will be attached to the more highly substituted carbon atom. Treating the halohydrin with base generates an alkoxide ion in situ, which displaces the bromide ion to form the epoxide.

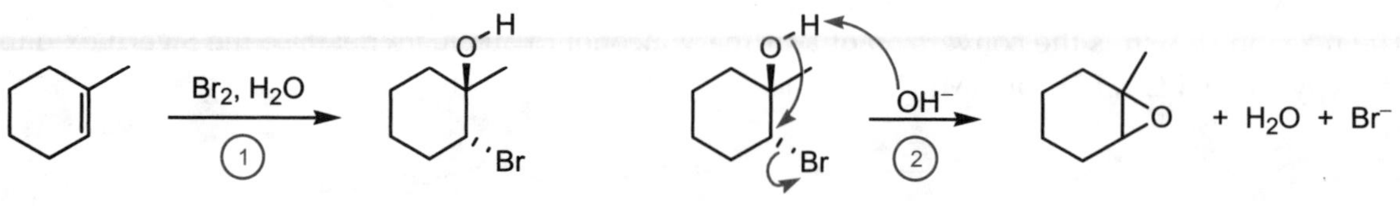

11.14. (continued)

11.15. To predict the products from various addition reactions, you have to recognize the reactivity properties of each reagent as well as the stereochemical consequences of its reactions. (Compare the reactions in this exercise with those in Exercise 9.25. and follow the procedures outlined in that exercise.) Reagents that are polar (electrophile and nucleophile) generally add to a double bond in a stepwise fashion, and addition occurs so that the electrophile becomes attached to the less highly-substituted carbon atom of the double bond (Markovnikov's rule). Some reagents cleave the double bond. Other reagents react with a double bond in a concerted manner, and the stereochemistry of those addition reactions is normally *syn*.

cis-2-Butene

The reagents that react via the electrophilic addition mechanism are shown in parts (a.) and (g.). Racemic products are formed because two new chiral centers are created in each instance by a stereospecific reaction that starts with an achiral reactant. For the reactions of *cis*-2-butene, the two carbon atoms in each product have the same configurations.

The reagents that cleave the double bond are those in parts (c.) and (d.). The symmetry of *cis*-2-butene means that two equivalents of the same product (acetaldehyde) are formed.

Hydrogenation and hydroboration are concerted processes. The stereochemistry of addition is not apparent in the products because only one chiral carbon atom at most is formed. Hydroboration yields a racemic mixture of 2-butanol.

11.15. (continued)

The remaining two reactions, which are shown in parts (f.) and (h.), generate three-membered ring products. These are concerted *syn* addition processes. Two new chiral centers are created in each transformation but the molecules have an internal mirror plane, so they are meso compounds. The methyl groups in these three-membered ring products are cis.

- -

***trans*-2-Butene**

The reactions with *trans*-2-butene and the listed reagents give the same products as those formed in the reactions of *cis*-2-butene. Any differences will manifest themselves in the stereochemistry of the products. For the reactions of the trans isomer with the electrophilic reagents, the two carbon atoms within each chiral product molecule will have opposite configurations.

The reagents that cleave the double bond will produce the same products from the trans isomer as those generated from the cis isomer.

Likewise, hydrogenation and hydroboration will produce the same products from the trans isomer as those generated from the cis isomer. Hydroboration yields a racemic mixture of 2-butanol.

The reactions that generate three-membered ring products will produce racemic mixtures of the products (only one enantiomer is shown for each) instead of a meso isomer. The methyl groups in these three-membered ring products are trans.

11.16. Follow the procedures outlined in the solution to Exercise 11.15. For methylenecyclohexane, all of the products are achiral.

Electrophilic addition:

Reagents that cleave the double bond:

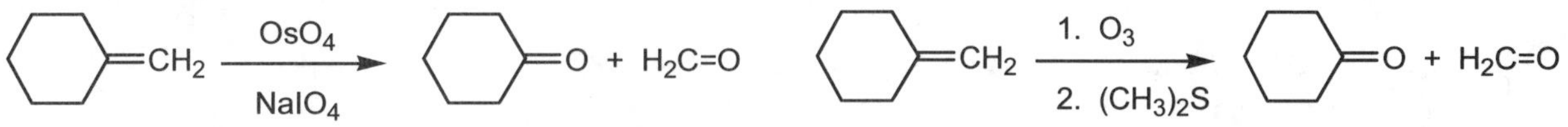

Hydrogenation and hydroboration:

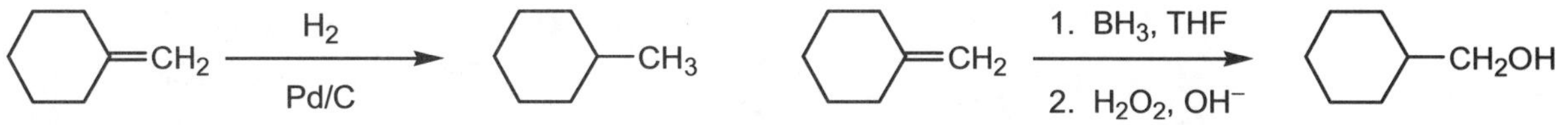

Reagents that generate three-membered ring products:

11.17. A tertiary alcohol is not susceptible to oxidation using metal oxide reagents. However, a tertiary allylic alcohol molecule that has been protonated can lose a molecule of water and rearrange to form a 2° allylic alcohol. All of these steps are reversible.

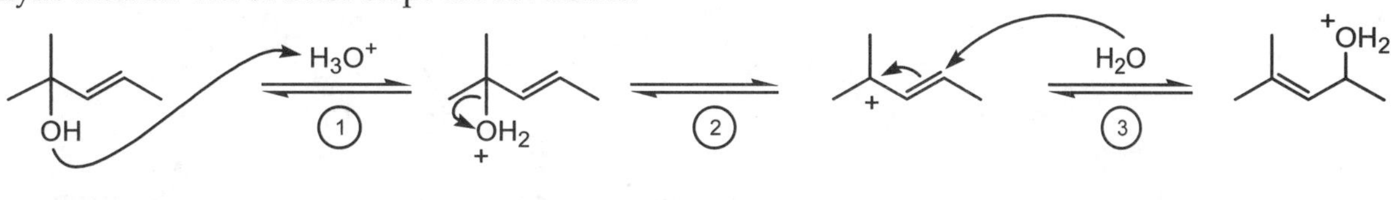

Unlike a tertiary alcohol, a secondary alcohol is readily oxidized to form a ketone. This reaction is irreversible when a good oxidant is present. Therefore, the secondary alcohol formed by the rearrangement process shown in steps 1–4 is oxidized, and this final step continually shifts the previous equilibria in favor of forming the ketone product.

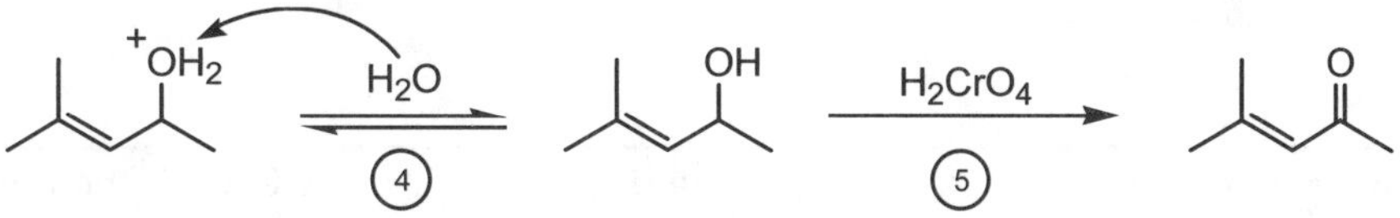

11.18. To determine the structure of the alkene that leads to the formation of carbonyl groups upon ozonolysis (with reductive workup), you only have to draw the molecule(s) with the carbonyl groups in proximity and facing each other, and then replace the C=O bonds with a single C=C bond, as shown in color for the following molecules. Finish by redrawing the alkene in its standard format.

a.

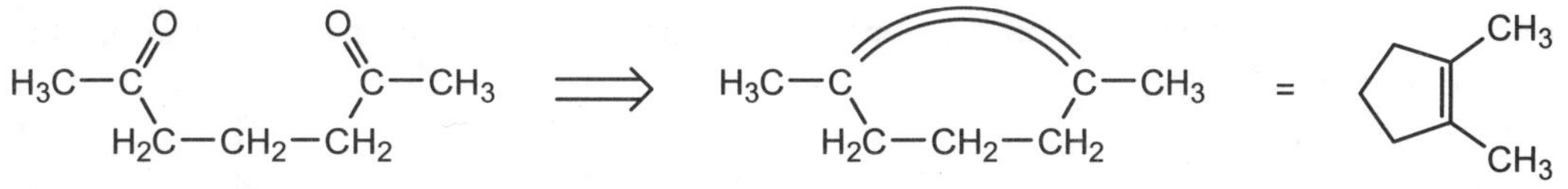

11.18. (continued)

b.

11.19. The dihydroxylation reaction catalyzed by osmium tetroxide yields a racemic mixture of products in which the two OH groups are cis. If an epoxide ring is opened with water, a racemic mixture of products is formed in which the OH groups are trans.

11.20. The involvement of DMSO as an oxidizing agent suggests that a species may be formed with a structure like that of the intermediate in the Swern oxidation. Because DMSO has a negatively-charged oxygen atom in one resonance form, substitution seems likely given that 1° alkyl halides undergo either substitution or elimination reactions with nucleophiles/bases. DMSO is not a strong base, so elimination is unlikely.

After formation of the sulfoxonium salt, the bicarbonate ion subsequently removes a proton from the adjacent carbon atom to form the carbon–oxygen double bond.

11.21. Among the molecular orbitals (MOs) of unsaturated molecules, the MO with the lowest energy is the one without nodes between the atoms. As the MOs increase in energy, an additional node appears. The highest energy MO has a node between each pair of atoms.

For the ozone molecule, the lowest energy MO has no nodes; the next higher energy MO has one node (at O2); the highest energy MO has two nodes, one between each pair of atoms (compare this diagram to that for the allyl system shown in Exercise 10.22). The four electrons are placed into the three orbitals, starting with the orbital at lowest energy; each orbital can hold two electrons. The middle MO is the HOMO because it is the orbital of highest energy that has electrons. For ozone, the orbital at the highest energy level is the LUMO.

11.21. (continued)

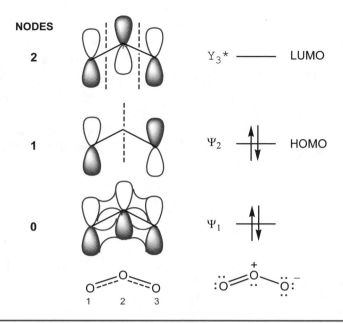

11.22. The structural framework and mechanism for the reaction between ozone and an alkene double bond are represented in the equation given directly below.

This reaction is facile because the symmetries at the termini of the reactants' MOs are the same for either combination that might occur. Notice that we can consider either the HOMO of the alkene reacting with the LUMO of ozone or the HOMO of ozone reacting with the LUMO of the alkene.

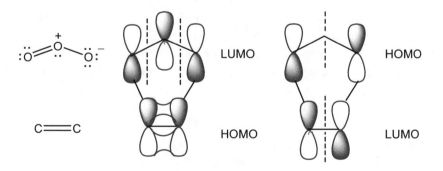

11.23. To predict the products formed by the oxidation or reduction reactions of alkenes and alkynes, you first must decide whether the π bond or both bonds (σ and π) will be broken. For the reagents described in this chapter, the results can be summarized as follows:

Only the π bond(s)	Product(s)	Both σ and π bonds	Product(s)
H_2, Pd/C	alkane	Ozone, $(CH_3)_2S$	aldehydes & ketones
H_2, Lindlar catalyst	alkene	Ozone, Zn and H_2O	aldehydes & ketones
OsO_4, NMO or t-BuOOH	gem-diol	Ozone, H_2O_2	carboxylic acids & ketones
RCO_3H	epoxide	OsO_4 with $NaIO_4$	aldehydes & ketones

11.23. (continued)

If only the π bond is broken, addition occurs with cis stereochemistry. As always, if new chiral carbon atoms are formed, either a racemic mixture or meso compound will be formed as the product(s).

If the oxidation process starts with an alcohol, then follow the procedure illustrated in the solution to Exercise 11.12.

a. The combination of OsO₄ and t-BuOOH leads to dihydroxylation of the double bond. The relative stereochemistry cannot be determined. A new chiral center is formed, however, so the product exists as a racemic mixture.

b. A peracid reacts with an alkene to form an epoxide. The cis stereochemistry of the double bond is retained. Two new chiral centers are formed, but the product is not symmetrical, so it is formed as a racemic mixture.

c. A peracid reacts with an alkene to form an epoxide. The stereochemistry of the double bond is not defined, so the product has only one chiral center and exists as a racemic mixture.

d. Ozonolysis with oxidative workup cleaves both σ and π bonds. If a carbon atom in the original double bond has a proton attached, then that end will be converted to a carboxylic acid group. Both carbon atoms in the double bond of this molecule have an attached H atom, so each end is converted to the COOH group. A cyclic alkene yields a diacid when the double bond is cleaved. This product is achiral.

e. An alkyne reacts with hydrogen in the presence of the Lindlar (or poisoned) catalyst to form a cis alkene. The product of this reaction has no chiral center.

11.23. (continued)

f. Ozonolysis with reductive workup cleaves both σ and π bonds. If a carbon atom in the original double bond has a proton attached, then that end will be converted to an aldehyde group. If a carbon atom that constitutes the double bond has two non-hydrogen groups attached, then that end becomes a ketone. In this transformation, both types of products—an aldehyde and a ketone—are formed.

g. An alkene reacts with hydrogen in the presence of a metal catalyst to form an alkane. The product of this reaction has no chiral center. These conditions do not affect the benzene ring.

h. The combination of OsO₄ and NaIO₄ cleaves both σ and π bonds. If a carbon atom in the original double bond has a proton attached, then that end will be converted to an aldehyde group. If a carbon atom that constitutes the double bond has two non-hydrogen groups attached, then that end becomes a ketone. In this transformation, both product molecules are aldehydes.

11.24. The Cope reaction occurs in the same way that the Swern oxidation does: a base within the same molecule removes a proton by way of a five-membered ring transition state. The electrons from the C–H bond displace the heteroatom bearing the positive charge, which creates the new π bond.

11.25. The fact that the deuterium atom becomes attached to the front face of the pyridine ring (as drawn) suggests that the substrate molecule will be situated such that the deuterium can approach the pyridinium ring from only that direction. A base site in the enzyme active site removes the proton attached to oxygen, and the D atom is transferred to the pyridinium ring. The new stereogenic center in the dihydropyridine molecule is (R).

11.25. (continued)

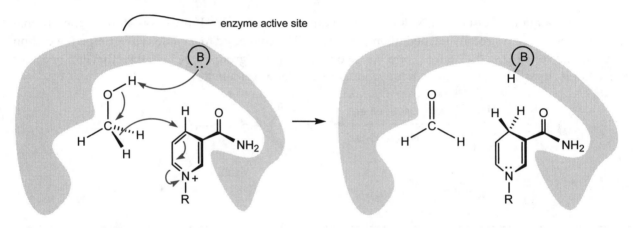

11.26. *Alcohol dehydrogenase* converts a 1° or 2° alcohol to the corresponding carbonyl compound. Formaldehyde is the product of methanol oxidation.

11.27. For the oxidation reactions of alcohols, follow the procedure illustrated in the solution to Exercise 11.12. For the cleavage of both σ and π bonds, make use of the results summarized in Exercise 11.23. For the oxidation reactions of nitrogen-containing compounds, make use of the data shown in Section 11.5.

a. The reaction shown is the conversion of an amine to a nitro compound. Dimethyldioxirane is the reagent used to carry out this transformation.

amine   →   nitro compound

b. The reaction shown is the conversion of a secondary, benzylic alcohol to a ketone: Use the Swern oxidation, CrO₃/pyridine, PCC, CrO₃ in aqueous acid, KMnO₄ in aqueous base, MnO₂, IBX, or DMP.

2° benzylic alcohol   →   ketone

11.27. (continued)

c.   The reaction shown is the conversion of a primary alcohol to a carboxylic acid: use CrO_3 in aqueous acid or $KMnO_4$ in aqueous base followed by acid workup.

1° alcohol   carboxylic acid

d.   The reaction shown is the dihydroxylation of an alkene π bond: use OsO_4 and a co-oxidant such as *t*-BuOOH or NMO.

alkene   vicinal diol

e.   This reaction results in the cleavage of a double bond, which forms two aldehyde molecules. Ozone followed by reductive workup or the combination of OsO_4 and $NaIO_4$ can be used to accomplish this transformation.

alkene   aldehydes (cleavage)

FREE RADICAL REACTIONS

12.1. The free radical chlorination of chloromethane occurs in the same way that chlorination of methane does. Three separate steps are involved: initiation, propagation, and termination (the termination steps are not shown here). The initiation step involves homolysis of the Cl–Cl bond by the action of UV light. The propagation steps take place by alternating steps of H-atom abstraction from chloromethane and Cl-atom abstraction from molecular chlorine. Termination occurs whenever any of the radicals shown in the following scheme combine to form a bond.

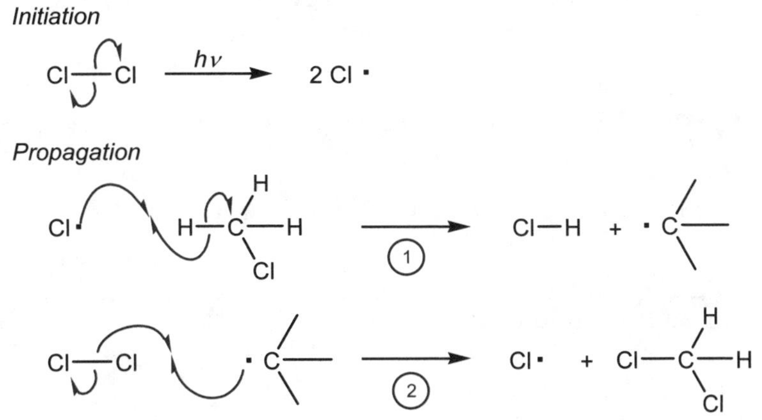

Initiation

Propagation

12.2. The ΔH values for the iodination reaction of methane are calculated from the formula, $\Delta H°_{rxn} = \Sigma(BDE_{broken}) - \Sigma(BDE_{made})$. The results of these calculations are shown below.

First, write equations for the two propagation steps and make sure they give the overall process when added (the radical species should cancel):

$$I\bullet + CH_4 \rightarrow HI + CH_3\bullet$$
$$\underline{CH_3\bullet + I_2 \rightarrow CH_3I + I\bullet}$$
$$CH_4 + I_2 \rightarrow CH_3I + HI$$

Next, tabulate the BDE values for each bond broken and made:

    Step (1): broken: CH_3-H (BDE = 104 kcal • mol^{-1}); made: $H-I$ (BDE = 71 kcal • mol^{-1})

    Step (2): broken: $I-I$ (BDE = 36 kcal • mol^{-1}); made: CH_3-I (BDE = 56 kcal • mol^{-1})

Third, calculate the $\Delta H°$ value of each step by subtracting the BDE values of the bonds made from those broken:

    Step (1): $\Delta H°_1 = BDE\ (CH_3-H) - BDE\ (H-I) = 104 - 71 = 33$ kcal • mol^{-1}

    Step (2): $\Delta H°_2 = BDE\ (I-I) - BDE\ (CH_3-I) = 36 - 56 = -20$ kcal • mol^{-1}

Finally, add the value of each step to obtain $\Delta H°_{rxn}$.

    $\Delta H°_{rxn} = \Delta H°_1 + \Delta H°_2 = 33$ kcal • mol^{-1} + $(-20$ kcal • mol$^{-1}) = +13$ kcal • mol^{-1}

12.2. (continued)

Conclusion: The overall enthalpy change for the free radical iodination reaction of methane is greater than zero, so the process is reactant favored. By comparison, the overall enthalpy change for the free radical bromination reaction of methane is exothermic, but not as much as the chlorination reaction is.

12.3. Attach the appropriate numbers of chlorine atoms to the carbon atoms of ethane. Then name each compound according to the procedures you learned in Chapter 1 (number the chain from each end to make certain that you have not drawn the same substance twice).

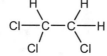

Chloroethane	**1,1-Dichloroethane**	**1,2-Dichloroethane**

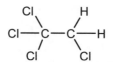 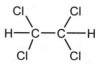

1,1,1-Trichloroethane	**1,1,2-Trichloroethane**	**1,1,1,2-Tetrachloroethane**	**1,1,2,2-Tetrachloroethane**

12.4. Follow the procedures outlined in the solution to Exercise 12.2.

a. $CH_3CH_2CH_3 + Cl\bullet \rightarrow HCl + $ 1-propyl radical

$\Delta H° = $ BDE (1° C–H) $-$ BDE (H–Cl) $= 98 - 103 = -5$ kcal $\bullet$ mol^{-1}

b. $CH_3CH_2CH_3 + Cl\bullet \rightarrow HCl + $ 2-propyl radical

$\Delta H° = $ BDE (2° C–H) $-$ BDE (H–Cl) $= 95 - 103 = -8$ kcal $\bullet$ mol^{-1}

c. $CH_3CH_2CH_3 + Cl_2 \rightarrow HCl + CH_3CH_2CH_2Cl$

$\Delta H°_1 = $ BDE (1° C–H) $-$ BDE (H–Cl) $= 98 - 103 = -5$ kcal $\bullet$ mol^{-1}

$\Delta H°_2 = $ BDE (Cl–Cl) $-$ BDE (1° C–Cl) $= 58 - 81 = -23$ kcal $\bullet$ mol^{-1}

$\Delta H°_{rxn} = \Delta H°_1 + \Delta H°_2$ $\qquad\qquad = -28$ kcal $\bullet$ mol^{-1}

d. $CH_3CH_2CH_3 + Cl_2 \rightarrow HCl + CH_3CHClCH_3$

$\Delta H°_1 = $ BDE (2° C–H) $-$ BDE (H–Cl) $= 95 - 103 = -8$ kcal $\bullet$ mol^{-1}

$\Delta H°_2 = $ BDE (Cl–Cl) $-$ BDE (2° C–Cl) $= 58 - 80 = -22$ kcal $\bullet$ mol^{-1}

$\Delta H°_{rxn} = \Delta H°_1 + \Delta H°_2$ $\qquad\qquad = -30$ kcal $\bullet$ mol^{-1}

12.5. Follow the procedures outlined in Example 12.2. First, draw all of the monochloro derivatives of 3-methylhexane. This molecule has no symmetry, so the substitution of chlorine for hydrogen at each carbon atom yields a unique isomer.

12.5. (continued)

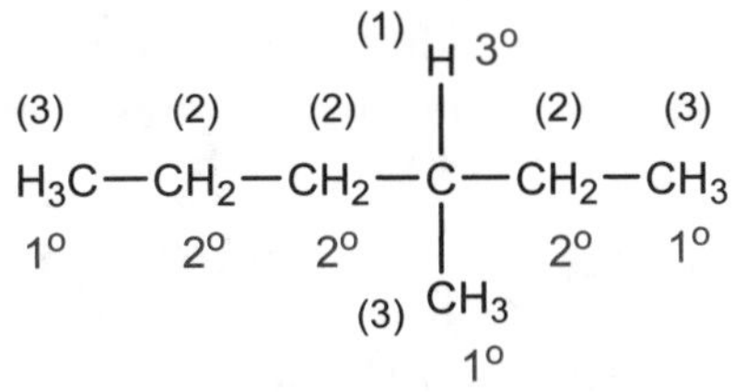

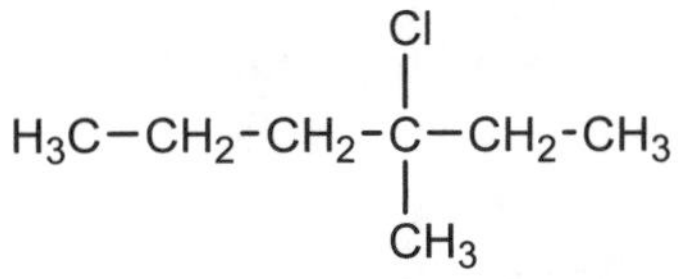

3-Chloro-3-methylhexane

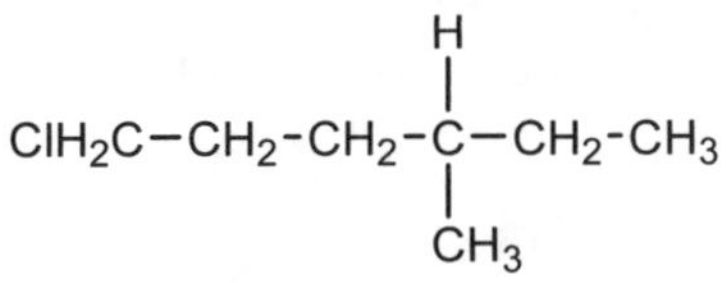

1-Chloro-4-methylhexane 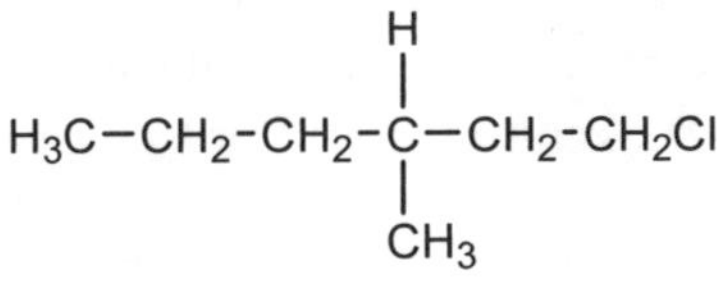   **1-Chloro-3-methylhexane** 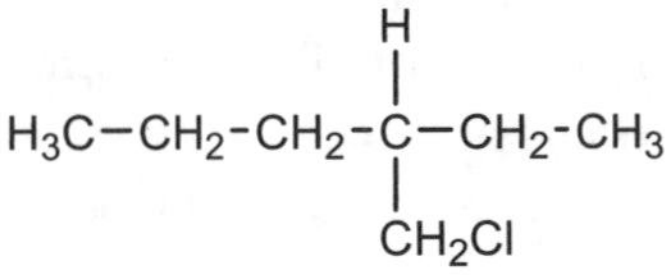   **3-(Chloromethyl)hexane**

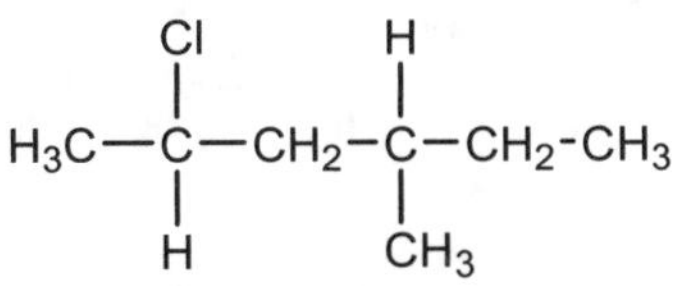

2-Chloro-4-methylhexane 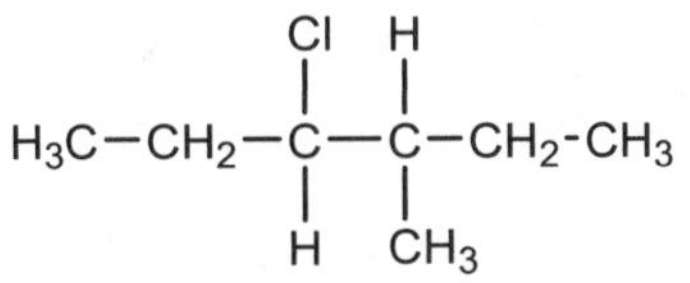   **3-Chloro-4-methylhexane** 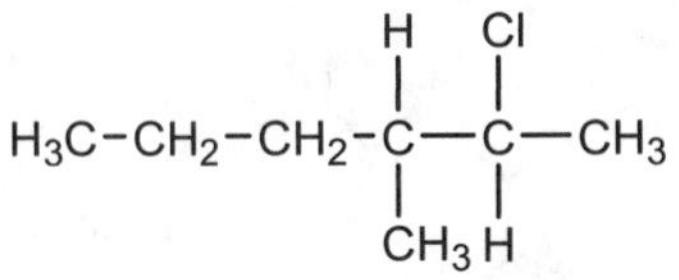   **2-Chloro-3-methylhexane**

Next, multiply the number of equivalent hydrogen atoms at each position being substituted (the numbers of hydrogen atoms are indicated in parentheses on the structure above at the left) by its reactivity quotient: 5 for tertiary H; 3.7 for secondary H; and 1 for primary H).

Finally, divide the number calculated for each hydrogen atom type by the total (36.2 for this molecule) and express the answer as a percentage.

3-Chloro-3-methylhexane	$1H \times 5$	$= 5$	$5 \div 36.2 = 0.138$ (13.8%)
1-Chloro-4-methylhexane	$3H \times 1$	$= 3$	$3 \div 36.2 = 0.083$ (8.3%)
1-Chloro-3-methylhexane	$3H \times 1$	$= 3$	$3 \div 36.2 = 0.083$ (8.3%)
3-(Chloromethyl)hexane	$3H \times 1$	$= 3$	$3 \div 36.2 = 0.083$ (8.3%)
2-Chloro-4-methylhexane	$2H \times 3.7$	$= 7.4$	$2 \div 36.2 = 0.204$ (20.4%)
3-Chloro-4-methylhexane	$2H \times 3.7$	$= 7.4$	$2 \div 36.2 = 0.204$ (20.4%)
2-Chloro-3-methylhexane	$2H \times 3.7$	$= \underline{7.4}$	$2 \div 36.2 = 0.204$ (20.4%)
		36.2 (total)	

The major products will be 2-chloro-4-methylhexane, 3-chloro-4-methylhexane, and 2-chloro-3-methylhexane, each of which will be formed with a yield of 20.4%. 3-Chloro-3-methylhexane will be formed with a yield of 13.8%. The three products in which a methyl group has been substituted will each be formed as 8.3% of the product mixture.

12.6. The relative reactivity order for hydrogen atoms in radical bromination reactions is benzylic, allylic > 3° > 2° > 1° > methyl. Primary and secondary C–H bonds will only react to a substantial degree if tertiary, allylic, and benzylic hydrogen atoms are absent. The starting material in this exercise has 1°, 2°, and 3° carbon atoms. Therefore, the reaction is expected to proceed by replacement of the 3° hydrogen atom.

12.7. In NBS bromination reactions, benzylic and allylic C–H bonds react preferentially. Within each of those types (benzylic and allylic), the reactivity order is 3° > 2° > 1°. The more stable the radical that can be formed by H-atom abstraction, the greater amount of the corresponding product will be formed. Therefore, classify each hydrogen atom type, and then choose the one(s) that will generate the most stable radical(s).

a.   The hydrogen atoms attached to the carbon atoms with sp^3 hybridization are all secondary. One set is allylic, one set is benzylic, and one set is both benzylic and allylic. The latter set should be most reactive because the resulting radical will be the best stabilized. When assessing free radical reactions of allylic systems, you must always consider its resonance forms because substitution can occur at either radical site. (A benzene ring will not be disrupted, so benzylic substitution always occurs adjacent to the benzene ring.)

b.   The hydrogen atoms attached to carbon atoms with sp^3 hybridization are either primary or tertiary, but several are also allylic. The tertiary, allylic hydrogen atom should be the most reactive because the resulting radical will be most stabilized. In this case, the allylic radical that is formed is symmetrical, so only one product is formed.

12.8. Hydrogenolysis reactions will cleave the bond between a heteroatom and a benzylic carbon atom with sp^3 hybridization. The bonds between heteroatoms and other types of carbon atoms are not normally affected.

12.9. Raney nickel reacts to remove sulfur and selenium atoms, replacing C–S and C–Se bonds with C–H bonds. The S or Se atom is converted to H_2S or H_2Se, respectively.

a.

12.9. (continued)

b.

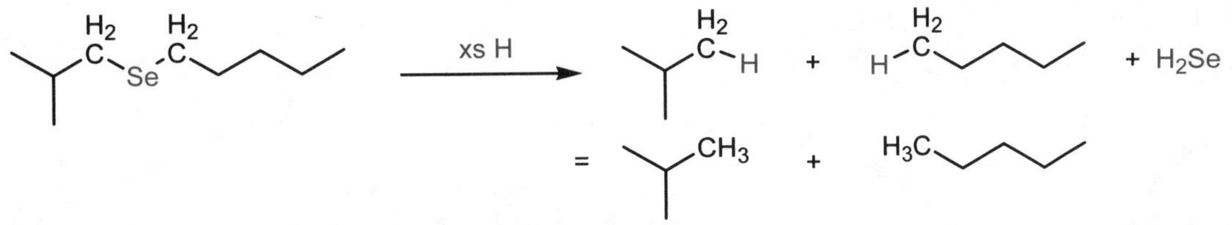

12.10. A trans double bond is usually formed when a triple bond is reduced using sodium in liquid ammonia containing an added proton source.

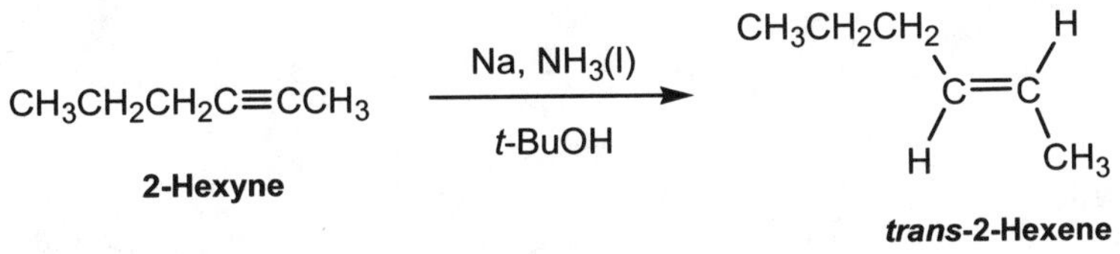

$$CH_3CH_2CH_2C{\equiv}CCH_3 \xrightarrow[t\text{-BuOH}]{Na,\ NH_3(l)}$$

2-Hexyne

trans-**2-Hexene**

12.11. An organohalide reacts with tributyltin hydride to form the product in which the halogen atom has been replaced by a hydrogen atom. If the molecule has a carbon-carbon double or triple bond within five atoms of the position at which a Cl, Br, or I atom is attached, then ring formation also has to be considered. Fluorine atoms are inert toward the reactions with tin hydride reagents.

a.

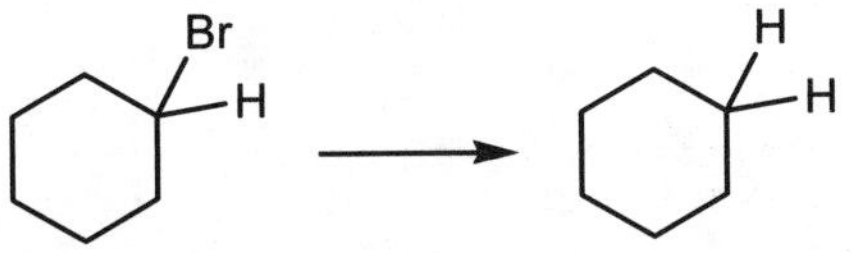

b.

12.12. Termination steps occur whenever any two radicals combine and form a bond. In the process of the radical addition of HBr to 1-hexene, there are two radical species that exist during the course of the reaction: bromine atoms and the 1-bromo-2-hexyl radical. These can combine as shown in the following equations.

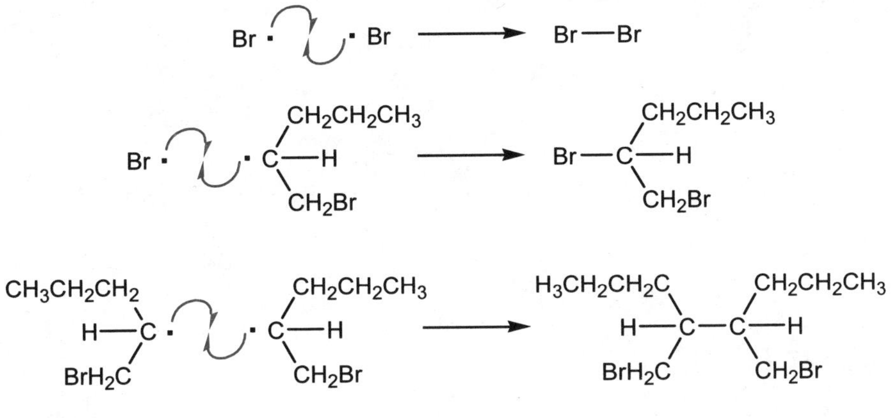

12.13. To calculate the $\Delta H°$ values for the given reactions, follow the examples outlined in the solutions of Exercises 12.2 and 12.4.

a. $RCH{=}CH_2 + Cl\bullet \rightarrow R(\bullet)CH{-}CH_2Cl$

$\Delta H° = BDE\ (C{=}C) - BDE\ (1°\ C{-}Cl) = 65 - 81 = -16\ kcal \bullet mol^{-1}$

12.13. (continued)

b.   RCH=CH₂ + Br • → R(•)CH– CH₂Br

    $\Delta H°$ = BDE (C=C) — BDE (1° C –Br) = 65 — 68 = –3 kcal • mol^{-1}

c.   RCH=CH₂ + I • → R(•)CH– CH₂I

    $\Delta H°$ = BDE (C=C) — BDE (1° C –I) = 65 — 53 = +12 kcal • mol^{-1}

The reaction coordinate diagram for a single step will have the reactants and products separated by a barrier that represents the free energy of activation. In these reactions the product is the radical intermediate. The addition of an iodine atom to an alkene is endothermic, but the addition of a bromine or chlorine atom is exothermic.

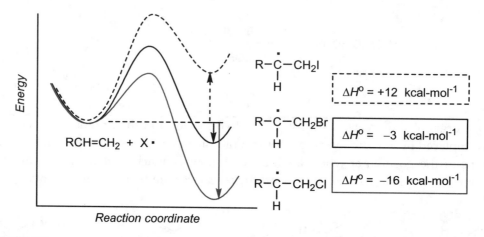

12.14. When styrene is converted to polystyrene with AIBN as the initiator, the first step involves decomposition of AIBN to form two 2-cyano-2-propyl radicals plus molecular nitrogen. This carbon radical adds to the π bond of styrene so as to form the benzylic radical. The resulting radical adds with the same regiochemistry to the π bond of another styrene molecule, a process that repeats over and over again.

12.15. The bromoalkene in this exercise has the 5-hexenyl type structure, so we expect cyclization to occur with formation of a five-membered ring product. The mechanism follows the usual pathway: Initiation (steps 1 and 2); abstraction of the halogen atom (step 3); addition to the π bond to generate the five-membered ring (step 4); and reduction of the carbon centered radical to form the hydrocarbon product with regeneration of the tributyltin radical (step 5).

Initiation

Halogen abstraction and addition

Reduction

Note that the cyclization step generates both cis and trans isomers.

12.16. The Bergman cyclization for the bicyclic molecule in this exercise creates a tricyclic system in which the central 6-membered ring is fused to a rigid diradical benzene ring and to a four-membered ring that has significant Baeyer strain. The geometric constraints imposed on the carbon skeleton by virtue of having three fused rings creates enough strain so as to prevent ring formation from occurring until the enediyne is heated to 100 °C.

12.17. If every hydrogen atom were abstracted at the same rate, then the product ratio would reflect only the total number of each hydrogen atom type. 2-Methylpropane has one tertiary hydrogen atom, and nine equivalent primary hydrogen atoms. The product ratio of 2-chloro-2-methylpropane to 1-chloro-2-methylpropane would therefore be 1:9.

12.17. (continued)

Ratio = 1:9

12.18. To predict whether a reaction will yield one principal product, you need to consider the relative reactivity ratios of each hydrocarbon group in the substrate molecule as well as the reactivity of each reagent. In radical halogenation reactions, bromine atoms react at any type of (sp^3)C–H bond, so we do not have to be concerned about the properties of the reagent itself. A benzylic C–H bond is more reactive than a 3° C–H bond, which in turn is more reactive than a 2° C–H bond. This order of reactivity allows us to predict which carbon atom(s) will undergo substitution. If there is a single type of C–H bond, then only one product will form. If there are several non-equivalent forms of a particular C–H bond type, then a mixture of products will be obtained, as in parts (b.) and (d.), below.

a. Cyclobutane has only secondary hydrogen atoms, all of which are equivalent. A single product will be formed in the free radical bromination process.

b. This substrate molecule has primary, secondary, and tertiary C–H bonds. For the free radical bromination procedure, we need only be concerned with the 3° centers that are present (cf. Exercise 12.6 for an illustration of this principle). There are two different tertiary C–H bonds in this molecule, so two products will be formed.

c. This substrate molecule has only primary and secondary hydrogen atoms, but two H atoms are also benzylic, so they will react preferentially. Because the benzylic hydrogen atoms are equivalent, a single product (ignoring stereochemistry) will be formed in the free radical bromination process.

d. This substrate has only primary and secondary hydrogen atoms, so the 2° centers will react preferentially (cf. Exercise 12.6). Three non-equivalent methylene groups are present in the starting material, so three monobromo products (ignoring stereoisomers) will be formed.

12.18. (continued)

12.19. To predict the structures of the products in each of the following transformations, you must learn the details about each reaction type, which can be found in the reaction summary section of the chapter. If a new chiral carbon center is formed, then both configurations will be formed in radical processes. Chiral centers that are away from the site of reaction are normally not affected by radical reactions.

a. NBS and benzoyl peroxide react to form bromine radicals, which lead to replacement of hydrogen atoms by bromine. The order of reactivity is benzylic, allylic > 3° > 2° > 1° > methyl. Free radical substitution produces both configurations at the carbon atom undergoing reaction because the radical intermediate is planar and can react with bromine at either its top or bottom face.

b. Raney nickel removes sulfur atoms from organic molecules, replacing them with two hydrogen atoms. The chiral carbon atom attached to sulfur in this molecule is no longer chiral after replacing the S atom with H. The other chiral carbon atom is not affected; therefore the product is chiral, not racemic.

c. Tributyltin hydride reacts with an organohalide to generate a radical at the carbon atom to which the halogen atom is attached. The resulting radical is reduced to yield the corresponding hydrocarbon. The stereochemistry at the halogen-substituted carbon atom is lost during the course of this transformation, but the other two chiral centers are retained.

d. In the presence of a peroxide co-reactant, HBr adds to double bonds with apparent anti-Markovnikov regiochemistry, meaning that the bromine atom becomes attached to the less highly-substituted end of the double bond. Addition of both H and Br can occur at either face of the double bond, so four stereoisomers are produced.

12.19. (continued)

e. Hydrogenolysis cleaves the bond between a benzylic carbon atom and a heteroatom. In this substrate, the bond between the benzylic carbon atom and oxygen atom is replaced with a C–H bond. Other types of carbon-oxygen bonds are not affected.

$$\text{(structure)} \xrightarrow{\text{H}_2,\ \text{Pd/C}} \text{(structure)} \quad \text{retention}$$

f. Allylic bromination reactions often give multiple products because the radical that is formed by H atom abstraction is delocalized:

Bromination therefore occurs at the two centers to yield three products (two are stereoisomers).

$$\text{(alkene)} \xrightarrow[\text{(PhCO}_2)_2]{\text{NBS, } \Delta} \underset{\text{achiral}}{\text{(Br product)}} + \underset{\text{(racemic)}}{\text{(Br product)}} + \text{(Br product)}$$

g-i. These three transformations compare the possible reduction reactions that triple bonds can undergo. Catalytic hydrogenation with hydrogen and Pd/C gives the corresponding alkane; the cis alkene is produced if a poisoned catalyst is employed (see Sections 11.2a and 11.2b). A dissolving metal reduction utilizes sodium in liquid ammonia to form trans alkenes from alkynes.

$$\text{(alkyne)}\ \text{—CH}_3 \xrightarrow{\text{H}_2,\ \text{Pd/C}} \text{(alkane)}$$

$$\text{(alkyne)}\ \text{—CH}_3 \xrightarrow[\text{quinoline poison}]{\text{H}_2,\ \text{Pd/BaSO}_3} \text{(cis alkene)} \text{CH}_3$$

$$\text{(alkyne)}\ \text{—CH}_3 \xrightarrow[\text{t-BuOH}]{\text{Na, NH}_3} \text{(trans alkene)} \text{CH}_3$$

12.20. To compare different mechanisms, the enthalpy values are calculated and graphically plotted.

a. The reaction coordinate diagram for the process of free radical chlorination starts with the hydrocarbon and chlorine, proceeds via formation of a radical intermediate, and yields the chlorinated product and HCl. The needed $\Delta H°$ values for methane chlorination were calculated in Section 12.2a.

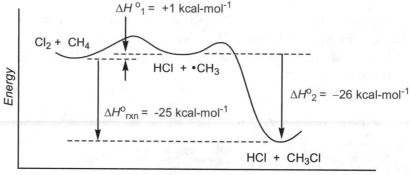

12.20. (continued)

b. For the alternate mechanism, first write the equations for the steps:

$$Cl \bullet + CH_4 \rightarrow CH_3Cl + H \bullet$$
$$H \bullet + Cl_2 \rightarrow HCl + Cl \bullet$$

$$\overline{CH_4 + Cl_2 \rightarrow CH_3Cl + HCl}$$

Then calculate the $\Delta H°$ values for each step:

Step (1): $\Delta H°_1$ = BDE (CH_3—H) — BDE (CH_3—Cl) = 104 — 84 = 20 kcal $\bullet$ mol^{-1}

Step (2): $\Delta H°_2$ = BDE (Cl—Cl) — BDE (H—Cl) = 58 — 103   = –45 kcal $\bullet$ mol^{-1}

Add the value of each step to obtain $\Delta H°_{rxn}$.

$\Delta H°_{rxn} = \Delta H°_1 + \Delta H°_2 = 20$ kcal $\bullet$ mol^{-1} + (–45 kcal $\bullet$ mol^{-1}) = –25 kcal $\bullet$ mol^{-1}

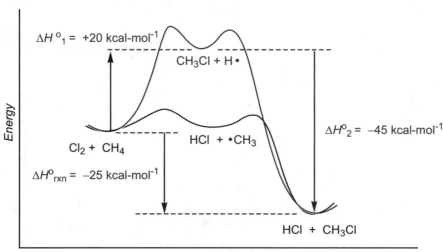

The original mechanism (via the methyl radical; black curve) is more reasonable because the alternate mechanism (via formation of the H atom; colored curve) requires significant amounts of energy to produce the first intermediate (even though this needed energy would be released in the second step). The $\Delta H°_{rxn}$ is the same for the two mechanisms, namely –25 kcal $\bullet$ mol^{-1}.

With the assumption that $\Delta G°_{rxn} = \Delta H°_{rxn}$, you can see that each mechanism gives the same numerical result. The standard free energy change for a reaction is based only on the energies of the reactants and products, not on the step-by-step mechanism for converting reactants to products.

12.21. This exercise is similar to Exercise 12.4; follow the procedure illustrated by its solution.

a. $CH_3CH_2CH_3 + Br \bullet \rightarrow HBr + 1$-propyl radical

$\Delta H°$ = BDE (1° C–H) — BDE (H–Br) = 98 — 88 = 10 kcal $\bullet$ mol^{-1}

b. $CH_3CH_2CH_3 + Br_2 \rightarrow CH_3CH_2CH_2Br + HBr$

$\Delta H°_1$ = BDE (1° C–H) — BDE (H–Br) = 98 — 88 = 10 kcal $\bullet$ mol^{-1}

$\Delta H°_2$ = BDE (Br–Br) — BDE (1° C–Br) = 46 — 68 = –22 kcal $\bullet$ mol^{-1}

$$\overline{\Delta H°_{rxn} = \Delta H°_1 + \Delta H°_2 \qquad\qquad\qquad = -12 \text{ kcal} \bullet \text{mol}^{-1}}$$

12.21.   (continued)

c. $CH_3CH_2CH_3 + Br\cdot \rightarrow HBr + $ 2-propyl radical

$\Delta H° = $ BDE (2° C–H) — BDE (H–Br) $= 95 — 88 = 7$ kcal • mol^{-1}

d. $CH_3CH_2CH_3 + Br_2 \rightarrow HBr + CH_3CHBrCH_3$

$\Delta H°_1 = $ BDE (2° C–H) — BDE (H–Br) $= 95 — 88 = 7$ kcal • mol^{-1}

$\Delta H°_2 = $ BDE (Br–Br) — BDE (2° C–H) $= 46 — 68 = -22$ kcal • mol^{-1}

$\Delta H°_{rxn} = \Delta H°_1 + \Delta H°_2$ $= -15$ kcal • mol^{-1}

12.22. The reaction coordinate diagram for the process of free radical bromination starts with the hydrocarbon and bromine, proceeds through formation of a radical intermediate, and yields the brominated product and HBr. Annotate such a diagram with the numerical data calculated in Exercise 12.21.

a. Reaction at the primary carbon atom

b. Reaction at the secondary carbon atom

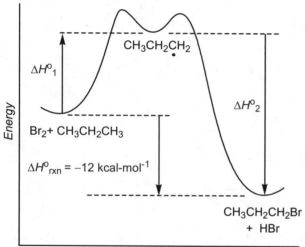

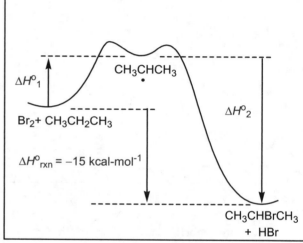

12.23.   Follow the procedure illustrated in the solution to Exercise 12.21.

a. $C_6H_5CH_3 + Br\cdot \rightarrow HBr + C_6H_5CH_2\cdot$

$\Delta H° = $ BDE ($C_6H_5CH_2$–H) — BDE (H–Br) $= 70 — 88 = -18$ kcal • mol^{-1}

b. $C_6H_6 + Br\cdot \rightarrow HBr + C_6H_5\cdot$

$\Delta H° = $ BDE (C_6H_5–H) — BDE (H–Br) $= 112 — 88 = 24$ kcal • mol^{-1}

c. $C_6H_5CH_3 + Br_2 \rightarrow C_6H_5CH_2Br + HBr$

$\Delta H°_1 = $ BDE (C–H) — BDE (H–Br) $= 70 — 88 = -18$ kcal • mol^{-1}

$\Delta H°_2 = $ BDE (Br–Br) — BDE (C–Br) $= 46 — 58 = -12$ kcal • mol^{-1}

$\Delta H°_{rxn} = \Delta H°_1 + \Delta H°_2$ $= -30$ kcal • mol^{-1}

12.23. (continued)

d. $C_6H_6 + Br_2 \rightarrow HBr + C_6H_5Br$

$\Delta H°_1 = \text{BDE (C–H)} - \text{BDE (H–Br)} = 112 - 88 = 24 \text{ kcal} \cdot \text{mol}^{-1}$

$\Delta H°_2 = \text{BDE (Br–Br)} - \text{BDE (C–Br)} = 46 - 82 = -36 \text{ kcal} \cdot \text{mol}^{-1}$

$\Delta H°_{rxn} = \Delta H°_1 + \Delta H°_2 \qquad\qquad = -12 \text{ kcal} \cdot \text{mol}^{-1}$

12.24. Follow the procedure outlined in the solution to Exercise 12.22. Both steps in the reaction of toluene are exothermic, but free radical bromination of benzene is highly endothermic in the first step.

a. Reaction at the benzylic carbon atom   b. Reaction at the aryl carbon atom

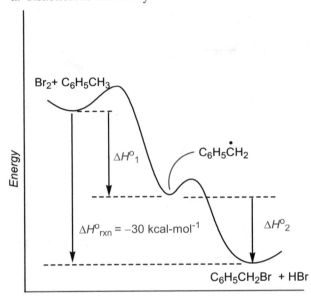

Reaction coordinate

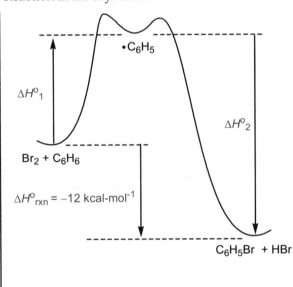

Reaction coordinate

12.25. In the free radical addition reactions of HBr, the bromine atom becomes attached to the less highly-substituted end of the π bond. The addition of one equivalent of HBr to an alkyne leads to formation of an alkene that can exist as both (*E*) and (*Z*) isomers.

$CH_3CH_2CH_2CH_2-C\equiv C-H$ $\xrightarrow{\text{HBr, ROOR, } \Delta}$

(*E*)-1-Bromo-1-hexene   +   (*Z*)-1-Bromo-1-hexene

12.26. Autoxidation can occur whenever a compound is exposed to molecular oxygen, which is a diradical species. Abstraction of a hydrogen atom from tetralin produces a benzylic radical (step 1) the formation of which is made possible by resonance stabilization.

Recombination of the two radicals generates the product (step 2).

12.26. (continued)

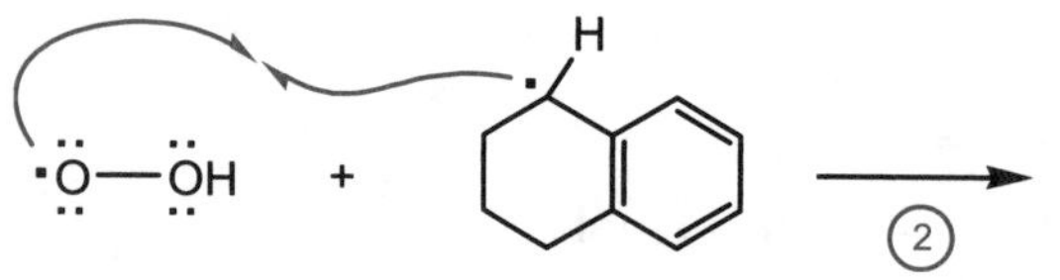

Alternatively, the radical formed in step 1 (above) can also combine with oxygen, as shown below (step 3). Hydrogen atom abstraction from another molecule of tetralin (step 4) establishes a chain reaction.

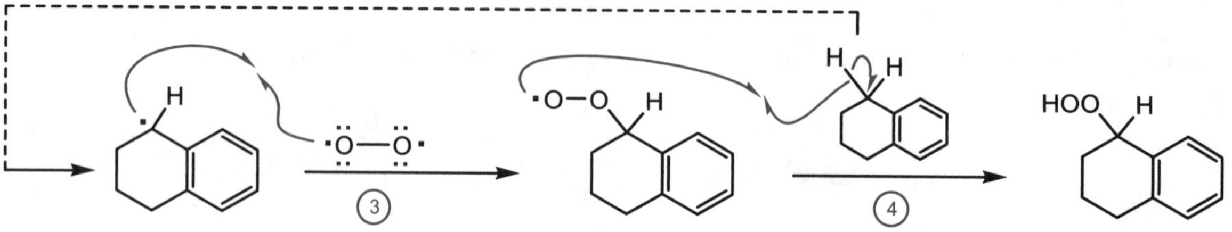

12.27. The radical reactions described in this chapter are of two general types: substitution and addition. To predict the structures of the products expected in each of the following transformations, you must know the details of each reaction type, which can be found in the reaction summary section of the chapter. When radical intermediates are involved in the formation of products having new chiral centers, the reactions are not stereospecific. Therefore, if more than one chiral center is created, a mixture of diastereomers will be formed.

a. The reaction of HBr with an alkene or alkyne and a peroxide reagent leads to "anti-Markovnikov" addition of H and Br to the double bond. A new chiral center is formed, so the product is racemic.

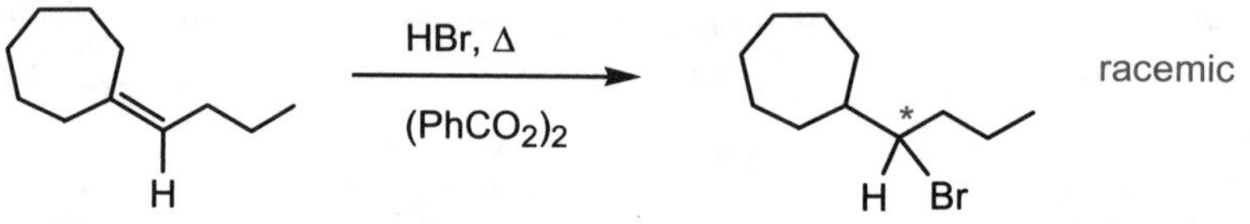

b. The combination of NBS with peroxide and heat is used for allylic and benzylic bromination procedures. The substrate molecule here has benzylic C–H bonds, so the substitution reaction should be facile. Even though the substrate molecule is chiral, the chiral center in the product is formed from an achiral radical intermediate, so the product molecule is racemic.

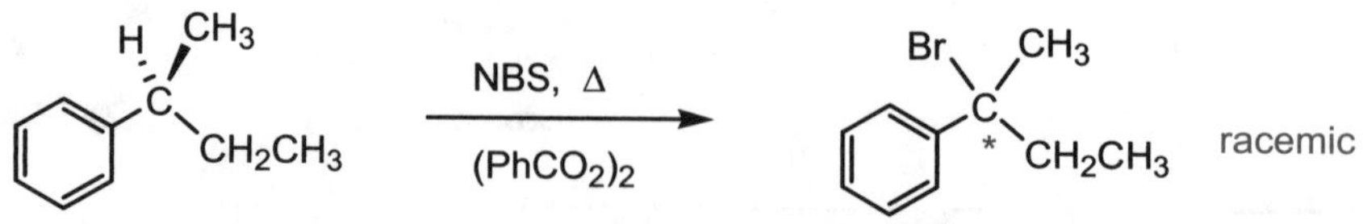

c. Tributyltin hydride, along with a radical initiator such as AIBN, reacts with organohalides to replace the halogen atom (except F) with a hydrogen atom. The chiral center in this substrate molecule is not affected by the reduction process, so the product has the same stereochemistry (unspecified) as the starting material.

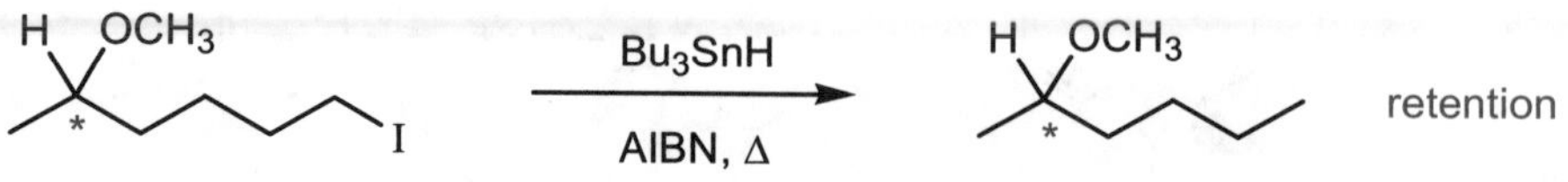

12.27. (continued)

d. Bromine reacts with hydrocarbons when irradiated with UV light to replace a hydrogen atom at the most highly substituted carbon atom. One chiral center that exists in the starting material is retained in the product, and a second chiral center in the product is formed via a radical inter-mediate that is planar at the reacting center. The product molecules are therefore diastereomers.

e. Sodium in liquid ammonia plus a proton source converts an alkyne to the corresponding trans [or (E)] alkene.

f. Hydrogenolysis cleaves the bond between a heteroatom and a benzylic carbon atom. Even though the starting compound is chiral, the radical intermediate involved in this reaction is achiral. Because a new chiral center is subsequently formed, the product is racemic.

g. Tributyltin hydride, along with a radical initiator such as AIBN, reacts with organohalides to replace the halogen atom (except F) with a hydrogen atom. If a double bond is separated by four carbon atoms from the halogen atom in the molecule, then cyclization can occur.

h. The combination of NBS with peroxide and heat is used for allylic and benzylic bromination procedures. This substrate molecule has eight equivalent allylic C–H bonds, so the substitution reaction should be facile. A new chiral center is formed, so the product is racemic.

i. Raney nickel replaces C–S bonds with C–H bonds. The sulfur-containing product is H₂S.

12.27. (continued)

j. When irradiated with UV light, chlorine replace hydrogen atoms in the structures of hydrocarbon molecules. The amounts of each product can be calculated by the process illustrated in the solution to Exercise 12.5. Note that there are six methyl protons, four each of two types of methylene protons, and two methylene protons directly across the ring from the point at which the methyl groups are attached. The major product of this chlorination reaction is 1-chloromethyl-1-methylcyclohexane.

| 6 x 5 = 30 | 4 x 3.7 = 14.8 | 4 x 3.7 = 14.8 | 2 x 3.7 = 7.4 |
| 30/67= 44.8% | 14.8/67 = 22.1% | 14.8/67 = 22.1% | 7.4/67 = 11% |

12.28. During pyrolysis of organic molecules, homolysis reactions will generally proceed by routes that generate hydrogen atoms and/or the most stable carbon radicals (benzylic, allylic > 3° > 2° > 1° > methyl).

a. The benzylic C–H and C–C bonds are the weakest because a benzyl radical formed by homolysis is resonance stabilized. For propylbenzene, cleavage can occur at either the C–H or C–C bond next to the ring.

b. 3,4-Dimethylpentane can lose a hydrogen atom to form a 3° radical. Cleavage of its central carbon-carbon bond generates two 2° alkyl radicals.

c. *tert*-Butylcyclohexane can lose a hydrogen atom to form a 3° radical or it can undergo homolysis of the exocyclic C–C bond to form a 2° alkyl radical and a 3° alkyl radical.

12.28. (continued)

d. 4-Methyl-1-pentene can lose a hydrogen atom to form an allylic radical. Cleavage of the carbon-carbon bond adjacent to the π bond will generate a 2° alkyl radical in addition to the allylic radical.

12.29. The tributyltin radical, generated from Bu₃SnH and AIBN (see the solution to Exercise 12.15 for the mechanism of its formation), abstracts the iodine atom from the 5-iodopentanal to generate a 1° alkyl radical (step 1). The carbonyl double bond intercepts this radical (step 2), and the resulting oxygen-centered radical is quenched with tin hydride, regenerating Bu₃Sn• to continue the chain reaction (step 3).

12.30. This exercise is a variation of Exercise 12.27. If you ignore the given product and instead predict what the expected major product will be, then you can deduce why the reaction does not occur as written.

a. An alkene will undergo allylic bromination when treated with NBS and a peroxide compound. The 3° allylic hydrogen atom is expected to be more reactive than the 1° allylic hydrogen atoms.

b. Organofluorides do not react with tributyltin radicals.

c. A haloalkene will form a five-membered ring if the double bond is four carbon atoms away from a radical center that develops (a radical of the 5-hexenyl type). The product is a spiro compound instead of the illustrated fused bicyclic compound.

5-hexenyl radical

12.31. The structure of the given substrate molecule is ideal for forming a five-membered ring by radical cyclization according to the following pathway:

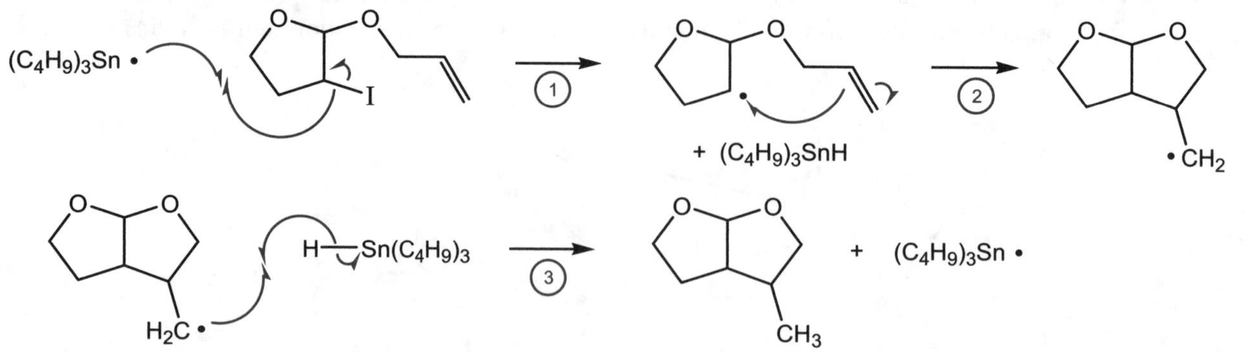

In part **B** of this exercise, the 1° alkyl radical formed in step 2 (above), reacts with molecular oxygen to form a carbon-oxygen bond (step 4), which subsequently forms the hydroperoxy group (step 5)

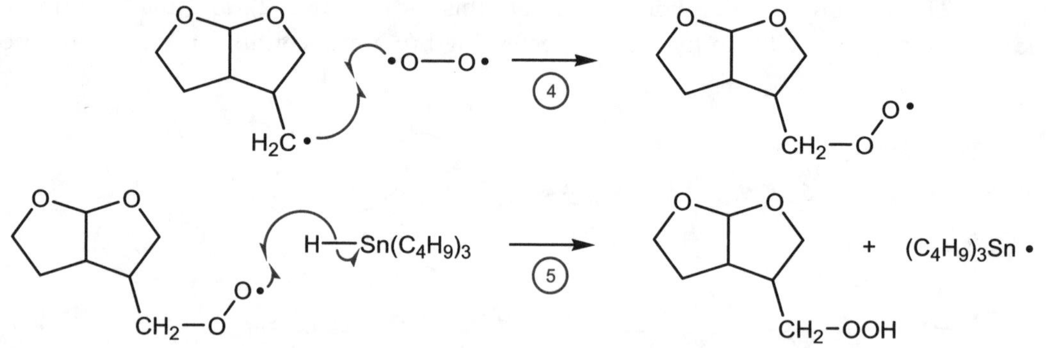

The hydroperoxy group becomes an alcohol functional group after cleavage of the oxygen-oxygen bond, which is quite weak. Tributyltin hydride is a good reducing agent, and it cleaves O—O bonds.

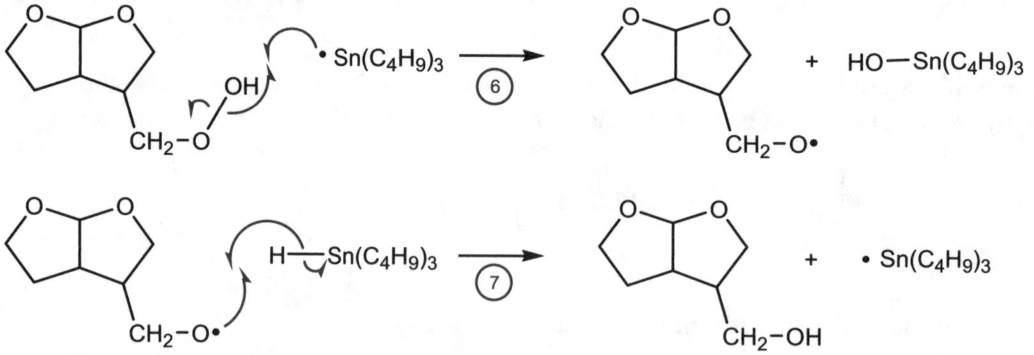

12.32. Reaction of an alkyl bromide with the tributyltin radical (look at the solution to Exercise 12.15 for the mechanism of its formation) generates an alkyl radical (step 1). Carbon monoxide can intercept this radical (step 2), and the resulting acyl radical abstracts a hydrogen atom from tributyltin hydride (step 3), which also regenerates the tributyltin radical and propagates the chain reaction.

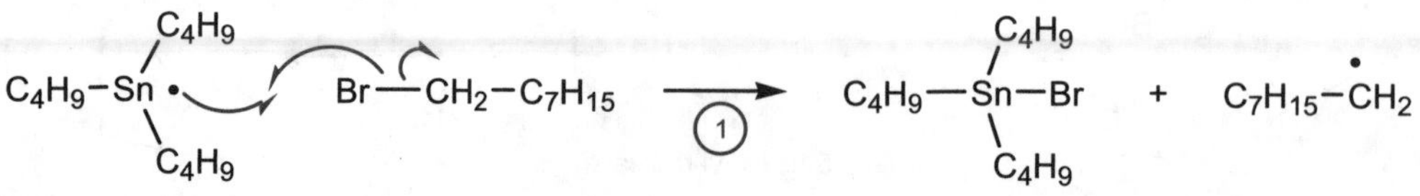

12.32. (continued)

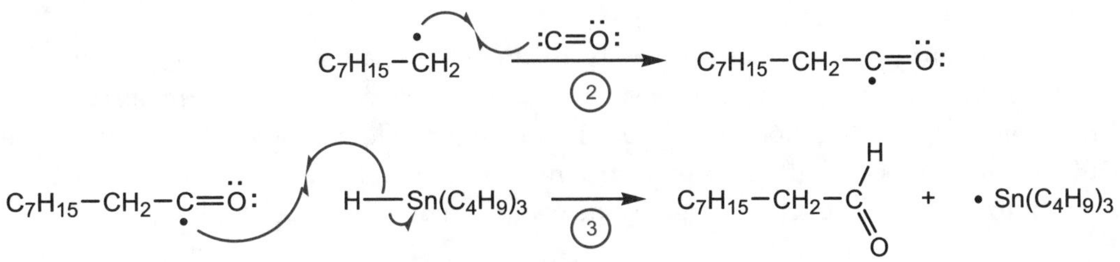

$$C_7H_{15}-\overset{\bullet}{C}H_2 \xrightarrow[\text{②}]{:C=\ddot{O}:} C_7H_{15}-CH_2-\overset{\bullet}{C}=\ddot{O}:$$

$$C_7H_{15}-CH_2-\overset{\bullet}{C}=\ddot{O}: \quad H-Sn(C_4H_9)_3 \xrightarrow[\text{③}]{} C_7H_{15}-CH_2-\overset{\overset{H}{|}}{\underset{\parallel}{C}}\underset{O}{} \quad + \quad \bullet Sn(C_4H_9)_3$$

12.33. Galvinoxyl has the structure shown below. This radical is stabilized by extensive delocalization of the unpaired electron over the atoms of the entire π system. Notice that the unpaired electron appears on three of the carbon atoms in the rings that are substituted, as well as on the O atoms, which are flanked by *tert*-butyl groups. The *tert*-butyl groups prevent two radical centers from getting close enough to form a bond, which would lead to dimerization and destruction of the radical center.

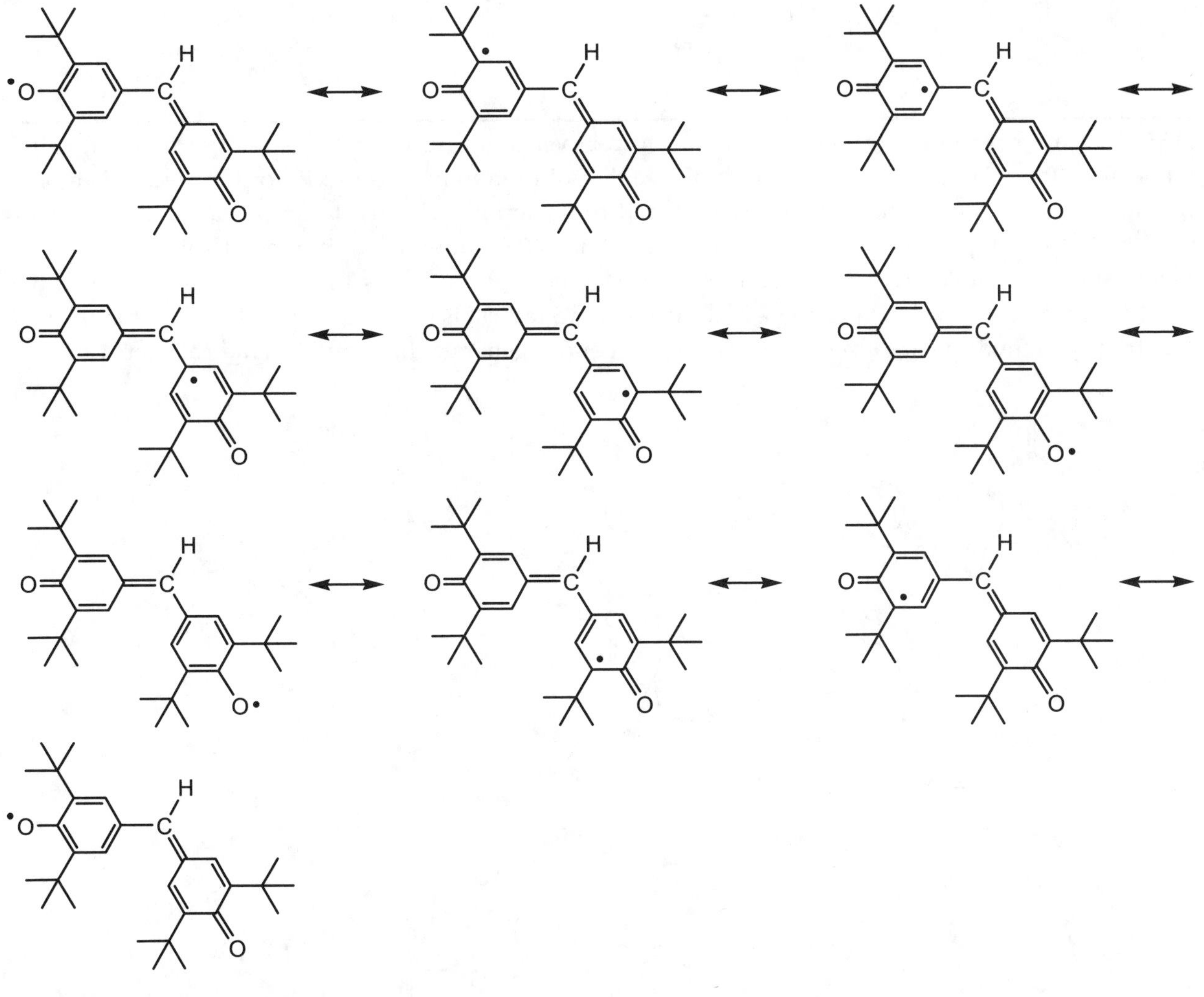

12.34. With its weak O—Cl bond, *tert*-butyl hypochlorite undergoes homolysis when irradiated, just as Cl₂ or Br₂ do when they are subjected to irradiation with UV light.

Initiation

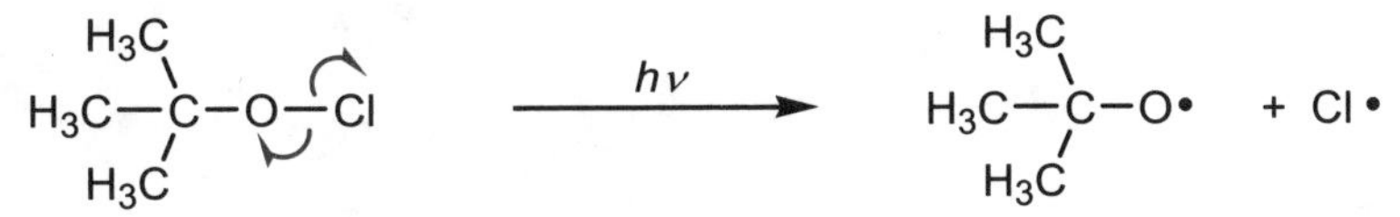

$$H_3C-\underset{\underset{CH_3}{|}}{\overset{\overset{CH_3}{|}}{C}}-O-Cl \xrightarrow{h\nu} H_3C-\underset{\underset{CH_3}{|}}{\overset{\overset{CH_3}{|}}{C}}-O\bullet \quad + \quad Cl\bullet$$

12.34. (continued)

A chain reaction subsequently ensues, starting with abstraction of an allylic hydrogen atom from cyclohexane (step 1). If there are two radical species present, as in this case, the H-atom abstraction step will occur with the atom or group that does not appear in the final product. The allylic radical formed in step 1 reacts with *tert*-butyl hypochlorite to form a carbon-chlorine bond (step 2). This second step also regenerates the *tert*-butoxy radical, which propagates the chain reaction.

Propagation

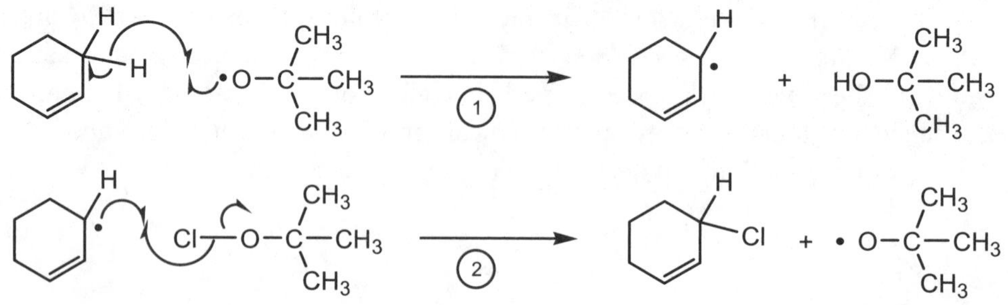

12.35. The mechanism for the given tandem radical cyclization reaction starts with abstraction of the iodine atom (step 1) by the tributyltin radical (see the solution to Exercise 12.15 for the mechanism of its formation). Cyclization creates the five-membered ring (step 2). The alkyl group bearing the radical is attached to the top side of the cyclopentanone ring (as drawn), so it most likely adds to the cyclopentanone double bond from the top as well, which forms the cis ring junction. A second five-membered ring is formed next (step 3), and the resulting radical is subsequently reduced by the reaction with tributyltin hydride to form the tricyclic product and regenerate the tributyltin radical (step 4).

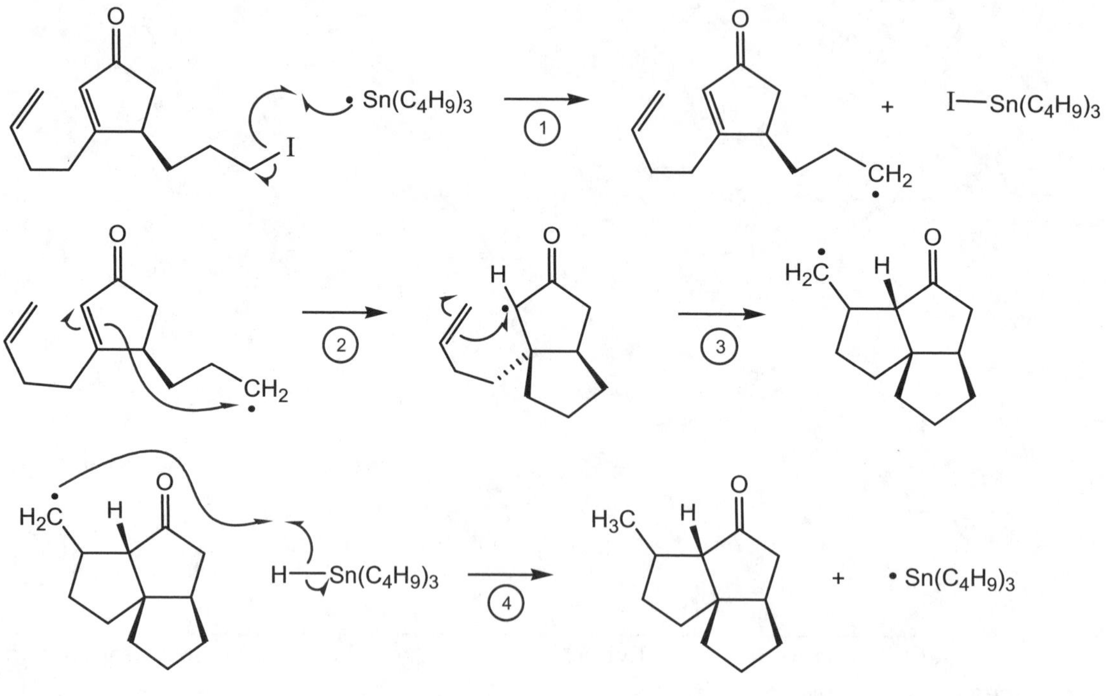

PROTON AND CARBON NMR SPECTROSCOPY

13.1. Follow the procedures outlined in Example 13.1.

a.

The six protons shown in color are equivalent.

b.

The four protons shown in color are equivalent.

c.

The eight protons shown in color are equivalent.

13.2. Follow the procedures outlined in Examples 13.2 and 13.3.

a. Aldehyde proton resonances normally appear at about 10 ppm, and alkene protons usually have chemical shift values between 5 and 6 ppm. The resonance for a methyl group attached to a double bond is expected to appear at δ 1.6 (Table 13.1).

b. All of the protons in this molecule are aliphatic ones, and the chemical shift values are estimated using the data in Table 13.1. The methylene group signal at δ 1.8 is estimated from the value of 1.7 ppm for the substructure –CH₂-C-C=C, assuming that a benzene ring has a slightly greater withdrawing effect than a double bond does.

c. Carboxylic acid proton resonances normally appear at about 12 ppm, and arene protons usually have chemical shift values between 7 and 8 ppm. The resonance for a methyl group attached to an aromatic ring is expected to appear at δ 2.3 (Table 13.1).

13.3. The integrated intensity value of a signal in the proton NMR spectrum is proportional to the number of protons producing that signal. For 1,4-dimethoxybenzene, you can see that it has only two types of protons: methyl and aromatic. A ratio of 6:4 is the same as 3:2, so you could assign the signals to the correct proton type based only on the integrated intensity values of the signals. It is worth noting, however, that aromatic protons have their resonances in the region around 7 ppm, and the methoxy groups will produce a resonance at about 4 ppm (Table 3.1).

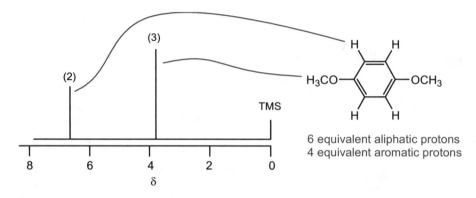

6 equivalent aliphatic protons
4 equivalent aromatic protons

13.4. Follow the procedures given in Example 13.5.

a. This molecule has only two types of protons: methyl and methylene. Because the molecule has two equivalent ethyl groups, the proton NMR spectrum for this molecule has the same appearance as a spectrum for a single ethyl group (Figure 13.11). The chemical shift values are estimated from the data in Table 13.1.

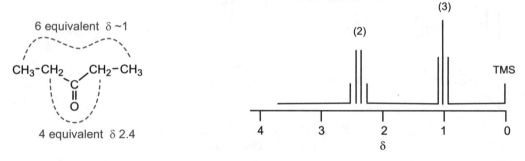

b. This molecule has three types of protons: a methine group and two different types of methyl groups. The spectrum should consist of a singlet for the methoxy group because the methyl protons are more than three bonds from any other protons. The remaining protons constitute the isopropyl group, which will manifest itself in the NMR spectrum as a doublet and septet (Figure 13.11) with a ratio of 6:1. The chemical shift values are estimated from the data in Table 13.1. The septet is shown as an expanded spectrum in the box.

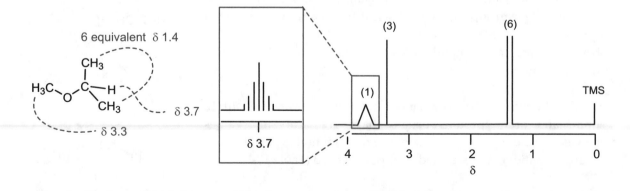

13.5. Follow the procedure outlined in Example 13.6.

The integrated intensity values are 1:3:6. Because the sum of these numbers (10) is the same as the number of hydrogen atoms in the given molecular formula, these numbers correspond to the absolute numbers of protons for each feature. This molecule has only aliphatic protons (all of the signals are between 0 and 3 ppm), and the observation of a doublet and a septet in a ratio of 6:1 means that an isopropyl group is present (Figure 13.11). The signal farthest downfield has a chemical shift value consistent with the presence of a CH fragment adjacent to a carbonyl group. Thus, this molecule has the $-CO-CH(CH_3)_2$ group.

The feature that produces the signal at δ 2.0 has an integrated intensity of 3, which means that it is likely produced by a methyl group. The chemical shift value suggests that a methyl group is adjacent to a carbonyl group, CH_3CO-, which is consistent with its appearance as a singlet. Combining the two fragments, we conclude that the molecule producing this spectrum is **3-methyl-2-butanone**.

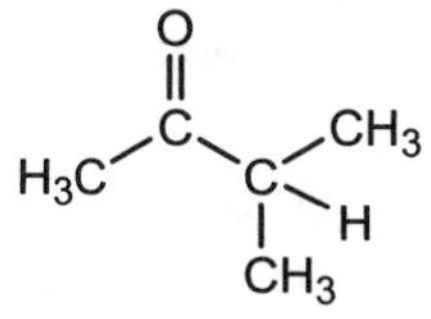

13.6. Follow the procedure outlined in the solution to Exercise 13.5.

The integrated intensity values are 1:2:2:2:3. Because the sum of these numbers (10) is the same as the number of hydrogen atoms in the given molecular formula, these numbers correspond to the absolute numbers of protons for each feature. This molecule has aliphatic protons (the signals between 0 and 3 ppm), aromatic protons (the signals between 7 and 8 ppm), and an aldehyde proton (most likely, based on the signal at 10 ppm). The observation of a triplet and a quartet with a ratio of 3:2 in the aliphatic region means that an ethyl group is present (Figure 13.11). The appearance of two doublets with a total integrated intensity value of 4 in the aromatic region is consistent with the presence of a 1,4-disubstituted benzene ring in this molecule.

Adding the atoms assigned so far ($C_6H_4 + CH_2CH_3 = C_8H_9$) accounts for all but one each of the carbon and hydrogen atoms plus an oxygen atom. These remaining atoms (C, H, and O) constitute the aldehyde group, which confirms that the signal at δ 10.0 is produced by an aldehyde group. Because the benzene ring has two substituents, we attach the ethyl and aldehyde groups to the ring. The signal for the methylene group should appear at about δ 2.7 (Table 13.1), as it does. The molecule that produces this spectrum is **4-ethylbenzaldehyde**.

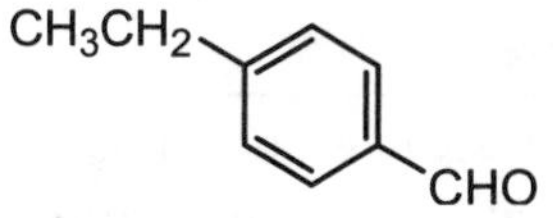

13.7. Follow the procedure outlined in the solution to Exercise 13.5.

The integrated intensity values are 3:1:6. Because the sum of these numbers (10) is the same as the number of hydrogen atoms in the given molecular formula, these numbers correspond to the absolute numbers of protons for each feature. This molecule has only aliphatic protons (all of the signals are between 0 and 4 ppm), and the observation of a doublet and a septet in a ratio of 6:1 means that an isopropyl group is present (Figure 13.11). The signal at δ 2.5 has a chemical shift value consistent with the presence of a CH fragment adjacent to a carbonyl group, probably that of an ester (Table 13.1). Thus, this molecule has the $-CO-CH(CH_3)_2$ group.

The feature that produces the signal at δ 3.7 has an integrated intensity of 3, which means that it is produced by a methyl group. The chemical shift value suggests the presence of a methyl group adjacent to the oxygen atom of the ester functional group, which means that the fragment CH_3O- is present in the molecule. A CH_3O group will produce a signal that appears as a singlet. Combining the two fragments, we conclude that the molecule producing this spectrum is **methyl 2-methylpropanoate**, which has the appropriate molecular formula and the structure shown at the right. (Do not be concerned if you did not name the molecule correctly; the nomenclature of esters will not be covered until Chapter 21.)

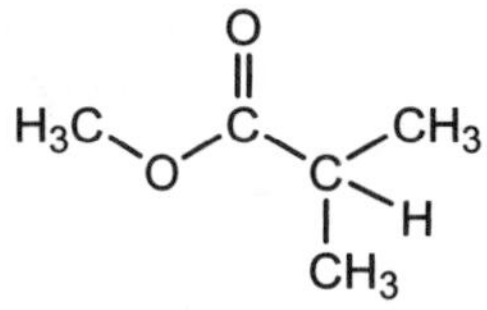

13.8. Follow the procedure outlined in the solution to Exercise 13.4.

   This molecule has three types of protons: a methine group, a methyl group, and a 1,4-disubstituted benzene ring. The aromatic portion should manifest itself as two doublets between 7 and 8 ppm; each doublet will have an integrated intensity value of 2. The methine proton (intensity = 1) should be split into a quartet by the adjacent methyl group (3 protons + 1 = 4 ⇒ quartet), and the methyl group protons (intensity = 3) should appear as a doublet because of coupling with the adjacent CH group (1 proton + 1 = 2 ⇒ doublet).

   The chemical shift values for the aliphatic protons are estimated from the data in Table 13.1 in the manner shown below. (Do not be concerned if the chemical shift values that you estimate differ from those given here; it is more important at this stage to draw the features with the correct splitting patterns and integrated intensity values and to position them in the correct *general* regions.)

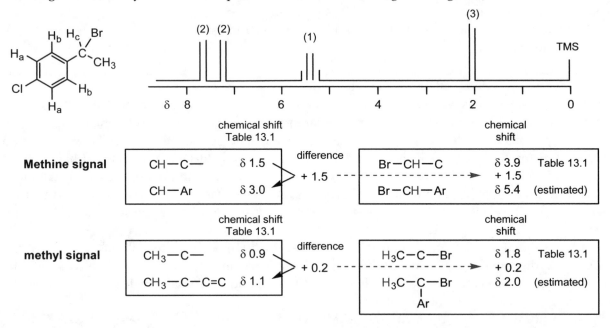

13.9. The isobutyl group should produce a spectrum with two doublets and a group of nine peaks. Assuming that X in the structure shown below is an electron-withdrawing group, the methylene protons (H_c) should produce the signal that is farthest downfield. This resonance will appear as a doublet because the CH_2 group is adjacent to a single proton (1 proton + 1 = 2 ⇒ doublet). The six protons of the two methyl groups are equivalent, and their signal will appear farthest upfield as a doublet because they are also within three bonds of the methine proton (1 proton + 1 = 2 ⇒ doublet). The upfield doublet should be about three times the height of the downfield doublet.

   The methine proton is three bonds away from eight aliphatic protons (2 methylene and 6 methyl protons). Therefore, its signal will be split into nine peaks (8 protons + 1 = 9).

13.10. Follow the procedure illustrated in the solution to Exercise 13.5.

The integrated intensity values are 2:2:2:3, the sum of which (9) is the same as the number of hydrogen atoms in the given molecular formula. This molecule has only aliphatic protons (all of the signals are between 0 and 3 ppm). Assuming that the four features represent one methyl and three methylene groups, the remaining elements (C and N) constitute the nitrile group. The signal at δ 2.5 has a chemical shift value consistent with the presence of an electron withdrawing group (the only option for which is the nitrile group), so this molecule most likely has the –CH₂–CN fragment. The feature at δ 2.5 is a triplet, so this methylene group adjacent to the nitrile group has a methylene group on its other side CH₂–CH₂–CN.

The feature that produces the signal at δ 0.95 has an integrated intensity of 3, which is associated with the presence of a methyl group. This feature is a triplet, so the methyl group producing it is adjacent to a methylene group CH₃–CH₂–.

Combining these two fragments, which account for all of the atoms in the molecular formula, we conclude that this molecule is pentanenitrile, CH₃–CH₂–CH₂–CH₂–CN. To confirm this assignment, verify that the splitting patterns for the signals of the internal methylene groups appear as expected.

$H_3C-CH_2-CH_2-CH_2-C\equiv N$

five adjacent protons + 1 = 6: The NMR signal for the protons in color will appear as a **sextet**.

$H_3C-CH_2-CH_2-CH_2-C\equiv N$

four adjacent protons + 1 = 5: The NMR signal for the protons in color will appear as a **quintet**.

The spectrum displays two triplets, a quintet, and a sextet, all as predicted for CH₃–CH₂–CH₂–CH₂–CN.

13.11. Follow the procedure illustrated in the solution to Exercise 13.10.

The integrated intensity values are 1:2:2:3, the sum of which (8) is the same as the number of hydrogen atoms in the given molecular formula. This molecule has three types of aliphatic protons (the signals are all between 0 and 3 ppm) and an aldehyde group (δ 9.8). Assuming that the three features in the aliphatic region represent one methyl and two methylene groups, the remaining elements (C, H, and O) are consistent with the presence of an aldehyde group. The signal at δ 2.5 has a chemical shift value consistent with the presence of an electron withdrawing group (the only option for which is the aldehyde group), so this molecule likely has the CH₂–CHO fragment.

The feature that produces the signal at δ 0.95 has an integrated intensity value of 3, which means that it is associated with a methyl group. This feature is a triplet, so associated methyl group producing it is adjacent to a methylene group CH₃–CH₂–.

Combining these two fragments, which account for all of the atoms in the molecular formula, we conclude that this molecule is butanal, CH₃–CH₂–CH₂–CHO. To confirm this assignment, verify that the splitting patterns for the remaining signals are as expected.

$CH_3-CH_2-CH_2-C$ (with =O and H)

one adjacent proton + 1 = 2: The NMR signal for the protons in color will appear as a **doublet** (J = 2 Hz).

two adjacent protons + 1 = 3: The NMR signal for the protons in color will appear as a **triplet** (J = 7 Hz).

The influence of two different types of neighboring protons means that the NMR signal for the protons in color will appear as a **doublet of triplets**.

$CH_3-CH_2-CH_2-C$ (with =O and H)

five adjacent protons + 1 = 6: The NMR signal for the protons in color will appear as a **sextet** (J = 7 Hz).

$CH_3-CH_2-CH_2-C$ (with =O and H)

two adjacent protons + 1 = 3: The NMR signal for the aldehyde proton will appear as a **triplet** (J = 2 Hz).

The spectrum displays a triplet (J = 2 Hz), a doublet of triplets (J = 2 Hz and 7 Hz), a sextet (J = 7 Hz), and a triplet (J = 7 Hz), all as predicted for CH₃–CH₂–CH₂–CHO.

13.12. Make use of the data in Figure 13.23 to estimate the expected chemical shift values for the indicated carbon atoms. The carbon atom in color in each structure is the one that will produce the signal that appears farther downfield.

a. $CH_3CH_2CH_2{-}C{\equiv}N$

$\delta \sim 120$

$\delta\ 30{-}40$

b. COOCH$_3$

$\delta \sim 165$

$\delta\ 40{-}60$

c. H_3C, CH_3, H, O

$\delta\ 20{-}30$

$\delta\ 110{-}150$

13.13. Make use of the data in Figure 13.23 to predict the type of carbon atom associated with the given chemical shift value for a specific resonance. Use the DEPT results to determine how many protons are attached to each carbon atom.

a. The compound that produces this spectrum has a carbonyl group, an aromatic ring, and two aliphatic carbon atoms. From its chemical shift value, the carbonyl group is either that of an aldehyde or ketone, but the resonance is not associated with a proton, so the molecule must be a ketone. The benzene ring must have two substituents (two carbon atoms have no attached H atom).

   The possible patterns for the carbon resonances associated with a disubstituted benzene ring are illustrated below (the different shapes represent different types of carbon atoms), and from them we can conclude that the ring is disubstituted in positions 1 and 4. The aliphatic carbon atoms comprise a methylene and a methyl group.

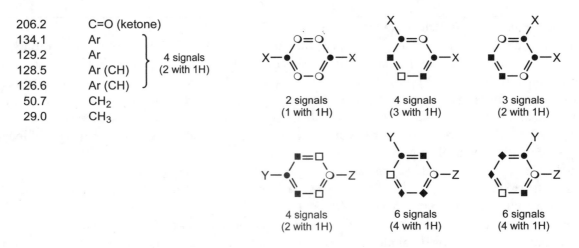

206.2	C=O (ketone)
134.1	Ar
129.2	Ar
128.5	Ar (CH)
126.6	Ar (CH)
50.7	CH$_2$
29.0	CH$_3$

4 signals (2 with 1H)

b. The compound that produces this spectrum has a carbonyl group and two aliphatic carbon atoms. From its chemical shift value, the carbonyl group is either that of an aldehyde or ketone. Because the carbonyl resonance is associated with one proton, the molecule must have the aldehyde group. The aliphatic carbon atoms comprise a methine and a methyl group.

13.14. Use the data given in Table 13.4 to estimate the chemical shift values for the associated carbon atoms. These are illustrated below as lines in the broad-band decoupled spectra (the values used for the resonance positions in these spectra are somewhat arbitrary, but the general regions are correct).

The expected DEPT spectra are indicated as the appropriate hydrocarbon group (or C atom) above each peak in the broad-band decoupled spectrum.

The expected patterns for off-resonance decoupled spectra are indicated by the letters below each peak in the broad-band decoupled spectrum (s = singlet, d = doublet, t = triplet, q = quartet).

13.14. (continued)

a. Example 13.7: **1-bromo-4-ethylbenzene**

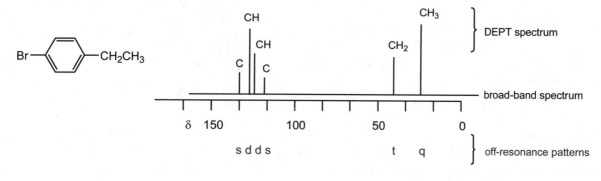

b. Example 13.8: **2-phenylethanol**

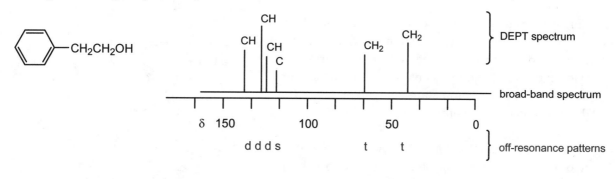

- - - - - - - - - - - - - - - - -

c. Example 13.9: **4-methylbenzaldehyde**

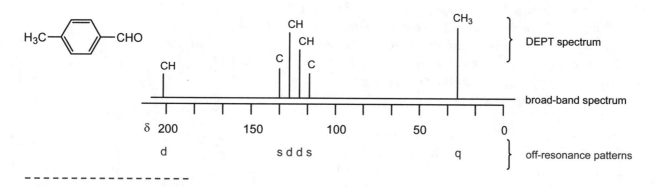

- - - - - - - - - - - - - - - - -

d. Example 13.10: **2-ethoxypropane**

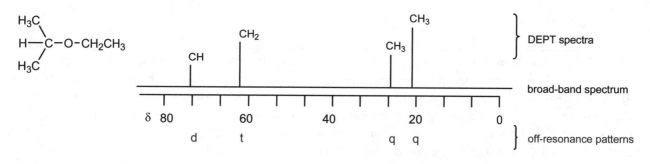

13.15. To identify chemically equivalent protons, start with one group and follow the procedures outlined in Example 13.1 and illustrated in the solution to Exercise 13.1. Continue until you have evaluated each set of protons.

a. This molecule has no symmetry elements (center, axis, or plane), so all six sets are different. The three protons of the methyl group are equivalent to each other, however.

b. This molecule has no symmetry elements (center, axis, or plane), so all five sets of protons are different. The three protons of each methyl group are equivalent to each other, as are the two protons of the methylene group.

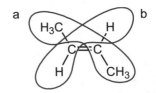

c. This molecule has a plane of symmetry that passes through the benzene ring and bisects the methyl group and the proton labeled "b," so there are three sets of protons. The protons attached to the ring next to the chlorine atoms are equivalent as are the three protons of the methyl group.

d. This molecule has a center of symmetry, so there are two different sets of protons. The two alkene protons are equivalent, as are all six protons that constitute the two methyl groups.

e. This molecule has a plane of symmetry that passes through the benzene ring and bisects the nitro and hydroxyl groups, so there are three sets of protons. The protons attached to the ring next to the hydroxyl group and those attached adjacent to the nitro group are equivalent to each other.

f. This molecule has a plane of symmetry that passes through the methine, methylene, and hydroxyl groups, so there are four sets of protons. All six protons that constitute the two methyl groups are equivalent, as are the two protons of the methylene group.

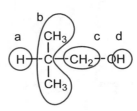

13.16. The chemical shift value for a particular proton is affected by its electronic environment. Magnetic anisotropy often has the greatest effect, especially if π bonds are present in a molecule. Inductive effects, created by the electronegativity properties of the substituents, influence chemical shift values as well. If you are evaluating the chemical shifts of protons in similar environments, pay attention to the electron-withdrawing or donating influences of neighboring groups. For more significant structural changes, use a compilation summary such as Figure 13.3 or Table 13.1.

13.16. (continued)

a.   The general structures of these molecules are the same, so the chemical shift values of the methine protons will be affected by the electron-withdrawing nature of the group attached to this center. The given values are from Table 13.1.

$CH_3CH_2-CH_2-\overset{\overset{\displaystyle Cl}{|}}{\underset{\underset{\displaystyle H}{|}}{C}}-CH_3$	$CH_3CH_2-CH_2-\overset{\overset{\displaystyle OH}{|}}{\underset{\underset{\displaystyle H}{|}}{C}}-CH_3$	$CH_3CH_2-CH_2-\overset{\overset{\displaystyle COOH}{|}}{\underset{\underset{\displaystyle H}{|}}{C}}-CH_3$
δ 2.5	δ 3.9	δ 4.0

b.   The structures of these molecules differ in the proximity of the methylene protons to the electron-withdrawing groups. Two of the values are calculated using data from Table 13.1.

c.   The protons that are being evaluated in these molecules are substantially different, so use Figure 13.1 to estimate their chemical shift values.

Br	CHO	COOH
δ 4	δ 10	δ 12

d.   The methylene protons that are being evaluated in these molecules differ in their proximity to the same electron-withdrawing atom. Their chemical shift values are obtained or estimated using the data in Table 13.1.

$CH_3CH_2CH_2CH_2Cl$	$CH_3CH_2CH_2CH_2Cl$	$CH_3CH_2CH_2CH_2Cl$	$CH_3CH_2CH_2CH_2Cl$
δ 0.9	δ ~1.2	δ 1.8	δ 3.3
Table 13.1	(estimated)	Table 13.1	Table 13.1

13.17. Make use of the data in Figure 13.3 and Table 13.1 to estimate the expected chemical shift values for the different types of protons. The proton shown in color in each structure is the one that will produce the signal that appears farther downfield.

a.

δ 2.4 δ 2.2

$$CH_3CH_2-CH_2-\overset{\overset{\displaystyle O}{\|}}{C}-CH_3$$

b.

H δ 7-8

δ 10

-CHO

c.

H   H

δ 7-8 δ 2.7

13.18. Make use of the data in Figure 13.23 to estimate the expected chemical shift values for the different types of carbon atoms. The carbon atom shown in color in each structure is the one that will produce the signal that appears farther downfield.

a.

δ 40-50 δ 30-40

$$CH_3CH_2-CH_2-\overset{\overset{\displaystyle O}{\|}}{C}-CH_3$$

b.

δ ~120

H

C

δ ~190

-CHO

c.

H   H

C   C

δ ~120 δ 30-40

13.19. The spin-spin splitting pattern observed for the resonance associated with a particular set of protons is related to the number of protons attached three bonds away. For n neighboring protons, the multiplicity of the resonance being observed is $n + 1$. For protons attached to a benzene ring, the predominant pattern is related to the number of protons attached to the adjacent carbon atoms. If different types of protons are three bonds away, then a more complex pattern will be observed (part c.).

It is usually helpful to expand a structure to be certain how many protons are three bonds away from the one(s) being evaluated. In the structures given below, the proton(s) being evaluated are shown in color and the neighboring protons are circled.

a.   There are two protons that are three bonds away from the methylene protons being evaluated. The methylene resonance will therefore appear as $2 + 1 = 3$ peaks, a **triplet**.

H   (H)   H

$$H-\overset{\overset{\displaystyle H}{|}}{C}-\overset{\overset{\displaystyle (H)}{|}}{C}-\overset{\overset{\displaystyle H}{|}}{C}-Br$$

H   (H)   H

b.   There are five similar protons (all aliphatic) that are three bonds away from the methylene protons being evaluated. The methylene resonance will therefore appear as $5 + 1 = 6$ peaks, a **sextet**.

(H)   H   (H)

$$(H)-\overset{\overset{\displaystyle }{|}}{C}-\overset{\overset{\displaystyle }{|}}{C}-\overset{\overset{\displaystyle }{|}}{C}-O-H$$

(H)   H   (H)

c.   There are six aliphatic protons plus one proton attached to the sulfur atom that are three bonds away from the methine proton being evaluated. The methylene resonance will therefore appear as $6 + 1 = 7$ peaks, a **septet** in addition to $1 + 1 = 2$, a **doublet**. Its overall appearance will therefore be a **doublet of septets**. If the J values for the two types of coupling are the same (say, 7 Hz), then we can group all of the neighboring protons together, and then the methine proton signal will appear as $(6 + 1) + 1 = 8$ peaks, an **octet**.

13.19. (continued)

d. There is only one proton within three bonds of the alkene proton being evaluated. Its resonance will therefore appear as 1 + 1 = 2 peaks, a **doublet**.

e. There is one proton that is three bonds away from the aromatic proton being evaluated. Its signal will therefore appear as 1 + 1 = 2 peaks, a **doublet**.

f. There is one proton that is three bonds away from the aromatic proton being evaluated. Its resonance will therefore appear as 1 + 1 = 2 peaks, a **doublet**.

13.20. The off-resonance spectrum will have peak patterns that are equal to one more than the number of protons attached to the carbon atom being evaluated.

a.　　　 2 protons

CH₃CH₂CH₂—Br

2 + 1 = 3
triplet

b.　　　 2 protons

CH₃CH₂CH₂—OH

2 + 1 = 3
triplet

c.　　　 1 proton

(CH₃)₂CH—SH

1 + 1 = 2
doublet

d.
(CH₃)₃C / OCH₃
0 proton

0 + 1 = 1
singlet

e.
1 proton
H₃C— —Br

1 + 1 = 2
doublet

f.
H₃C—C— —H
Br   H
0 proton

0 + 1 = 1
singlet

13.21. Follow the procedure illustrated in the solutions to Exercises 13.4 and 13.8.

a. This molecule has four types of protons: a methyl group, two methylene groups, and a 1,4-disubstituted benzene ring. The aromatic portion should manifest itself as two doublets between 7 and 8 ppm; each doublet will have an integrated intensity value of 2. One methylene group (intensity = 2) should produce a singlet. The ethyl group should manifest itself as a quartet and triplet (intensity ratio 2:3). (The chemical shift values for the aliphatic protons are estimated by using the data in Table 13.1 and are only approximate.)

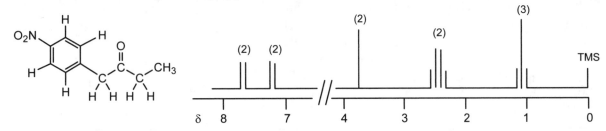

13.21. (continued)

b. This molecule has an isobutyl group and a methyl group, both of which are attached to a carbonyl group. The methyl group (intensity = 3) should produce a singlet, and the isobutyl group will produce two doublets and a nine-peak pattern (see Exercise 13.9). (The chemical shift values for the protons are estimated by using the data in Table 13.1.)

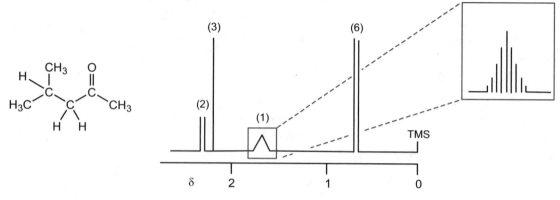

---

c. This molecule has four types of protons: a methyl group, a methylene group, and two different protons of a cis alkene. The alkene proton resonances will appear between 5 and 6 ppm, and the aliphatic proton signals will be observed between 1 and 3 ppm. The latter are part of an ethyl group, which should manifest itself as a quartet and triplet (intensity ratio 2:3) except that the methylene group is also three bonds away from one of the alkene protons. If we assume that all of the coupling constants are 7 Hz (as shown directly below), then we can group all of the neighboring protons together. In that case, the methylene group will appear as a quintet [(3 + 1) protons + 1 = 5 peaks], and H_a will appear as a quartet [2 + 1 protons + 1 = 4 peaks]. No matter what the J values are, the resonance for the methyl group will appear as a triplet, and the signal for H_b will appear as a doublet.

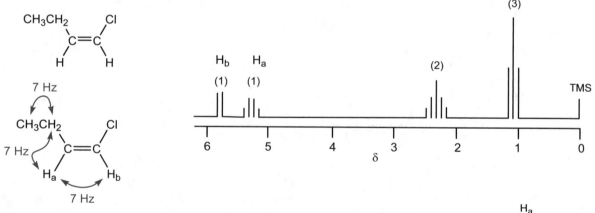

If the coupling constants are different (the most likely difference will be in the coupling between the alkene protons because J values among aliphatic protons are typically 7 Hz), then the signal for H_a will appear as a doublet of triplets, as shown at the right. Such a pattern will be more noticeable as the magnitudes of the J values become increasingly different.

13.22. Follow the procedures outlined in the solution to Exercise 13.14. (The chemical shift values are only approximate.)

a.

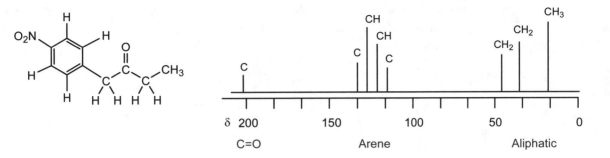

b.

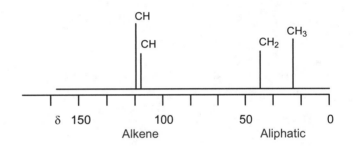

c.

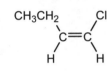

13.23. There can be many subtle differences that arise because of the variations of chemical shift values with structure. To discriminate between isomers using proton NMR spectra, however, it is most helpful to focus on the more substantive differences associated with the distinct types of protons that are present in each molecule, especially protons within functional groups or aliphatic protons adjacent to functional groups.

a.

HC≡C—CH₂CH₂CH₂OH

δ ~1.2   δ 3.6

1H, exchanges with D₂O

H₂C=C⟨CH₂CH₂CHO / H

δ 5-6, 3H   δ 10   δ 2.4

b.

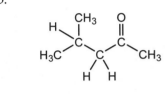 CH₂Br   δ ~5, 2H

δ 7-8, 5H

CH₃   δ 2.6, 3H

Br

δ 7-8, 4H, two doublets

13.24. There can be many subtle differences that arise because of the variations of chemical shift values with structure. To discriminate between isomers using ^{13}C NMR spectra, however, it is most helpful to focus on the more substantive differences associated with the distinct types of carbon atoms that are present in each molecule, especially those that constitute functional groups. DEPT spectra may vary because of different ways the protons are distributed. The resonances associated with aliphatic carbon atoms often do not vary appreciably, but the DEPT spectra can differentiate between them.

a.

b.

d 110-150 4 resonances
3 with 1H (DEPT)
1 with 0H (DEPT)

δ 110-150 4 resonances
2 with 1H (DEPT)
2 with 0H (DEPT)

13.25. The same criteria described in the solution to Exercise 13.23 apply in cases such as the ones in this exercise: Focus on the substantive structural changes that occur.

a.   The ethyl groups within each molecule—starting material or product—are equivalent. This equivalence is manifest in the integrated intensity values (given in the labels as 2H, 3H, 6H, etc.).

b.   The aromatic rings in all three molecules will likely produce similar signals, so the focus should be on the side chain substructures.

13.26. The same criteria described in the solution to Exercise 13.24 apply in cases such as the ones in this exercise: Focus on the substantive structural changes that occur.

a.   The ethyl groups within each molecule—starting material or product—are equivalent. The carbon atom attached to the oxygen atom in each molecule will be the most different.

13.26. (continued)

b. The aromatic rings in all three molecules will likely produce similar signals, so the focus should be on the side chain substructures.

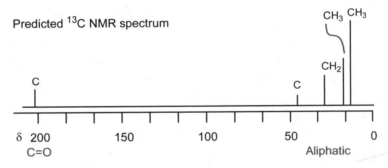

δ 10-20	δ 30-40	δ 30-40	δ 40-70
3H (DEPT)	2H (DEPT)	2H (DEPT)	3H (DEPT)

13.27. The proton NMR spectrum in this exercise displays only singlets, so the carbon atoms bearing protons must be isolated from the others by quaternary carbon atoms, heteroatoms, or functional groups that do not have protons. The integrated intensity values of 2, 3, and 9 (left to right) suggest the presence of a methylene group, a methyl group, and a *tert*-butyl group. Subtracting those elements from the molecular formula ($C_7H_{14}O - CH_2 - CH_3 - C_4H_9$) leaves C and O, a carbonyl group. The methylene and methyl groups cannot be adjacent or spin coupling would occur, so com-bining these fragments is possible in only one way, as shown at the right.

To support this conclusion using ^{13}C NMR spectroscopy, we would expect to see three resonances in the δ 20-30 region for the two different types of methyl groups and the methylene group, a peak in the δ 40-60 region for the quaternary carbon atom, and a signal around δ 205 for the ketone carbonyl carbon atom.

Predicted ^{13}C NMR spectrum

δ 200   150   100   50   0
C=O   Aliphatic

13.28. The proton NMR spectrum in this exercise has only singlets, so the carbon atoms bearing protons must be isolated from the others by quaternary carbon atoms, heteroatoms, or functional groups that do not have protons. The integrated intensity values of 5, 3, and 6 (left to right) suggest the presence of a monosubstituted benzene ring (δ 7.3), a methylene group, and two equivalent methyl groups.

Subtracting these elements from the molecular formula ($C_9H_{13}N - C_6H_5 - CH_2 - CH_3 - CH_3$) leaves only a nitrogen atom. The methylene and methyl groups cannot be adjacent or spin coupling would occur, so combining these fragments is possible in only one way:

13.29. The integrated intensity values are 1:2:2:6:3, the sum of which (14) is the same as the number of hydrogen atoms in the given molecular formula. This molecule has only aliphatic protons (all of the signals are between 0 and 3 ppm), and the appearance of a septet (1H, δ 2.6) and doublet (6H, δ 1.1) is consistent with the presence of an isopropyl group. The signal farthest downfield has a chemical shift value consistent with the presence of a CH fragment adjacent to a carbonyl group. Thus, this molecule likely has the $-CO-CH(CH_3)_2$ group.

13.29. (continued)

The remaining three features comprise a triplet, sextet, and triplet with integrated intensity ratios of 2:2:3, which are consistent with the presence of a propyl group (compare Figure 13.19). The triplet at δ 2.45 (2H) is likely a methylene group adjacent to a carbonyl group; so combining these two fragments with the assumption that the carbonyl group is shared by each, we conclude from the proton NMR spectrum that this molecule is **2-methyl-3-hexanone**.

We can now confirm this conclusion by evaluating the ^{13}C NMR spectrum. For 2-methyl-3-hexanone, we expect to see a resonance for the ketone carbon atom, one for each of the methylene groups, one for the methine group, one for the isopropyl methyl groups and one for the methyl group of the propyl fragment. The resonance for the methine and one of the methylene carbon atoms should be downfield from the others because they are adjacent to the electron-withdrawing carbonyl group. As expected, all six peaks are observed at the appropriate chemical shift values with the correct numbers of protons.

^{13}C NMR and DEPT spectra

δ 214.77	C=O	no proton (ketone)
δ 42.26	CH	(adjacent to the C=O group)
δ 40.80	CH₂	(adjacent to the C=O group)
δ 18.23	CH₃	
δ 17.24	CH₂	
δ 13.81	CH₃	

13.30. The integrated intensity values are 2:1:6. Because the sum of these numbers (9) is the same as the number of hydrogen atoms in the molecular formula, these numbers correspond to the absolute numbers of protons for each feature. This molecule has only aliphatic protons (all of the signals are between 0 and 3 ppm), and the observation of two doublets in addition to a feature with nine peaks (expanded region) suggests that the isobutyl group is present (Exercise 13.9)

Subtracting the elements of the isobutyl group from the total (C_5H_9N – C_4CH_9) leaves the elements C and N, which likely constitute the nitrile group. Combining these two fragments (the isobutyl and nitrile groups), we conclude from the evaluation of the proton NMR spectrum that this molecule is **3-methylbutanenitrile**.

We can confirm our conclusion by evaluating the ^{13}C NMR spectrum. For 3-methylbutanenitrile, we expect to see a resonance for the nitrile carbon atom, one for the methylene group, one for the methine group, and one for the equivalent methyl groups. As expected, all four peaks are observed at the appropriate chemical shift values with the correct numbers of protons.

^{13}C NMR and DEPT spectra

δ 118.89	CN	no proton
δ 26.09	CH₂	
δ 25.94	CH	
δ 21.77	CH₃	

13.31. This proton NMR spectrum has absorption peaks only in the aromatic region, and the integrated intensity values are 1:1:1, the sum of which equals the number of protons in the given molecular formula. Therefore, the benzene ring is trisubstituted and has three protons.

13.31. (continued)

There are only three ways in which to attach three protons to a benzene ring, and the dominant splitting pattern observed for each proton reflects the number of protons attached to its immediate neighboring atoms. Thus, a proton will appear as a singlet if both adjacent carbon atoms are substituted by a group other than H; as a doublet if only one of the adjacent carbon atoms has a proton; or as a triplet if both adjacent carbon atoms bear a proton. These possibilities are summarized as follows:

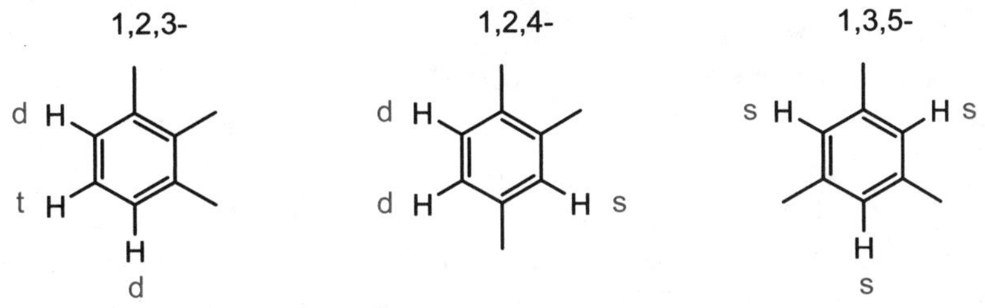

The proton NMR spectrum for this unknown compound displays a singlet and two doublets in its aromatic region. Therefore, it has substituents (two chlorine atoms and a nitro group) in the 1, 2, and 4-positions. (The molecule producing this spectrum is 3,4-dichloronitrobenzene).

13.32. Both spectra have absorptions in two regions: δ 7-8 and δ 1-3; therefore, the molecules have both aromatic and aliphatic protons. Reading left to right, the integrated intensity values for both are 2:2:2:3:3, the sum of which (12) equals the number of protons in the given molecular formula.

Both aromatic regions display the two-doublet pattern (integrated intensity = 4, total) that is associated with a 1,4-disubstituted benzene ring (C_6H_4). The structural differences between the two isomers are therefore in the aliphatic portion.

The features closest to δ 3 (2H, quartet) and at δ 1 (3H, triplet) indicate that each molecule contains an ethyl group (Figure 13.11). Furthermore, each spectrum displays a singlet (3H), which means that each molecule also has a methyl group. Subtracting these fragments from the molecular formula ($C_{10}H_{12}O$ – C_6H_4 – CH_2CH_3 – CH_3) leaves C and O, the carbonyl group.

There are only two ways to assemble these fragments into stable molecules, and these are shown below. Using the data in Table 13.1, we can predict where the resonances are likely to appear for the protons adjacent to the electron-withdrawing groups (carbonyl or aromatic ring); these values are listed in color, below.

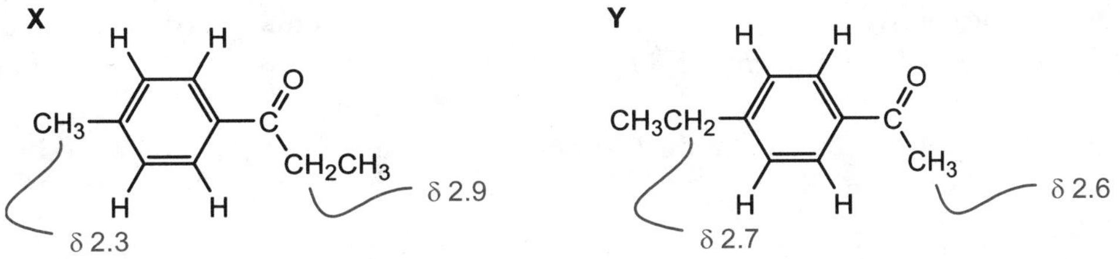

Looking now at the actual spectra in the textbook, we see that the resonances between δ 2 and 3 are at 2.4 and 2.9 ppm in spectrum (a.) and at 2.6 and 2.7 ppm in spectrum (b.). Therefore, spectrum (a.) corresponds to compound **X**, and spectrum (b.) corresponds to compound **Y**.

13.33. This spectrum has absorption peaks in two regions: δ 7-8 and δ 2-4; therefore, the molecule has both aromatic and aliphatic protons. Reading left to right, the integrated intensity values are 5:2:2:2, the sum of which (11) equals the number of protons in the given molecular formula.

The aromatic region has an integrated intensity value of 5, which means that the ring is monosubstituted (C_6H_5).

13.33. (continued)

The aliphatic region has three features with equal integrated intensity values (each 2), which suggests the presence of three methylene groups. Because each methylene group displays spin coupling, they likely constitute the 1,3-propylene fragment: $-CH_2CH_2CH_2-$.

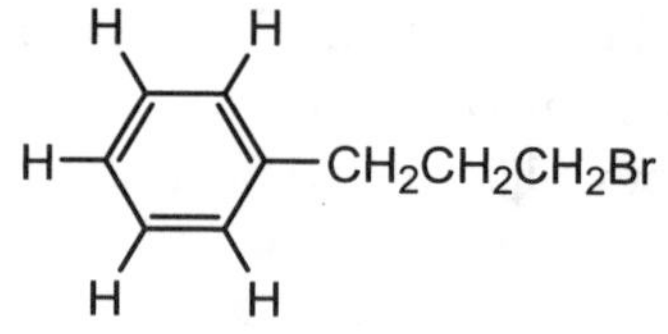

Subtracting these fragments from the molecular formula ($C_9H_{11}Br - C_6H_5 - CH_2CH_2CH_2$) leaves only the Br atom. Assembling all of the fragments leads us to conclude that the molecule is **1-bromo-3-phenylpropane**.

We can confirm this conclusion by evaluating the ^{13}C NMR spectrum. For 1-bromo-3-phenyl-propane, we expect to see four resonances for the benzene carbon atoms, one of which should have no attached proton and three of which will have 1 proton. There should also be three resonances, one for each of the methylene carbon atoms (2H attached to each carbon atom). As expected, all seven resonances are observed at appropriate chemical shift values with the correct numbers of protons.

^{13}C NMR and DEPT spectra

δ 140.45	Ar	no proton
δ 128.45	Ar	1H
δ 128.41	Ar	1H
δ 126.07	Ar	1H
δ 34.14	CH_2	
δ 33.96	CH_2	
δ 32.98	CH_2	

13.34. This spectrum has absorption peaks in three regions: δ 9.5, δ 6-7 and δ 1-3. This molecule likely has an aldehyde group, an alkene portion, and an alkyl group. Reading left to right, the integrated intensity values are 1:1:1:2:3, the sum of which equals the number of protons in the given molecular formula.

Working right to left, we can see that the protons in the aliphatic region (2H and 3H) could signify the presence of an ethyl group. The methylene portion is more complex than a simple quartet however. In fact, it is split into a double of quartets (an approximate quintet), which means it is being influenced by another proton in addition to the methyl group. Let us assume for now that this molecule contains the CH_2CH_3 group.

The aldehyde proton signal appears as a doublet, which means that it is adjacent to a single proton. Subtracting CH_2CH_3 and CHO from C_5H_8O leaves C_2H_2, which represents the alkene portion of the molecule.

There are three ways to construct a disubstituted alkene and these are illustrated below along with the expected spin-spin splitting patterns:

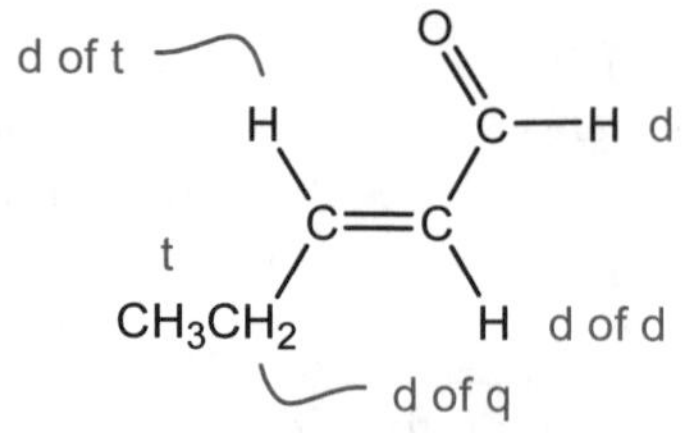

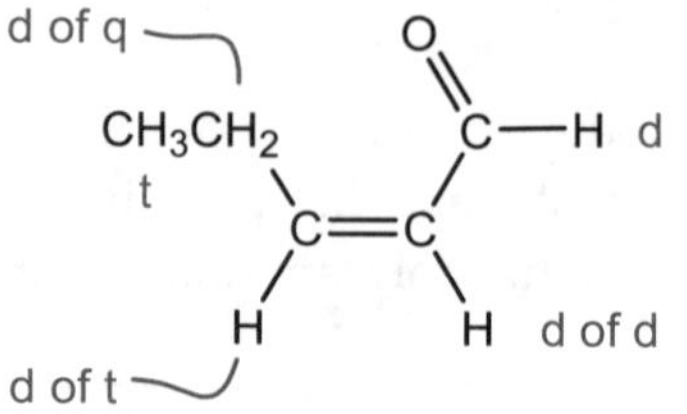

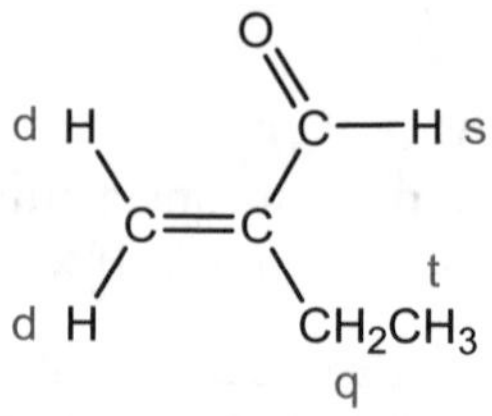

The aldehyde proton resonance in this spectrum appears as a doublet, so we can eliminate structure **C** from consideration.

13.34. (continued)

If we consider what the alkene proton region would look like for structures **A** and **B**, we can create the following splitting diagrams. The *J* values of 16 Hz for coupling between trans alkene protons and 8 Hz for cis alkene protons are typical, as is the alkane/alkene *J* value of 7 Hz. The aldehyde/alkene *J* value of 7 Hz can be determined by measuring the splitting of the aldehyde proton signal at δ 9.5. With these data, we construct the following diagrams:

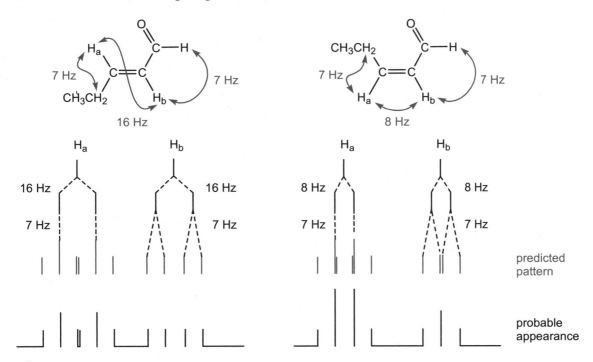

The four equally sized peaks at δ ~6.1 in the actual spectrum matches the predicted pattern for the trans isomer. The signal for H$_b$ in the cis isomer would probably look more like a triplet. Furthermore, the two highest peaks in the H$_a$ signal will be 16 Hz apart in the trans isomer, but only 8 Hz apart in the cis isomer. This separation of ~16 Hz is observed for the signal at δ ~6.9 in the actual spectrum.

We can confirm the structure (except for its stereochemistry) by evaluating the ^{13}C NMR spectrum. For *trans*-2-pentenal, we expect to see a resonance for the aldehyde carbon atom (with 1H), two resonances for the alkene carbon atoms (each with 1 H), and one resonance each for the methylene and methyl carbon atoms. As expected, all five resonances are observed at appropriate chemical shift values with the correct numbers of protons.

^{13}C NMR and DEPT spectra

δ 194.04	C=O	1H (aldehyde)
δ 160.06	C=C	1H
δ 132.07	C=C	1H
δ 25.87	CH$_2$	
δ 11.96	CH$_3$	

DETERMINING THE STRUCTURES
OF ORGANIC MOLECULES

14.1. Follow the procedures outlined in Examples 14.1 and 14.2.

$$35.9 \text{ g C} \left(\frac{1 \text{ mol C}}{12.011 \text{ g C}} \right) = 2.9889 \text{ mol C}$$

$$5.83 \text{ g H} \left(\frac{1 \text{ mol H}}{1.008 \text{ g H}} \right) = 5.7837 \text{ mol H}$$

$$34.6 \text{ g O} \left(\frac{1 \text{ mol O}}{16.00 \text{ g O}} \right) = 2.1625 \text{ mol O}$$

$$23.7 \text{ g S} \left(\frac{1 \text{ mol S}}{32.07 \text{ g S}} \right) = 0.7390 \text{ mol S}$$

$$\frac{2.9889 \text{ mol C}}{0.7390} = 4.04 \qquad \frac{5.7837 \text{ mol H}}{0.7390} = 7.83 \qquad \frac{2.1625 \text{ mol O}}{0.7390} = 2.93 \qquad \frac{0.7390 \text{ mol S}}{0.7390} = 1.00$$

$$C_{4.04}H_{7.83}O_{2.93}S_{1.00}$$

The empirical formula is $C_4H_8O_3S$.

14.2. To calculate the number of π bonds and rings in a molecule—collectively called the "sites of unsaturation"— apply the formula given below:

$$\text{No. of sites of unsaturation} = \frac{(2n + 2) - m - p + q}{2} \qquad \begin{cases} n = \text{number of C atoms} \\ m = \text{number of H atoms} \\ p = \text{number of halogen atoms} \\ q = \text{number of N atoms} \end{cases}$$

For the molecule under consideration,

$$C_6H_{13}NO_2 \qquad \begin{cases} n = 6 \\ m = 13 \\ q = 1 \end{cases}$$

$$\text{No. of sites of unsaturation} = \frac{2(6) + 2 - 13 + 1}{2} = 0 \text{ sites of unsaturation}$$

This molecule has no rings or π bonds.

14.3. To calculate the number of sites of unsaturation, follow the procedure outlined in the solution to Exercise 14.2. From the actual structures of these molecules, count the number of π bonds and rings to see that the results are identical. You may have to expand the structure to make certain that all of the π bonds are obvious (do not overlook π bonds between C and heteroatoms: C=O, C≡N, etc.)

14.3. (continued)

a. This molecule has a ring and 4 π bonds (three in the benzene ring and the C=O bond of the carboxylic acid group).

4 π bonds + 1 ring = 5 sites of unsaturation

$C_{10}H_{11}BrO_2$ $\dfrac{2(10) + 2 - 11 - 1}{2}$ = 5 sites of unsaturation

b. This molecule has no rings but 4 π bonds (two in the alkyne group, one in the alkene bond, and the C=O bond of the aldehyde group).

4 π bonds = 4 sites of unsaturation

C_7H_8O $\dfrac{2(7) + 2 - 8}{2}$ = 4 sites of unsaturation

c. This molecule has two rings and 2 π bonds (the C=O bond of the ketone and the C=C of the alkene).

2 π bonds + 2 rings = 4 sites of unsaturation

$C_9H_{12}ClNO$ $\dfrac{2(9) + 2 - 12 - 1 + 1}{2}$ = 4 sites of unsaturation

---

14.4. To assign the isotopic distributions of elements in a compound, first specify its composition. The formula for 2-butanone is C_4H_8O. Next consider what common isotopes exist for each element present. These are listed in Table 14.1 in the text. Finally, calculate the ratios of the isotopes that combine to give the observed m/z value. For 2-butanone, which has a mass of 72, the (M+2) peak corresponds to a value of m/z = 74. The possible combinations are as follows:

$(^{12}C)_2(^{13}C)_2(^1H)_8{}^{16}O$ $(^{12}C)_4(^1H)_6(^2H)_2{}^{16}O$

$(^{12}C)_3(^{13}C)(^1H)_7(^2H)^{16}O$ $(^{12}C)_4(^1H)_7(^2H)^{17}O$

$(^{12}C)_3(^{13}C)(^1H)_8{}^{17}O$ $(^{12}C)_4(^1H)_8{}^{18}O$

14.5. To calculate exact mass values, apply the data in Table 14.2 to the given formulas and compare these calculated values with the observed value (122.078 in this exercise).

$C_4H_{10}O_4$	$4(12.0000) + 10(1.007825) + 4(15.9949)$	=	122.0579
$C_8H_{10}O$	$8(12.0000) + 10(1.007825) + 15.9949$	=	122.0732
$C_7H_{10}N_2$	$7(12.0000) + 10(1.007825) + 2(14.0031)$	=	122.0845

The probable molecular formula is $C_8H_{10}O$.

14.6. Make use of the data given in Figure 14.3 to discern the presence of the heteroatoms Br, Cl, and S. The fact that the M and M + 2 peaks in this spectrum have approximately the same sizes indicates that the molecule producing this spectrum contains bromine.

14.7. The molecule being evaluated for this exercise is an alcohol. Referring to Table 14.3, you can see that alcohols tend to fragment by losing a molecule of water and by cleavage of the bonds to the carbon atom attached to the OH group. The three peaks listed in this exercise are associated with the molecular ion (the parent compound minus one electron), the fragment that has lost a molecule of water, and the fragment that bears a positive charge formed by breaking the C_α–C_β bond:

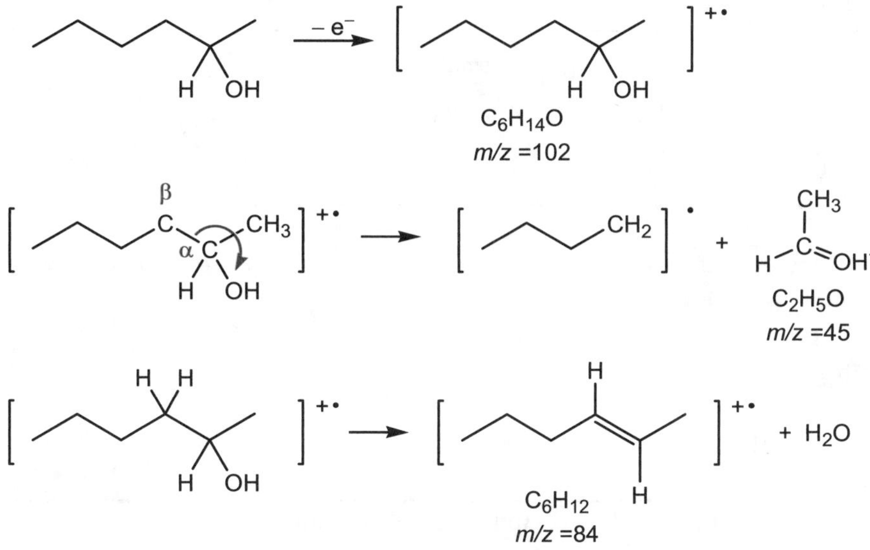

14.8. Follow the procedure outlined in Example 14.6. The intense band at 1715 cm^{-1} is the absorption associated with the C=O stretching vibration of the carbonyl group. By applying the criteria listed in Table 14.5, all but the ketone functional group can be ruled out. (The identification of a ketone usually results from eliminating all of the other possibilities.)

Functional group	Characteristic	Conclusion
Acid chloride	νC=O ~1800 cm^{-1}	The observed νC=O ~1715 cm^{-1} is too low in frequency.
Aldehyde	νC–H ~2720 cm^{-1}	No such band observed.
Amide	νC=O <1680 cm^{-1}	The observed νC=O ~1715 cm^{-1} is too high in frequency.
Carboxylic acid	νO–H ~2500-3500 cm^{-1}	No such band observed.
Ester	νC–O ~1200 cm^{-1}	No such band observed.
Ketone	νC=O ~1715 cm^{-1}	**The observed νC=O is in the correct region.**

14.9. Follow the procedures outlined in Examples 14.7-14.10.

(1) The MW of this unknown can be obtained from the mass spectrum (M = 152); furthermore we know that this molecule has a bromine atom because the M+2 peak has about the same intensity as that of the M peak.

(2) Examine the IR spectrum. The strong absorption at 1725 cm^{-1}, along with the strong and broad band between 2500 and 3500 cm^{-1} indicates that this molecule has a carboxylic acid group.

(3) Examine the proton NMR spectrum. There are three features, one of which must correspond to the resonance for the carboxylic acid proton (COOH). The NMR data are summarized below.

14.9. (continued)

Proton NMR data for Compound **5**.

Chemical shift (ppm)	Integrated intensity	Assignment	Multiplicity	No. adjacent protons (n)
12	1	COOH	singlet	0
3.6	2	CH_2	triplet	2
3.0	2	CH_2	triplet	2

ethylene group

(4) Use the multiplicity and integrated intensity of each feature to establish the connectivity between pairs of hydrocarbon groups (see Figure 13.11).

Compound **5** is **3-bromopropanoic acid**.

$$Br-CH_2-CH_2-COOH \qquad C_3H_5BrO_2 \quad FW = 152.2$$

14.10. Follow the procedures outlined in the solutions to Exercises 14.1 and 14.5.

$$47.17\,g\,C\left(\frac{1\ mol\ C}{12.011\ g\,C}\right) = 3.9272\ mol\ C \qquad 8.63\,g\,H\left(\frac{1\ mol\ H}{1.008\ g\,H}\right) = 8.5615\ mol\ H$$

$$20.91\,g\,O\left(\frac{1\ mol\ O}{16.00\ g\,O}\right) = 1.3068\ mol\ O \qquad 23.29\,g\,Cl\left(\frac{1\ mol\ Cl}{35.45\ g\,Cl}\right) = 0.6570\ mol\ Cl$$

$$\frac{3.9272\ mol\ C}{0.6570} = 5.98 \qquad \frac{8.5615\ mol\ H}{0.6570} = 13.03 \qquad \frac{1.3068\ mol\ O}{0.6570} = 1.99 \qquad \frac{0.6570\ mol\ Cl}{0.6570} = 1.00$$

$$C_6H_{13}ClO_2 \qquad FW = 153.20 \qquad Exact\ mass\ = 152.060$$

14.11. Follow the procedures outlined in the solutions to Exercises 14.1 and 14.5.

$$61.3\,g\,C\left(\frac{1\ mol\ C}{12.011\ g\,C}\right) = 5.1037\ mol\ C \qquad 10.2\,g\,H\left(\frac{1\ mol\ H}{1.008\ g\,H}\right) = 10.1190\ mol\ H$$

$$28.5\,g\,N\left(\frac{1\ mol\ N}{14.007\ g\,N}\right) = 2.0347\ mol\ N$$

$$\frac{5.1037\ mol\ C}{2.0347} = 2.508 \qquad \frac{10.119\ mol\ H}{2.0347} = 4.973 \qquad \frac{2.0347\ mol\ N}{2.0347} = 1.00$$

$$C_{2.51}H_{4.97}N_{1.00} = C_5H_{10}N_2 \qquad FW = 98.15 \qquad Exact\ mass\ = 98.084$$

14.12. Follow the procedure illustrated in the solution to Exercise 14.8.

a. The strong band at 1750 cm^{-1} is the absorption associated with the C=O stretching vibration of the carbonyl group. Applying the criteria listed in Table 14.5, we conclude that the molecule producing this spectrum is a carboxylic acid ester.

Functional group	Characteristic	Conclusion
Acid chloride	νC=O ~1800 cm^{-1}	The observed νC=O ~1750 cm^{-1} is too low in frequency.
Aldehyde	νC–H ~2720 cm^{-1}	No such band observed.
Amide	νC=O <1680 cm^{-1}	The observed νC=O ~1750 cm^{-1} is too high in frequency.
Carboxylic acid	νO–H ~2500-3500 cm^{-1}	No such band observed.
Ester	νC–O ~1200 cm^{-1}	**A strong band is observed in this region. The νC=O value is also appropriate (see Figure 14.10).**
Ketone	νC=O ~1715 cm^{-1}	The observed νC=O ~1750 cm^{-1} is too high in frequency.

- .

b. The strong band at 1710 cm^{-1} is the absorption associated with the C=O stretching vibration of the carbonyl group. Applying the criteria listed in Table 14.5, we conclude that the molecule producing this spectrum is a carboxylic acid.

Functional group	Characteristic	Conclusion
Acid chloride	νC=O ~1800 cm^{-1}	The observed νC=O ~1710 cm^{-1} is too low in frequency.
Aldehyde	νC–H ~2720 cm^{-1}	No such band observed (but it could be obscured by the absorption band between 2500 and 3500 cm^{-1}).
Amide	νC=O <1680 cm^{-1}	The observed νC=O ~1710 cm^{-1} is too high in frequency.
Carboxylic acid	νO–H ~2500-3500 cm^{-1}	**A broad, strong band is observed in this region. The νC=O value is also appropriate (see Figure 14.10).**
Ester	νC–O ~1200 cm^{-1}	The bands in this region are not intense enough.
Ketone	νC=O ~1715 cm^{-1}	The observed νC=O ~1710 cm^{-1} is appropriate, but the broad absorption between 2500 and 3500 cm^{-1} is good evidence for the presence of a COOH group.

14.13. When you perform a reaction in which you know what the product will be (or is expected to be), you want to focus on those spectroscopic features that are unique for the starting material and product. Although many subtle changes will undoubtedly occur in many parts of the spectrum, it helps to look for a limited number of specific changes.

a. This transformation is an oxidation reaction that converts a secondary alcohol to a ketone. The functional group change in the IR spectrum should be quite clearly observed. In the ^{13}C NMR spectrum, the resonance for the carbon atom of the principal functional group will also be quite different. In the ^{1}H NMR spectrum, the alcohol will display no resonances between 2 and 3 ppm. On the other hand, the ketone will have two peaks in this same region for the neighboring aliphatic groups. The alcohol will have a signal at δ 3.9 for the CH–OH proton, which will be farther downfield than any signal in the spectrum of the product.

14.13. (continued)

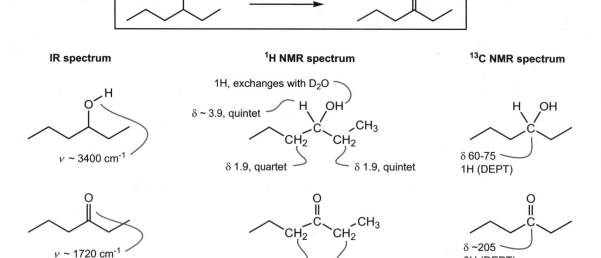

IR spectrum   **¹H NMR spectrum**   **¹³C NMR spectrum**

b. In this substitution reaction, the functional group change (alcohol to nitrile) should be readily detected by IR spectroscopy. In the ¹³C NMR spectrum, the resonance for the cyano carbon atom will be prominent because it is farther downfield than every other signal. In the ¹H NMR spectrum, the methylene group adjacent to the functional group will be easily observed because it is downfield from the other signals. Its chemical shift changes substantially in going from reactant to product.

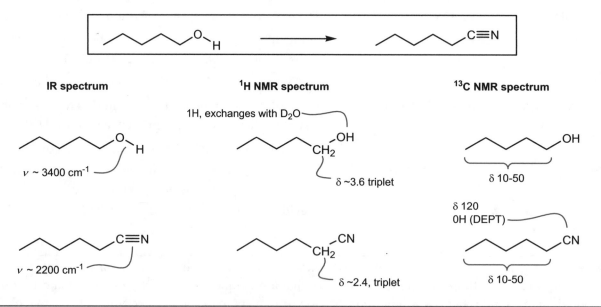

IR spectrum   **¹H NMR spectrum**   **¹³C NMR spectrum**

14.14. The mass spectrum peaks mentioned in this problem indicate that chlorine is present in the structure of compound **X**. The mass value of 92 (³⁵Cl) is consistent with the composition C₄H₉Cl, which has no sites of unsaturation. The absence of absorption bands in the region around 1600 cm⁻¹ confirms that no carbon–carbon double bonds are present in this unknown compound. The possible structures for C₄H₉Cl are shown below.

14.15. This molecule contains an oxygen atom and the IR spectrum displays a strong band around 3400 cm^{-1} that can be assigned as an O–H stretching vibration. This molecule is a saturated alcohol (there are no sites of unsaturation). You are told that the OH group is attached to a secondary carbon atom, so the possible structures for compound **Y** are:

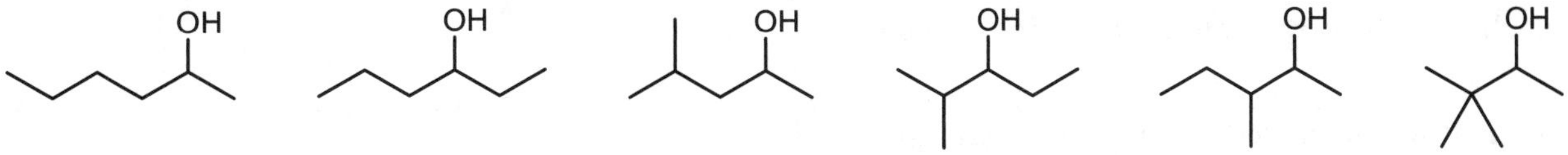

14.16. Follow the procedure outlined in the solution to Exercise 14.5.

$C_6H_{14}O_2$	$6(12.0000) + 14(1.007825) + 2(15.9949)$	=	118.2245
$C_6H_{14}S$	$6(12.0000) + 14(1.007825) + 31.9721$	=	118.0816
$C_5H_{12}NO_2$	$5(12.0000) + 12(1.007825) + 14.0031 + 2(15.9949)$	=	118.0868

The probable molecular formula is $C_6H_{14}S$.

14.17. Follow the procedure outlined in the solution to Exercise 14.5.

$C_7H_{14}S$	$7(12.0000) + 14(1.007825) + 31.9721$	=	130.0816
$C_7H_{11}Cl$	$7(12.0000) + 11(1.007825) + 34.9689$	=	130.0550
$C_6H_{12}NS$	$6(12.0000) + 12(1.007825) + 14.0031 + 31.9721$	=	130.0691

The probable molecular formula is $C_7H_{11}Cl$.

14.18. This molecule has three types of protons: aromatic (3H), aliphatic (6H), and hydroxyl (1H), which account for all of the protons in the molecular formula. The assignment of the resonance at δ 4.9 to an OH group is not an obvious one. Upon examination of the IR spectrum, however, we see a broad, medium-intensity band at about 3300 cm^{-1} that can be attributed to the OH stretching vibration. If this molecule is an alcohol or phenol, we would expect to see a resonance for that proton in the ^{1}H NMR spectrum, hence the assignment of the δ 4.9 signal as the OH resonance.

The benzene ring has a total of three protons, and we can deduce its substitution pattern by comparing the splitting patterns in the aromatic portion of the NMR spectrum with those illustrated in Figure 13.12. The triplet, doublet, doublet pattern means that the ring is substituted in the 1, 2, and 3 positions, and the substituents comprise two methyl groups and an OH group, all of which display observable signals. These fragments ($C_6H_3 + CH_3 + CH_3 + OH = C_8H_{10}O$) account for all of the atoms.

There are only two ways to attach the three groups:

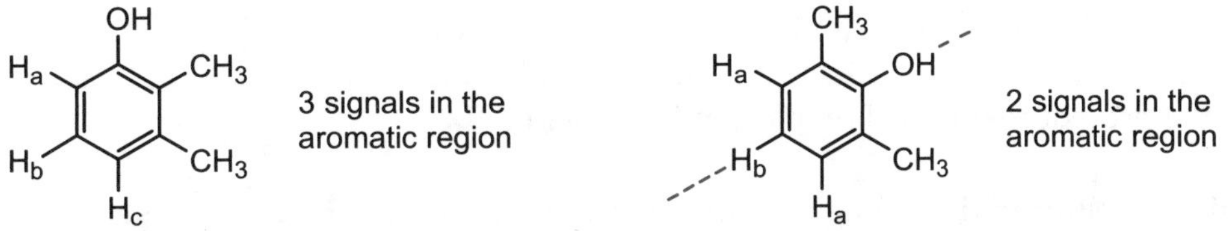

The structure at the right (above) has a plane of symmetry, so it would only display two signals in the aromatic region of the ^{1}H NMR spectrum (in a ratio of 2:1 H_a:H_b). Furthermore, the methyl groups are also equivalent when they flank the OH group, so the aliphatic region would have only one peak with an integrated intensity value of 6. The actual compound, **2,3-dimethylphenol** (above, left), will have three aromatic proton resonances (ratio 1H, 1H, 1H) and two methyl signals (3H, 3H).

14.19. The strong band at 1800 cm^{-1} is clearly an absorption band associated with the C=O stretching vibration of a carbonyl group. Applying the criteria listed in Table 14.5, we conclude that this spectrum was obtained on a carboxylic acid chloride because the frequency of the absorption is so high.

To confirm the presence of the acid chloride functional group using ^{13}C NMR spectra, we would look for a signal between 165 and 185 ppm (Table 13.4), and the DEPT spectrum would reveal that no proton is attached to the carbon atom producing that signal.

14.20. Follow the procedure outlined in Example 14.7.

(1) From the given molecular formula, $C_5H_{10}O$, calculate the number of unsaturation sites: [2(5) + 2 – 10]/2 = 1 site of unsaturation. This means that compound **R** contains a double bond or a ring.
(2) Examine the IR spectrum. The medium intensity absorption at 3400 cm^{-1} can be attributed to the OH stretching vibration of an alcohol or phenol.
(3) Examine the proton NMR spectrum. There are three general features: alkene protons, aliphatic protons, and the OH proton. These data are summarized in the following table.

Proton NMR data for Compound **R**

Chemical shift (ppm)	*Integrated intensity*	*Assignment*	*Multiplicity*	*No. adjacent protons (n)*	
4.88	1	=CH	singlet	0	
4.78	1	=CH	singlet	0	
3.70	2	CH_2	triplet	2	ethylene group
2.30	2	CH_2	triplet	2	ethylene group
2.12	1	OH	singlet	0	
1.78	3	CH_3	singlet	0	

(4) The resonance at 1.78 ppm suggests that the methyl group is attached to the double bond. There are three possible structures that we can draw, and two require that the signal at 1.78 ppm would be split into a doublet. Because the signal associated with the methyl group is a singlet, the alkene must be disubstituted at the same end rather than at each end of the π bond.

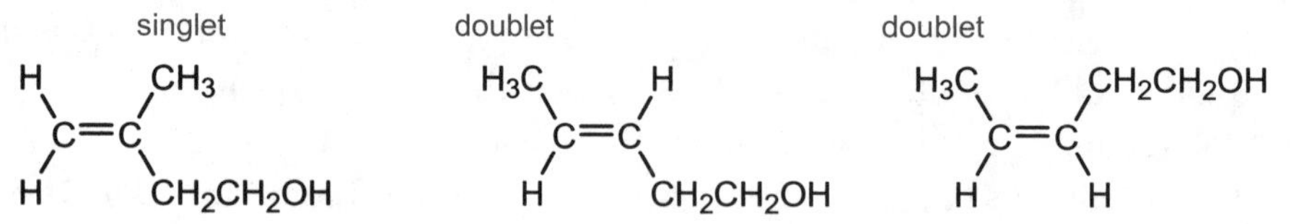

The molecule that produces these spectra is **3-methyl-3-butene-1-ol** (above, left).

14.21. Follow the procedure outlined (in part) in Example 14.9.

(1) Examine the ^{13}C NMR to determine the number and types of carbon atoms present. There are three types: C=O, aromatic, and aliphatic. A benzene ring has 6 carbon atoms, plus there are 6 resonances in the rest of the spectrum. Start by assuming that the compound has 12 carbon atoms.
(2) Examine the integrated intensity values in the proton NMR spectrum. The total is 16. Start with the assumption that the compound has 16 hydrogen atoms.
(3) Use the MW = 192 to see if it can be fit with a formula of the type $C_{12}H_{16}O_n$. The answer is yes, when n = 2. We will assume that the molecular formula is $C_{12}H_{16}O_2$ (and revise it if necessary).

14.21. (continued)

(4) Calculate the number of unsaturation sites: $[2(12) + 2 - 16]/2 = 5$ sites of unsaturation. The NMR data reveal that this molecule has a benzene ring, which accounts for 4 sites of unsaturation (3 double bonds and a ring). Therefore, Compound **S** contains one double bond or one ring *in addition to the benzene ring*. The carbon NMR spectrum shows [step (1)] that this molecule has a carbonyl group, which is the "fifth" site of unsaturation.

(5) Examine the IR spectrum. The strong band at 1730 cm^{-1} is attributed to a C=O stretching vibration. The equally strong absorption at 1200 cm^{-1} suggests the presence of an ester functional group.

(6) Examine the proton NMR spectrum. There are seven features with a total integration of 16 protons. The proton NMR data are summarized in the following table. The aromatic region has an integrated intensity value of 5, so the benzene ring has only one substituent. The multiplicities and integrated intensity values for the aliphatic protons are consistent with the presence of an ethylene group and a propyl group. The structural fragments to assemble are therefore as follows:

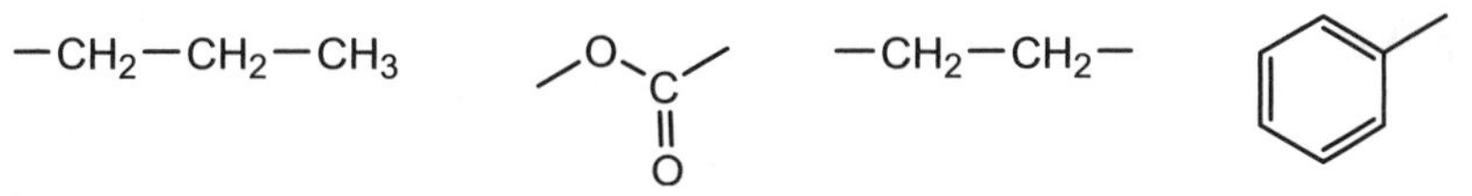

Proton NMR data for Compound **S**

Chemical shift (ppm)	Integrated intensity	Assignment	Multiplicity	No. adjacent protons (n)
7.25	5	ArH	multiplet	n.a.
4.3	2	CH$_2$	triplet	2
2.9	2	CH$_2$	triplet	3
2.3	2	CH$_2$	triplet	2
1.6	2	CH$_2$	sextet	5
0.9	3	CH$_3$	triplet	2

There are two ways to put these fragments together:

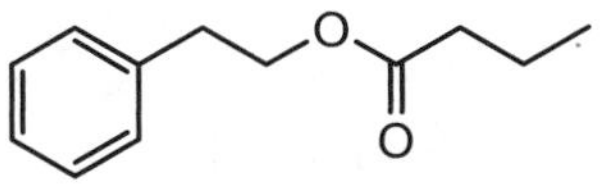

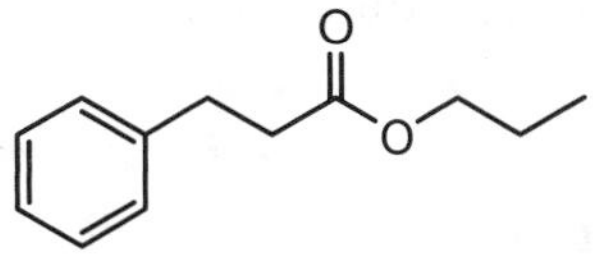

To differentiate between these two molecules is not a simple task. We know that the central methylene group in the propyl portion of the molecule produces the resonance at δ 1.6 because it is split into a sextet pattern.

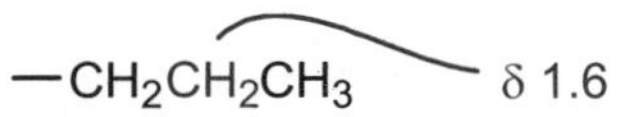

For structure **II**, however, we would expect to see the signal for that methylene group at δ 1.9, based on the data for that substructure given in Table 13.1.

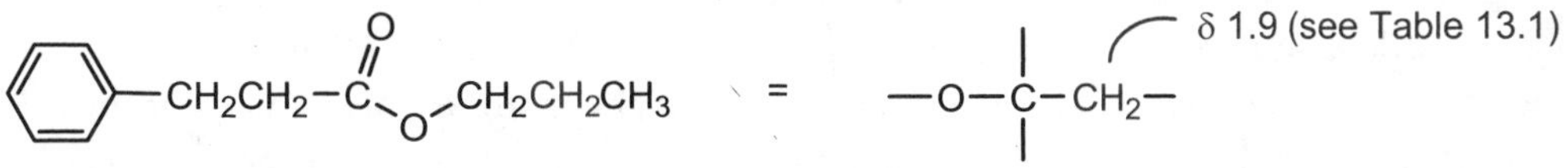

Therefore, the propyl group is more likely attached to the carbonyl group rather than the oxygen atom, and we choose structure **I** as the answer (which incidentally is correct).

14.22. The molecule being evaluated in this exercise is a carboxylic acid. Referring to Table 14.3, you can see that carboxylic acids tend to undergo fragmentation by cleavage of the bond that is one carbon atom removed from the carbonyl oxygen atom as well as by the McLafferty rearrangement. The four peaks listed in this exercise are associated with the molecular ion (the parent compound minus one electron), the fragment that bears a positive charge formed by breaking each of the two bonds between C_α and the β atoms, and the fragment that bears a positive charge formed by the McLafferty rearrangement.

$C_6H_{12}O_2$
$m/z = 116$

$C_6H_{11}O$
$m/z = 99$

$m/z = 45$

$C_3H_6O_2$
$m/z = 74$

14.23. Follow the procedure outlined in Example 14.9.

(1) Examine the ^{13}C NMR to determine the number and types of carbon atoms present. There are three types (C=O, alkene, and aliphatic) consisting of six peaks. Start with the assumption that the compound has 6 carbon atoms.

(2) Examine the integrated intensity values in the proton NMR spectrum. The total is 10 (1:1:2:3:3, left to right). Start with the assumption that the compound has 10 hydrogen atoms.

(3) Use the high-resolution data to determine the molecular formula with the initial assumption that there are 6 C atoms, 10 H atoms and some number of oxygen atoms. For **C$_6$H$_{10}$O**, the calculated high resolution MW is 98.073 (observed MW = 98.071). This formula is consistent with the data.

(4) Calculate the number of unsaturation sites: $[2(6) + 2 - 10]/2 = 2$ sites of unsaturation. Thus, compound **T** contains two double bonds, a double bond and a ring, two rings, or a triple bond. The ^{13}C NMR spectrum shows [step (1)] that this molecule has a carbonyl group and an alkene double bond, which account for the two unsaturation sites.

(5) Examine the IR spectrum. The strong absorption at 1700 cm^{-1} is attributed to a C=O stretching vibration (ketone or aldehyde) that is conjugated with the double bond. The carbonyl resonance in the ^{13}C NMR spectrum indicates that this carbonyl group is a ketone (no attached H).

(6) Examine the proton NMR spectrum. There are seven features with a total integration of 10 protons. The proton NMR data are summarized in the following table.

14.23. (continued)

Proton NMR data for Compound **T**

Chemical shift (ppm)	Integrated intensity	Assignment	Multiplicity	No. adjacent protons (n)
6.85	1	=CH	d of q	1 and 3
6.12	1	=CH	doublet	1
2.56	2	CH$_2$	quartet	3
1.90	3	CH$_3$	doublet	1
1.10	3	CH$_3$	triplet	2

(2.56, 1.10 marked as ethyl group)

The multiplicities and integrated intensity values for the aliphatic protons are consistent with the presence of an ethyl group. The chemical shift value associated with the other methyl group indicates that it is attached to the double bond.

There are two ways to put these groups together:

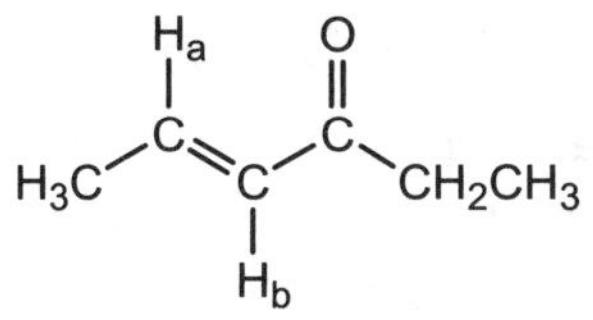

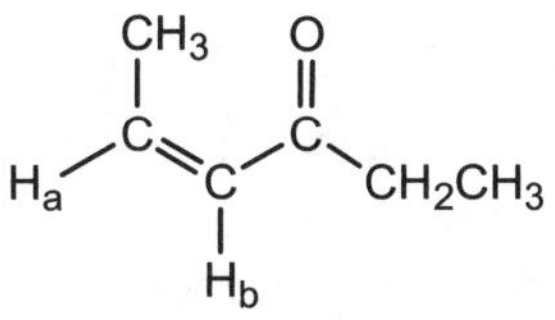

The alkene splitting pattern is relatively simple to interpret because H$_b$ is coupled only to H$_a$ (H$_a$ is coupled both to H$_b$ and to the methyl group). We can readily see that the J value for the alkene coupling interaction is 15 Hz, which means that the double bond has the trans geometry (**I**).

Compound **T** is ***trans*-4-hexene-3-one** (structure **I**, above).

14.24. Follow the procedures outlined in Examples 14.8 and 14.10.

(1) Use the analytical data to determine the empirical formula (section 14.1b): **C$_8$H$_9$Br**.
(2) Calculate the number of unsaturation sites: $[2(8) + 2 - 9 - 1]/2 = 4$ sites of unsaturation.
(3) Examine the ^{13}C NMR to determine the types of carbon atoms present. There are six carbon resonances in the aromatic region, which accounts for the 4 sites of unsaturation, and there are two aliphatic carbon resonances. The 8 peaks account for all of the carbon atoms.
(4) Examine the proton NMR spectrum. The benzene ring is trisubstituted and there are two methyl groups, which must be attached to the ring because they show no spin-spin splitting. The proton NMR data are summarized below.

Proton NMR data for Compound **U**

Chemical shift (ppm)	Integrated intensity	Assignment	Multiplicity	No. adjacent protons (n)
7.36	1	ArH	doublet	1
7.03	1	ArH	doublet	1
6.90	1	ArH	triplet	2
2.37	3	CH$_3$	singlet	0
2.30	3	CH$_3$	singlet	0

14.24. (continued)

(5) The pattern of peaks in the aromatic region indicates that the ring is substituted in three adjacent positions. In fact, the pattern (two doublets and a triplet) is one you have seen before—in Exercise 14.18. This molecule is **3-bromo-1,2-dimethylbenzene**.

14.25. Follow the procedure outlined in the solution to Exercise 14.9.

(1) The MW of this unknown can be obtained from the mass spectrum (M = 150); furthermore we know that this molecule has a bromine atom because the M+2 peak has about the same intensity as that of the M peak.

(2) Subtracting the mass of ^{79}Br from 150, we obtain m/z 71, which corresponds to the mass of C_5H_{11}. The sum of the integrated intensity values in the proton NMR spectrum (2 + 3 + 6 = 11) confirms the number of hydrogen atoms, which in turn is consistent with the number of carbon atoms.

(3) Examine the proton NMR spectrum. There are three features, as summarized below. Two of these constitute an ethyl group. The signal with an integrated intensity of 6 is likely two methyl groups, and they must be attached to a carbon atom that has no protons.

Proton NMR data for Compound **V**

Chemical shift (ppm)	Integrated intensity	Assignment	Multiplicity	No. adjacent protons (n)	
1.80	2	CH_2	quartet	3	⎫
1.65	6	CH_3	singlet	0	⎬ ethyl group
1.10	3	CH_3	triplet	2	⎭

(4) Assembling the pieces [–Br, –CH₂CH₃, and –C(CH₃)₂] generates the structure of Compound **V**, which is **2-bromo-2-methylbutane**.

FW = 150.23

ORGANOMETALLIC REAGENTS
AND CHEMICAL SYNTHESIS

15.1. A reaction between the cyanide ion and a compound with a good leaving group yields the corresponding organonitrile as a substitution product. This transformation normally follows the S$_N$2 pathway, leading to inversion of configuration if the leaving group is attached to a stereogenic carbon atom. An alcohol OH group can be replaced with the cyanide ion by first preparing its alkyl sulfonate ester derivative (mesylate or tosylate).

a. The first step in this transformation involves free radical bromination of the methyl group, which has benzylic hydrogen atoms. The bromide ion is then replaced with the cyanide ion.

b. The first step in this transformation involves preparation of the mesyl (or methanesulfonate) derivative of the alcohol.

15.2. A Grignard reagent has the same general structure as the corresponding organohalide from which it is made, except that MgX replaces X.

15.3. A Grignard reagent reacts with an aldehyde to form a secondary alcohol after workup. A new chiral center is formed in this reaction, so the product is a mixture of enantiomers.

15.4. A Grignard reagent reacts with an epoxide at the less highly substituted carbon atom of the three-membered ring, and an alcohol is formed after workup. In this transformation, a 2° alcohol is the product. A chiral carbon atom in the epoxide retains its configuration if the Grignard reagent reacts at the *other* carbon atom of the epoxide ring. If the Grignard reagent reacts with an epoxide molecule at a chiral carbon atom, the configuration of that stereogenic center is inverted.

15.5. Groups that interfere with Grignard reagent formation include those with acidic protons and those with a double or triple bond between carbon and a heteroatom (carbonyl and nitrile groups). Carbon-carbon double and triple bonds (except for the terminal alkyne group, which has an acidic proton) do not interfere.

reactive multiple bond   *forms a Grignard reagent readily*   *acidic proton*

15.6. Groups that interfere with Grignard formation include those with acidic protons and those with a double or triple bond between carbon and a heteroatom (carbonyl and nitrile groups). Carbon-carbon double and triple bonds (except for the terminal alkyne group, which has an acidic proton) do not interfere.

acidic proton   *acidic proton*   *reactive multiple bond*

15.7. A terminal alkyne reacts with a strong base such as the amide ion (step 1) to form an alkynyl carbanion, which can be alkylated with a primary alkyl halide (step 2).

Steps 3 and 4 lead to the addition of H and OH groups to the triple bond (see Section 9.4c).

15.7. (continued)

The resulting vinyl alcohol undergoes tautomerism to form the corresponding ketone.

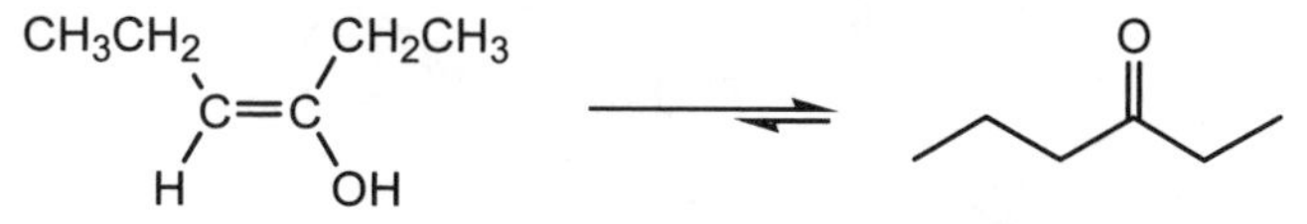
CH3CH2
CH2CH3
C=C
H
OH
O

15.8. A terminal alkyne reacts with a Grignard reagent to form an alkynyl Grignard reagent. Ethylene oxide undergoes ring-opening upon reaction with the alkynyl Grignard reagent, and the product of the overall transformation is a 1° alcohol (after hydrolysis) with two carbon atoms more than the starting alkyne.

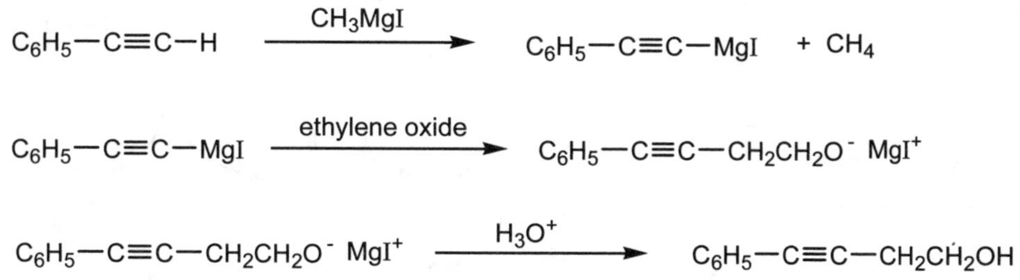
C6H5—C≡C—H
CH3MgI
C6H5—C≡C—MgI + CH4
C6H5—C≡C—MgI
ethylene oxide
C6H5—C≡C—CH2CH2O⁻ MgI⁺
C6H5—C≡C—CH2CH2O⁻ MgI⁺
H3O⁺
C6H5—C≡C—CH2CH2OH

15.9. Organocuprates—both Gilman as well as higher-order reagents—react to replace the halogen atom of organobromides and iodides with the organic group bonded to copper (shown in color in the equation directly below).

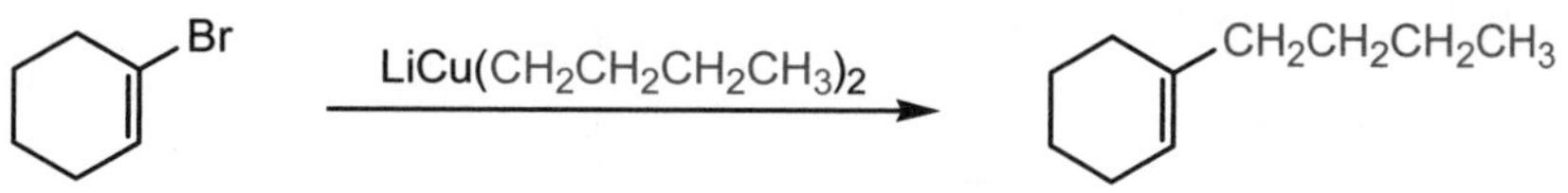
Br
LiCu(CH2CH2CH2CH3)2
CH2CH2CH2CH3

15.10. Higher-order organocuprates react to replace the halogen atom of organobromides and iodides with the organic group bonded to copper (shown in color in the equations below). A higher-order cuprate reacts with an epoxide by opening of the three-membered ring, and the transferred group attaches to the less highly-substituted carbon atom of the original epoxide ring.

a.

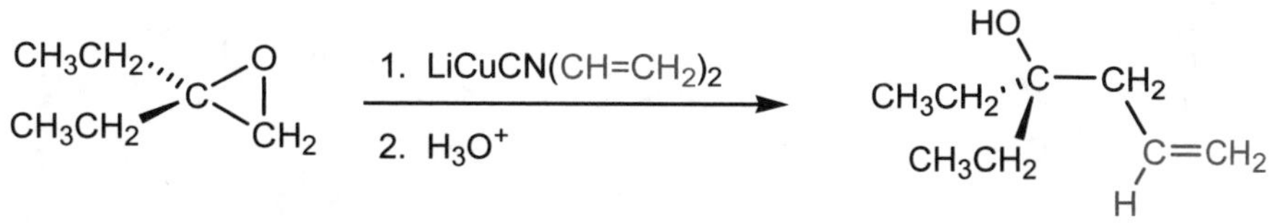
CH3CH2
CH3CH2
O
C
CH2
1. LiCuCN(CH=CH2)2
2. H3O⁺
HO
CH3CH2
C—CH2
CH3CH2
C=CH2
H

b.

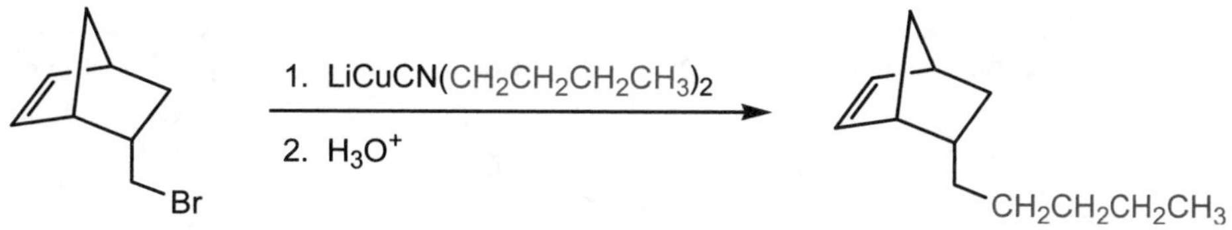
Br
1. LiCuCN(CH2CH2CH2CH3)2
2. H3O⁺
CH2CH2CH2CH3

15.11. Structures **A** through **C** are deduced by applying the descriptions and examples in the accompanying discussion of the text. After two equivalents of triphenylphosphine dissociate to generate $Pd[Ph_3P]_2$, oxidative addition of iodobenzene takes place to form **A**. The butyl group (R′ in the following scheme) is transferred from *B*-butyl-9-BBN, producing **B**. After dissociation of the iodide ion, reductive elimination occurs to form the product, and bis(triphenylphosphine)-palladium(0) is regenerated.

15.11. (continued)

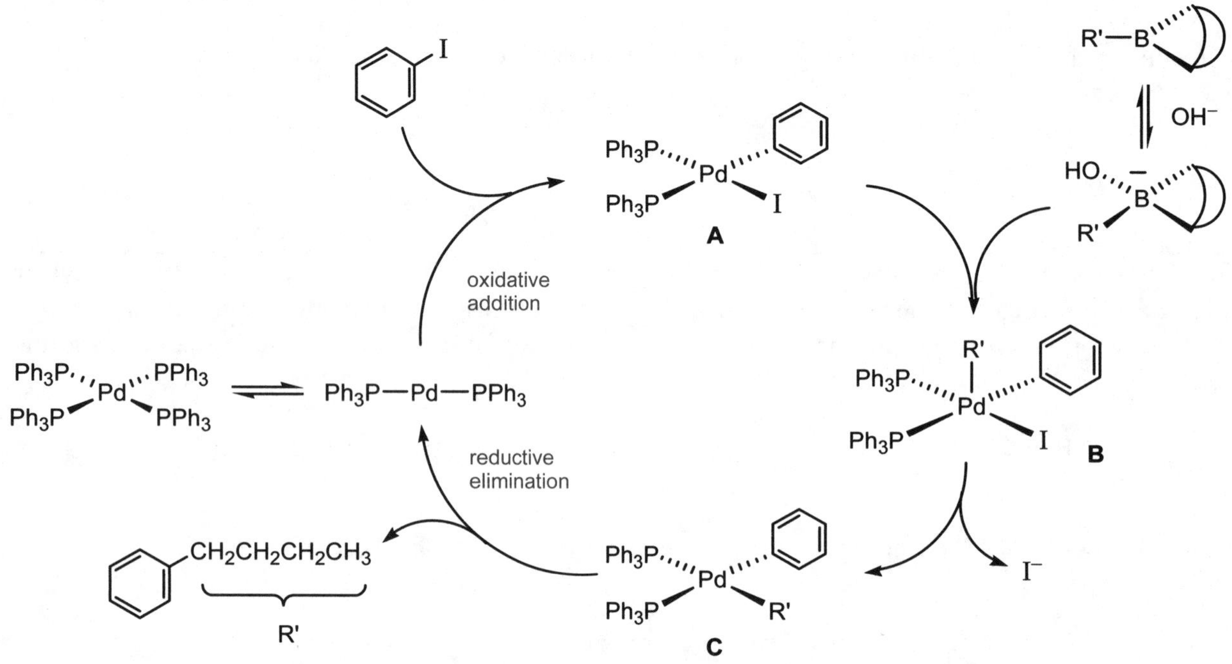

15.12. Catecholborane is prepared from the reaction between borane and catechol, which liberates two molecules of hydrogen. To deduce which starting material is used to make the desired organoborane, remove the boron-containing group and a hydrogen atom from C2, replacing those two groups with a π bond.

The starting materials used to make the required organoboranes are the alkynes shown below at the right.

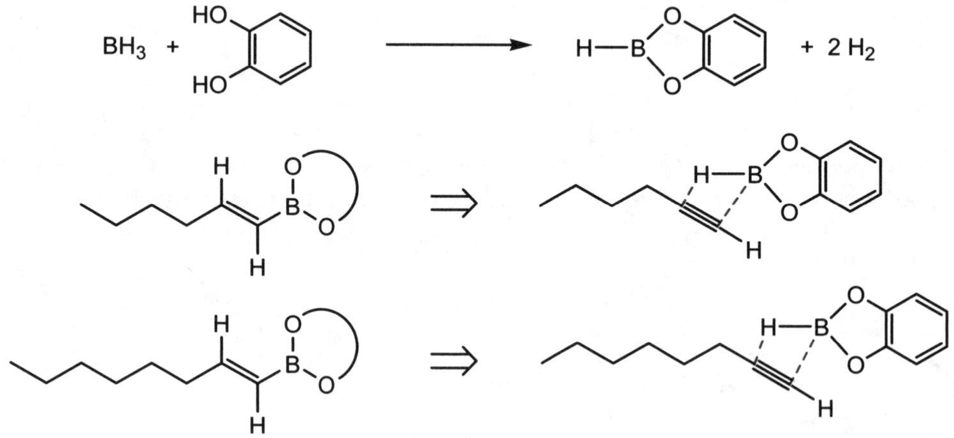

15.13. The equation showing the synthesis of 3-heptanol from bromoethane and pentanal is as follows:

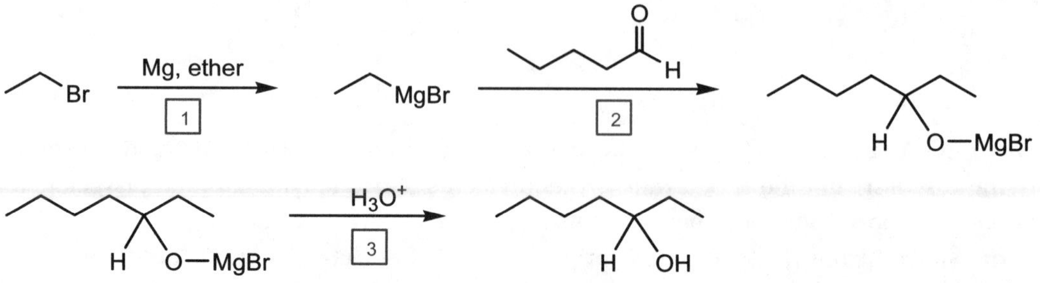

15.14. Alcohols are readily prepared by addition reactions between Grignard reagents and suitable carbonyl compounds or epoxides. For the retrosynthesis, break the molecule into fragments of approximately equal size, each having five or fewer carbon atoms. Bromobenzene is also an allowed starting material. The starting compounds are shown in color. (There may be other ways to synthesize the molecules shown in this exercise; only one method is presented for each.)

a. The product is a 2° alcohol, so the reaction between an aldehyde and a Grignard reagent is needed. The fragment with the oxygen atom is the one derived from the aldehyde reactant.

Retrosynthesis

Synthesis

b. The product is a 1° bromide, which is easily prepared from a 1° alcohol. The reaction between a Grignard reagent and either formaldehyde or ethylene oxide will form the desired alcohol molecule. Because the product has seven carbon atoms, we can break the molecule into a five-carbon and a two-carbon fragment in the retrosynthesis, and then make use of the Grignard reaction with an epoxide.

Retrosynthesis

Synthesis

15.15. Because you were given the retrosynthesis steps, you only have to suggest which reagents are needed to carry out each transformation in the normal or "synthesis" order.

a. Cyclohexene is converted to cyclohexanol by addition of H and OH to the double bond (this transformation can also be accomplished with H_3O^+). Oxidation of the secondary alcohol produces cyclohexanone.

Cyclohexanone reacts with the Grignard reagent derived from bromobenzene. The resulting alcohol is dehydrated using strong acid to yield the alkene product.

15.15. (continued)

b. Cyclohexene is converted to an organoborane derivative using 9-BBN-H. The Suzuki reaction is then used to attach the benzene ring. Free radical bromination is used to create the tertiary organobromide, and strong base promotes dehydrohalogenation to form the alkene product.

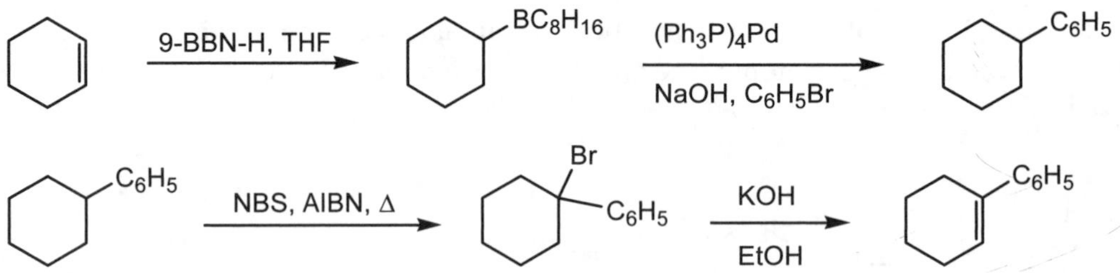

15.16. The bond disconnection next to an alcohol functional group is a reasonable retrosynthetic choice that reveals structures of the alkyl halide and aldehyde reactants needed to make the 2° alcohol product.

Retrosynthesis

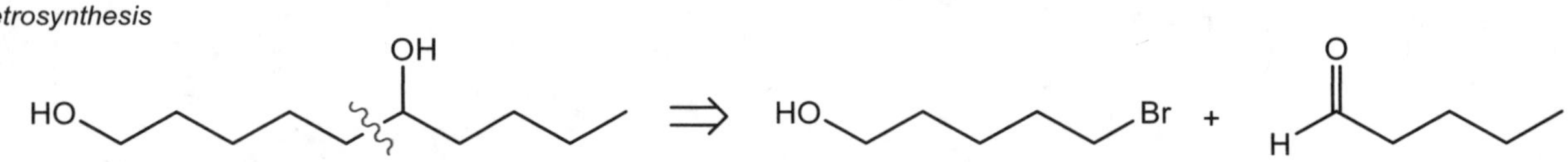

The bromoalkane that is needed has a functional group (alcohol) that is incompatible with the formation of a Grignard reagent, so a protecting group will be required before the Grignard reagent is made. After adding the Grignard reagent to the aldehyde, the protecting group is removed using fluoride ion to yield the diol product. The principal starting materials are shown in color.

Synthesis

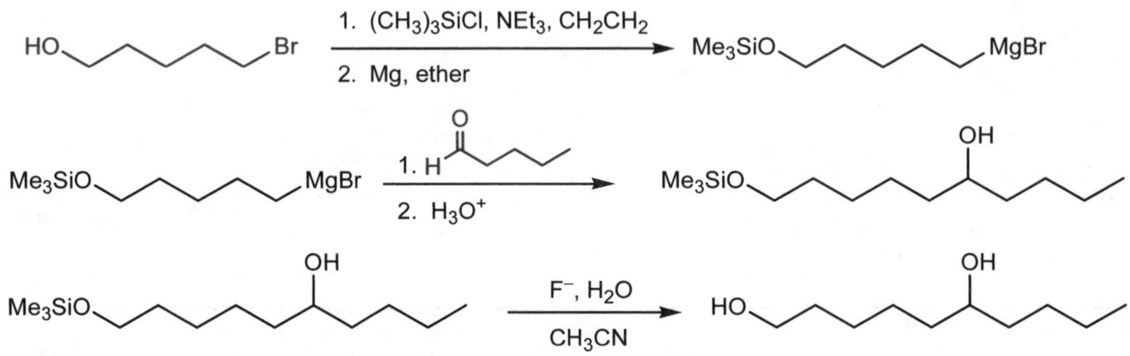

15.17. To draw the structure of the Grignard reagent derived from a specific organohalide, simply insert "Mg" into the carbon–halogen bond (see the procedure outlined in the solution to Exercise 15.2). For parts d and e, draw each structure in its dimensional form to incorporate the given stereochemistry.

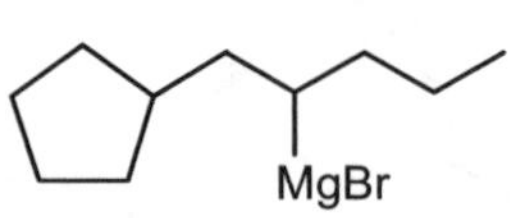

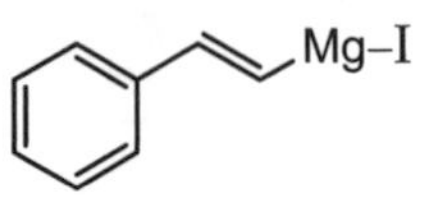

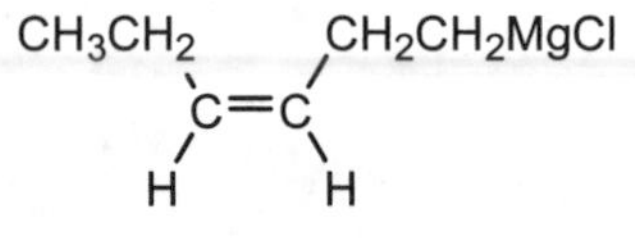

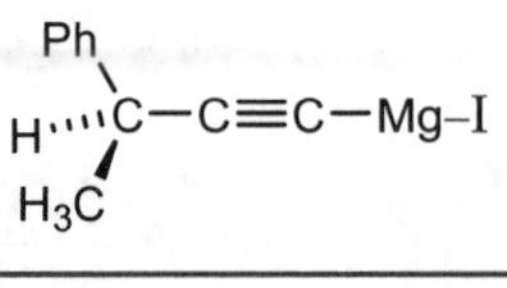

15.18. Groups that interfere with Grignard formation include those with acidic protons and those with carbonyl, nitrile, or nitro groups. Alkyl fluorides do not normally react with magnesium metal to form Grignard reagents. The organohalide in (b.) will form a Grignard reagent.

a.

acidic proton

b.

c.

unreactive halogen

d.

acidic proton

15.19. A Grignard reagent reacts with formaldehyde to generate a primary alcohol after hydrolysis. To decide which alkyl halide is needed, remove the CH_2OH group: the remaining organic fragment, bonded to Cl, Br, or I, is an appropriate starting material.

a.

1. Mg, THF
2. CH_2O
3. H_3O^+

b.

1. Mg, THF
2. CH_2O
3. H_3O^+

c.

1. Mg, THF
2. CH_2O
3. H_3O^+

15.20. A Grignard reagent reacts with an aldehyde other than formaldehyde to yield a secondary alcohol after hydrolysis. To decide which alkyl halide is needed, break the molecule at one of the carbon-carbon bonds adjacent to the OH group. The portion containing the OH group corresponds to the aldehyde needed. The other organic fragment, bonded to Cl, Br, or I, is the starting material from which the Grignard reagent is made. There are often two ways to disconnect the carbon skeleton; only one possible answer is given here.

a.

C_2 aldehyde

1. Mg, THF
2. CH_3CHO
3. H_3O^+

b.

C_4 aldehyde

1. Mg, THF
2. $CH_3CH_2CH_2CHO$
3. H_3O^+

c.

C_3 aldehyde

1. Mg, THF
2. CH_3CH_2CHO
3. H_3O^+

15.21. A Grignard reagent reacts with a ketone to produce a tertiary alcohol after hydrolysis with aqueous ammonium chloride solution. To decide which alkyl halide is required, break the molecule at a carbon-carbon bond adjacent to the OH group. The portion having the OH group corresponds to the required ketone. The other organic fragment, bonded to Cl, Br, or I, is the starting material from which the Grignard reagent is made. There are often three ways to disconnect the carbon skeleton; only one possible answer is given here. Rarely will a bond within a ring be broken in the retrosynthesis.

a.

b.

c.

15.22. Phenylmagnesium bromide reacts as any Grignard reagent does: It adds to the carbon-oxygen double bond of aldehydes and ketones and it opens epoxide rings. It reacts by acid-base reactions with substances that have protons with pK_a values < 40. The products of an acid-base reaction in this case are benzene and the conjugate base of the acidic reactant. If there are no reactive groups, then no reaction occurs.

a.   This is an acid-base reaction; the acidic proton in the reactant is shown in color (alcohol pK_a ~ 15). The proton that is removed results in the formation of benzene from the Grignard reagent.

b.   This is an acid-base reaction; the acidic proton in the reactant is shown in color (pK_a alkyne ~ 25). The proton that is removed results in the formation of benzene from the Grignard reagent.

c.   This is an addition reaction. The phenyl group of the Grignard reagent adds to the carbonyl group of the aldehyde.

15.22. (continued)

d. This is an acid-base reaction; the acidic proton in the reactant is shown in color (pK_a amine ~ 40). The proton that is removed results in the formation of benzene from the Grignard reagent.

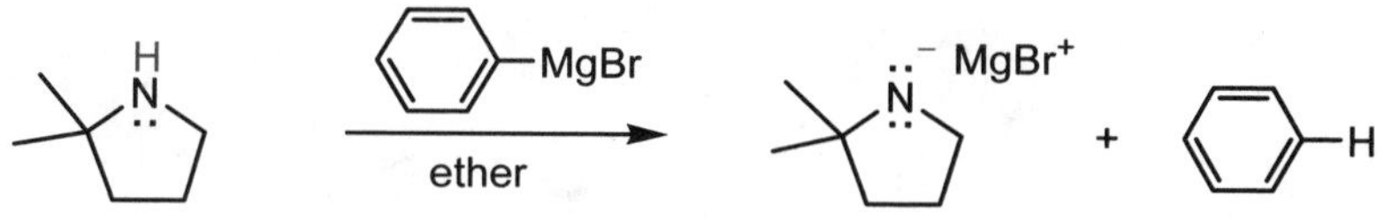

e. This is an acid-base reaction; the acidic proton in the reactant is shown in color (pK_a thiol ~ 10). The proton that is removed results in the formation of benzene from the Grignard reagent.

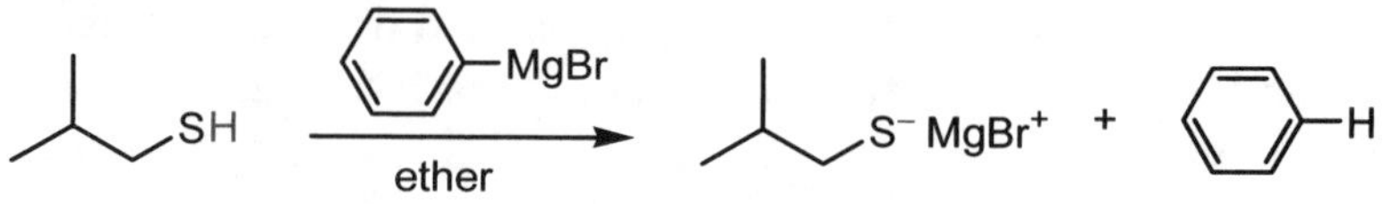

15.23. 1-Bromo-*cis*-2-butene is an allylic halide and it forms a Grignard reagent. Like many allylic species, it can react at two different carbon atoms because the π bond can move with formation of a new bond. In the first step, the ketone carbonyl oxygen atom complexes with the magnesium ion:

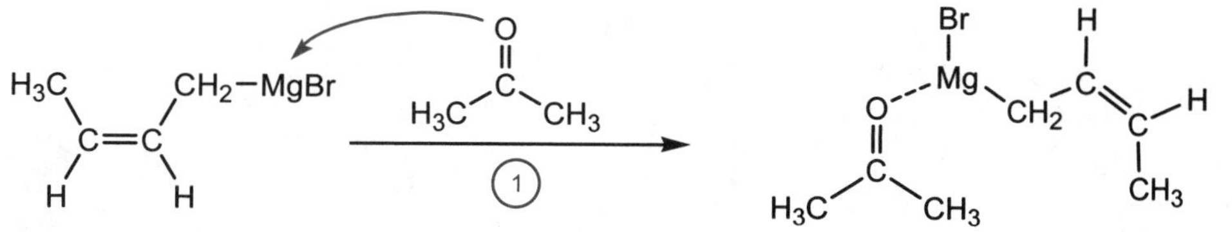

In step 2 (or 2′) the electrons react with the carbonyl carbon atom to form the new bond. In the second pathway shown below (step 2′), the π electrons of the carbon–carbon bond are also involved in the addition reaction. Hydrolysis (steps 3 and 3′) yields the alcohol products.

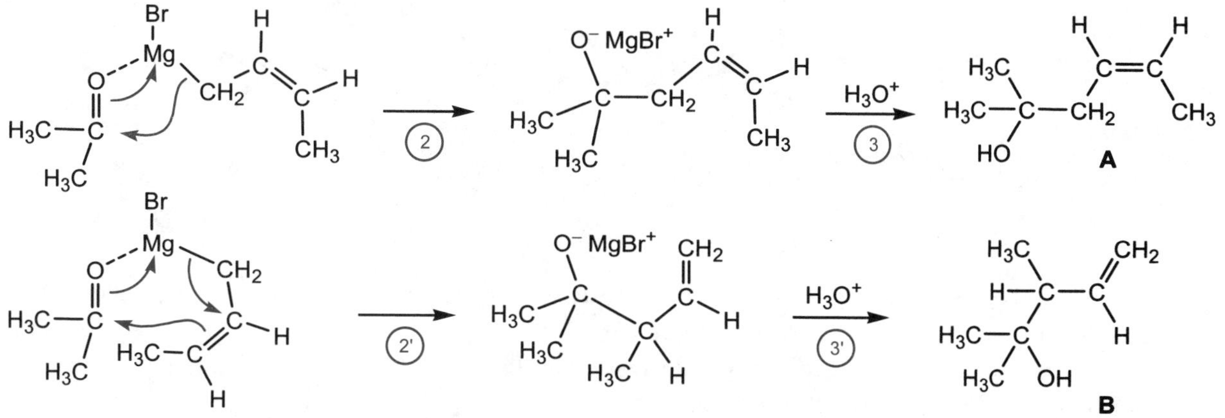

15.24. Protection of an alcohol OH group with the trimethylsilyl group involves stirring the alcohol with chlorotrimethylsilane and a non-nucleophilic base, usually triethylamine. Deprotection of the trimethylsilyl ether involves stirring this silyl compound with fluoride ion.

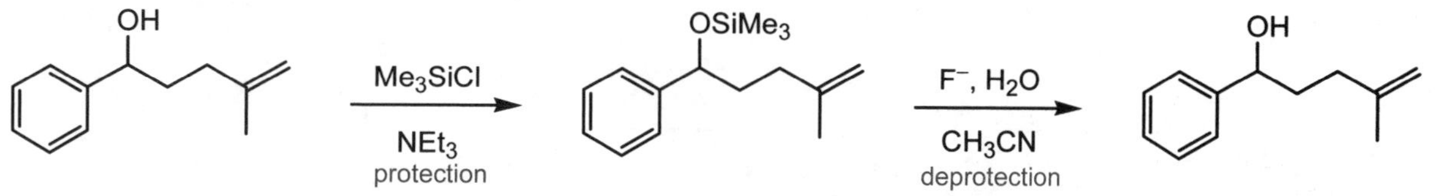

15.24. (continued)

Protection of an alcohol OH group as its benzyl ether requires formation of the alcohol's conjugate base, followed by alkylation with benzyl bromide. Deprotecting the benzyl ether is done using hydrogenolysis.

15.25. An alcohol molecule is prepared using the reaction between a Grignard reagent and a suitable carbonyl compound or epoxide. When the starting material is specified, then the retrosynthetic analysis should be done to generate compounds that are structurally similar to the indicated starting material. The starting compounds are shown in color.

a.   2-Cyclohexylethanol from bromocyclohexane. The product is a 1° alcohol that has two carbon atoms more than the given starting material. The reaction between ethylene oxide and a Grignard reagent permits such a transformation to be carried out.

b.   1-Hexanol from 1-pentene. The product, a 1° alcohol, has one carbon atom more than the starting material, so the reaction between a Grignard reaction and formaldehyde is appropriate. The bromide required is made from the specified starting material, 1-pentene by anti-Markovnikov (radical) addition of HBr.

c.   1-Cyclohexylethanol from cyclohexanol. The product is a 2° alcohol with the same ring skeleton that appears in the starting material. An aldehyde corresponding to the acyclic portion of the product is required to react with a Grignard reagent derived from bromocyclohexane, which is prepared from cyclohexanol by means of a substitution reaction.

15.25. (continued)

d.   3-Heptanol from 1-iodobutane. The product is a 2° alcohol that has three carbon atoms more than the given starting material. The reaction between propanal and the Grignard reagent generated from 1-iodobutane will produce the desired heptanol isomer.

Retrosynthesis

Synthesis

15.26. To predict the products of reactions that involve organometallic compounds, you first must know what general constraints exist. For instance, most organometallic compounds react with acidic groups to remove the proton in an acid-base reaction. For the reactions described in this chapter, the following summary can be made:

- Grignard reagents and organolithium compounds are prepared from organohalides and Mg and Li, respectively.

- Alkynyl Grignard reagents are prepared from alkynes plus another Grignard reagent by an acid-base reaction.

- Alkynyl metal compounds are prepared from the reaction of alkynes with a strong base (LDA in THF) and these organometallic compounds will react with primary alkyl halides to form new carbon–carbon bonds.

- Grignard reagents (and organolithium compounds) add to the carbonyl group of aldehydes and ketones to form alcohols after workup with aqueous acid.

- Grignard reagents (and organolithium compounds) add to the carbonyl group of carbon dioxide to form carboxylic acids after workup with aqueous mineral acid.

- Grignard reagents, organolithium compounds, and organocuprates open epoxide rings at the less highly substituted carbon atom to form alcohols after workup with aqueous acid.

- Organocuprates are prepared from organolithium compounds by treatment with a copper(I) halide or copper(I) cyanide salt.

- Organocuprates react with primary and secondary alkyl halides to form new carbon–carbon bonds.

- Organoboranes react with aryl halides in the presence of Pd(0) catalysts to form alkyl and alkenyl benzene derivatives.

As for stereochemistry, the creation of new chiral centers means that the product(s) are either racemic or meso. When a chiral epoxide ring is opened, the configuration of the carbon atom at which the reaction takes place is inverted; the configuration of the other carbon atom is retained.

a.   The first step generates the Grignard reagent. Reaction with ethylene oxide generates a primary alcohol with two additional carbon atoms. This product is achiral.

achrial

15.26. (continued)

b. The first step generates the lithium derivative of the alkyne. Alkylation occurs in step 2. The third step is partial reduction of the triple bond to form the corresponding cis alkene.

1. LDA, THF

2. CH_3CH_2Br
3. H_2, Lindlar catalyst

achrial

c. The first step generates the Grignard reagent. Reaction with the epoxide molecule generates a chiral secondary alcohol. Reaction of the Grignard reagent occurs at the primary carbon atom of the epoxide. The chiral center therefore retains its stereochemical configuration.

1. Mg, THF

2.

3. H_3O^+

retention

d. The organocuprate reacts with the primary alkyl iodide (the carbon skeleton is shown in color) to couple the two hydrocarbon fragments together.

$Li[(CH_3)_2Cu]$

achrial

e. An organolithium compound adds to the carbonyl group of a ketone to form a tertiary alcohol after workup with aqueous acid. A new chiral carbon atom is generated, so a racemic mixture is obtained.

1. BuLi

2. H_3O^+

racemic

f. Wilkinson's catalyst is used to hydrogenate alkene double bonds. This product is achiral.

$RhCl(Ph_3P)_3$, H_2

benzene, ethanol

achrial

g. The first step generates an organoborane derivative of the alkene. The second step is the Suzuki reaction, which couples the two hydrocarbon units together. This product is achiral.

1. 9-BBN-H

2. PhBr, $Pd(Ph_3P)_4$
   NaOH

achrial

h. The first step generates the Grignard reagent. Reaction with carbon dioxide generates a carboxylic acid after workup with aqueous acid. This product is achiral.

1. Mg, THF

2. CO_2
3. H_3O^+

achrial

15.26. (continued)

i. The first step generates the organolithium compound. Reaction with copper(I) cyanide generates a higher order organocuprate reagent. This product is achiral.

$$CH_3CH_2CH_2{-}Br \quad \xrightarrow[\text{2. CuCN}]{\text{1. Li, hexane}} \quad (CH_3CH_2CH_2)_2Cu(CN)Li \quad \text{achrial}$$

j. The first step generates the Grignard reagent. Reaction with the aldehyde generates a secondary alcohol after workup with aqueous acid. A new chiral carbon atom is generated, so a racemic mixture is obtained. The fourth step produces the trimethylsilyl derivative of the alcohol. This reaction does not affect the stereochemistry of the chiral carbon atom, so the final product is also racemic.

15.27. With constraints on the number of carbon atoms that each starting material can have, the retrosynthesis should break the molecule into pieces that have appropriate sizes within the given parameters. Using organocuprate reagents to couple two groups is often straightforward if one portion is derived from an organolithium compound and the other reactant is a 1° alkyl bromide or iodide. The starting compounds are shown in color.

a. The retrosynthetic disconnection creates two fragments, one with five carbon atoms (do not ignore the methyl group when counting carbon atoms) and one with six. The portion without the methoxy group is chosen as the one from which the organocuprate is derived because the organometallic reactant will then have the metal ion associated with a 2° carbon atom and the other portion can be 1° iodoalkane.

Retrosynthesis

Synthesis

b. The retrosynthetic disconnection creates two fragments, each with six carbon atoms. Making the organocuprate from the benzene ring allows the other fragment to be a primary iodoalkane.

Retrosynthesis

Synthesis

15.27. (continued)

c.   The retrosynthetic disconnection creates two fragments, each with five carbon atoms. Making the organocuprate from the cyclic portion allows the other fragment to be a primary iodoalkane.

Retrosynthesis

Synthesis

15.28. This transformation, which is similar to the reaction that takes place between a Grignard reagent and an epoxide, provides a way to extend a carbon chain by three carbon atoms. Ring opening occurs because the four-membered ring has significant strain energy. Furthermore, the magnesium ion associates with the oxygen atom of the oxetane ring to form a good leaving group that facilitates ring-opening. Even so, this transformation requires fairly high temperatures to be successful, so it is not as general as the reaction that epoxides undergo.

15.29. To assess the atom economy of a pair of reactions in a semi-quantitative sense, write a complete equation for each (if possible) and consider what by-products are formed in each process. Addition reactions normally have the best atom economy, especially if all of the atoms of a reactant add to the organic substrate. For the transformations in each part of this exercise, the reaction in the colored box has the better atom economy of each pair being considered.

a.   Oxymercuration is an inherently poor procedure because mercury is formed as a product. Disposal of this highly toxic material is costly.   Markovnikov addition of water is highly efficient.

b.   Hydroboration is a fairly benign process, but it does generate boric acid as a by-product. Electrophilic addition of water is highly efficient.

15.29. (continued)

c.   Only one-third equivalent of PBr₃ is needed to convert an alcohol to its alkyl bromide derivative.

d.   Catalytic hydrogenation is highly efficient because the catalyst can be recovered and recycled.

$$CH_3CH_2CH_2-C{\equiv}CH \xrightarrow[\text{2 } tert\text{-BuOH}]{\text{2 Na, liquid NH}_3} CH_3CH_2CH_2-CH{=}CH_2 \quad + \text{ 2 NaO-}t\text{-Bu}$$

$$CH_3CH_2CH_2-C{\equiv}CH \xrightarrow[\text{Lindlar catalyst}]{\text{H}_2} CH_3CH_2CH_2-CH{=}CH_2$$

15.30. When you perform a reaction in which you know what the product will be (or is expected to be), you want to focus on those spectroscopic features that are unique for the starting material and product. Although many subtle changes will undoubtedly occur in several parts of the spectrum, it helps to look for a limited number of specific changes that will be readily observed.

a.   The reaction between a Grignard reagent and ethylene oxide adds two carbon atoms to the starting organohalide and yields a primary alcohol. The calculated analytical data are listed along with the observed data given in parentheses, and the results show that the expected product was likely formed. The starting material has few spectroscopic properties that differ from those of the product molecule—that is, the absorption bands associated with the methoxy group and benzene ring will be similar for both reactant and product. Thus, the data that need to be evaluated will be the new bands that appear in the hydroxyethyl group of the product, which are listed below.

$C_9H_{12}O_2$ C, 71.03%; H, 7.95%
(C, 71.15%; H, 7.87%)

IR spectrum   **¹H NMR spectrum**   **¹³C NMR spectrum**

15.30. (continued)

b. The reaction between a Grignard reagent and an aldehyde yields a secondary alcohol. The calculated analytical data are listed along with the observed data given in parentheses. The results show that the expected product was likely formed.

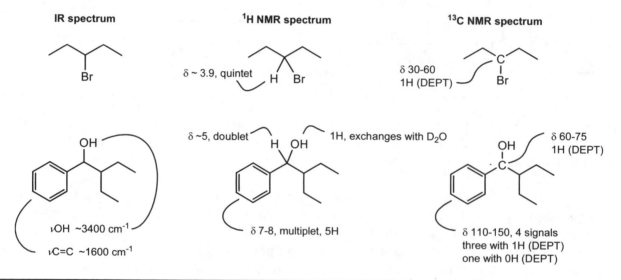

$C_{12}H_{18}O$ C, 80.3%; H, 10.1%
(C, 80.8%; H, 10.3%)

The starting material has few spectroscopic properties that differ from those of the product molecule. The addition of a benzene ring and an OH group to the starting compound accounts for the most obvious new absorption bands.

IR spectrum	^{1}H NMR spectrum	^{13}C NMR spectrum

δ ~ 3.9, quintet

δ 30-60
1H (DEPT)

δ ~5, doublet   1H, exchanges with D_2O

δ 60-75
1H (DEPT)

νOH ~3400 cm^{-1}
νC=C ~1600 cm^{-1}

δ 7-8, multiplet, 5H

δ 110-150, 4 signals
three with 1H (DEPT)
one with 0H (DEPT)

15.31. By looking at the integrated intensity values of the given proton NMR spectrum, you will see that the by-product molecule has 10 protons, which is one more than the starting organobromide has. Knowing that Grignard reagents often engage in acid-base reactions, it is likely that this by-product has been formed by the reaction of the Grignard reagent with a proton source.

$$CH_3-C\equiv C-CH_2CH_2CH_2MgBr \xrightarrow{H_3O^+} CH_3-C\equiv C-CH_2CH_2CH_3$$

If the acid-base reaction has occurred, then the byproduct will have a propyl group, which should manifest itself as two triplets and a sextet at appropriate chemical shift values. In fact, such peaks are seen in the spectrum, as summarized below, so we conclude that the by-product is 2-hexyne.

Chemical shift (ppm)	Integrated intensity	Assignment	Multiplicity	No. adjacent protons (n)	
2.1	2	CH_2	triplet	2	
1.8	3	CH_3	singlet	0	propyl group
1.4	2	CH_2	sextet	5	
0.95	3	CH_3	triplet	2	

15.32. Because you do not know the structure of the aldehyde used in your reaction, the product should be treated the same as you would for an unknown compound. You do know that it contains a phenyl group because you started with the Grignard reagent made from bromobenzene. Follow the procedure outlined in the solution to Exercise 14.23.

(1) Examine the ^{13}C NMR to determine the number and types of carbon atoms present. There are three types (C=O, aromatic, and aliphatic) consisting of nine peaks, four of which represent the benzene ring. Start with the assumption that the compound has 11 carbon atoms (6 for the benzene ring plus one for the carbonyl group and four aliphatic carbon atoms).

(2) Examine the integrated intensity values in the proton NMR spectrum. The total is 14 (5:2:2:2:3, left to right). Start with the assumption that the compound has 14 hydrogen atoms.

(3) Use the high-resolution data to determine the molecular formula with the initial assumption that there are 11 C atoms, 14 H atoms, and some number of oxygen atoms. For $C_{11}H_{14}O$, the calculated high resolution MW is 162.104 (observed MW = 162.105). This formula is consistent with the data.

(4) Calculate the number of unsaturation sites: $[2(11) + 2 - 14]/2 = 5$ sites of unsaturation. Four sites are in the benzene ring and one is the carbonyl group, which accounts for all of the sites.

(5) Examine the IR spectrum. The strong absorption at 1700 cm^{-1} is attributed to a C=O stretching vibration (ketone or aldehyde) that is conjugated with the aromatic ring. The carbonyl resonance in the ^{13}C NMR spectrum indicates that this carbonyl group is a ketone (no attached H).

(6) Examine the proton NMR spectrum. You already know that the molecule contains a benzene ring, so you can focus on the aliphatic region. The data, summarized below, indicate that a butyl group is present in addition to a monosubstituted benzene ring.

Chemical shift (ppm)	Integrated intensity	Assignment	Multiplicity	No. adjacent protons (n)	
7.4-8	5	ArH	multiplet	--	
2.95	2	CH$_2$	triplet	2	
1.70	2	CH$_2$	quintet	4	butyl group
1.40	2	CH$_2$	sextet	5	
0.90	3	CH$_3$	triplet	2	

Assembling these fragments (C=O, C$_6$H$_5$, CH$_2$CH$_2$CH$_2$CH$_3$) generates the structure of this unknown compound, **1-phenyl-1-pentanone**.

15.33. Follow the procedure outlined in the solution to Exercise 15.32.

(1) Examine the ^{13}C NMR to determine the number and types of carbon atoms present. There are two types (aromatic and aliphatic) consisting of eight peaks, four of which represent a benzene ring. Start with the assumption that the compound has 10 carbon atoms (6 for the benzene ring plus four aliphatic carbon atoms).

(2) Examine the integrated intensity values in the proton NMR spectrum. The total is 14 (5:2:2:2:3, left to right). Start with the assumption that the compound has 14 hydrogen atoms.

(3) Use the high-resolution data to determine the molecular formula with the initial assumption that there are 10 C atoms, 14 H atoms, and some number of O atoms. For $C_{10}H_{14}$, the calculated high resolution MW is 134.110 (observed MW = 134.110). This formula is consistent with the data.

(4) Calculate the number of unsaturation sites: $[2(10) + 2 - 14]/2 = 4$ sites of unsaturation, which are in the benzene ring.

15.33.   (continued)

(5) Examine the proton NMR spectrum. You already know that the molecule contains a benzene ring, so you can focus on the aliphatic region. The data, summarized below, indicates that a butyl group is present in addition to a monosubstituted benzene ring.

Chemical shift (ppm)	Integrated intensity	Assignment	Multiplicity	No. adjacent protons (n)
7.1-7.3	5	ArH	multiplet	--
2.60	2	CH₂	triplet	2
1.60	2	CH₂	quintet	4
1.35	2	CH₂	sextet	5
0.90	3	CH₃	triplet	2

Assembling these fragments (C_6H_5 and $CH_2CH_2CH_2CH_3$) generates the structure of this unknown compound, butylbenzene.

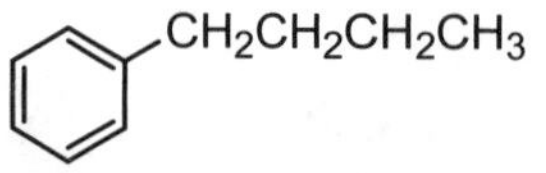

To synthesize this compound from bromobenzene and an alkene, you can make use of the Suzuki reaction. Treating 1-butene with catecholborane yields one reactant. Bromobenzene couples with this organoborane to yield the desired butylbenzene molecule. The starting compounds are shown in color.

Retrosynthesis

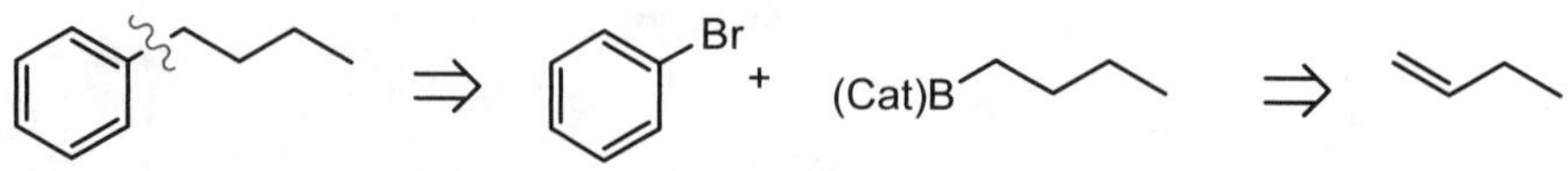

Synthesis

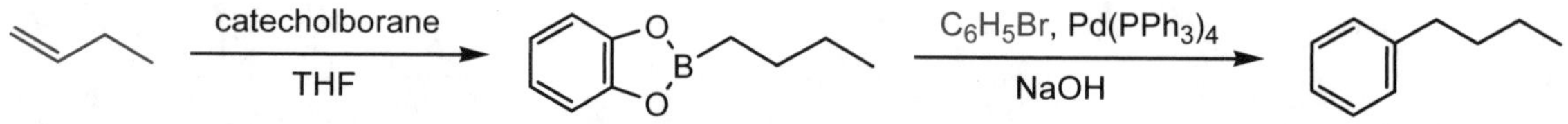

15.34. For many molecules, the alcohol functional group defines a place at which to make a logical disconnection in the retrosynthesis. If an OH group is not present, consider how the functional group that does exist can be made from an alcohol. For instance, alcohols are oxidized to form carbonyl compounds, and alcohols are dehydrated to form alkenes.

If the synthetic goal has a ring as part of the carbon skeleton, consider ways to include the ring as a starting material. For instance, benzene derivatives are often made conveniently using the phenyl Grignard reagent. The Diels-Alder reaction (Section 10.4b) is a useful way to make six-membered rings from acyclic starting materials.

There are many ways to prepare the molecules shown below. Only one route for each is presented. The starting compounds are shown in color.

a.   The alkene can be made by dehydrating an alcohol; the alcohol in turn is accessible by using a Grignard reagent with a ketone.

Retrosynthesis

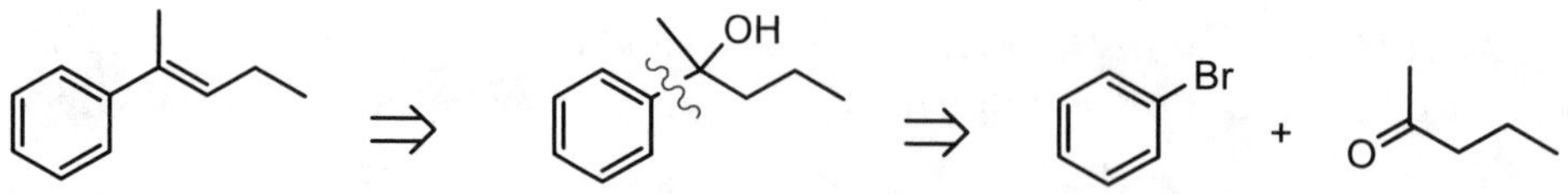

15.34. (continued)

Synthesis

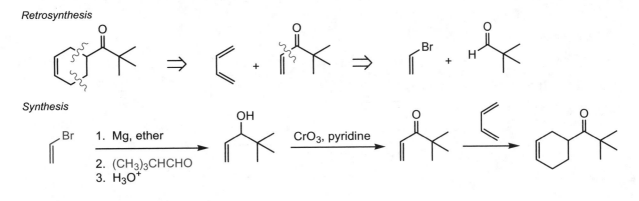

b. The ketone and aldehyde groups can be made by oxidizing alcohols. The diol in turn is accessible by using a Grignard reagent with an aldehyde having a protected alcohol group.

Retrosynthesis

Synthesis

c. The cyclohexane ring can be made by means of the Diels-Alder reaction (Section 10.4b). The dienophile is made by oxidizing an alcohol made from a Grignard reaction with an aldehyde.

Retrosynthesis

Synthesis

d. This alcohol can be made by opening an epoxide ring with a Grignard reagent. Whenever a carbon-containing substituent is attached to the carbon atom *adjacent* to a carbon atom bearing an OH group, consider using the epoxide ring-opening reaction as a way to prepare the alcohol.

Retrosynthesis   *Synthesis*

15.34. (continued)

e. The alcohol product can be made by the reaction of a Grignard reagent with an aldehyde; the needed alkyl bromide is made using a Grignard reagent with acetaldehyde followed by conversion of the alcohol to the bromoalkane.

Retrosynthesis

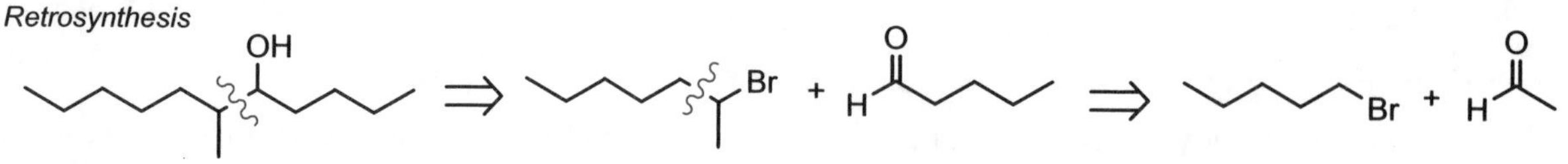

Synthesis

- - - - - - - - - - - - - - - - - - - -

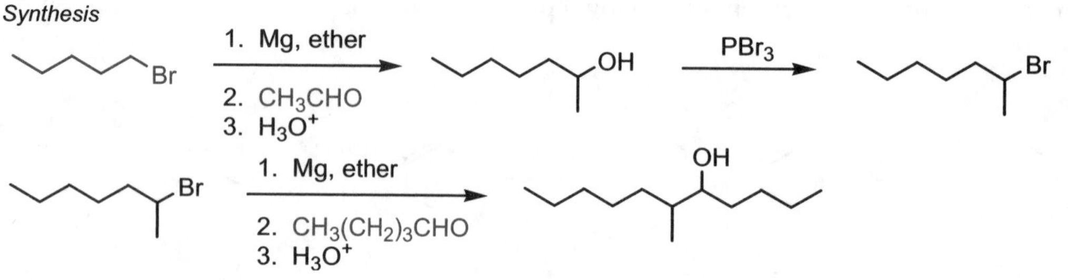

f. This alcohol can be made by opening an epoxide ring with a Grignard reagent. Whenever a carbon-containing substituent is attached to the carbon atom adjacent to a carbon atom bearing an OH group, consider using the epoxide ring-opening reaction as a way to prepare the alcohol.

Retrosynthesis   *Synthesis*

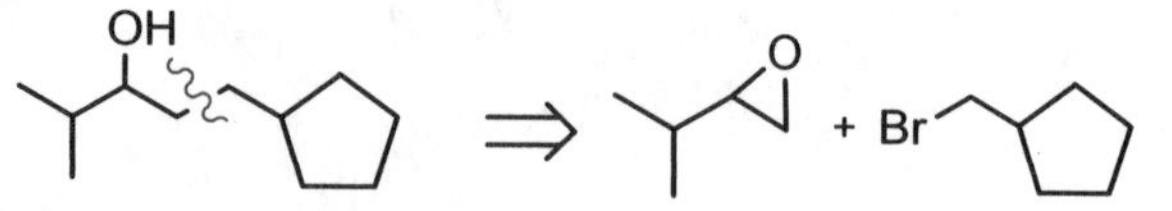

- - - - - - - - - - - - - - - - - - - -

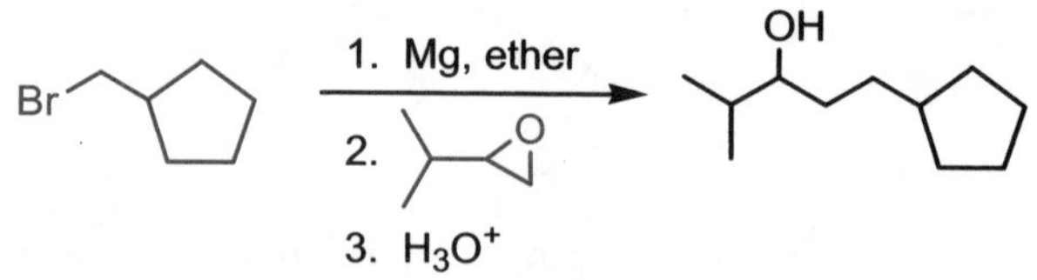

g. The alkene can be made by dehydrating an alcohol; the alcohol in turn is accessible by using a Grignard reaction with an aldehyde.

Retrosynthesis

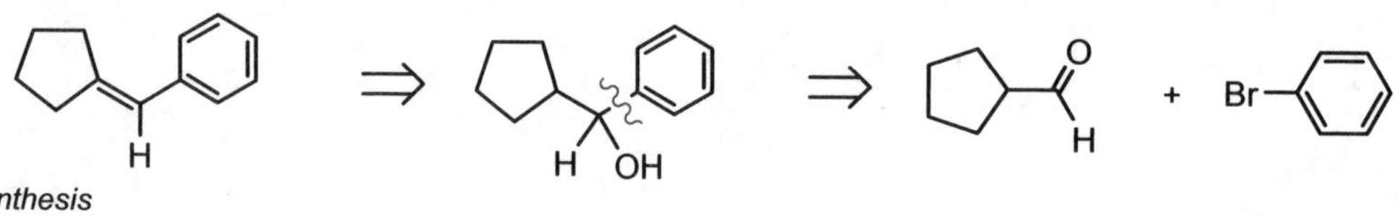

Synthesis

- - - - - - - - - - - - - - - - - - - -

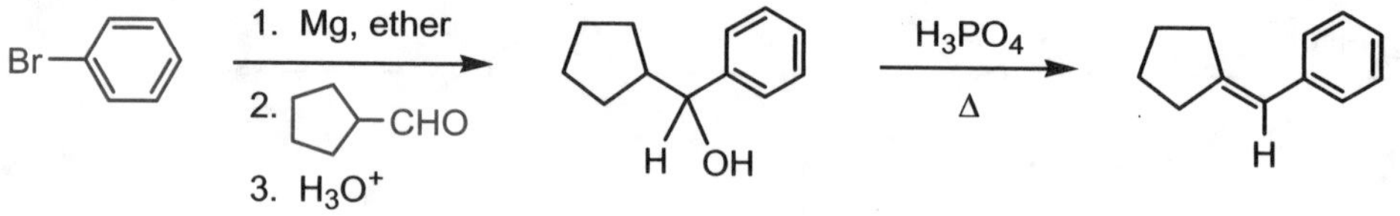

h. The azidoalkane can be made via a substitution reaction from an alcohol; the alcohol in turn is accessible by the reaction of a Grignard reagent with an aldehyde.

Retrosynthesis

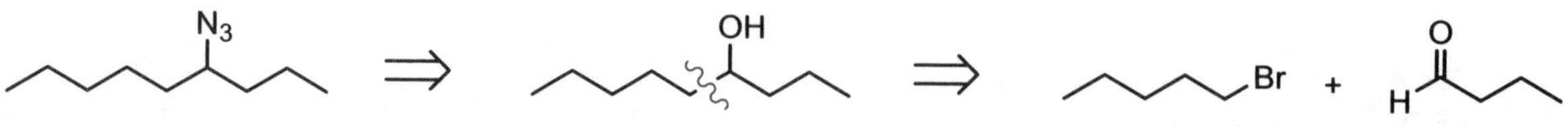

Synthesis

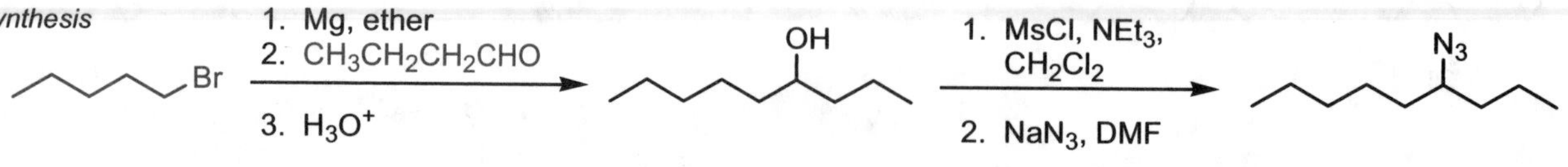

Solutions to Exercises for CHAPTER 16

ASYMMETRIC REACTIONS AND SYNTHESIS

16.1. Brucine has seven stereogenic centers, and quinine has five (it is helpful to expand the structures to show the protons): They are carbon and nitrogen atoms with sp^3 hybridization having four different groups attached (an unshared electron pair on N counts as a group). In the following structures, these centers are marked with colored asterisks. Nitrogen atoms that are chiral must be locked into a specific orientation to prevent pyramidal inversion (Section 4.2f). Note that a nitrogen atom adjacent to a carbonyl group has sp^2 hybridization and is not chiral.

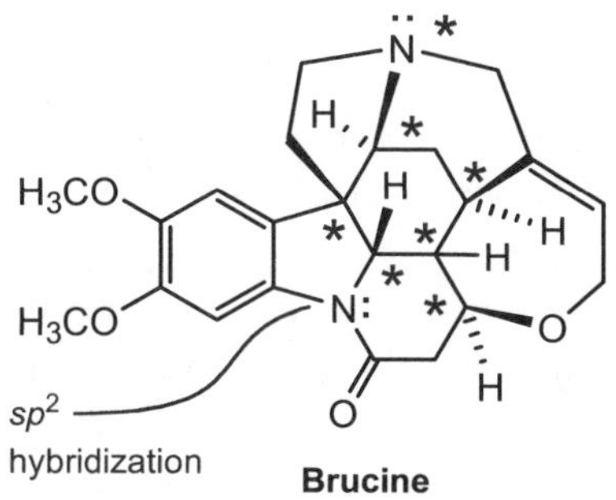

Brucine

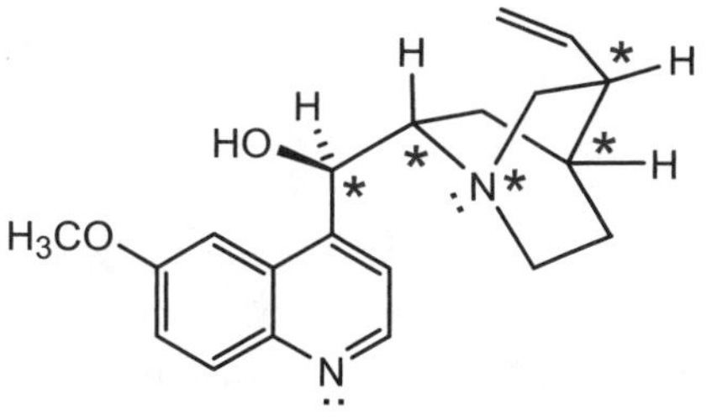

Quinine

16.2. The resolution of a racemic mixture of amine molecules is carried out by forming diastereomeric salts with the chiral sulfonic acid shown below. Its chiral carbon atom has the (*R*) configuration. Note that this chiral carbon atom is attached to four other carbon atoms, so to assign its configuration, you have to consider the next shell of substituent atoms, which is listed for each position in the expanded structure (middle).

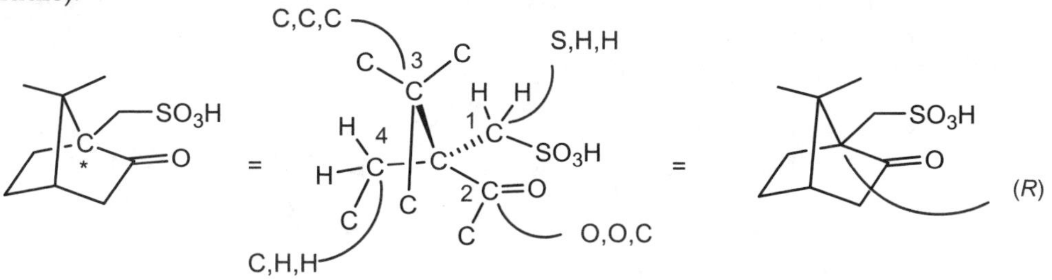

In the resolution process, the amine mixture reacts with the acid to form diastereomeric salts, which are then subjected to fractional crystallization.

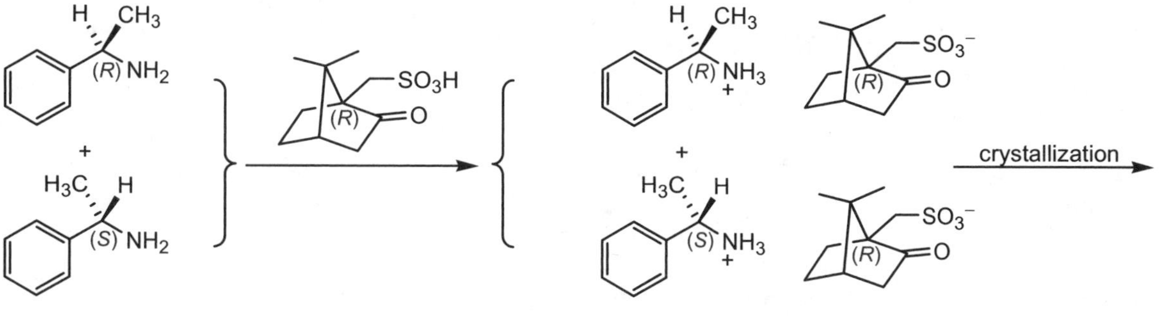

16.2. (continued)

The separated diastereomers are each then treated with aqueous base to liberate the corresponding chiral amines. This step yields the sodium salt of the chiral sulfonic acid, which can be recovered by acidifying it with aqueous hydrochloric acid.

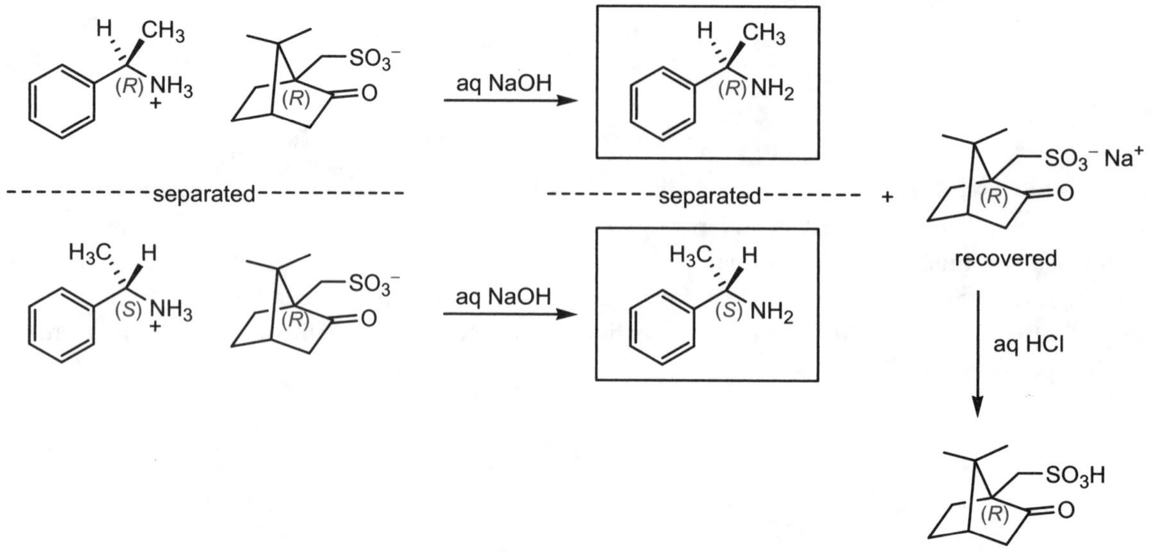

16.3. The assignment of pro-(R) and pro-(S) is made by looking at the configuration of the new stereogenic center in each product molecule. The hydrogen atom that was replaced to form the new (R) center is pro-(R), and the other one is pro-(S).

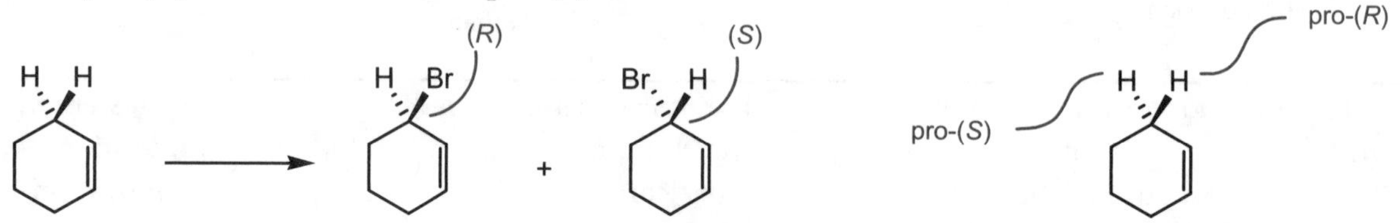

16.4. The assignment of pro-(R) and pro-(S) is made by looking at the configuration of the new stereogenic center in each product molecule. The double bond face at which addition yielded the new (R) center is pro-(R), and the other one is pro-(S).

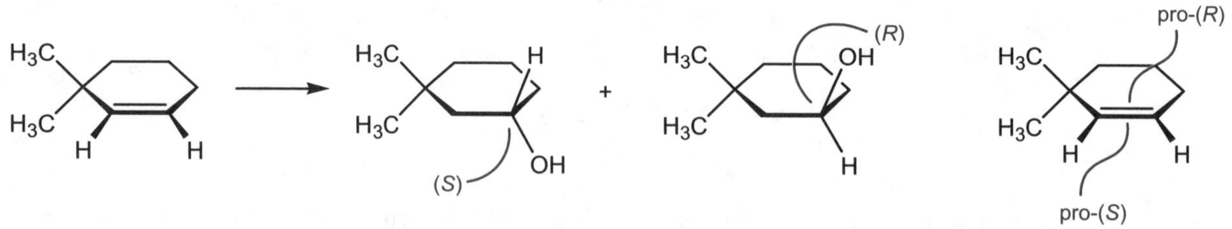

16.5. The de is the numeric difference between the amounts of each diastereomer, expressed as a percentage. The major diastereomer is 80% of the mixture, and the minor one is 20%. The de is therefore $80\% - 20\% = 60\%$.

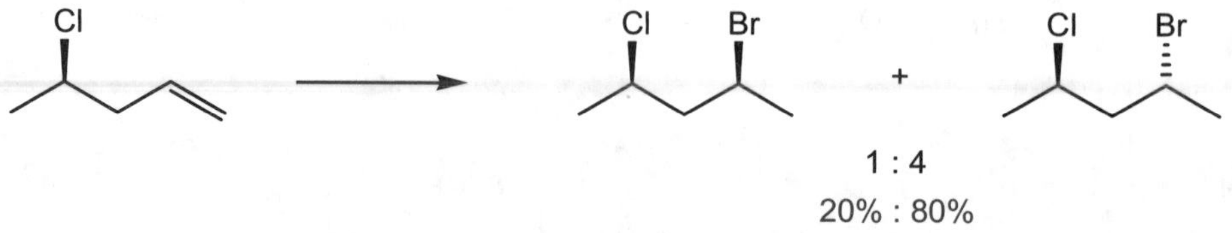

1 : 4

20% : 80%

16.6. The assignments of the configurations are carried out as described in the solutions to Exercises 4.7 and 4.9.

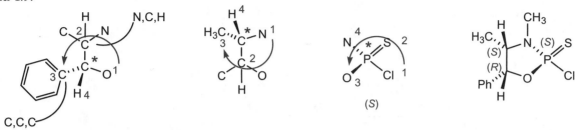

(S) as drawn, but the priority 4 group is coming forward, so the actual configuration is (R).

(R) as drawn, but the priority 4 group is coming forward, so the actual configuration is (S).

16.7. From the integrated intensity values, calculate the intensity ratio. The amount of each divided by the sum of their amounts is equal to the percentage of each. The difference between these percentage values is the ee.

13.30

3.16

$$\frac{13.3}{3.16} = \frac{4.20}{1.00}$$

$$\frac{4.20}{4.20 + 1.00} = 81\%$$

$$\frac{1.00}{4.20 + 1.00} = 19\%$$

$$81 - 19 = \boxed{62\% \text{ ee}}$$

16.8. Follow the procedure outlined in Example 16.1. This needed alkene is an unsaturated carboxylic acid, so Noyori's chiral ruthenium catalyst is a suitable choice (Table 16.2).

(BINAP*)Ru(OAc)₂

100 atm H₂, CH₃OH

16.9. Follow the procedure outlined in Example 16.2. The needed alkene in part (b.) should be cis in order to achieve a high degree of enantioselectivity.

a.

1. (Ipc*)BH₂
2. crystallize

3. 4 equiv CH₃CHO
4. H₂O₂, OH⁻

16.9. (continued)

b.

16.10. To determine the structure of the needed alkene, replace the oxygen atom of the epoxide with a carbon–carbon double bond. To choose the reagent needed to prepare the chiral epoxide, make use of the data in Table 16.2.

a.

b.

c.

16.11. Grignard reagents react with an epoxide at the less highly-substituted end of the three-membered ring after complexation between the oxygen and magnesium atoms. The configuration of the carbon atom at which RMgX reacts undergoes inversion; the configuration of the other carbon atom is retained.

16.12. The first step in the given synthesis is a hydroboration reaction, which gives the organoborane derivative of methylcyclohexene. Crystallization (step 2) occurs with formation of the dimeric organoborane species, which is a single diastereomer.

16.12. (continued)

Step 3, which consists of treating the pure diastereomer with acetaldehyde, yields the enantiomerically pure boronate derivative. The chiral pinene molecules are liberated in this step and can be recovered.

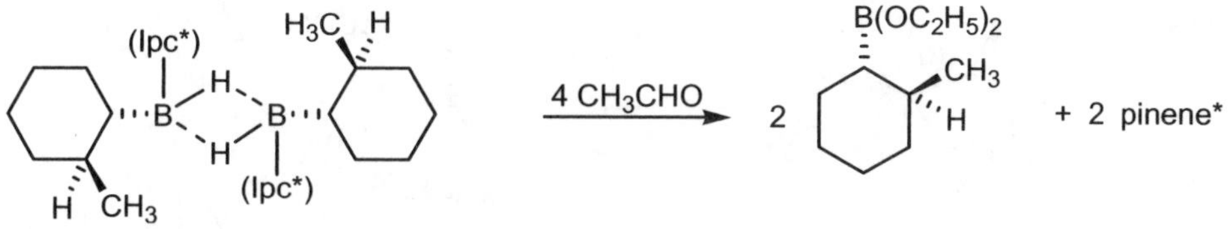

The boronate is treated with hydrogen peroxide and base, which yields the chiral alcohol, and the Swern oxidation procedure converts the secondary alcohol to the corresponding ketone.

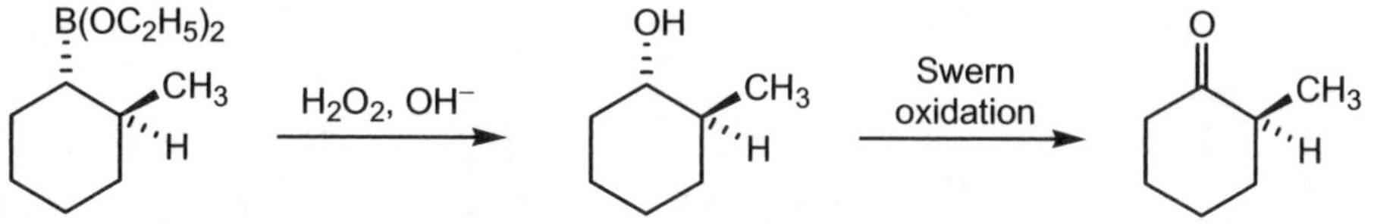

16.13. Follow the procedures outlined in Examples 16.4-16.7. The starting compounds are shown in color.

a. The nitrile can be prepared from an alcohol using a substitution reaction, and chiral alcohols are readily prepared using chiral organoborane reagents. The substitution reaction will employ S$_N$2 conditions in order to control the stereochemistry of the reaction, therefore the alcohol prepared from the alkene must have a configuration opposite that of the product.

Retrosynthesis

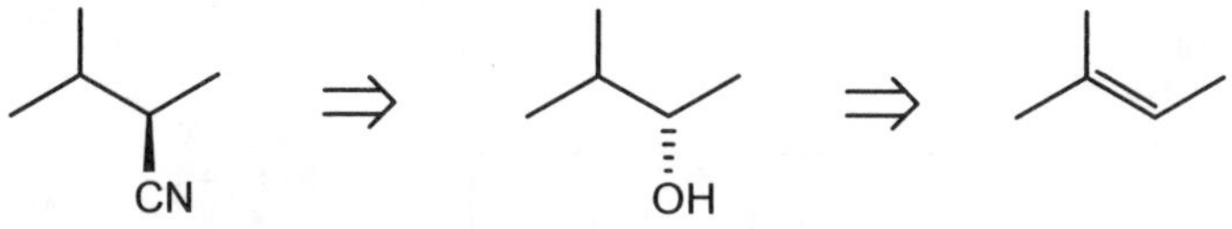

Synthesis

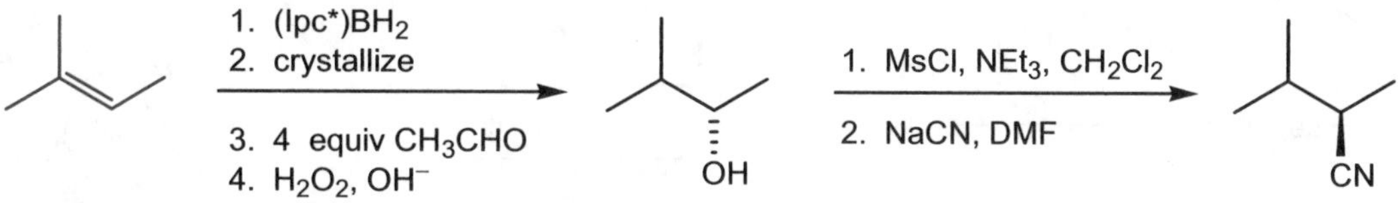

b. The dimethoxy compound in this exercise can be made by alkylation of the corresponding diol, which in turn can be made in chiral form via the Sharpless AD procedure.

Retrosynthesis

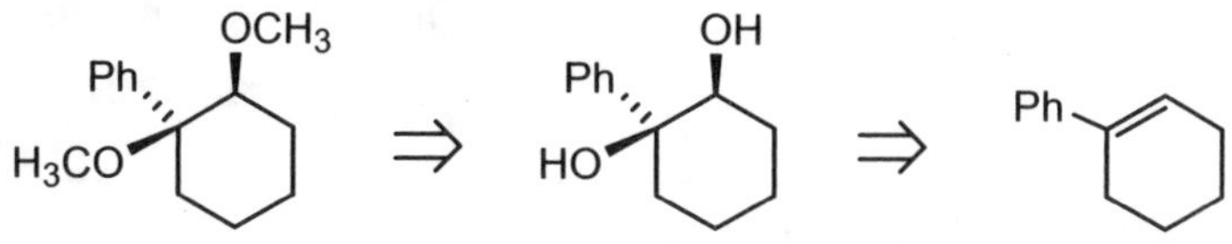

Synthesis

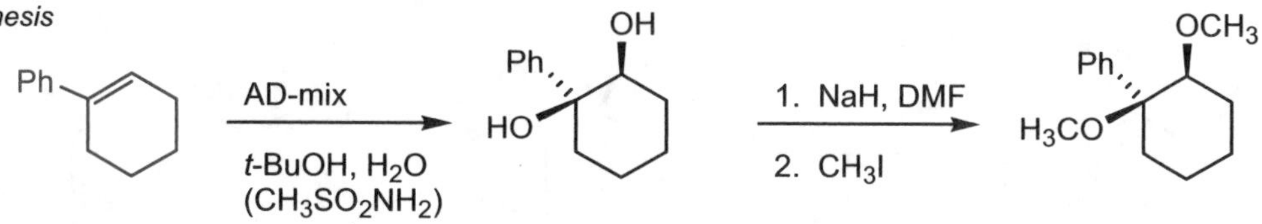

16.14. The resolution of racemic carboxylic acid molecules is carried out by forming salts with a chiral amine. The one chosen in this example (any chiral amine can be used) has two chiral atoms. In the first stage, diastereomeric salts are formed.

Next, the salts are crystallized to separate one diastereomer (as a solid) from the other, which stays in solution (in the ideal situation).

Finally, the carboxylic acid enantiomers are regenerated by treating them separately with aqueous HCl. This step also yields the amine hydrochloride salt, which can be converted to the free base amine and recovered after treatment with aqueous base.

16.15. Follow the procedure given in the solutions to Exercises 16.3 and 16.4.

a.

16.15. (continued)

b.

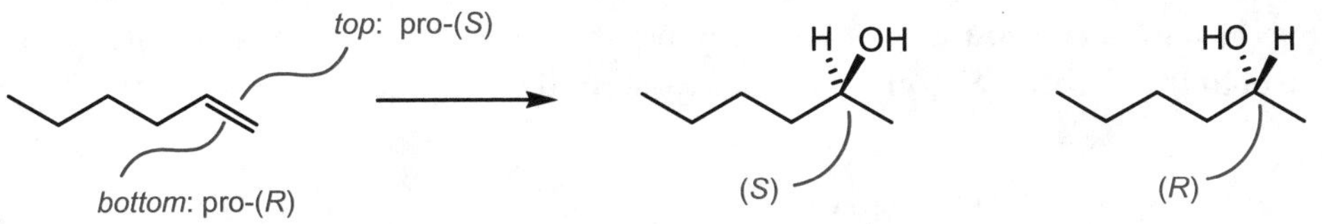

top: pro-(S)

bottom: pro-(R)

(S)   (R)

c.

top: pro-(S)

bottom: pro-(R)

(S)   (R)

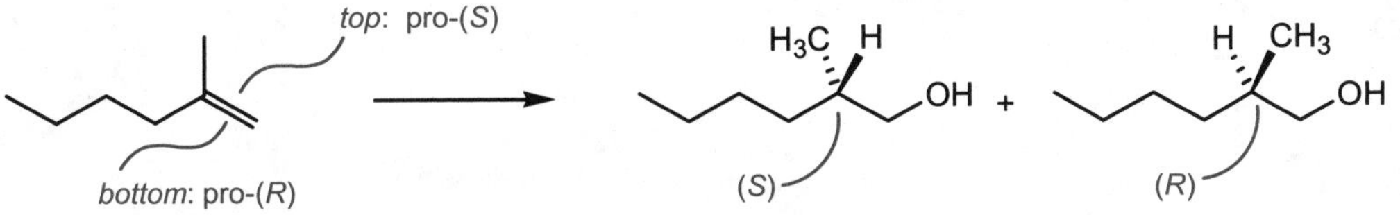

16.16. Follow the procedure outlined in the solution to Exercise 16.3. If replacing the protons in turn at a given carbon atom yields enantiomers, then the hydrogen atom is enantiotopic. If diastereomers are formed by the replacement processes, then the replaced H is diastereotopic. The protons in methyl groups are homotopic and not prochiral.

Arachidonic acid

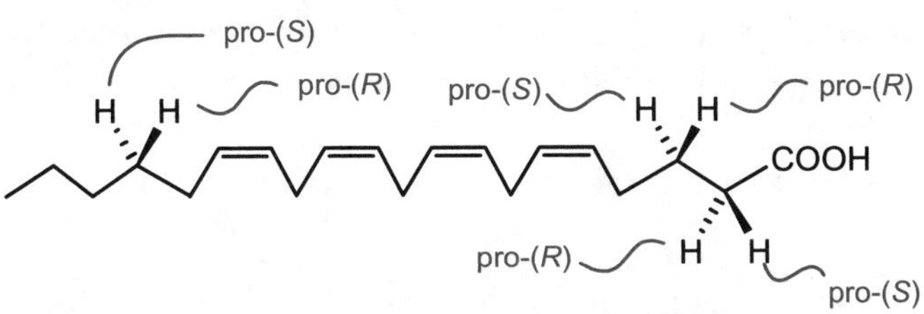

All of the indicated hydrogen atoms are enantiotopic.

Lauric acid

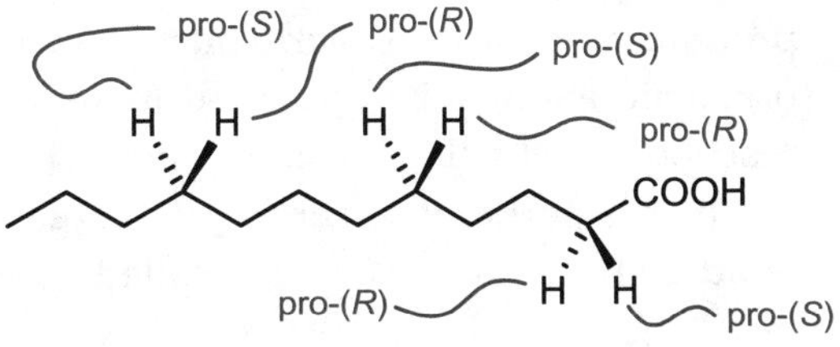

All of the indicated hydrogen atoms are enantiotopic.

Oxalosuccinate

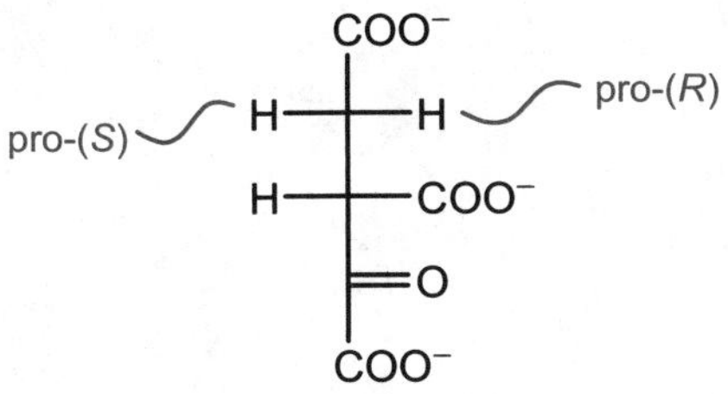

The indicated hydrogen atoms are diastereotopic.

Citrate is an unusual molecule because it is achiral, yet replacing the indicated hydrogen atoms will produce diastereomers (the carbon atom bearing the OH group becomes chiral when either of the indicated protons are substituted).

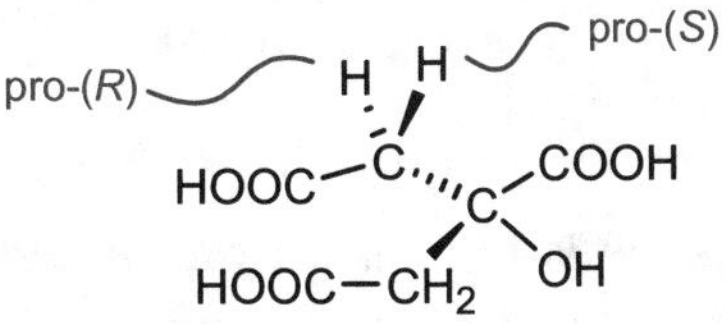

The indicated hydrogen atoms are diastereotopic.

L-Phenylalanine

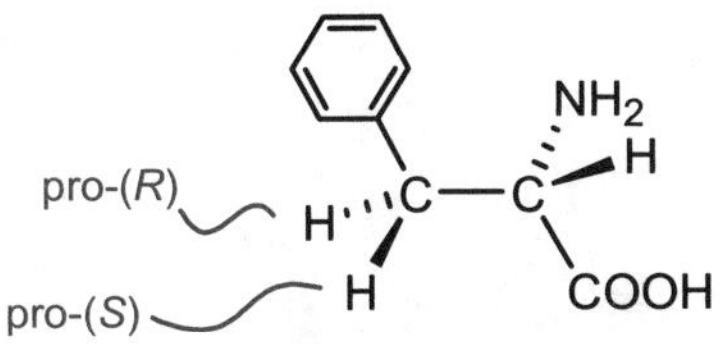

The indicated hydrogen atoms are diastereotopic.

Pyruvic acid

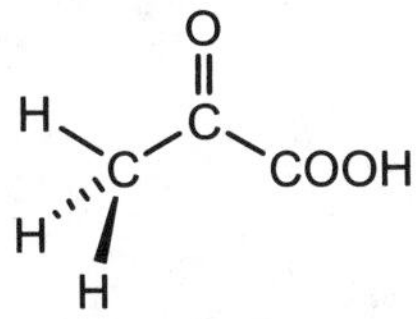

The methyl protons are homotopic.

16.17. Given both the names and structures of the molecules involved in this transformation, we can draw each with its specified stereochemistry. The top face of the double bond in aconitate is the one to which the incoming H and OH groups add, but saying that the top face of aconitate is pro-(*S*) is ambiguous, because both the (*R*) and (*S*) configurations exist in the product.

(2R,3S)-Isocitrate

Therefore, we have to modify our designation to reflect which chiral center in the product is referred to. The top face of aconitate is the pro-(3*S*) [or pro-(2*R*)] face with respect to hydration.

16.18. Follow the procedures illustrated in the solution to Exercise 16.9. The structures of the needed starting compounds are shown in color.

a. Because you are essentially told to use a hydroboration procedure to create the chiral center in the product, you have to ask yourself how an organoborane reactant can be exploited. Either the organoborane will be converted to an alcohol, which can be modified subsequently by substitution reactions, or else the Suzuki reaction will be employed.

In this part of the exercise, two organic rings are brought together to form a new carbon–carbon bond, so the Suzuki reaction is called for.

Retrosynthesis

Synthesis

- -

b. An azidoalkane is prepared from an alcohol via its alkyl sulfonate ester derivative (Section 7.1b). The needed chiral alcohol, the configuration of which must be opposite that of the product, is made from a cis alkene using hydroboration followed by oxidative hydrolysis.

Retrosynthesis

Synthesis

16.18. (continued)

c.   An alkene can be prepared by the E2 reaction of alkyl sulfonate ester, which in turn is prepared from an alcohol.   The E2 reaction requires that the leaving group is trans to the proton that is also removed.   The presence of the methyl group ensures that the elimination step proceeds as needed, which leaves the chirality set at the carbon atom bearing the methyl group.

Retrosynthesis

Synthesis

16.19. The degree of ee or de of a reaction is the numeric difference between the amounts of each chiral isomer (ee for enantiomers; de for diastereomers), expressed as a percentage.

a.

b.

16.20. The signals in this spectrum are assigned by the procedures outlined in Exercises 13.3-13.9, making use of the data in Table 13.1, the integrated intensity values (the numbers of protons), and the spin-spin splitting patterns.   The methyl group next to the carbonyl group will produce a singlet with an intensity of 3H; the methyl group at the opposite end of the chain will generate a signal that appears as a triplet because it is next to a methylene group; and the methyl group adjacent to the methine proton will generate a doublet signal.   The methine proton produces the signal farthest downfield, and it will appear as a sextet because it is adjacent to both methyl and methylene groups (3 + 2 protons + 1 = 6, a sextet).

The remaining two signals in the spectrum must be produced by the methylene protons.   These are diastereotopic because replacing each in turn generates diastereomers (the methine carbon atom is chiral).   Therefore, the protons labeled H_a and H_b are not equivalent, and they will each be split by the combination of the methyl (end of the chain) and methine protons (3 + 1 protons + 1 = 5, a quintet), as well as by each other (geminal coupling).   Assuming J values of 7 and 3.5 Hz (the choice of 3.5 Hz is somewhat arbitrary) creates the following splitting patterns, which are not unlike the ones that are actually observed.

16.20. (continued)

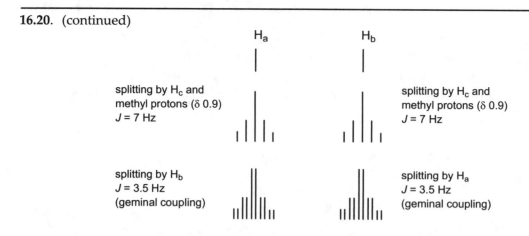

16.21. Follow the procedures illustrated in the solution to Exercise 16.13. The starting compounds are shown in color.

a. The ketone can be prepared from an alcohol by oxidation, and chiral alcohols are readily prepared from alkenes using chiral organoborane reagents.

Retrosynthesis

Synthesis

1. (Ipc*)BH₂
2. crystallize
3. 4 equiv CH₃CHO
4. H₂O₂, OH⁻

1. DMSO, ClCOCOCl
   −60°C
2. NEt₃

- -

b. The sulfide can be prepared from an alcohol by using a substitution reaction of the corresponding alkyl sulfonate ester, and the chirality of the adjacent carbon atom is set during the hydroboration reaction that serves in the preparation of the required alcohol group.

Retrosynthesis

Synthesis

1. (Ipc*)BH₂
2. crystallize
3. 4 equiv CH₃CHO
4. H₂O₂, OH⁻

1. MsCl, NEt₃, CH₂Cl₂
2. NaSPh, DMF

- -

c. The chiral alcohol product can be prepared with a ring-opening reaction between a Grignard reagent and epoxide. The chiral epoxide is made using the dioxirane reagent, FR*.

Retrosynthesis

16.21. (continued)

Synthesis

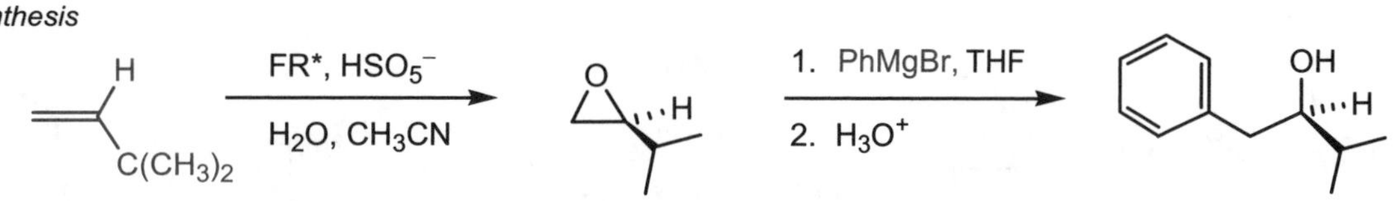

d. This carboxylic acid can be prepared by hydrogenation of the corresponding unsaturated acid, which in turn is made by carboxylation of the Grignard reagent prepared from an alkenyl bromide.

Retrosynthesis

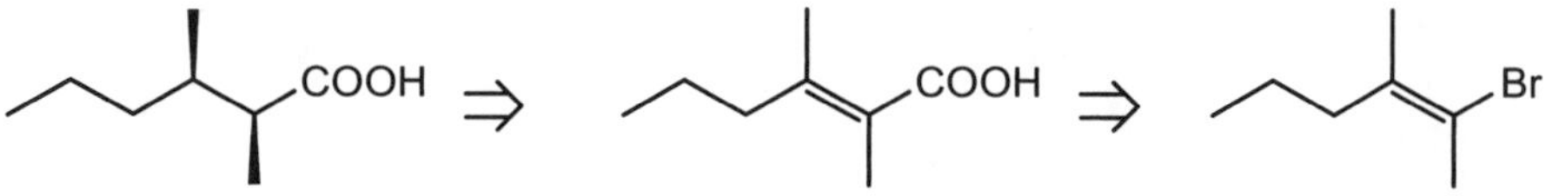

Synthesis

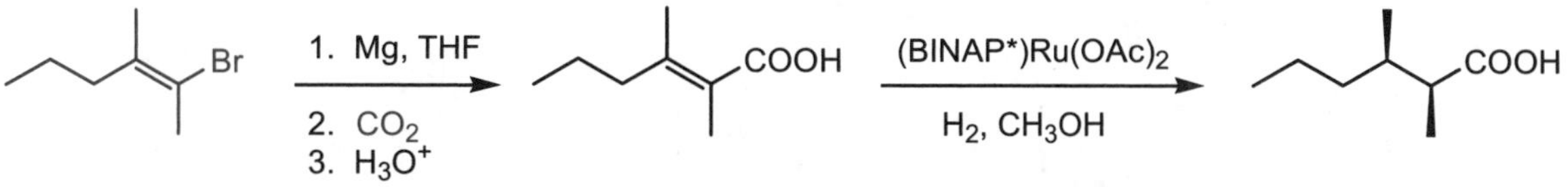

e. The ketone can be prepared from an alcohol by oxidation, and the required chiral diol is prepared by an asymmetric dihydroxylation reaction of a trisubstituted alkene. The tertiary alcohol is not affected by the oxidation step.

Retrosynthesis

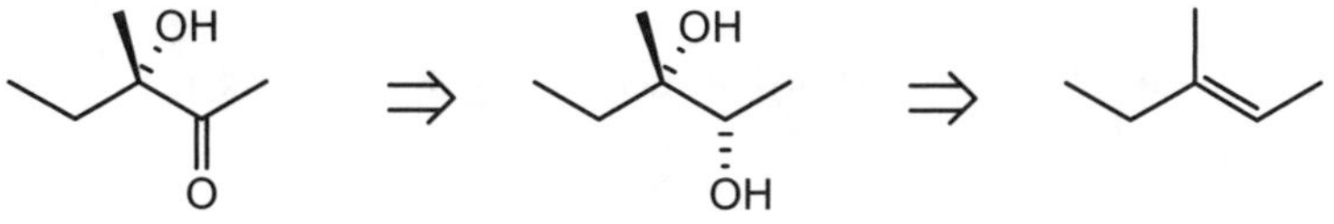

Synthesis

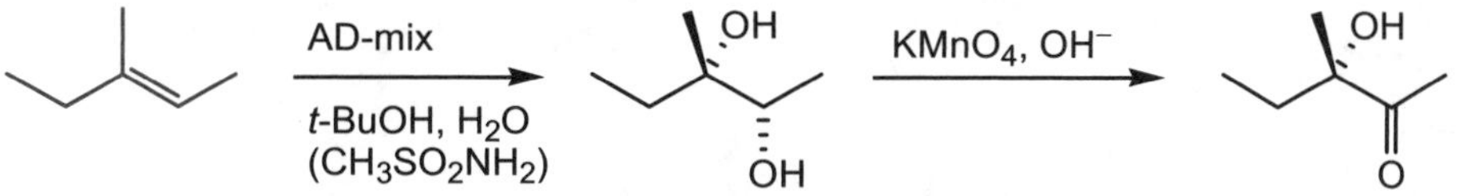

f. This alkyl benzene derivative can be prepared using the Suzuki coupling. The chiral organoborane is made from an alkene, taking advantage of the steric bulk of the *tert*-butyl group to direct the boron atom to the carbon atom farther away from the *tert*-butyl group.

Retrosynthesis

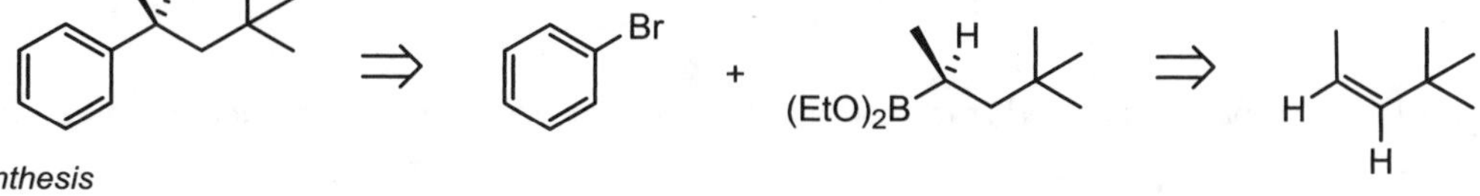

Synthesis

hindered by the presence of the *tert*-butyl group

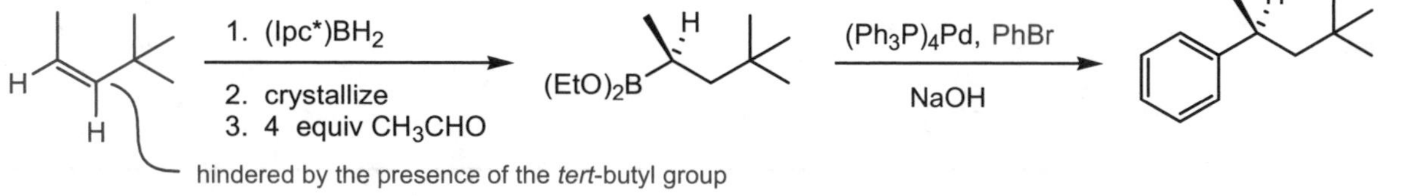

16.22. The two diastereomers formed from the reaction of the racemic mixture of amines with the phosphorus reagent differ in their configurations at the carbon atom bearing the amino group.

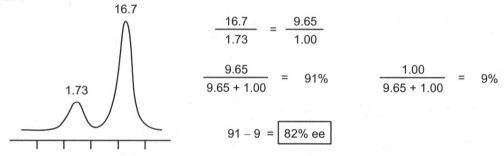

The ee of the associated amination reaction is calculated using the procedure outlined in the solution to Exercise 16.7.

$$\frac{16.7}{1.73} = \frac{9.65}{1.00}$$

$$\frac{9.65}{9.65 + 1.00} = 91\% \qquad \frac{1.00}{9.65 + 1.00} = 9\%$$

$$91 - 9 = \boxed{82\% \text{ ee}}$$

16.23. The first step in the transformation presented in this exercise converts the alcohol to its sulfonate ester derivative.

MsCl, NEt₃, CH₂Cl₂

If the methanethiolate ion were to react in step 2 at the carbon atom bearing the leaving group, then the configuration of the chiral carbon atom would be retained.

– OMs⁻

The fact that inversion of configuration occurs must mean that the epoxide ring must be involved.

You learned previously (Section 7.2e) that epoxides are susceptible to reactions with nucleophiles, so considering that possibility, you can see that ring opening will create an alkoxide nucleophile that can subsequently displace the mesylate group. This step inverts the configuration of the chiral carbon atom.

– OMs⁻

This second step is likely to be a concerted (rather than stepwise) process, in which case the mechanism is written as follows:

– OMs⁻

16.24. First, draw the predicted products based on the given transformation (hydrogenation, hydroboration, and so on) and taking into account the relative stereochemistry associated with each reaction (syn addition, anti addition, etc.) Assign the configuration to each chiral carbon atom, and then draw its enantiomer by switching the configuration of every chiral center. Finally, calculate the quantity of each enantiomer from the given ee value by applying the following equation: $ee = X - (100 - X)$, where X is the amount of the major enantiomer. This equation simplifies to: $X = (100 + ee\ value)/2$.

a. The AD reaction occurs by syn addition, but only one new chiral carbon atom is created in the product molecule. $X = (100 + 91)/2 = 95.5$.

Ratio = 95.5 : 4.5 = 91 %ee (S) > (R)

b. This asymmetric hydrogenation reaction occurs by syn addition, but only one new chiral carbon atom is created in the product molecule. $X = (100 + 96)/2 = 98$.

Ratio = 98 : 2   96 %ee (S) > (R)

c. The AD reaction occurs by syn addition, but only one new chiral carbon atom is created in the product molecule. $X = (100 + 84)/2 = 92$.

Ratio = 92 : 8   84 %ee (R) > (S)

16.25. First summarize the reactions described in the paragraph.

Compound **A** is an alkene because it has one site of unsaturation (Section 14.1c) and reacts with ozone and epoxidizing reagents.

Compound **B** is an aldehyde, the structure of which can be deduced from the given proton NMR data. The resonance at δ 9.9 is in the correct range for the CHO group, and it is split into a triplet, which means it must be within three bonds of 2 other protons: –CH₂–CHO. The splitting of the methylene protons signal (doublet of quartets with J values of 2 and 7 Hz) means that the methylene group is also adjacent to a methyl group: CH₃–CH₂–CHO. Because only compound **B** is formed by ozonolysis of compound **A**, the alkene must be a symmetric compound, 3-hexene.

Compound **C** is the epoxide derivative of 3-hexene. The fact that it is meso means that **A** must be *cis*-3-hexene. We can now replace the letters with structures, as follows.

16.25. (continued)

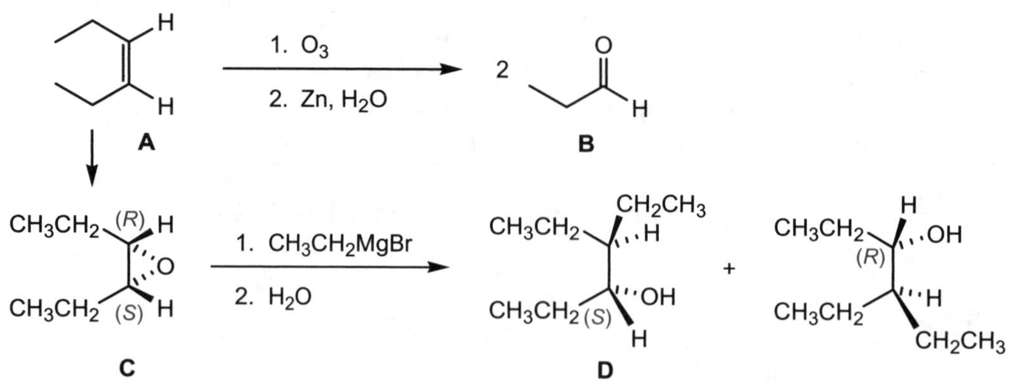

The reaction of the Grignard reagent with the meso epoxide yields a racemic mixture because the ethyl group can react at either carbon atom of the epoxide ring. The carbon atom that retains its configuration is the only chiral center in compound **D** because the other carbon atom has two ethyl groups attached.

16.26. An enantioselective reaction occurs whenever the possible transition states leading to each of the enantiomeric products differ in their free energy of activation values, as illustrated in Figure 16.1. An enzyme active site is chiral because its constituents are amino acids, which themselves are chiral. A prochiral starting material interacts with the chiral active site to produce diastereomeric transition states.

16.27. The active site of the enzyme that converts citrate to aconitate and isocitrate is chiral, and citrate itself is prochiral, having two CH_2COO^- groups attached to the central carbon atom. When citrate binds within the active site by means of hydrogen bonds and/or electrostatic interactions, the prochiral CH_2COO^- groups are differentiated by their proximities to the base that promotes dehydration. Furthermore, each hydrogen atom of the methylene group is also prochiral, and each of those is also positioned with a different proximity to this base (denoted as B: in the structure at the left, below). Therefore, the reaction to remove a specific hydrogen atom from only one of the methylene groups has a lower free energy of activation than does the reaction with any of the other three methylene protons. The result is that only one proton can react to form the specific isomer of aconitate shown and that subsequently generates (2R, 3S)-isocitrate.

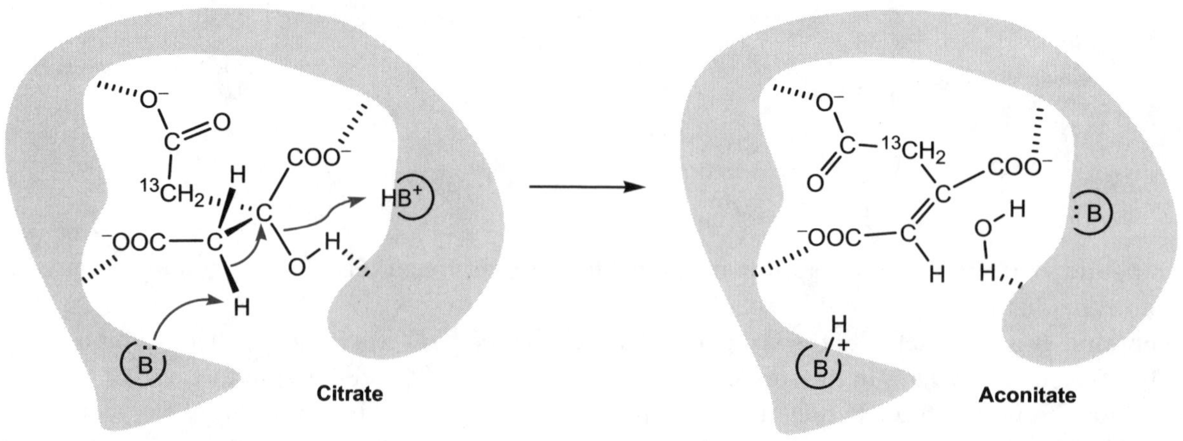

Citrate   Aconitate

THE CHEMISTRY OF BENZENE
AND ITS DERIVATIVES

17.1. The relative energy levels of the molecular orbitals in compounds that are thought to be aromatic can be generated by extending horizontal lines from each vertex of the ring after orienting the ring with one vertex pointing down. For cyclooctatetraene, there are five levels, and all but the lowest and highest energy levels would have two orbitals if the molecule were planar and permitted overlap between all eight orbitals. Placing the eight electrons into the orbitals would lead to the presence of two unpaired electrons.

Instead, cyclooctatetraene undergoes distortion to remove the degeneracy among the MO energy levels; and its shape is a boat form, which has four π bonding orbitals, all of which are filled. There are also four π* antibonding orbitals, all of which are unfilled. This orbital arrangement allows the eight electrons to be paired.

17.2. The energy levels for the MOs of a seven-membered ring are generated as described in the solution to Exercise 17.1. The cation has six electrons, the radical has seven electrons, and the anion has eight electrons. Only the cation has all of its electrons paired. Because all of the bonding orbitals are filled, it is the only species of the three shown below that is aromatic.

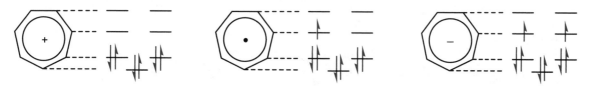

17.3. According to the Hückel formalism, a planar, cyclic molecule with an uninterrupted π system will be aromatic when it has 2, 6, 10, 14, 18, etc. electrons. A π bond counts as two electrons and an unshared pair of electrons counts as two electrons if the pair is in a *p* orbital rather than in a hybrid orbital. The four most common cases that involve unshared electron pairs are illustrated directly below.

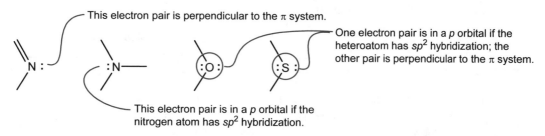

Among the molecules shown in this exercise, those in parts (a.) and (c.) are expected to be aromatic. An atom with *sp*³ hybridization prevents a molecule from being aromatic because it disrupts the π system, as in part (d.).

17.3. (continued)

a.

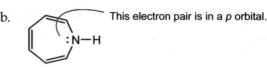

10 π electrons

b.

This electron pair is in a *p* orbital.

8 π electrons

c.   H₃C—N   N :

This electron pair is in a *p* orbital and is part of the π system.

This electron pair is perpendicular to the π system.

6 π electrons

d.

This carbon atom has *sp*³ hybridization, which disrupts the continuity of the π system.

—NH

4 π electrons

17.4. Follow the procedure outlined in the solution to Exercise 1.12.

a. **3-Chloro-2-methylphenol**

phenol   a benzene ring with the OH group attached;
                        its attachment point defines C1 of the ring
3-chloro   chlorine atom attached at C3 of the ring
2-methyl   CH₃ group attached at C2 of the ring

b. ***m*-Hydroxybenzoic acid**

benzoic acid   a benzene ring with the –COOH group attached;
                        its attachment point defines C1 of the ring
m-hydroxy   OH group attached at C3 of the ring (meta)

c. ***o*-Dichlorobenzene**

benzene   a benzene ring
o-dichloro   two chlorine atoms attached to adjacent carbon
                        atoms of the ring (ortho or 1,2)

d. **2,6-Dimethoxytoluene**

toluene   a benzene ring with the CH₃ group attached;
                        its attachment point defines C1 of the ring
2,6-dimethoxy   two OCH₃ groups, one attached at C2 and one
                        attached at C6

e. **α,α′-Dibromo-*m*-xylene**

m-xylene   a benzene ring with CH₃ groups
                        attached at C1 and C3
α,α′-dibromo   two Br atoms, one attached at
                        each methyl group

***m*-Xylene**

17.5. Follow the procedures outlined in the solution to Exercise 1.17.

a. This compound has a methyl group attached to the benzene ring, so the parent compound is toluene. The point of attachment of the methyl group defines C1 of the ring.

   A bromine atom is attached at C3, which is meta to the methyl group. The name is either **3-bromotoluene** or ***m*-bromotoluene**.

17.5. (continued)

b. This compound has a carboxylic acid group attached to the benzene ring, so the parent compound is benzoic acid. The point of attachment of the COOH group defines C1 of the ring.

    A methyl group is attached at C2, which is ortho to the carboxylic acid group. The name is either **2-methylbenzoic acid** or *o*-**methylbenzoic acid**. (A common name for this molecule is *o*-**toluic acid** because it derives from toluene.)

c. This compound has an OH group attached to the benzene ring, so the parent compound is phenol. The point of attachment of the OH group defines C1 of the ring.

    An amino group is attached at C3, and a chlorine atom is attached at C4. The name is **3-amino-4-chlorophenol**.

d. This compound has an aldehyde group attached to the benzene ring, so the parent compound is benzaldehyde. The point of attachment of the CHO group defines C1 of the ring.

    An isopropyl group is attached at C4, which is para to the aldehyde group. The name is either **4-isopropylbenzaldehyde** or *p*-**isopropylbenzaldehyde**.

17.6. The assignment of resonances in the aromatic region of a proton NMR spectrum to the specific protons in a known structure makes use of both the integrated intensity values and splitting patterns. The splitting patterns observed for aryl protons are dominated by coupling with the other protons that are ortho to the one giving the signal, so only a singlet (no ortho proton), doublet (one ortho proton), or triplet (two ortho protons) will be observed.

    Benzoic acid has a plane of symmetry (indicated below as the dashed line), so there are three types of aromatic protons: two that are ortho to the COOH group, two that are meta to the COOH group, and one that is para. There is only one proton para to the COOH group, so the signal with an integrated intensity of 1H is assigned to H_c (as labeled in the structure) and it appears as a triplet (two protons are ortho to H_c. Protons H_a have only one proton that is ortho, so its signal is the doublet appearing farthest downfield. The H_b protons produce the triplet signal (two ortho protons—H_a and H_c) that appears farthest upfield.

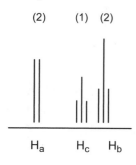

(2)   (1) (2)

H_a H_c H_b

integrated intensity = 1H
2 adjacent ortho protons
triplet

integrated intensity = 2H
1 adjacent ortho proton
doublet

integrated intensity = 2H
2 adjacent ortho protons
triplet

17.7. Chlorine reacts with aluminum chloride to generate Cl^+, an electrophile. Benzene reacts with this electrophile to form a cation intermediate (step 1) and the tetrachloroaluminate ion removes a proton from the intermediate to regenerate the aromatic system (step 2). Aluminum chloride is regenerated with the concomitant formation of HCl.

17.7. (continued)

17.8. When a benzenesulfonic acid derivative is treated with aqueous sulfuric acid, the ring can be protonated at the carbon atom bearing the SO_3H group (step 1). Water deprotonates the sulfonic acid group to regenerate the aromatic ring with formation of sulfur trioxide (step 2). Water serves another purpose as well: it reacts with SO_3 to form diluted sulfuric acid (step 3). This last step is crucial in preventing the formation of the electrophile SO_3H^+, which is needed to sulfonate the ring. Thus, the reaction occurs in the direction written below, which leads to removal of the SO_3H substituent.

$$SO_3 + H_2O \xrightarrow{\quad 3 \quad} H_2SO_4 \text{ (diluted)}$$

17.9. In the pathway leading to the non-rearranged product, 1-bromopropane forms a cationic complex by its reaction with aluminum bromide (step 1a), and this intermediate is intercepted by benzene (step 2a). Remember that primary carbocations are not stable, so dissociation of $AlBr_4^-$ accompanies the reaction between benzene and the cationic complex. The tetrabromoaluminate ion functions as a base to regenerate the aromatic ring in the last step (step 3a).

To form the rearranged product, the intermediate produced in step 1a, above, undergoes hydride ion migration to form the isopropyl carbocation (step 1b, below). The isopropyl carbocation is then intercepted by benzene (step 2b), followed by regeneration of the aromatic ring (step 3b).

17.10. As is the case for all electrophilic aromatic substitution reactions, the first step involves formation of the electrophile: Protonation of the alkene double bond generates a carbocation (step 1). Benzene intercepts this electrophile (step 2), and the aromatic ring is regenerated by deprotonation (step 3).

17.11. For Friedel-Crafts acylation reactions, retrosyntheses break the bond between the benzene ring and the carbonyl group of the side chain. If the substituent is an alkyl group, the retrosynthesis should include a step that shows the acyl derivative. The starting compounds are shown in color.

a. The bonds of this molecule are disconnected to give benzene and the corresponding acid chloride. The synthesis includes a hydrolysis step, the purpose of which is to remove aluminum chloride from its complex with the carbonyl group of the product.

Retrosynthesis   *Synthesis*

b. The bonds of this molecule are disconnected to give benzene and the acid chloride after including the acyl derivative that corresponds to the alkyl product molecule. The synthesis includes hydrolysis after the acylation step and reduction of the acyl group to form the alkyl group.

Retrosynthesis

Synthesis

17.12. The methyl group is an ortho/para director, so the products will consist of a mixture of *o*-nitrotoluene and *p*-nitrotoluene. The electrophile is NO_2^+, which is formed from the reaction between nitric acid and sulfuric acid.

17.13. The intermediate carbocation with the electrophile attached to the carbon atom that is para to the methoxy group has the same general resonance forms that can be drawn for the ortho isomer.

17.13. (continued)

In particular, the positive charge can be delocalized onto the oxygen atom of the methoxy group (note that the oxygen atom still has an octet of electrons).

17.14. Phenol is a highly activated aromatic compound, so it can undergo three successive nitration reactions even at room temperature. The OH group is an ortho/para director, so the three nitro groups will be attached to the ring at C2, C4, and C6. Picric acid is explosive, as is the related trinitro compound TNT. A methyl group, which is also an ortho/para director, is not as activating as the OH group, so the preparation of TNT requires the use of harsher conditions (heat and more concentrated acids).

2,4,6-Trinitrophenol (picric acid)   2,4,6-Trinitrotoluene (TNT)

17.15. As is the case for any benzene derivative bearing a substituent with an electron pair on the atom adjacent to the ring, additional stabilization (structures in the colored boxes) can take place when the incoming electrophile attaches at the ortho (C2) or para (C4) positions, but not at C3.

17.16. As is the case for any benzene derivative with a substituent having a positive or partial positive charge on the atom adjacent to the ring, destabilization (structures in the colored boxes) take place when the incoming electrophile (Br^+ in this exercise) attaches at the ortho (C2) or para (C4) positions.

17.16. (continued)

In one structure, adjacent atoms have positive charges; in the other, the positive charge is on the highly electronegative oxygen atom, which has only six electrons (a positive charge on oxygen is no problem if it also has an octet—see the solution to Exercise 17.13). The electrophile goes to C3 by default.

17.17. As noted in the solution to Exercise 17.16, an intermediate with a partial positive charge on the atom attached directly to the ring is destabilized (structures in the colored boxes) when the incoming electrophile attaches at the ortho (C2) or para (C4) positions. The electrophile reacts at C3 by default.

17.18. Follow the procedures outlined in Examples 17.1-17.4.

a. The substituents on the benzene ring are part of another ring, but we can still evaluate the effects of these groups by focusing on each separately. One is an acyl group (a meta director) and one is an alkyl group (an ortho/para director, represented by "R" in the structure below).

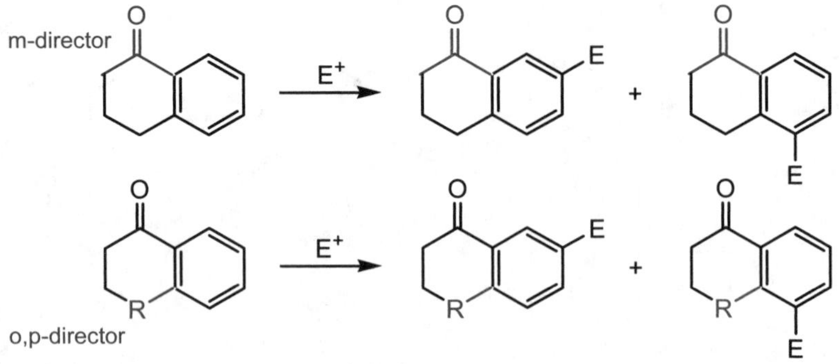

Each substituent directs the incoming electrophile (Br$^+$) to the same two positions, so the expected products are the ones shown below.

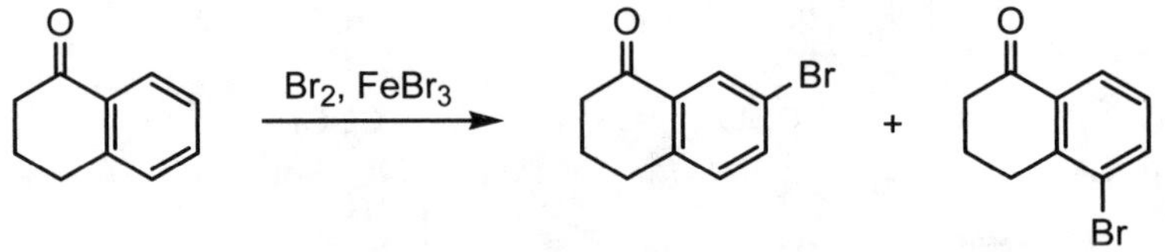

b. The substituents on the benzene ring are both ortho/para directors, so there are three possible products, which appear at the same positions.

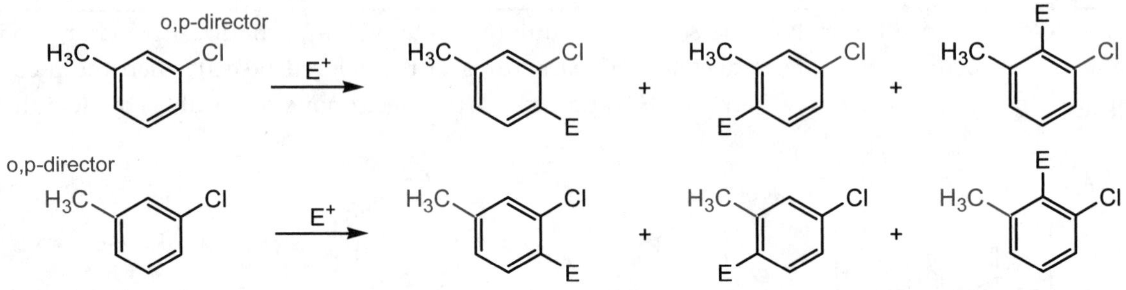

An incoming electrophile will not enter between two substituents, so we can ignore the structures at the far right, above. The expected products of this nitration procedure are those shown below.

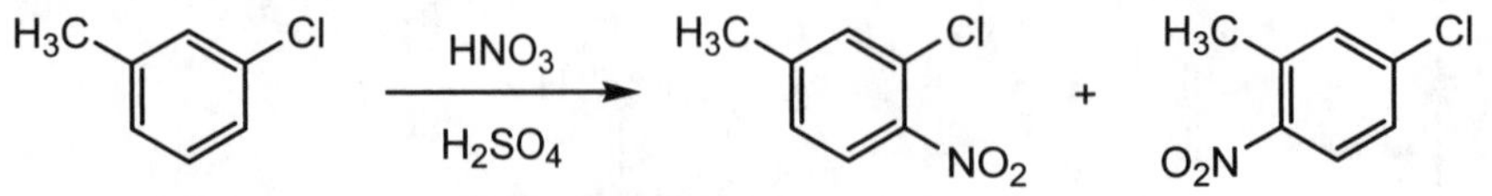

17.19. The total of the integrated intensity values for signals in the aromatic region equals 4, so the ring is disubstituted. From the data shown in Figure 17.5, we deduce that the two substituents are *ortho*, because two doublets and two triplets are observed. All of the signals have chemical shifts upfield from δ 7.25, so we can conclude that both of the otherwise unknown substituents are activating groups.

17.20. An alkyl group attached to a benzene ring can be oxidized to the carboxylic acid group if at least one benzylic hydrogen atom is present.

a. The methyl group is oxidized to the COOH group; the *tert*-butyl group does not have a benzylic proton, so it does not react.

17.20. (continued)

b. Both methyl and isopropyl groups are oxidized to COOH groups.

17.21. In planning the synthesis of benzene derivatives, it is helpful first to consider which groups can be readily introduced by electrophilic reactions and which require the use of a diazonium intermediate. Other methods are also available, including oxidation, Grignard reactions, and reduction.

Electrophilic: Br, Cl, acyl, alkyl, NO₂, SO₃H

Diazonium: Br, Cl, I, F, OH, CN

Other: COOH (ArMgX + CO₂; oxidation of alkyl or acyl groups) and NH₂ (reduction of NO₂)

a. Both groups (OH and I) can be attached via diazonium derivatives, but because both are ortho/para directors, we cannot start with phenol or iodobenzene. Instead, an amino group will have to be placed at each ring position, but not at the same time.   By starting with aniline, we can prepare *m*-nitroaniline by using an electrophilic substitution reaction: The strong acid conditions protonate the amino group, making it into the ammonium group, which is a meta director. Diazotization is used to make the iodo compound from the original amino group. Then the nitro group is reduced and converted to the phenol OH group via a diazonium intermediate as well.

Retrosynthesis

Synthesis

b. The cyano group is introduced via use of a diazonium ion. The bromine atom is an ortho/para director, so we can nitrate bromobenzene and subsequently prepare the amino group by reduction (after separating the ortho and para isomers of bromonitrobenzene).

Retrosynthesis

17.21. (continued)

Synthesis

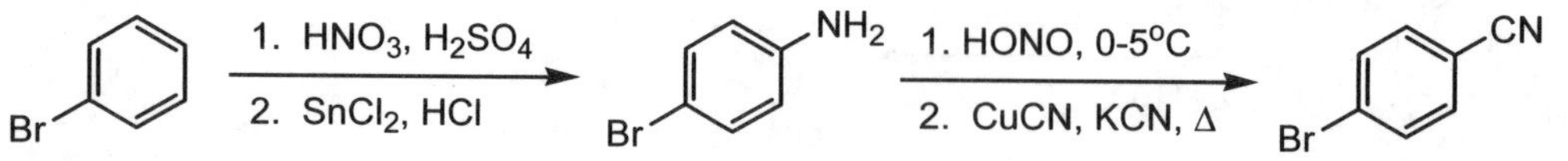

c. The phenol group is attached via use of the diazonium ion. The bromine atom is an ortho/para director, so bromobenzene is first nitrated. After separating the ortho and para isomers of bromo-nitrobenzene, the nitro group is reduced to form the aniline derivative. The amino group is converted to an OH group via the diazonium ion.

Retrosynthesis

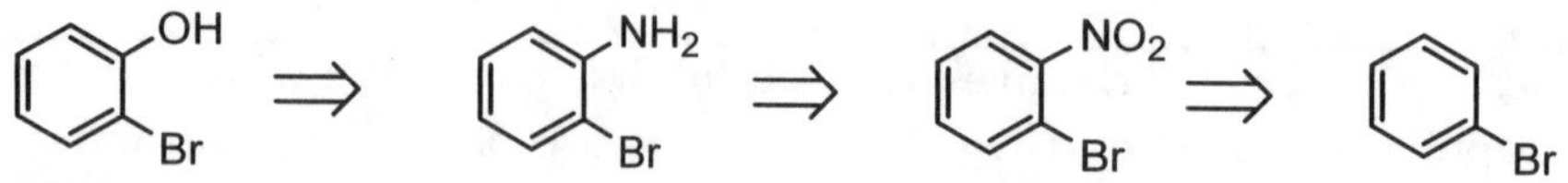

Synthesis

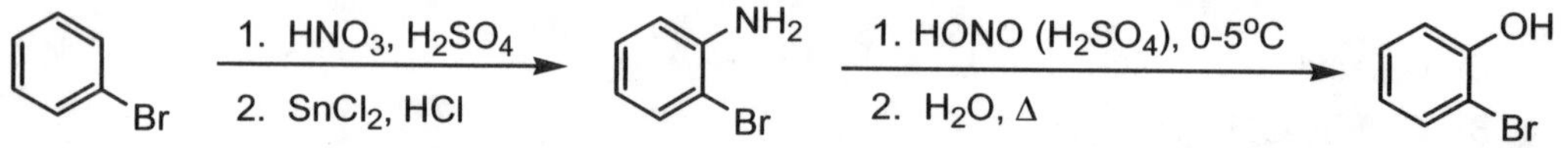

17.22. To make Alizarin Yellow R, the diazonium ion derivative of *p*-nitroaniline will be treated with salicylic acid. To make *p*-nitroaniline, aniline must be converted to its acetamide derivative first, otherwise the meta isomer will be formed (see the solution to Exercise 17.21a). After nitration, the amide is hydrolyzed to form the aniline derivative. Diazotization and coupling with salicylic acid produces the desired dye. The starting compounds are shown in color.

Retrosynthesis

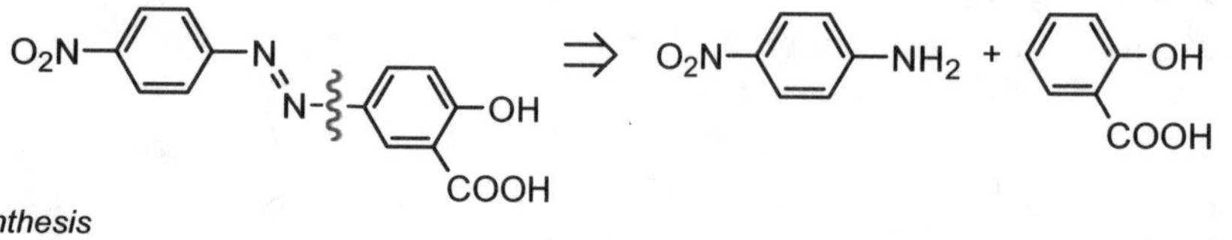

Synthesis

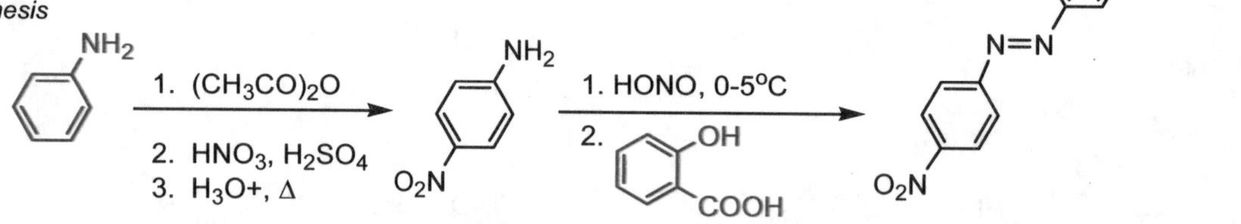

17.23. The mechanism of this reaction follows the same course observed for the para-disubstituted analog. The nucleophile reacts at the carbon atom bearing the halogen atom. Elimination of the halide ion subsequently regenerates the aromatic system. The intermediate is stabilized by delocalization of the electrons within the ring and into the nitro group, as shown in the expanded box below.

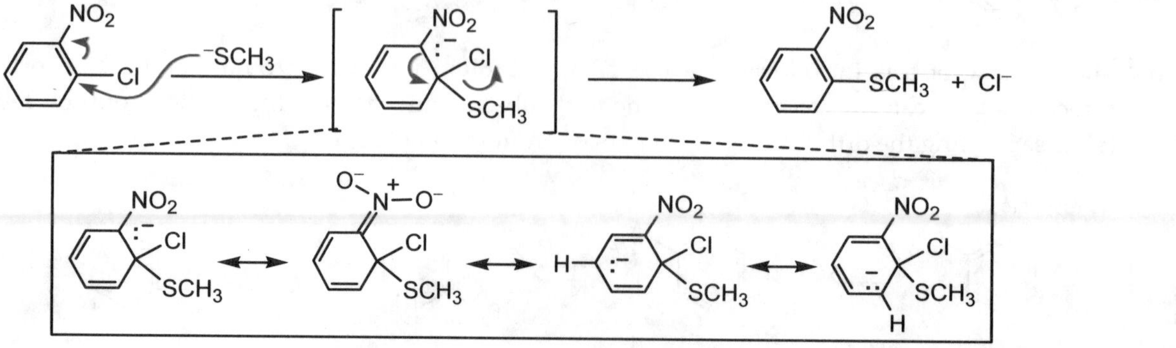

17.24. Follow the examples illustrated in the solutions to Exercise 17.5

a.   This compound has an aldehyde group attached to the benzene ring, so the parent compound is benzaldehyde. The point of attachment of the CHO group defines C1 of the ring. A methoxy group is attached at C2, and a fluorine atom is attached at C5. The name is **5-fluoro-2-methoxybenzaldehyde**.

- -

b.   This compound has no readily identifiable functional group or substituent, so we designate as C1 the carbon atom attached to the bromine atom. The root word is benzene. Nitro groups are attached at C2 and C4, so the name of this molecule is **1-bromo-2,4-dinitrobenzene**.

- -

c.   This compound has an OH group attached to the benzene ring, so the parent compound is phenol. The point of attachment of the OH group defines C1 of the ring. A methyl group is attached at C2 and a nitro group is attached at C5. The name is **2-methyl-5-nitrophenol**.

- -

d.   This compound has a methyl group attached to the benzene ring, so the parent compound is toluene. The point of attachment of the methyl group defines C1 of the ring. Two chlorine atoms are attached to the ring, one at C3 and one at C5. The name is **3,5-dichlorotoluene**.

- -

e.   This compound has a cyano group attached to the benzene ring, so the parent compound is benzonitrile. The point of attachment of the cyano group defines C1 of the ring. A bromine atom is attached to the ring at C4.   The name is either **4-bromobenzonitrile** or **p-bromobenzonitrile**.

17.25. Follow the examples illustrated in the solutions to Exercise 17.4.

a. **p-Bromoaniline**

aniline	a benzene ring with the NH₂ group attached at C1
p-bromo	a Br atom attached at C4 of the ring

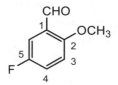

b. **2,3-Dibromo-5-nitrophenol**

phenol	a benzene ring with the OH group attached at C1
2,3-dibromo	two Br atoms, one each attached at C2 and C3
5-nitro	a NO₂ group attached at C5 of the ring

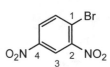

c. **3-[(R)-1-Hydroxyethyl]benzoic acid**

benzoic acid	a benzene ring with the COOH group attached at C1
3-[X]	a substituent X attached to the ring at C3

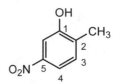

X = substituent: **(R)-1-Hydroxyethyl**

ethyl	2 carbon atoms; by convention attached through its C1
1-hydroxy	OH group attached at C1 of the substituent chain
(R)	the substituent's chiral center has the (R) configuration

17.25. (continued)

d. **3-Chloro-2-methylbenzaldehyde**

benzaldehyde	a benzene ring with the CHO group attached at C1
3-chloro	a Cl atom attached at C3 of the ring
2-methyl	a CH₃ group attached at C2 of the ring

e. **m-Nitroaniline**

aniline	a benzene ring with the NH₂ group attached at C1
m-nitro	a NO₂ group attached at C3 of the ring

17.26. According to Hückel's rule, a cyclic compound with an uninterrupted π system that is planar and has $4n + 2$ electrons will be aromatic. All of the compounds shown below are aromatic, except for the cycloheptatrienyl anion in part (c).

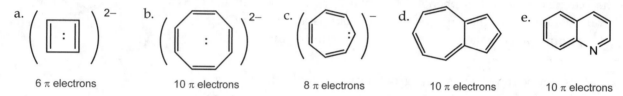

a.	b.	c.	d.	e.
6 π electrons	10 π electrons	8 π electrons	10 π electrons	10 π electrons

17.27. To assess how reactive an arene is toward electrophilic substitution, look at the substituents attached to the ring, and then use the data in Figure 17.10 to evaluate their activating or deactivating influences.

a. A methyl group is a stronger activating group than a halogen atom.

        more reactive

b. These two xylene derivatives have the same groups attached to the ring, so their activation levels should be about the same.

        same

c. A chlorine atom is less deactivating than the nitro group.

        more reactive

d. A methoxy group is a stronger activating group than a methyl group.

        more reactive

17.28. A substituent that functions as an ortho/para director is either an alkyl group or has an electron pair on the atom adjacent to the ring. Only the aldehyde and ester groups, shown in the boxes below, do not function as ortho/para-directing substituents.

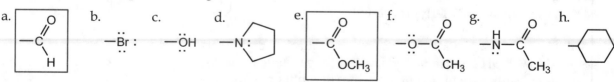

17.29. There are two issues to consider when evaluating the number of isomeric products that may be formed: 1) how many unique positions are present; 2) what are the directing influences of the substituents. For the isomers of xylene, the numbers of hydrogen atom types attached to the ring are three, two, and one, respectively. The dashed lines in the structures shown below represent planes of symmetry that exist in each molecule.

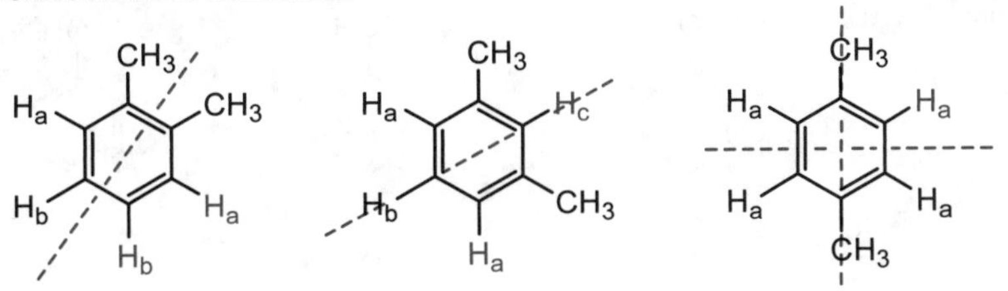

The methyl group is an ortho/para director, so there are two possible positions to which an incoming electrophile will be directed in the ortho and meta isomers (the hydrogen atoms to be substituted are shown in color, above), but only one position in the para disubstituted compound. Therefore the simple answer to this exercise is that *p*-xylene will yield only one monosubstituted product.

Looking at the equations for the reaction of the meta isomer, you can see that *m*-xylene will *effectively* produce only one isomer because of the constraint that an incoming electrophile can attach between two substituents only with difficulty.

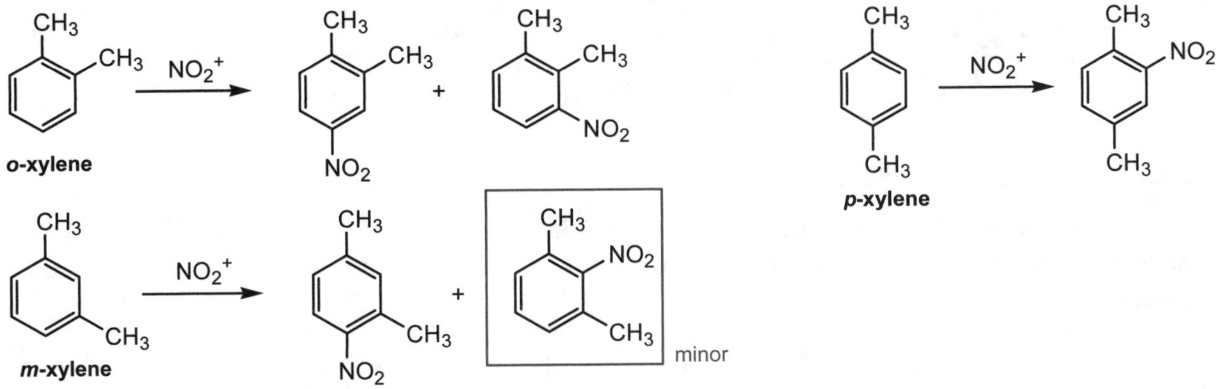

17.30. Follow the examples illustrated in the solutions to Exercise 17.18.

a. One substituent on the benzene ring is an ortho/para director and the other is a meta director, but each directs the incoming electrophile—NO$_2^+$ in this exercise—to the same two positions.

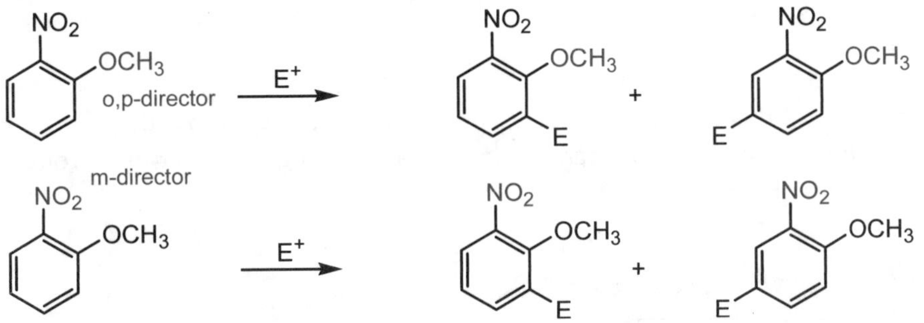

The expected products of this nitration procedure are the ones shown in the following scheme.

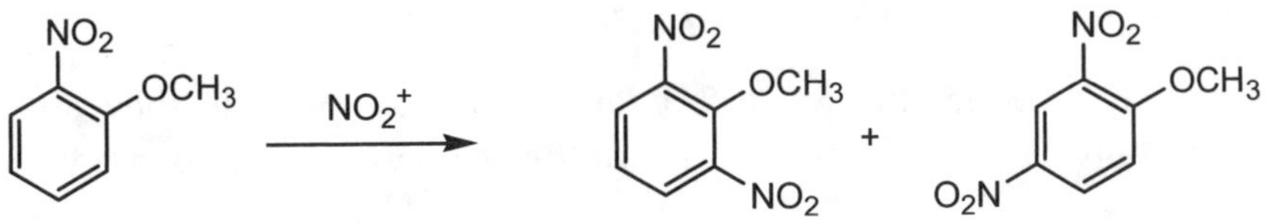

17.30. (continued)

b. Both substituents on the benzene ring are ortho/para directors. Because they are meta to one another, each directs the incoming electrophile—NO_2^+ in this exercise—to the same three positions.

Because an incoming electrophile will not attach between two substituents, we can ignore the 1,2,3-trisubstituted isomer. The expected products of this nitration procedure are as follows:

- -

c. Both substituents on the benzene ring are ortho/para directors. Because they are ortho to one another, each directs the incoming electrophile—NO_2^+ in this exercise—to a different pair of positions.

In a case such as this, the major products will be those directed by the more potent of the substituents. A methoxy group is more activating than a methyl group, so those product isomers are predicted to be the ones obtained.

17.31. Follow the examples illustrated in the solutions to Exercise 17.30.

a. Both substituents on the benzene ring are ortho/para directors. Because they are ortho to one another, each directs the incoming electrophile—Br^+ in this exercise—to an apparently different pair of positions.

17.31. (continued)

In a case such as this, in which the substituents are identical, the seemingly different product isomers are actually equivalent (see the solution to Exercise 17.29). This reaction yields two products.

- -

b. One substituent on the benzene ring is an ortho/para director and the other is a meta director, but each directs the incoming electrophile—Br⁺ in this exercise—to the same two positions.

The expected products of this bromination reaction are the following:

- -

c. Both substituents on the benzene ring are ortho/para directors. Because they are meta to one another, each directs the incoming electrophile to the same three positions.

17.31. (continued)

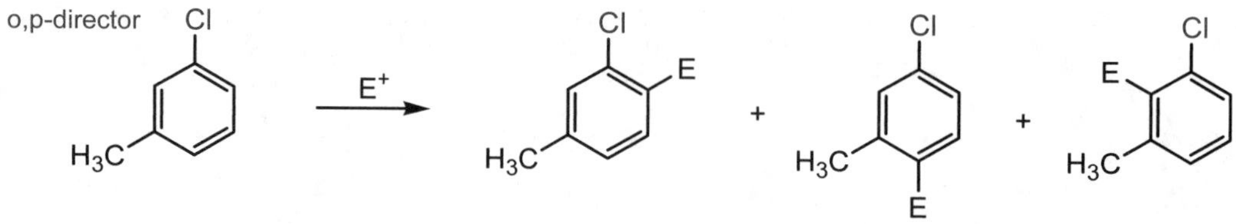

Because an incoming electrophile will not enter between two substituents, we can ignore the 1,2,3-trisubstituted isomer. The expected products of this nitration procedure are as follows:

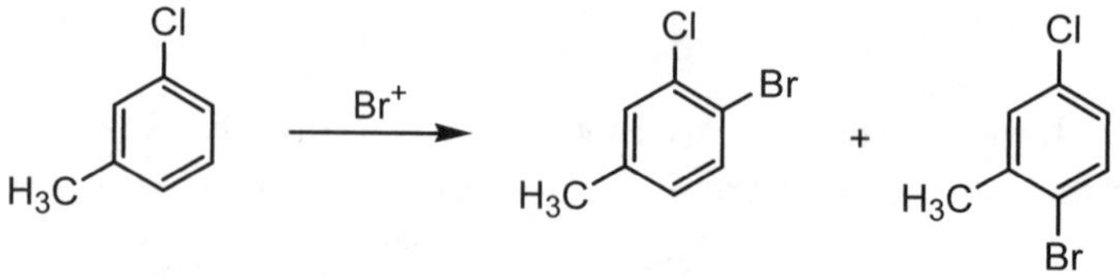

17.32. Follow the examples illustrated in the solutions to Exercises 17.30 and 17.31.

a. Both substituents on the benzene ring are ortho/para directors, and they are equivalent. Because they are ortho to one another, each directs the incoming electrophile—HSO_3^+ in this exercise—to equivalent positions as described in Exercise 17.31a.

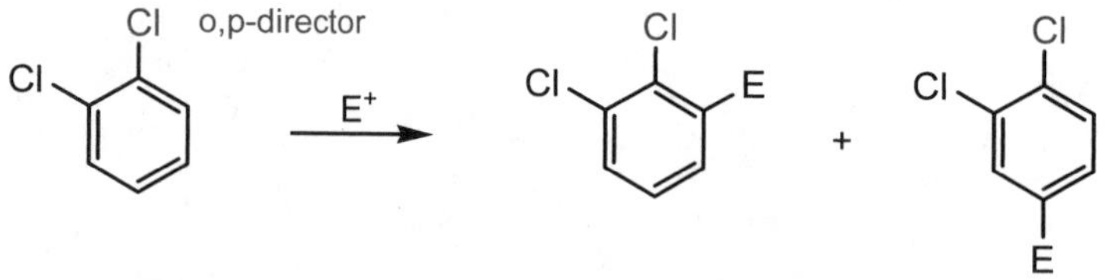

The expected products of this sulfonation reaction are as follows:

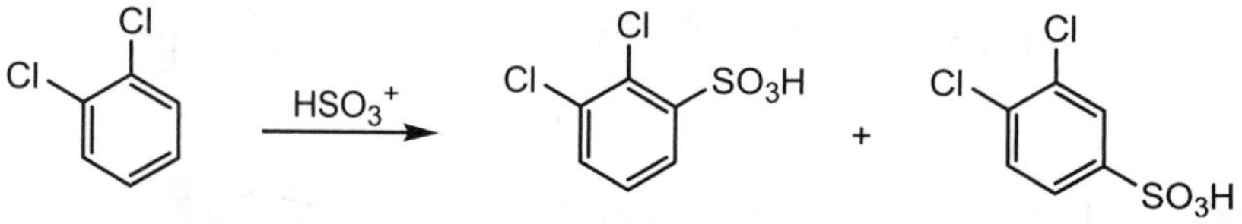

- -

b. Both substituents on the benzene ring are ortho/para directors, and they are equivalent. Because they are meta to one another, each directs the incoming electrophile—HSO_3^+ in this exercise—to the same positions, two of which are equivalent because of symmetry (compare this result with the one described in Exercise 17.29).

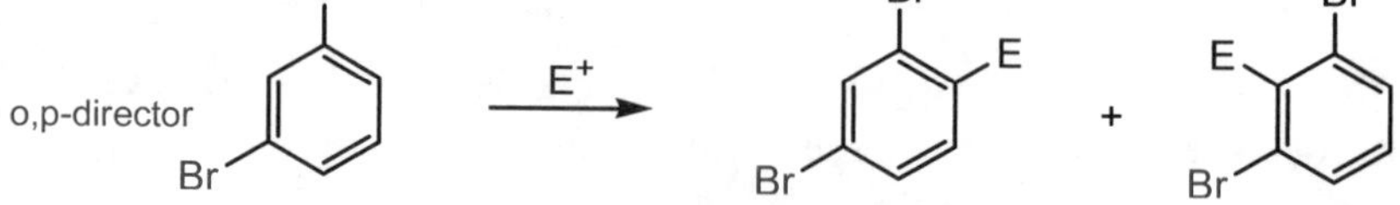

Because an incoming electrophile will not attach between two substituents, we can ignore the 1,2,3-trisubstituted isomer. The expected product of this sulfonation reaction is the following:

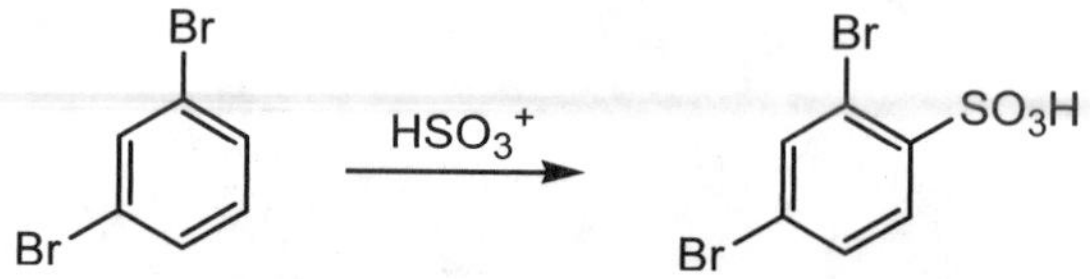

- -

17.32. (continued)

c. One substituent on the benzene ring is an ortho/para director and the other is a meta director, but each directs the incoming electrophile—Br⁺ in this exercise—to the same position.

The expected product of this sulfonation reaction is the following:

17.33. A "major product of an electrophilic substitution reaction" means that the predominant directing influences of the substituents attached to the ring should activate the position in which the new substituent appears in the product. Because ortho/para directors often lead to the formation of more than one isomer, we will assume that either one constitutes a major product (in most cases).

To assess whether a particular compound can be prepared, disconnect a substituent that can be introduced as an electrophile (these are shown in color in this exercise and the next). If the remaining substituents activate the position from which the substituent was removed, then the synthesis reaction should be successful. If not, then treat the problem as the synthesis of an unknown compound and consider other methods.

a. Both the chlorine atom and a nitro group can be attached using electrophilic substitution reactions. Disconnect each in turn and decide whether the forward reaction will be successful. Two isomers will be formed in each reaction, but either option is viable.

- - - - - - - - - - - - - - - - - -

b. The OH group cannot be introduced using an electrophilic substitution reaction, but a bromine atom can. Because the OH group is a potent ortho/para director, it will direct the incoming electrophile more than the bromine atom will. In this case, the desired isomer should be the only product.

17.33. (continued)

c. Each group can be introduced using electrophilic substitution reactions, but only by introducing the nitro group can the desired product be made from a disubstituted benzene precursor. Friedel-Crafts reactions cannot have two deactivating groups on the ring.

This ring is too deactivated to undergo Friedel-Crafts reactions.

Both substituents are meta directions, so reactions will occur at the position in color if the ring is not too deactivated.

The only viable option is the following:

17.34. Follow the procedures outlined in the solution to Exercises 17.33.

a. The methyl and nitro groups can be introduced using electrophilic substitution reactions.

The Friedel-Crafts reactions may work because the deactivating effect of the nitro group will be offset by the presence of the strongly activating methoxy group. Aluminum chloride can complex with the oxygen atom of the methoxy group, however, which may negate its activating effects.

Friedel-Crafts methylation is a difficult reaction to control in any case, so the nitration route is preferable. Nitration of *o*-methylanisole will yield the desired product.

- -

b. Neither the COOH nor the methoxy group can be introduced using an electrophilic substitution reaction, but the nitro group can. Because the OCH₃ group is a potent ortho/para director, it will direct the incoming electrophile more than the COOH group will. Two isomers will be formed, including the desired one.

17.34. (continued)

c. The methyl group and chlorine atom can be introduced using electrophilic substitution reactions. The substituents are all ortho/para directors, however, so the desired isomer will not be formed.

Instead, an amino group is used to direct the introduction of the chlorine atoms, and it is subsequently removed via its diazonium derivative.

17.35. The nitroso group has an unshared pair of electrons on the atom adjacent to the ring, which is the common feature of most ortho/para-directing substituents. Draw the resonance structures according to the procedures outlined in the solution to Exercise 17.15. The ones in the boxes contribute added stability to the cationic intermediate.

Nitrosobenzene differs from nitrobenzene because the nitroso group has an electron pair adjacent to the ring whereas the nitro group has a positive charge on the atom adjacent to the ring.

17.36. Styrene reacts with sulfuric acid to form the stable benzylic cation (step 1), which can undergo Markovnikov addition to the π bond of another molecule (step 2).

17.36. (continued)

This second carbocation is subsequently intercepted by the benzene ring of the first unit in a Friedel-Crafts alkylation sequence.

17.37. In the Gattermann-Koch reaction, carbon monoxide, HCl, and aluminum chloride react to form the formyl cation stabilized by complexation with the tetrachloroaluminate ion.

This cation subsequently reacts with toluene by the sequence that takes place in any Friedel-Crafts acylation reaction: The cation is intercepted by the benzene ring (step 2), and the cationic intermediate is converted to the aromatic product by deprotonation (step 3).

(Only the para isomer of the product is shown in the preceding equation, but the methyl group is an ortho/para director, so the ortho isomer is also formed.)

17.38. Iodosuccinimide has an electrophilic iodine atom, so its reaction with an activated benzene derivative follows the normal two-step process, namely reaction of the arene with the electrophile (step 1) and regeneration of the aromatic ring (step 2).

17.39. Follow the examples illustrated in the solutions to Exercise 17.13.

a. An alkyl aryl ketone can be prepared by the Friedel-Crafts acylation reaction. With a methyl group on the ring to start, both ortho and para products will be formed, but the para isomer should predominate because the two substituents are relatively large.

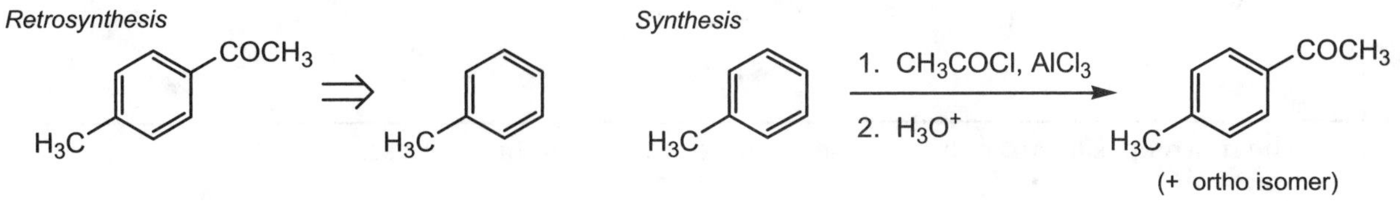

Retrosynthesis   *Synthesis*

(+ ortho isomer)

b. A benzene derivative with a primary alkyl group as a substituent can be prepared by the Friedel-Crafts acylation reaction followed by reduction. An acyl group is a meta director, so the bromine atom is attached by an electrophilic substitution reaction at that stage of the synthesis.

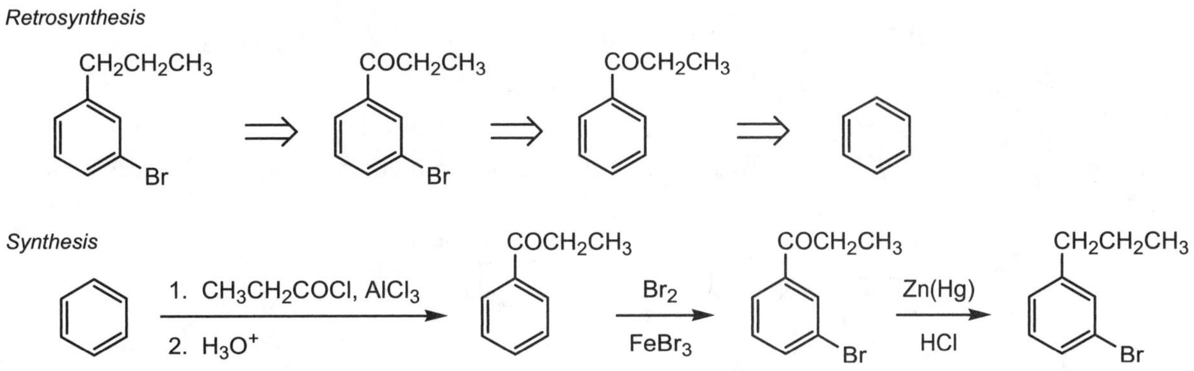

Retrosynthesis

Synthesis

c. Sulfonic acid derivatives of benzene can be prepared by direct sulfonation. To obtain a reasonable amount of the ortho isomer, the para position should be blocked with a meta director. The nitro group will serve this purpose, and it can be removed via reduction of a diazonium derivative.

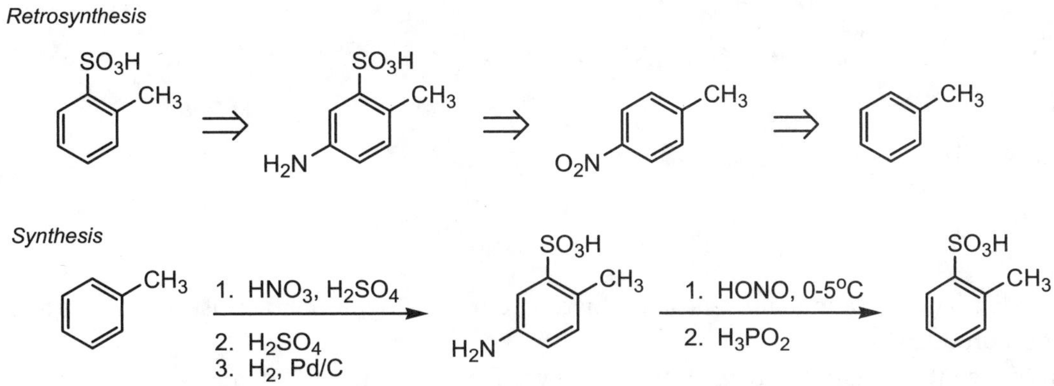

Retrosynthesis

Synthesis

d. The carboxylic acid group can be prepared by oxidation of an alkylbenzene derivative. The subsequent nitration of benzoic acid yields the desired isomer. Benzoic acid can also be prepared from benzene by bromination, formation of a Grignard reagent, carboxylation, and hydrolysis.

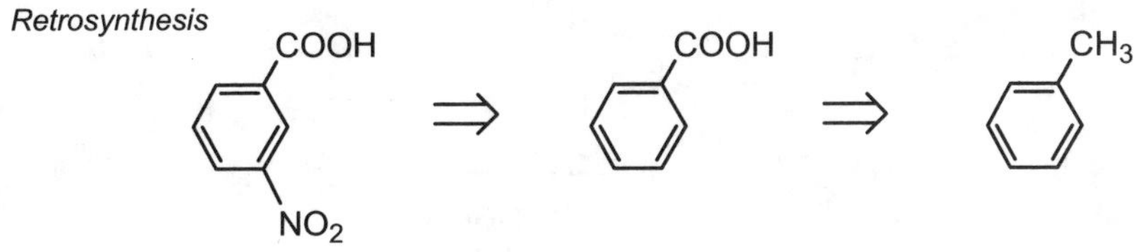

Retrosynthesis

17.39. (continued)

Synthesis

$$\text{toluene (CH}_3\text{)} \xrightarrow[\text{2. H}_3\text{O}^+]{\text{1. KMnO}_4,\ \text{OH}^-} \text{benzoic acid (COOH)} \xrightarrow{\text{HNO}_3,\ \text{H}_2\text{SO}_4} \text{3-nitrobenzoic acid (COOH, NO}_2\text{)}$$

17.40. The first step is to summarize the reactions described in this exercise:

$$\underset{C_7H_7Br}{\textbf{X}} \xrightarrow{\text{HNO}_3,\ \text{H}_2\text{SO}_4} \underset{C_7H_6BrNO_2}{\textbf{Y + Z}} \xrightarrow{\text{Zn, HOAc}} \underset{C_7H_8BrN}{\textbf{P + R}} \xrightarrow{\text{Br}_2\ (\text{xs})} \underset{C_7H_7Br_2N}{\textbf{Q + T}}$$

Next, consider the possible structures for compound **X**, which based on its formula, is a bromo derivative of toluene.

(structures: benzyl bromide (CH_2Br); o-bromotoluene; m-bromotoluene; p-bromotoluene)

You are told that nitration of **X** yields two isomers of its nitro product, and the possible structures can be predicted:

(nitration schemes for benzyl bromide, o-bromotoluene, m-bromotoluene, and p-bromotoluene shown with their respective nitro products)

From these predictions, we eliminate *o*-bromotoluene from consideration because it can form four possible nitration products.

Reduction of the nitro groups yields aniline derivatives **P** and **R** (replace the nitro groups with amino groups in the foregoing structures). You now only need to consider which of the aniline compounds will react with excess bromine to form products having only two bromine atoms. Remember that an amino group is highly activating, so any open positions that are ortho and para to the amino group will be substituted if excess bromine is added.

$$\underset{\textbf{P}}{\text{(o-amino benzyl bromide)}} + \underset{\textbf{R}}{\text{(p-amino benzyl bromide)}} \xrightarrow{\text{Br}_2\ (\text{xs})} \text{(tribromo products)}\quad C_7H_6Br_3N$$

17.40. (continued)

Only the meta isomer undergoes bromination to form dibromo products. Therefore, compound **X** is *m*-bromotoluene.

17.41. The diazotization reaction of anthranilic acid is straightforward and yields the diazonium ion substituent ortho to the carboxylic acid group. A base (chloride ion as shown; or water) starts the movement of electrons that leads to formation of carbon dioxide and molecular nitrogen, which generates benzyne.

17.42. The reactions described in this chapter can be classified within three categories: electrophilic substitution reactions, diazonium reactions, and other. To predict the structures of the products in each of the following transformations, you must know the details of each reaction type, which can be found in the reaction summary section of the chapter. Every product molecule is achiral.

a. The first step of this transformation is a Friedel-Crafts alkylation reaction that attaches an isopropyl group to the ring. Chlorine is an ortho/para director, so two isomers are formed. The second step leads to the oxidation of the isopropyl groups to carboxylic acid groups, so the product comprises a mixture of *o*- and *p*-chlorobenzoic acids.

b. This transformation is an electrophilic sulfonation reaction and the ethyl group is an ortho/para director, so two isomers are formed.

c. This is a diazonium reaction in which a chlorine atom replaces the amino group

17.42. (continued)

d. The first step of this transformation is a reduction reaction that converts the nitro to an amino group. Next, the diazonium ion derivative is prepared. Boiling water reacts with an aryl diazonium ion to form a phenol. The overall procedure converts the nitro group to a hydroxyl group.

e. This electrophilic chlorination reaction will yield two isomers. The acyl group is a meta director, and the alkyl group is an ortho/para director. This reaction is similar to the one shown in Exercise 17.18a.

f. This transformation is an electrophilic nitration reaction. A methyl group is an ortho/para director, and the cyano group is a meta director, so the same positions are activated. Because of symmetry, only one product is formed.

g. This transformation is an example of a nucleophilic aromatic substitution reaction. A halogen atom ortho or para to a nitro group is replaced by a nucleophile (OH⁻) in the first step. In the second step, the nitro group is reduced to form the amino group.

h. This transformation is an example of the Friedel-Crafts acylation reaction. Both the methyl group and the bromine atom are ortho/para directors, so two isomers are formed.

i. This transformation involves two electrophilic substitution reactions. The bromine atom is an ortho/para director, so two isomers of the sulfonation product (step 2) are formed.

h. The first step of this transformation is an example of nucleophilic aromatic substitution. The nitro group is then reduced to form an amino group, and diazotization and reduction replaces the amino group with a proton.

17.42. (continued)

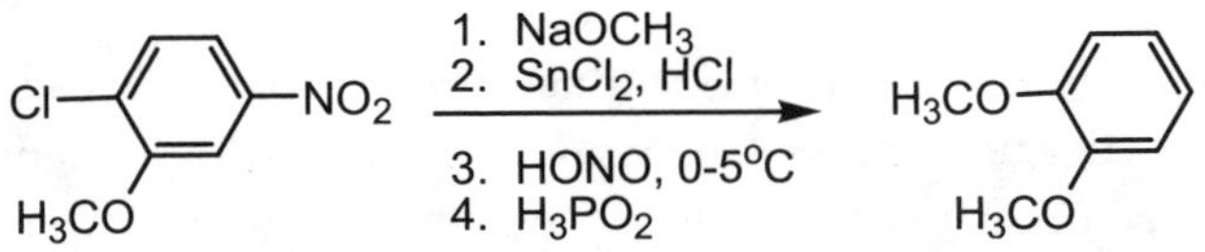

17.43. NO⁺ is an electrophile that is formed from HONO. As an electrophile, NO⁺ will react with the π electrons of an activated aromatic ring (step 2). Regeneration of the aromatic system completes the substitution mechanism.

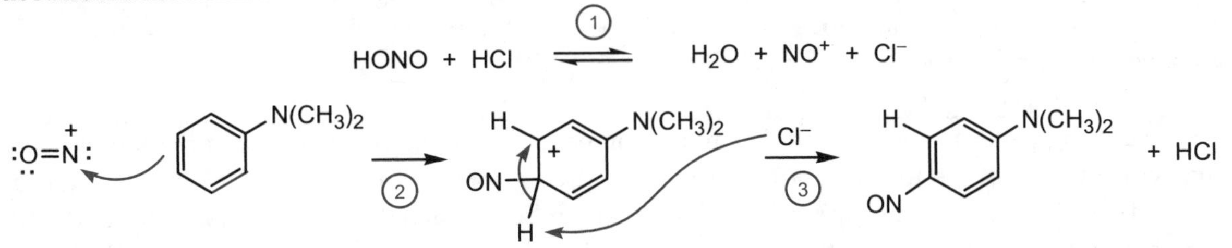

17.44. Follow the procedure outlined in the solution to Exercise 14.23.

(1) Examine the ^{13}C NMR to determine the number and types of carbon atoms present. There are three types (C=O, arene, and aliphatic) consisting of nine peaks. Start with the assumption that the compound has 9 carbon atoms (six for the benzene ring plus three others).

(2) Examine the integrated intensity values in the proton NMR spectrum. The total is 10 (1:1:1:1:3:3, left to right). Start with the assumption that the compound has 10 hydrogen atoms.

(3) Use the high-resolution data to determine the molecular formula with the initial assumption that there are 6 C atoms, 10 H atoms, and some number of oxygen atoms. For $C_9H_{10}O_3$, the calculated high resolution MW is 166.063 (observed MW = 166.068). This formula is consistent with the data.

(4) Calculate the number of unsaturation sites: $[2(9) + 2 - 10]/2 = 5$ sites of unsaturation, four of which are the benzene ring. The carbonyl group (identified in step 1) is the fifth site.

(5) Examine the IR spectrum. The strong absorption at 1700 cm^{-1} is attributed to a C=O stretching vibration (ketone or aldehyde) that is conjugated with the double bond. The carbonyl resonance in the ^{13}C NMR spectrum indicates that this carbonyl group is an aldehyde (1 H attached). The proton NMR spectrum also shows a signal for the aldehyde proton. The proton NMR data are summarized as follows.

Chemical shift (ppm)	Integrated intensity	Assignment	Multiplicity	No. adjacent protons (n)	
10.4	1	CHO	singlet	0	
7.30	1	ArH	singlet	0	
7.15	1	ArH	doublet	1	1,2,4 substitution
6.90	1	ArH	doublet	1	
3.90	3	OCH$_3$	singlet	0	
3.80	3	OCH$_3$	singlet	0	

There are three ways to arrange the groups on a 1,2,4-trisubstituted benzene ring, and these are as follows:

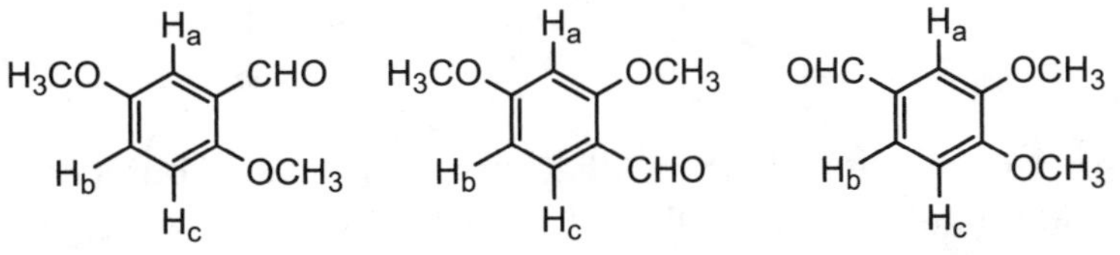

I   **II**   **III**

17.44. (continued)

To differentiate among these isomers using the chemical shift data is not trivial. We can rule out isomer **II** because H_a is between two strongly activating groups, so it should be shifted significantly upfield from δ 7.25. This is not the case however; in fact the signal for H_a is slightly downfield from δ 7.25. To evaluate structures **I** and **III**, it is necessary to know that the signal for a proton ortho to a carbonyl-containing substituent is farther downfield from δ 7.25 than are signals for a proton that is para to a carbonyl group. Structure **III** has two protons ortho to the aldehyde substituent, whereas compound **I** has only one proton ortho to the CHO group. Because only one signal in the aromatic region is downfield from δ 7.25, we would choose structure **I**. (In fact, the proton NMR spectrum for compound **III** does have two signals downfield from δ 7.25.)

17.45. The splitting patterns observed for benzene protons in a proton NMR spectrum are dominated by coupling with other protons that are ortho to the one giving the signal. This spectrum has aromatic protons, and the integrated intensity value of 4 (total) indicates that the ring is disubstituted. The observation of two doublets and two triplets means that the ring is ortho disubstituted (Figure 17.5).

Subtracting the atoms of the disubstituted benzene fragment (C_6H_4) from those in the molecular formula (C_6H_6BrN) leaves Br and NH_2, which constitute the two substituents.

The amino group is a strongly activating group, so the signals assigned to the protons ortho and para to the amino group should be upfield from δ 7.25, as they are. Protons ortho and para to the bromine atom should be around δ 7.25 (slightly upfield or downfield) since this substituent is a weak deactivator but an ortho/para director.

The protons of the amino group generate the broad peak that appears at approximately δ 3.8.

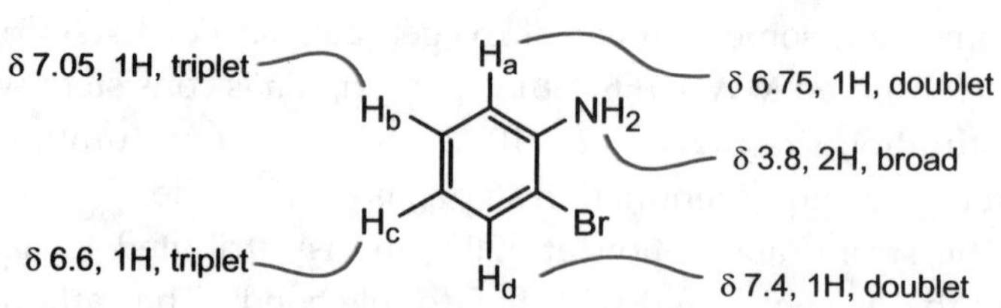

17.46. The appearance of a sizable $M + 4$ peak in the mass spectrum of a compound usually means that two halogen atoms are present. We know from the proton NMR spectrum that a trisubstituted benzene ring is present, and we also know that the molecule contains nitrogen, because the MW is an odd number. We can also surmise, based on the positions of the resonances, one of which is significantly downfield from δ 7.25, that a nitro group is present. Adding the masses for the principal isotopes of C_6H_3 and NO_2 gives 121, which is 70 less than the molecular mass of 191. With this assumption, we conclude that two ^{35}Cl atoms are also present to account for the mass of the molecular ion.

Looking at the splitting patterns of the peaks in the proton NMR spectrum reveals that the ring is 1,2,4-trisubstituted (Figure 13.12). There are three ways to arrange a nitro group and two chlorine atoms in this substitution pattern, and they are as follows:

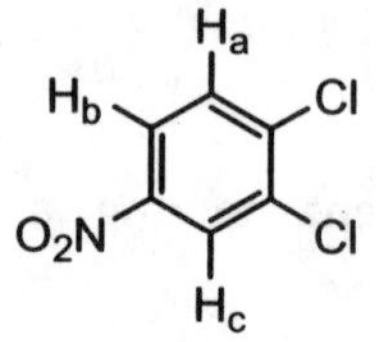

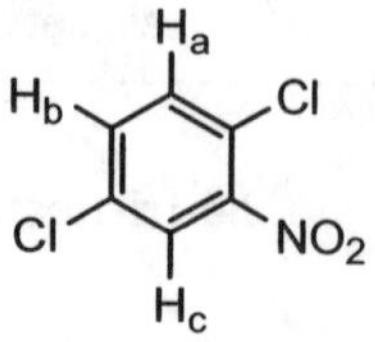

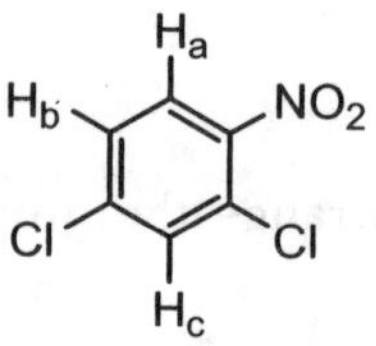

Proton H_c will appear as a singlet because there is no other proton ortho to it; but in structures **I** and **II**, H_c is ortho to the nitro group and should appear farthest downfield. Therefore, we can conclude that this unknown compound is **2,4-dichloronitrobenzene (III)**.

NUCLEOPHILIC ADDITION REACTIONS
OF ALDEHYDES AND KETONES

18.1. Cyanide ion is a good nucleophile, and it adds to the ketone carbonyl group by reacting with the electrophilic carbon atom (step 1). The resulting alkoxide ion reacts with HCN, forming the cyanohydrin derivative of acetone and regenerating cyanide ion (step 2).

18.2. Recall from Section 3.2a that cyclopropane rings are highly strained. Even with bent bonds, the internal $\angle$C–C–C values are approximately $102°$, which is compressed from the normal value of $109.5°$ expected for a carbon atom with sp^3 hybridization. If a carbonyl group is present in a three-membered ring, then the disparity between the theoretical ($120°$) and actual ($102°$) bond angles is even greater. By reacting with water, the carbonyl carbon atom changes from sp^2 to sp^3 hybridization, which relieves some of the strain energy associated with having a small ring along with an sp^2–hybridized atom.

The bond angles around unconstrained carbon atoms with sp^2 and sp^3 hybridization are $120°$ and $109°$, respectively.

The bond angle at each vertex of a cyclopropane ring is about $102°$, which is closer to $109°$ than to $120°$.

18.3. Markovnikov addition of water to the double bond of 1-bromocyclohexene occurs by protonation of the π bond (step 1), followed by reaction of a water molecule at the electrophilic carbon atom (step 2).

Removal of a proton from the cation intermediate yields a geminal bromohydrin (step 3). Such species are unstable and readily lose HX (HBr in this case) to form the ketone product (step 4).

18.3. (continued)

18.4. Organometallic reagents undergo addition with the carbonyl groups of ketones and aldehydes: The R group of RLi or RMgX attaches to the carbon atom of the original carbonyl group, and the metal ion becomes associated with the oxygen atom. The corresponding alcohol is the product after aqueous acid workup. For both of the reactions in this exercise, a new chiral center is created by the addition process, so a racemic mixture of products is obtained.

18.5. An alcohol molecule is the product expected from the reaction between a metal hydride reagent (LiAlH$_4$ or NaBH$_4$) and the carbonyl group of an aldehyde or ketone. Both carbon and oxygen atoms of the original carbonyl group obtain a hydrogen atom during these reduction reactions. An aldehyde forms a primary alcohol, and a ketone forms a secondary alcohol. For the reactions in this exercise, a new chiral center is not formed, so the product molecules are achiral.

18.6. Unsymmetrical ketones in which the carbonyl group is attached to a large group (usually an aryl ring) and a small alkyl group can be reduced enantioselectively using the oxazaborolidine reagents described in the text. As noted in Example 18.1, the (S,S) reagent yields the (R)-alcohol when the aryl group has a higher Cahn-Ingold-Prelog priority than the alkyl group, which is the case for both of the substrate molecules in this exercise.

18.7. Unsymmetrical ketones in which the carbonyl group is attached to a large group (usually an aryl ring) and a small alkyl group can be reduced enantioselectively using the Noroyi catalysts described in the text. The (S,S) reagent yields the (R)-alcohol when the aryl group has a higher Cahn-Ingold-Prelog priority than the alkyl group. In this reaction, the alkyl group has a higher priority than the phenyl ring, so the (R,R) reagent will produce the (R)-alcohol.

18.8. The Friedel-Crafts acylation of benzene with benzoyl chloride yields benzophenone. Clemmensen reduction of this ketone produces the hydrocarbon diphenylmethane.

Considering a retrosynthesis of diphenylmethane that works back to bromobenzene and benzaldehyde, we consider that the starting material will likely be used in the Grignard reaction, the product of which will be the corresponding alcohol. A benzylic alcohol can be deoxygenated by hydrogenolysis (Section 12. 2a). The retrosynthesis and synthesis (starting compounds are shown in color) are as follows:

Retrosynthesis

Synthesis

18.9. Chromium and manganese-based oxidants that function in aqueous solution will convert aldehydes to carboxylic acids. The same types of oxidants in non-aqueous solvents do not oxidize aldehydes. Ketones are inert to oxidation by these reagents under most conditions.

a. Aqueous permanganate ion is potent enough to oxidize aldehydes. This reagent is used in basic solution, so the carboxylate salt is the first product, which yields the carboxylic acid after acid workup.

b. Chromium oxide in pyridine and dichloromethane will convert primary alcohols to aldehydes, but it will not oxidize the aldehyde product further.

c. Ketones are inert to reactions with chromium and manganese-based oxidants under most conditions.

18.10. The Baeyer-Villiger reaction occurs when a ketone or aldehyde is treated with a peracid. The product is the ester formed by inserting an oxygen atom into the bond between the carbonyl group and the substituent that is more highly substituted (aryl groups migrate more readily than alkyl groups). If the migrating group is chiral at the atom attached to the carbonyl group, it will migrate with retention of configuration, although neither of these substrate molecules is chiral. A cyclic ketone undergoes ring expansion.

a.

b.

18.11. Follow the procedure outlined in the solution to Exercise 1.34.

a. **(Z)-3-Chloro-2-pentenal**

pent	5 carbon atoms
en	a double bond at C2
al	aldehyde functional group; its presence defines C1 of the carbon chain
3-chloro	a Cl atom is attached at C3
(Z)	the higher priority groups at the ends of the double bond are on the same side

b. **2-Bromo-3-nitrobenzaldehyde**

benzaldehyde	the aldehyde derivative of benzene; the attachment point of the CHO group defines C1 of the ring
2-bromo	a Br atom is attached at C2
3-nitro	a NO_2 group is attached at C3

c. **(R)-3-Phenyl-4-pentene-2-one**

pent	5 carbon atoms
en	a double bond at C4
one	a ketone; the carbonyl group is at C2
3-phenyl	a C_6H_5 group at C3
(R)	the configuration of C3 is (R)

d. **4,4-Dimethyl-2-cyclopentenecarbaldehyde**

cyclopent	5 carbon atoms in a ring
en	double bond starting at C2
carbaldehyde	an aldehyde group is attached to the ring; its point of attachment defines C1
4,4-dimethyl	two CH_3 groups at C4

18.12. Follow the procedures outlined in the solution to Exercises 13.21 and 13.22.

a. This molecule has four types of protons: a methyl group, a methylene group, an alkene proton, and an aldehyde proton. The aldehyde and alkene protons are three bonds apart and will couple with each other, producing a doublet signal with an intensity value corresponding to 1H. The ethyl group should manifest itself as a quartet and triplet (intensity ratio 2:3). The chemical shift values for the protons are estimated using the data in Table 13.1 and Figure 13.3 and are approximate.)

The carbon NMR spectrum should have the signals shown; these are estimated by using the data in Table 13.4.

18.12. (continued)

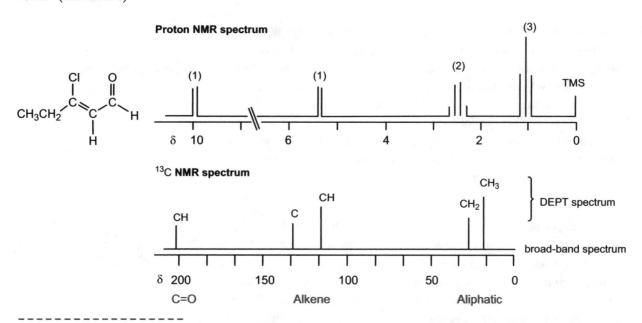

- - - - - - - - - - - - - - - - - - -

b. This molecule has only aryl and aldehyde protons. Spin-spin splitting will occur only between the protons attached to the benzene ring, and the pattern will consist of two doublets and a triplet (Figure 13.12). Because all of the substituents on the ring are electron-withdrawing, the aryl proton signals will appear downfield from δ 7.25 (Section 17.3g).

The carbon NMR spectrum should have the signals shown; these are estimated by applying the data in Table 13.4.

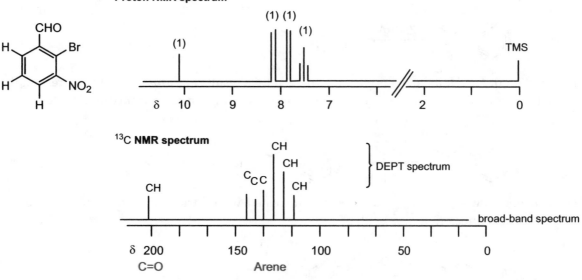

18.13. Follow the procedure outlined in the solution to Exercise 1.32.

a. This compound is a ketone with three carbon atoms in the chain, which has no double or triple bond: prop/an/one = propanone.

The principal functional group (ketone) is at C2. The substituent (cyclopentyl) defines the order of numbering so as to give it the lower number (1 instead of 3). The name of this molecule is **1-cyclopentyl-2-propanone**.

18.13. (continued)

b. This compound has seven carbon atoms in a chain without double or triple bonds. There are two functional groups, one of which (aldehyde) has to be at the end of the chain. The ketone can be anywhere in the chain. Numbering starts with the aldehyde because that gives a lower set of numbers (1 and 5 versus 3 and 7): hept/an/al = heptanal.

   The ketone functional group is not the principal one, so it is treated as a substituent at C5 using the prefix oxo. The name of this molecule is **5-oxoheptanal**.

----

c. This compound is a ketone with four carbon atoms in the chain, which has no double or triple bond: but/an/one = butanone.

   The principal functional group (ketone) is at C2. Two methyl groups are attached at C3, and a phenyl group and chlorine atom are attached at C4, creating a chiral center. The name of this molecule is **(S)-4-chloro-3,3-dimethyl-4-phenyl-2-butanone**.

18.14. Ketones reacts with a variety of reagents by undergoing addition of nucleophiles to their carbon–oxygen double bonds. The benzene ring of 1-phenyl-1-propanone (propiophenone) is also susceptible to reactions with electrophiles as described in Chapter 17.

a. The combination of Zn(Hg) and HCl constitutes the Clemmensen reduction, which converts a ketone carbonyl group to a methylene group.

b. The reaction of NaBH$_4$ in aqueous ethanol with a ketone results in formation of the corresponding 2° alcohol. A new chiral center is formed, so the product comprises a racemic mixture.

c. The benzene ring will undergo nitration when treated with nitric and sulfuric acids. The ketone group is a meta director.

d. The reaction of LiAlH$_4$ with a ketone results in formation of the corresponding 2° alcohol after hydrolytic workup. A new chiral center is formed, so the product comprises a racemic mixture.

18.15. A general method for making aldehydes is by oxidation of the corresponding primary alcohol molecule. Hydroboration and substitution reactions can be used to make 1° alcohols that are subsequently oxidized. Ozonolysis is a good way to make an aldehyde if the starting material can be converted to an alkene and has more carbon atoms than the desired product.

a. From 1-butanol: oxidation of a primary alcohol

b. From 1-bromobutane: substitution to make a primary alcohol followed by oxidation

c. From 1-butene: hydroboration followed by oxidative hydrolysis and oxidation

d. From 1-pentene: ozonolysis followed by reductive workup

18.16. For the synthesis of many molecule types, the alcohol functional group defines a logical place at which to make a disconnection in a retrosynthesis. If the synthetic goal has a ring as part of the carbon skeleton, consider ways to include the ring within one of the starting materials. For instance, benzene derivatives are often made using the phenyl Grignard reagent. Several ways to prepare each molecule can be conceived; only one route for each is given. The starting compounds are shown in color.

a. This secondary alcohol molecule can be made by the Grignard reaction with an aldehyde.

Retrosynthesis

Synthesis

b. This tertiary alcohol can be made by the Grignard reaction with a ketone.

Retrosynthesis

18.16. (continued)

Synthesis

c. This tertiary alcohol can be made by the Grignard reaction with a ketone, but in this case, the aldehyde group must either be protected or created after the Grignard reaction has been accomplished. Ozonolysis is used to make the aldehyde from an alkene.

Retrosynthesis

Synthesis

18.17. In the pinacol rearrangement, **2,3-dimethyl-2,3-butanediol** (pinacol) reacts with acid and is protonated at one of its equivalent OH groups (step 1). Dissociation of a molecule of water (step 2) yields a carbocation, and a methyl group migrates to generate the more stable hydroxy-substituted cation (step 3). Deprotonation yields pinacolone, **3,3-dimethyl-2-butanone** (step 4).

2,3-Dimethyl-2,3-butandiol

resonance stabilized   **3,3-Dimethyl-2-butanone**

18.18. The reactions described in this chapter comprise addition, reduction, and oxidation processes. To predict the structures of the products in each of the following transformations, learn the details of each reaction type, which can be found in the reaction summary section of the chapter. It is also important to know which reagents do not react with aldehydes and ketones.

a. This transformation involves an addition reaction between a ketone and a Grignard reagent to form a 3° alcohol molecule. A new chiral center is formed, so the product is a racemic mixture.

racemic

18.18. (continued)

b. Ketones do not undergo addition reactions with organocuprates, but organoiodides react via substitution. The ethyl group of the organocuprate reagent replaces the iodine atom of the reactant.

c. This transformation is an addition reaction between an aldehyde and a Grignard reagent. The product is a secondary alcohol. A new chiral center is formed, but the starting material has a chiral center to start, so the product comprises a mixture of two diastereomers.

d. Aldehydes react with HCN and the cyanide ion to form cyanohydrins. A new chiral center is formed, but the starting material has three chiral centers to start, so the product comprises a mixture of two diastereomers.

e. Ketones and aldehydes react with sodium borohydride to form alcohols. The product is achiral.

f. Ketones (aryl ketones in particular) are reduced completely when treated with amalgamated zinc in hydrochloric acid. The carbonyl group is converted to the methylene group.

g. Ketones are formed from secondary alcohols upon treatment with most types of oxidizing agents.

h. This transformation involves the addition reaction between an aldehyde and a Grignard reagent to form a secondary alcohol molecule. A new chiral center is formed, so the product is a racemic mixture.

18.18. (continued)

i.   Aldehydes are oxidized to carboxylic acids by a variety of chromium and manganese reagents in aqueous solution.

j.   Ketones react with peracids to form esters in a transformation called the Baeyer-Villiger reaction. The migration aptitude of the carbon-containing substituent follows the order, aryl > 4° > 3° > 2° > 1° carbon atom.

18.19. The reaction between a ketone and a peracid constitutes the Baeyer-Villiger reaction, which yields an ester as the product. The migration aptitude of the carbon-containing substituent follows the order, aryl > 4° > 3° > 2° > 1° carbon atom. A chiral center migrates with retention of configuration.

a.   b.   c.

18.20. When the cyanide ion is present with a ketone or aldehyde molecule, the nucleophilic carbon atom of CN⁻ reacts with the electrophilic carbon atom of the carbonyl group (step 1). The silicon–cyanide bond is weaker than a silicon–oxygen bond, so the alkoxy group reacts with cyanotrimethylsilane to displace the cyanide ion and form the trimethylsilyl derivative of the cyanohydrin (step 2). The cyanide ion is thereby regenerated and so fulfills its role as a catalyst in this transformation.

a.   Fluorine has an extremely high affinity for silicon. Therefore, the reaction of a trimethylsilyl cyanohydrin with HF leads to removal of the trimethylsilyl group and protonation of the oxygen atom. A cyanohydrin, however, is stable toward weak acids, so no further reaction occurs.

b.   A trimethylsilyl ether is stable toward base, so no reaction occurs.

c.   Fluorine has an extremely high affinity for silicon. Therefore, the fluoride ion reacts with the trimethylsilyl group, which generates the conjugate base of the cyanohydrin. Displacement of the cyanide ion leads to formation of the ketone carbonyl group.

18.21. The hydroxide ion reacts with HN_3 to form water and the azide ion, which is a good nucleophile. The azide ion adds to the carbonyl group (step 2), and the resulting alkoxide ion reacts with HN_3 to regenerate the azide ion (step 3). Steps 2 and 3 can repeat until the reactants are depleted.

18.22. The reactions described in this chapter mainly comprise addition, reduction, and oxidation reactions. To predict the structures of the products in each of the following transformations, learn the details of each reaction type, which can be found in the reaction summary section of the chapter.

a. The first step of this sequence is an enantioselective reduction reaction. The (S,S)-oxazaborolidine reagent produces the (R)-alcohol from an aryl alkyl ketone when the aryl group has a higher Cahn-Ingold-Prelog priority than the alkyl chain. The second step converts the alcohol to its trimethylsilyl ether derivative.

b. The first transformation in this sequence is an enantioselective reduction reaction, the results of which are determined by comparison with the example given in the chapter. The second step of this sequence converts the alcohol to its alkoxide derivative, which subsequently displaces the chloride ion to form the cyclic ether (see Section 7.2c).

c. The first step of this sequence involves the electrophilic bromination of the benzene ring; the ketone substituent is a meta director. The second transformation is an enantioselective reduction reaction. The (S,S)-catalyst produces the (R)-alcohol from an aryl alkyl ketone when the aryl group has a higher priority than the alkyl chain as it does here.

d. This transformation involves the Grignard reaction with a ketone to form a tertiary alcohol product. The starting material also has a carboxylic acid group, so an excess of the Grignard reagent is needed to react first with this acidic proton. The carboxylate ion does not react further, and the acid workup protonates the carboxylate ion to regenerate the carboxylic acid group.

18.23. For many molecule types, the alcohol functional group defines a place at which to make the disconnection in the retrosynthesis. If the synthetic goal has a ring as part of the carbon skeleton, consider ways to include the ring in one of the starting materials. There are several ways to prepare the molecules shown below. Only one route for each is presented. The starting compounds are shown in color.

a. A tertiary alcohol is the product of the reaction between a ketone and a Grignard reagent. The alkyl halide needed to make the Grignard reagent is made by the anti-Markovnikov addition of HBr to an alkene. The ketone is prepared by oxidation of a secondary alcohol, which is prepared by hydration of an alkene.

Retrosynthesis

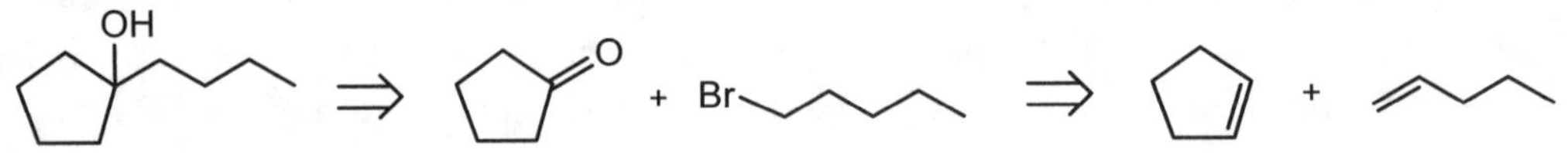

Synthesis

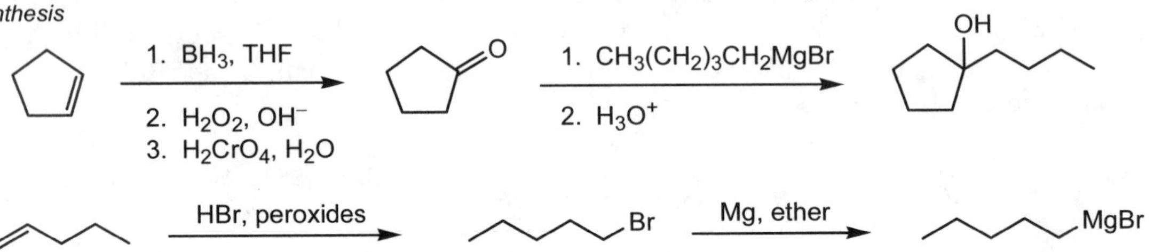

b. A cyclic ester is prepared from a ketone using the Baeyer-Villiger reaction. The ketone is prepared by oxidizing a 2° alcohol, which is made by anti-Markovnikov hydration of the cyclic alkene.

Retrosynthesis

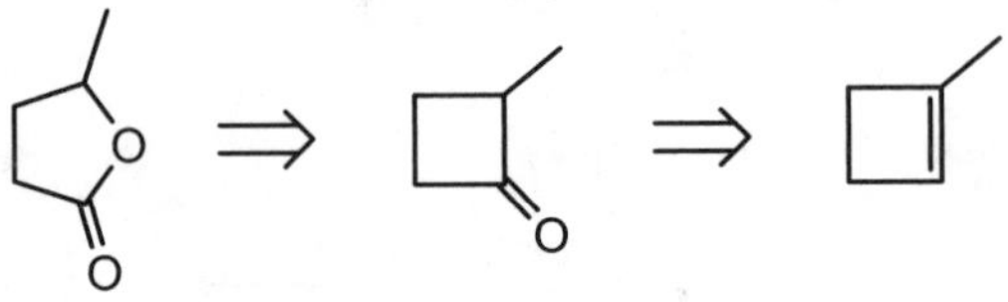

Synthesis

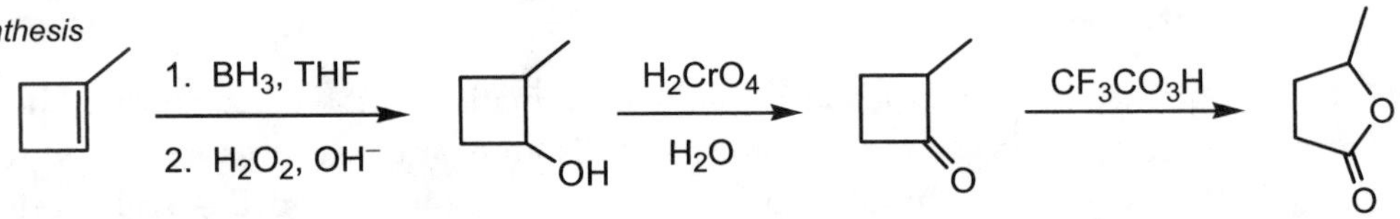

18.24. You learned several ways in Chapter 16 for making chiral alcohols, and the enantioselective reduction of ketones provides another strategy. For the first step of the retrosynthesis, consider which ketone can be reduced to form the specified alcohol, and then decide how the ketone can be prepared.

a. An aryl alkyl ketone is reduced in an enantioselective fashion by oxazaborolidine reagents and by hydrogenation procedures. Such ketones are readily prepared by the Friedel-Crafts acylation reaction.

Retrosynthesis

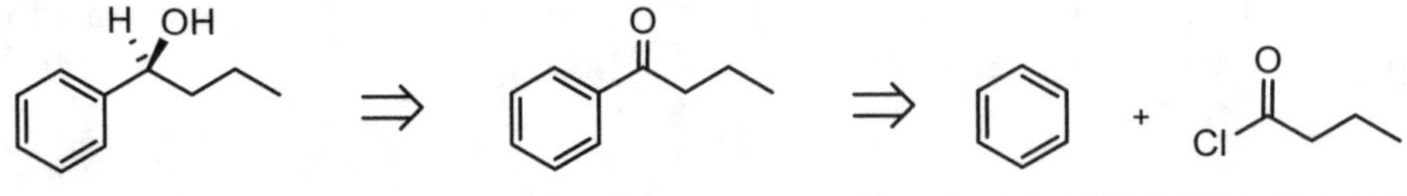

Synthesis

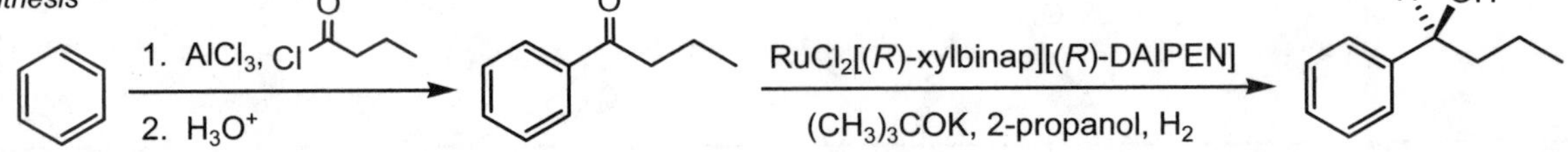

18.24. (continued)

b. A dialkyl ketone is reduced enantioselectively by oxazaborolidine reagents if the two alkyl groups are significantly different in size. The ketone can be prepared by oxidizing a secondary alcohol, which in turn can be made as a racemic mixture by the Grignard reaction with an aldehyde.

Retrosynthesis

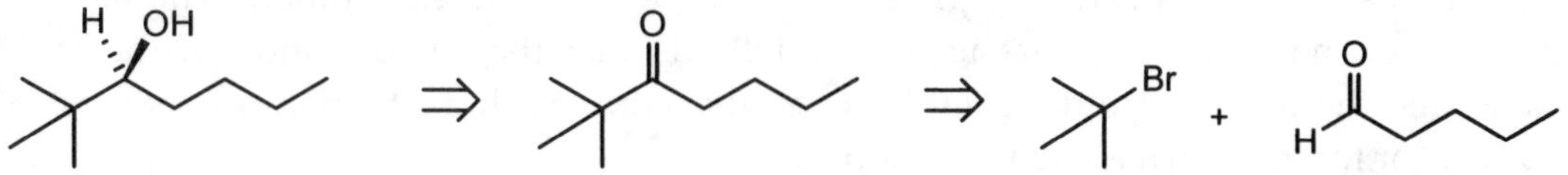

Synthesis

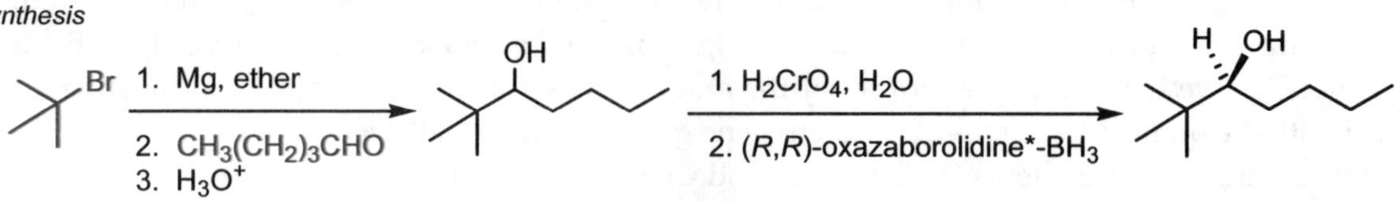

18.25. When *B*-allyl-9-BBN is treated with benzaldehyde, an acid-base reaction takes place between the boron atom, a Lewis acid, and the oxygen atom of benzaldehyde, a Lewis base. Such reactions are common for boron-containing substances.

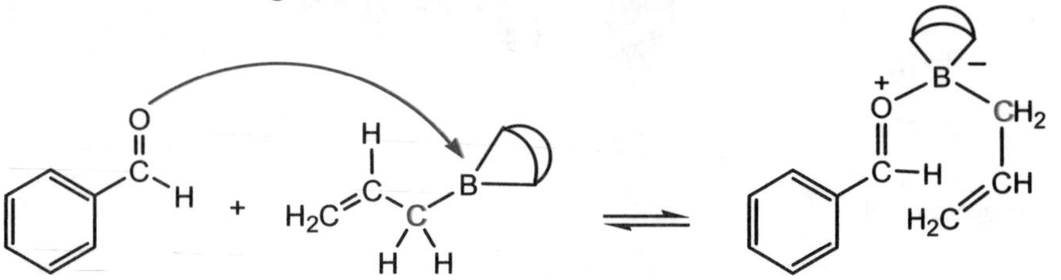

In the next step, electrons move through a concerted six-membered ring transition state to form the new carbon-carbon bond between the terminus of the allyl group and the carbonyl carbon atom. The ^{13}C atom is shown in color.

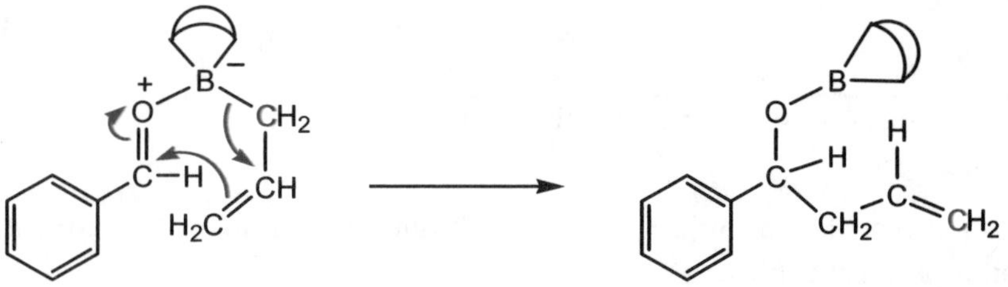

18.26. The first step is to summarize the reactions described in the accompanying paragraph of this exercise.

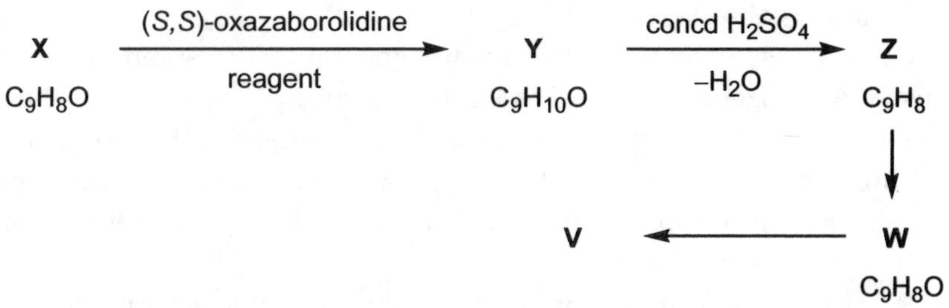

Next, use the spectroscopic data to identify one or more of the unknowns. The IR spectrum of **X** displays a strong absorption band at 1705 cm⁻¹, which can be interpreted as a C=O stretching vibration. Looking at the appropriate portions of the spectrum summarized in Table 14.4, we conclude that the compound is a ketone because diagnostic bands for the other types of carbonyl groups are absent.

18.26. (continued)

The proton NMR spectrum shows that the compound is a disubstituted aromatic compound (δ 7-8) because the total integrated intensity for this region is 4H. From the splitting pattern in the aromatic region (two doublets and two triplets), we conclude that the two substituents are ortho.

In addition, compound **X** has two methylene groups (δ 3.1, 2H and δ 2.65, 2H). These signals exhibit spin-spin splitting, so the two methylene groups must be adjacent to each other. The fact that these splitting patterns do not appear as classic triplets tells us that the conformation of the molecule is constrained so that the protons are not completely equivalent. Such a situation occurs most often for protons in a ring (or for those that are diastereotopic).

Using the given molecular formula, we calculate that there are six sites of unsaturation—four are in the benzene ring and one is the ketone carbonyl group, so there must be one additional double bond or ring. Because we have accounted for all of the atoms, this sixth site of unsaturation must be a ring, in accord with the qualitative results of the methylene group splitting patterns.

Assembling these fragments (o-C_6H_4, C=O, and CH_2CH_2) generates 1-indanone.

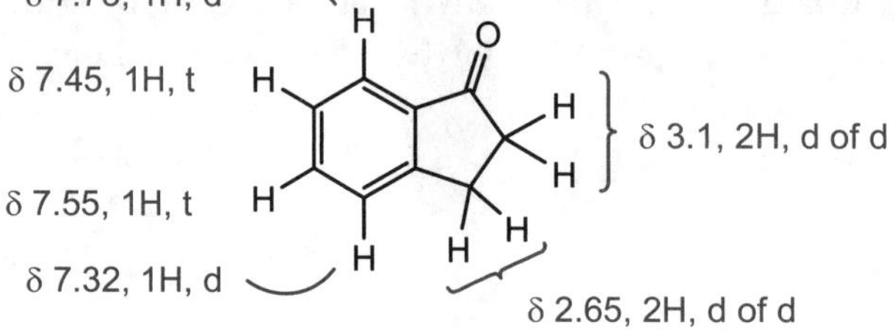

δ 7.73, 1H, d
δ 7.45, 1H, t
δ 7.55, 1H, t
δ 7.32, 1H, d
δ 3.1, 2H, d of d
δ 2.65, 2H, d of d

The compounds therefore have the following structures.

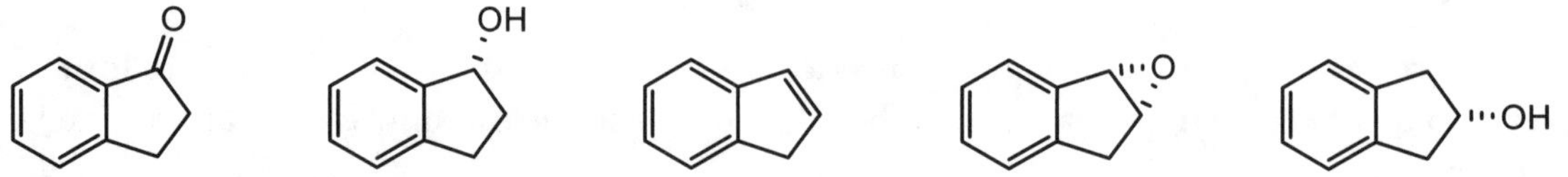

X   **Y**   **Z**   **W**   **V**

18.27. Follow the procedures outlined in the solution to Exercise 14.21.

a. **Compound 27A**

   (1) From the given molecular formula, C_7H_5ClO, calculate the number of unsaturation sites: $[2(7) + 2 - 6]/2 = 5$ sites of unsaturation.

   (2) Examine the IR spectrum. The IR spectrum of compound **27A** displays a strong absorption band at 1700 cm^{-1} that is assigned as a carbonyl stretching vibration. Looking at the appropriate portions of the spectrum, we conclude that the compound is an aldehyde because a band for the C–H stretching vibration is observed at 2710 cm^{-1}.

   (3) Examine the proton NMR spectrum. There are two general features: an aldehyde proton and aromatic protons. A benzene ring and a carbonyl group account for the five sites of unsaturation. The integrated intensity values and splitting pattern in the aromatic region reveals that the benzene ring has two substituents that are para to each other (two doublets). One is the aldehyde group. The other is a chlorine atom, which is the only element not accounted for by the other data.

   (4) The carbon NMR spectra confirm the presence of the aldehyde group and a disubstituted benzene ring.

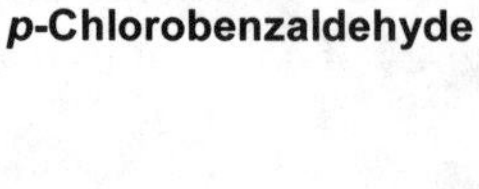

Cl—⟨ ⟩—CHO

***p*-Chlorobenzaldehyde**

18.27. (continued)

b. **Compound 27B**

(1) From the given molecular formula, which is $C_6H_{12}O$, calculate the number of unsaturation sites: $[2(6) + 2 - 12]/2 = 1$ site of unsaturation.

(2) Examine the IR spectrum. The IR spectrum of compound **27B** displays a strong absorption band at 1720 cm^{-1} that is assigned as a carbonyl stretching vibration. Looking at the appropriate portions of the spectrum, we conclude that the compound is a ketone because diagnostic bands for other types of carbonyl groups are absent.

(3) Examine the proton NMR spectrum. This molecule has only aliphatic protons: the resonances are attributable to the presence of a methyl group adjacent to a carbonyl group, a methylene group adjacent to a carbonyl and a methine group, and two methyl groups that are adjacent to a methine group. This combination is diagnostic for the isobutyl group (Exercise 13.9), although this assignment is made indirectly because the splitting pattern of the CH group is not readily discerned.

Chemical shift (ppm)	Integrated intensity	Assignment	Multiplicity	No. adjacent protons (n)	
2.35	2	CH$_2$	doublet	1	
2.1	1	CH	multiplet	several	Based on the splitting patterns, these groups must be adjacent to the CH group.
2.1	3	CH$_3$	singlet	0	
0.95	6	CH$_3$ (2)	doublet	1	

The compound is **4-methyl-2-pentanone**.

4-Methyl-2-pentanone

ADDITION–SUBSTITUTION REACTIONS OF ALDEHYDES AND KETONES; CARBOHYDRATE CHEMISTRY

19.1. One factor that accounts for the instability of a hemiacetal is in the strength of the carbonyl double bond that is formed when a molecule of water is removed. If the process occurs in a stepwise fashion, the alkoxy oxygen atom if first protonated, and then the OH group is deprotonated, concomitant with displacement of the alcohol molecule.

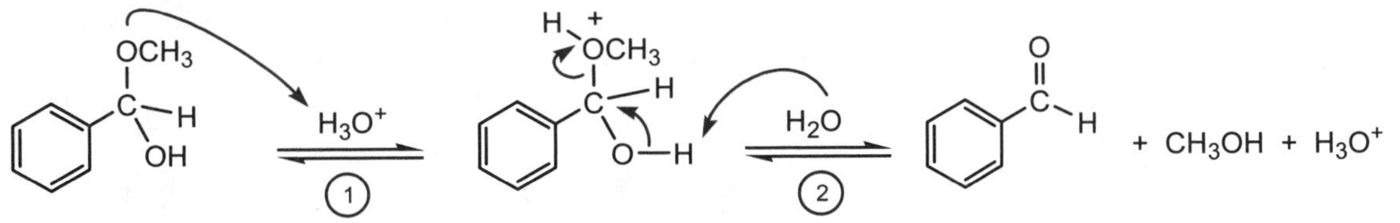

The overall process may be concerted, however, in which case it occurs as shown by the following equation.

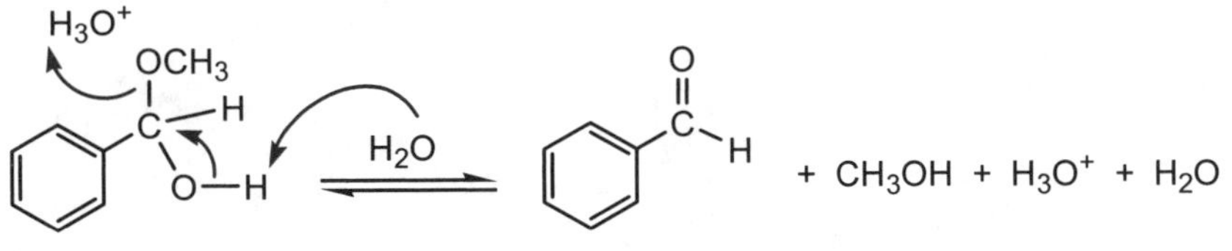

In either case, the aldehyde is formed by loss of a molecule of alcohol.

19.2. In the reaction between benzaldehyde and ethylene glycol, the carbonyl group of benzaldehyde is protonated by its reaction with *p*-toluenesulfonic acid (step 1), and then an oxygen atom of ethylene glycol acts as a nucleophile and reacts at the carbon atom of the cationic species (step 2).

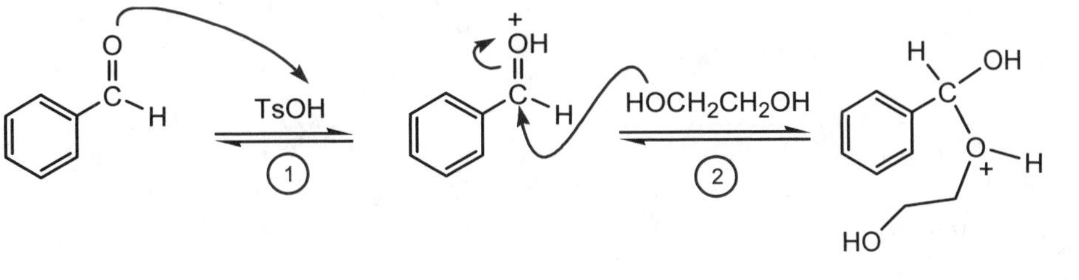

Next, a proton is transferred from one oxygen atom to the other (step 3). This step is shown here as an intramolecular process, but it more likely occurs between different molecules of the cationic intermediate, which would rely on the acid and base properties of the solvent and/or *p*-TsOH. Loss of water generates another carbocation (step 4).

19.2. (continued)

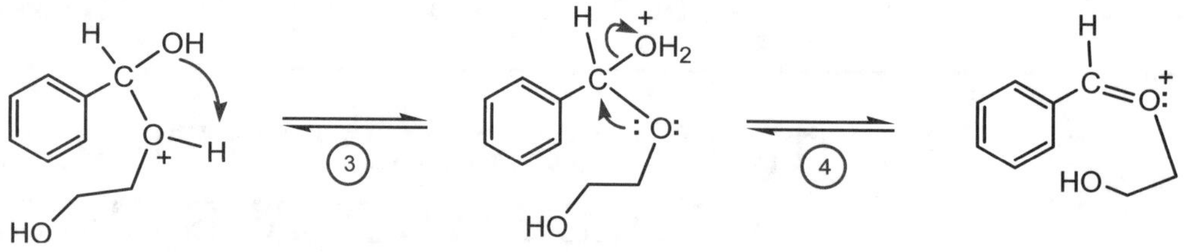

The other oxygen atom of ethylene glycol molecule can now react with the electrophilic carbon atom (step 5), and a final acid-base reaction (step 6) yields the ethylene acetal of benzaldehyde. Another name for this product is 2-phenyl-1,3-dioxolane (1,3-dioxalane is a five-membered ring with oxygen atoms at positions 1 and 3).

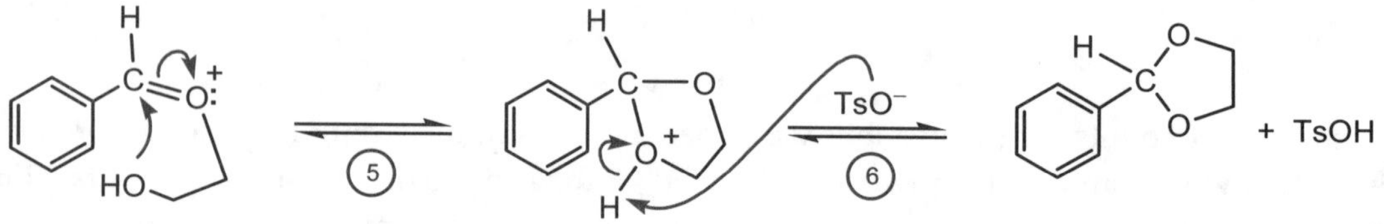

19.3. In the formation of an acetal from a vinyl ether molecule, the alkene double bond is protonated by its reaction with sulfuric acid (step 1). The proton adds in the Markovnikov fashion, generating a resonance-stabilized cation (the resonance form is not shown). The nucleophilic oxygen atom of the alcohol molecule intercepts this cation (step 2), and an acid-base reaction occurs to form the product (step 3).

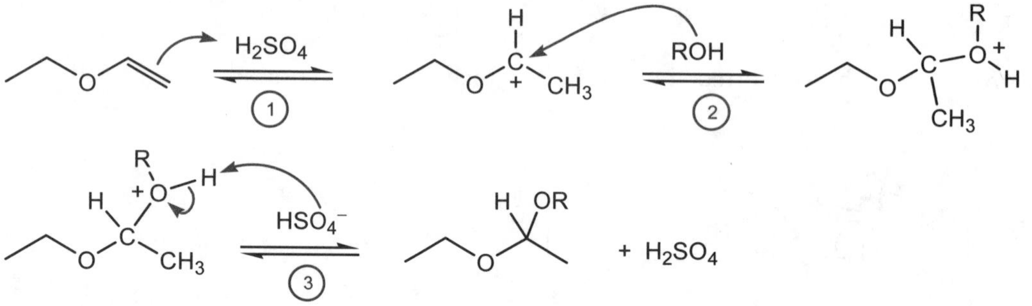

19.4. The formation of a carbonyl compound from a hemiacetal was depicted in the solution to Exercise 19.1. Shown below is the concerted version of the mechanism.

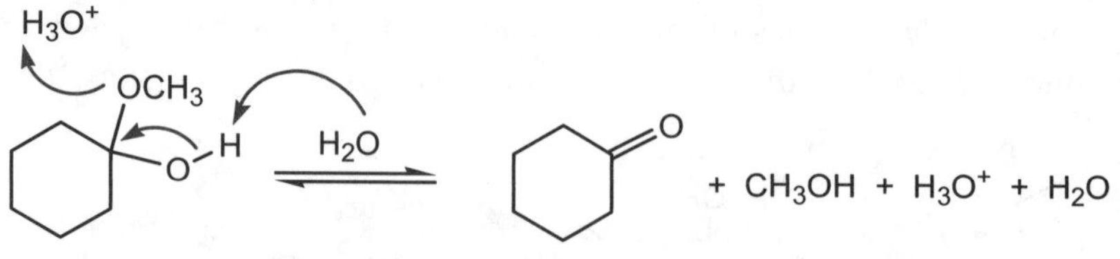

19.5. The mechanism for the hydrolysis reaction of a THP ether molecule is the same as that for the hydrolysis of any acetal compound. The exocyclic (outside of the ring) oxygen atom is protonated (step 1) to generate a good leaving group. Next, an electron pair on the endocyclic (within the ring) oxygen atom displaces the alcohol molecule (step 2).

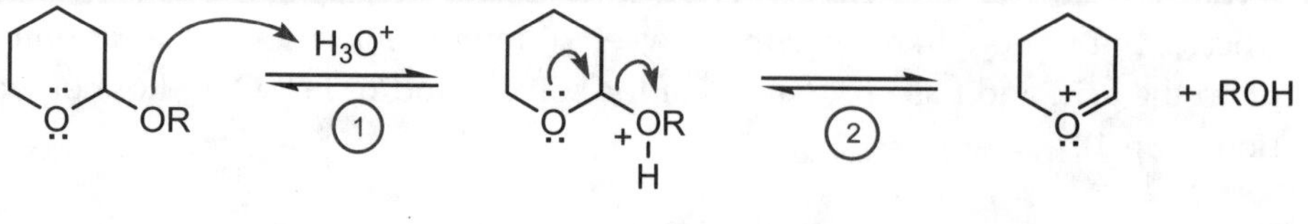

19.5. (continued)

A molecule of water reacts with this cation species (step 3), and an acid-base reaction takes place to yield the neutral hemiacetal (step 4).

19.6. The MOM derivative of an alcohol is made by using the Williamson ether synthesis (Section 7.2b). The alkoxide ion is first generated by treating the alcohol with NaH in DMF, and then chloromethyl methyl ether is added. An S_N2 reaction yields the protected alcohol derivative.

19.7. In this transformation, the OH group of the hemithioacetal reacts with zinc chloride to generate a good leaving group (step 1). Dissociation of the leaving group produces an electrophilic species (step 2) that is intercepted by the sulfur atom of a thiol molecule (step 3). The hydroxide ion associated with zinc chloride removes a proton to produce the thioacetal (step 4) and regenerate the Lewis acid catalyst, $ZnCl_2$.

19.8. Analogous to the reaction depicted in the solution to Exercise 19.1 for the decomposition of a hemiacetal, a hemithioacetal is deprotonated to form the carbonyl group and a thiolate ion. An acid-base reaction with carbonic acid regenerates the bicarbonate ion and produces the thiol.

19.9. The first two steps of this sequence constitute the Friedel-Crafts acylation procedure (Section 17.2e). Formation of the thioacetal derivative occurs in the third step, which proceeds by the mechanism illustrated in the solution to Exercise 19.7. Raney nickel replaces the sulfur atoms with hydrogen atoms, converting the thioacetal carbon atom to a methylene group. Ethane and hydrogen sulfide are two by-products of the last step.

19.10. Make use of the examples in Section 19.3a along with the structures shown in Figure 19.1 to draw the structures that correspond to the given names.

a.   To draw the structure of an L-carbohydrate, first draw the structure of the D-enantiomer, and then switch the positions of the H atom and the OH group at each chiral carbon atom.

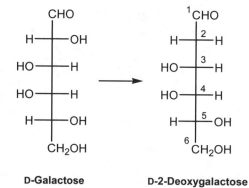

D-Arabinose   L-Arabinose

b.   To draw the structure of a deoxy sugar, first draw the structure of the parent carbohydrate, and then replace the OH group at each specified position with a hydrogen atom.

D-Galactose   D-2-Deoxygalactose

c.   To draw the structure of a deoxy L-carbohydrate, first draw the structure of the parent carbohydrate, and then replace the OH group at the specified position(s) with an H atom. Finally, switch the positions of the H atom and the OH group at each chiral carbon atom.

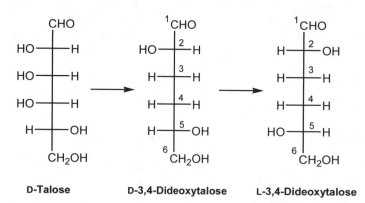

D-Talose   D-3,4-Dideoxytalose   L-3,4-Dideoxytalose

19.11. Follow the procedures outlined in Example 19.1. Remember that in the conversion of a Fischer projection to the Haworth structure, a group on the **left** (Fischer) is **up** (Haworth) and one on the **right** (Fischer) is **down** (Haworth).

a.   The β anomer has the OH group to the left in the Fischer projection and up in the Haworth. The furanose suffix means that the cyclic form has a five-membered ring.

19.11. (continued)

D-Arabinose   β-D-Arabinofuranose

b. The α anomer has the OH group to the right in the Fischer projection and down in the Haworth. The pyranose suffix means that the cyclic form has a six-membered ring.

D-Galactose   α-D-Galactopyranose

c. For L-carbohydrates, the α anomer has the OH group to the left in the Fischer projection. The furanose suffix means that the cyclic form has a five-membered ring. For an L-carbohydrate, the anomeric carbon atom (C1) is at the left side of the Haworth structure, and the correlations between the Fischer and Haworth structures are reversed from those of D-carbohydrates. Thus, for L-carbohydrates, a group on the **right** (Fischer) is **up** (Haworth) and one on the **left** (Fischer) is **down** (Haworth).

D-Ribose   L-Ribose   α-L-Ribofuranose

d. The β anomer has the OH group to the left in the Fischer projection. The furanose suffix means that the cyclic form has a five-membered ring. Note that an aldotetrose cannot form a six-membered ring. Furthermore, the furanose form uses the terminal carbon atom of an aldotetrose to generate the hemiacetal form, so the customary CH2OH group next to the endocyclic oxygen atom is not present.

19.11. (continued)

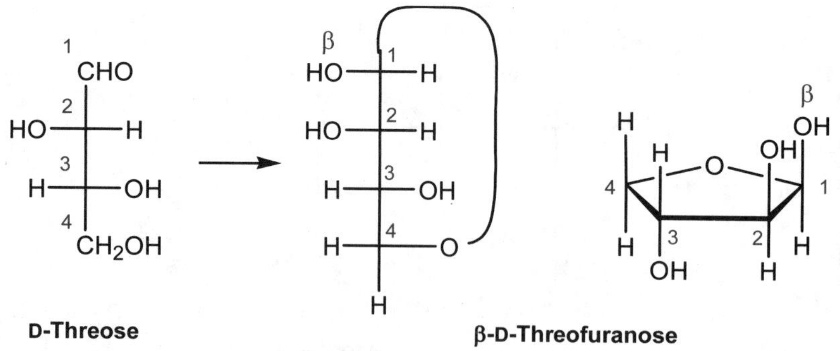

D-Threose → β-D-Threofuranose

e. The β anomer has the OH group to the left in the Fischer projection. The furanose suffix means that the cyclic form has a five-membered ring.

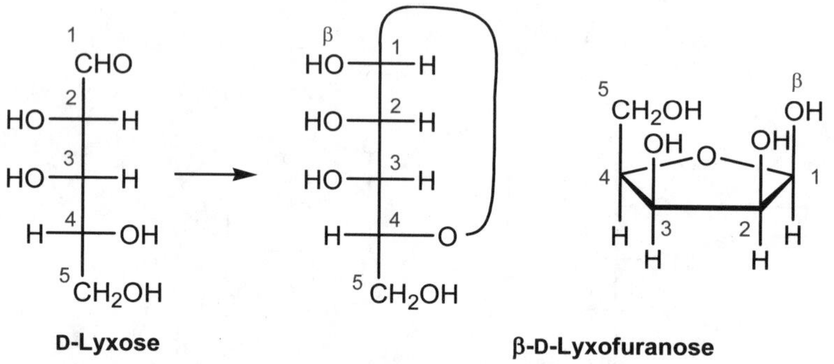

D-Lyxose → β-D-Lyxofuranose

f. The α anomer has the OH group to the right in the Fischer projection. The pyranose suffix means that the cyclic form has a six-membered ring.

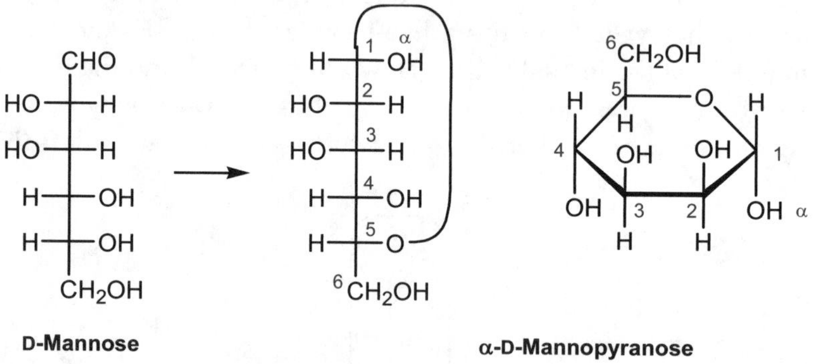

D-Mannose → α-D-Mannopyranose

19.12. The hemiacetal forms of carbohydrates in solution exist as an equilibrium mixture with their open-chain forms. This process is facilitated in aqueous solution because protons can be shuttled as needed.

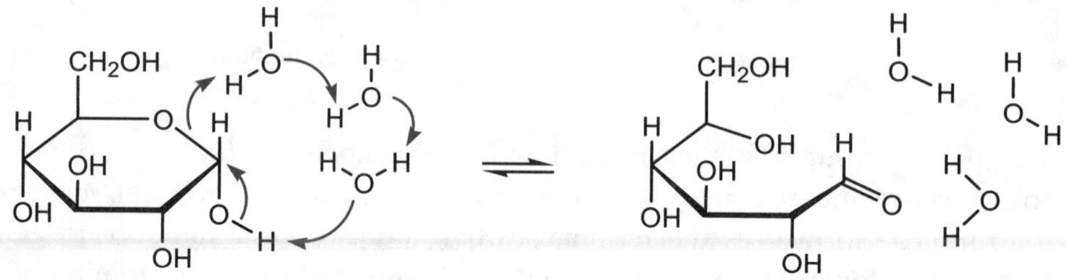

Redrawing the open form in the more familiar Fischer projection,

19.12. (continued)

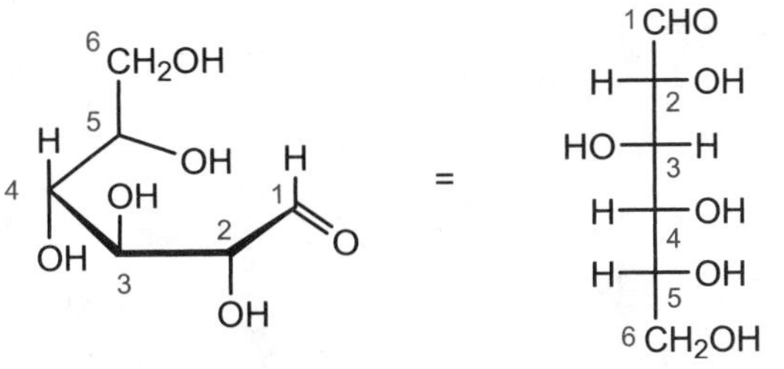

19.13. To interpret the name of a carbohydrate with the suffix "–oside," first recognize that the sugar is a glycoside (an acetal). The unnumbered prefix is the group bonded to the exocyclic oxygen atom at the anomeric carbon atom. Otherwise, follow the examples illustrated in the solution to Exercise 19.11.

a. The β anomer has the OR group (R = ethyl) to the left in the Fischer projection. The furanoside suffix means that the cyclic form has a five-membered ring. The parent carbohydrate is a 2-ketose; therefore, the anomeric carbon atom will have the OR group as well as the CH₂OH group.

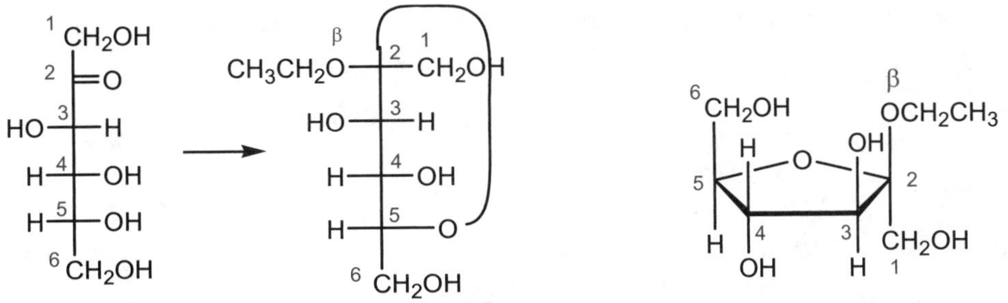

D-Fructose

Ethyl β-D-Fructofuranoside

b. The α anomer has the OR group (R = isopropyl) to the right in the Fischer projection. The pyranoside suffix means that the cyclic form has a six-membered ring.

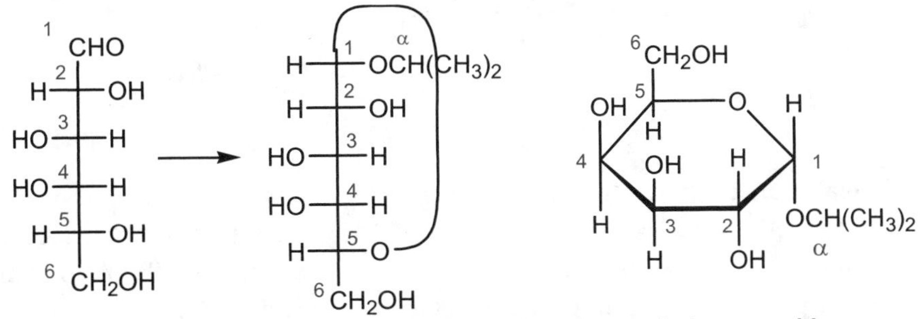

D-Galactose

Isopropyl α-D-galactopyranoside

c. The α anomer has the OR (R = methyl) group to the right in the Fischer projection. The furanoside suffix means that the cyclic form has a five-membered ring.

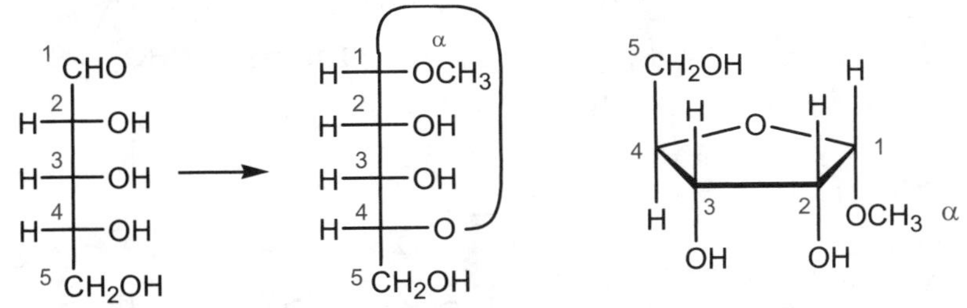

D-Ribose

Methyl α-D-ribofuranoside

19.13   (continued)

d.   The β anomer has the OR (R = phenyl) group to the left in the Fischer projection. The pyranoside
suffix means that the cyclic form has a six-membered ring.

D-Allose   **Phenyl β-D-allopyranoside**

19.14. The anomer of a carbohydrate is an isomer in which only the configuration of the hemiacetal
carbon atom is inverted. Two anomeric carbon atoms are present in lactose. The one that exists in the
hemiacetal form is at the lower right end of the structure (shown in color). β-Lactose has the anomeric
OH group in the equatorial position. Anomers interconvert via the open form (middle, bottom).

α-Lactose   **β-Lactose**

19.15. A bacterial cell wall can be hydrolyzed in strong acid by protonation of the exocyclic acetal
oxygen atom (step 1). The endocyclic oxygen atom subsequently displaces the polysaccharide alcohol
group (step 2), and then a molecule of water intercepts this cation species (step 3). A final acid-base step
produces the hemiacetal form of the other half of the polysaccharide chain (step 4).

19.15. (continued)

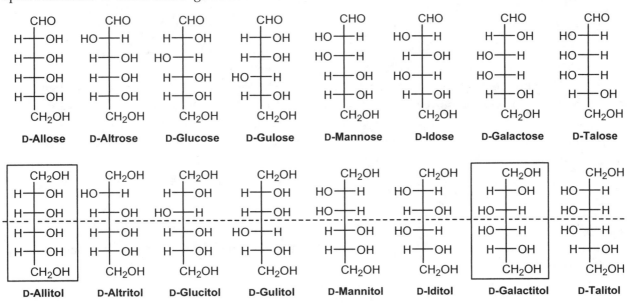

19.16. In order for cell wall cleavage to occur in the reaction catalyzed by lysozyme, an acid (Glu-35) is required to protonate the acetal oxygen atom. After dissociation of the alcohol fragment, a carbocation forms. This intermediate is stabilized partially by the proximity of the negative charge of carboxylate ion of Asp-52. If both carboxylic acid side chains (Asp and Glu) are protonated, then no carboxylate ion is present to stabilize the cation intermediate. If both side chains are deprotonated, then a proton is not available to initiate the first step of the reaction sequence. At the optimum pH value, both reactive centers—a carboxylic acid group and a carboxylate ion—are present in the active site of the enzyme.

19.17. If a D-aldohexose is known to yield a meso alditol upon reduction, then the OH groups attached to C2 & C5 and to C3 & C4 must have mirror-image relationships to each other (the mirror plane is shown as a dashed, colored line in the structures below). Only two D-aldohexoses form meso alditols upon reduction: D-allose and D-galactose.

19.18. Because the lactone groups that are formed in this oxidized sugar exist as five-membered rings, the ring junction between these two rings should be cis. For the oxidation product of glucose, a dilactone can form because the configurations of C3 and C4 are opposite each other. Thus, the orientations are correct to form two, five-membered rings without inducing significant strain into the molecule. (If you have difficulty seeing how this structure forms, it will help to make a model.)

19.19. If a D-aldopentose yields a meso diacid upon oxidation of its terminal groups, then the OH groups attached at C2 and C4 must have a mirror relationship to each other (a mirror plane passes through C3). Only two aldopentoses yield meso aldaric acids: D-ribose and D-xylose.

D-Ribose   D-Arabinose   D-Xylose   D-Lyxose

D-Ribaric acid   D-Arabaric acid   D-Xylaric acid   D-Lyitaric acid

19.20. To name a carbohydrate, first identify the root word by comparing the structure of the compound in its Fischer projection with the structures illustrated in Figures 19.1 and 19.2.

- The name must include D or L according to the criteria described in Section 19.3a.
- If the molecule is cyclic, a suffix is appended to the root word to denote whether the compound exists in a five- (furanose) or six-membered (pyranose) ring.
- If the compound is an acetal, which is identified by the presence of an –OR group attached to the anomeric carbon atom, then the suffix is changed from "ose" to "oside" and the name of the substituent group of the acetal appears as a separate, unnumbered prefix.
- For cyclic forms, the configuration of the anomeric carbon atom must be specified as either α or β (Section 19.3b).
- The absence of OH groups is denoted by the prefix "deoxy" along with a number to indicate which carbon atom lacks the OH group.

A reducing sugar is one that has either an aldehyde or hemiacetal functional group (keto sugars are also reducing in many instances because they isomerize to aldoses). Compounds that are glycosides (acetals) are not reducing sugars.

a. This carbohydrate is a six-membered ring acetal (pyranoside) with the OR group up (β). The R group is "methyl." Breaking the ring and putting the OH groups in the appropriate positions of a Fischer projection (Haworth "up" = Fischer "left"), we see that this molecule has the L-configuration (the chiral center farthest from the carbonyl group is to the left). The molecule is **methyl β-L-lyxopyranoside**. This molecule is a non-reducing sugar.

L-Lyxose

19.20. (continued)

b. This carbohydrate is not cyclic, and the chiral center farthest from the carbonyl group has the OH group on the right, so the sugar has the D-configuration. Comparing the structures in Figure 19.1, we see that the molecule is **D-gulose**, which is a reducing sugar.

c. This carbohydrate is a five-membered ring hemiacetal (furanose) with the anomeric OH group up (β). Breaking the ring and putting the OH groups in their appropriate positions in the Fischer projection (Haworth "up" = Fischer "left"), we see that this molecule has the D-configuration (the chiral center farthest from the carbonyl group has the OH group to the right).   C2 lacks an OH group, so this compound is a 2-deoxy sugar. There are two possible names: the molecule is either **β-D-2-deoxyribofuranose** or **β-D-2-deoxyarabinofuranose**. This molecule is a reducing sugar.

D-2-Deoxyribose

D-2-Deoxyarabinose

d. This carbohydrate is a five-membered ring hemiacetal (furanose) with the anomeric OH group on the left (β). We can see that this molecule has the D-configuration (the chiral center farthest from the carbonyl group has the OH group on the right). The molecule is **β-D-lyxofuranose**. This molecule is a reducing sugar.

D-Lyxose

e. This carbohydrate is a five-membered ring acetal (furanoside) with the OR group to the right (α). The R group is "methyl." We can see that this molecule has the D-configuration (the chiral center farthest from the carbonyl group has the OH group on the right). The molecule is **methyl β-D-gulofuranoside**. This molecule is a non-reducing sugar.

D-Gulose

19.21. Follow the examples illustrated in the solution to Exercise 19.10.

a. To draw the structure of an L-carbohydrate, first draw the structure of the D-enantiomer, and then switch the positions of the H atom and OH group at each chiral carbon atom.

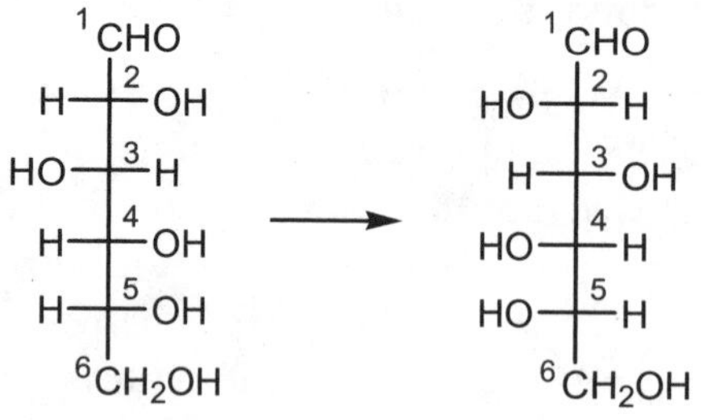

D-Glucose → L-Glucose

b. To draw the structure of a deoxy sugar, first draw the structure of the parent carbohydrate, and then replace the OH group at the specified position with an H atom.

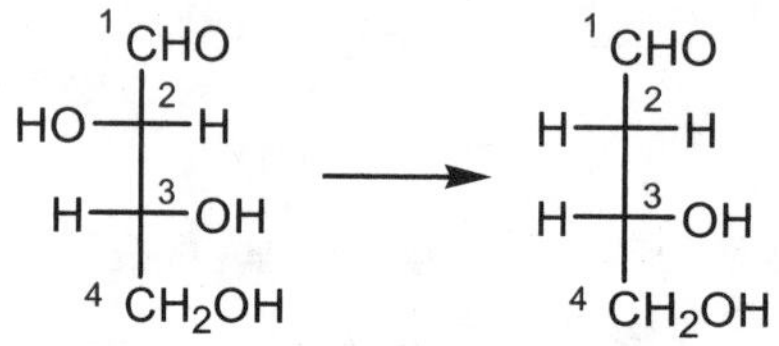

D-Threose → D-2-Deoxythreose

c. To draw the structure of a deoxy L-carbohydrate, first draw the structure of the parent carbohydrate, and then replace the OH group at the specified position with an H atom. Finally, switch the positions of the H atom and OH group at each chiral carbon atom.

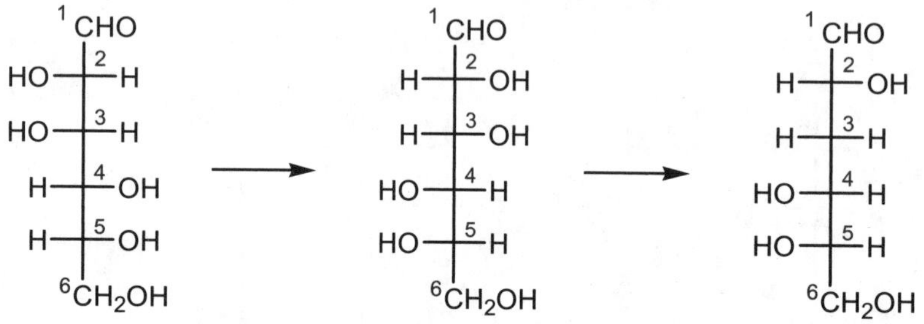

D-Mannose → L-Mannose → L-3-Deoxymannose

19.22. In carbohydrate names, the D or L designation specifies the stereochemistry of the stereogenic carbon atom farthest from the carbonyl group. An aldehyde sugar is an aldose, and a keto sugar is a ketose. The total number of carbon atoms identifies the carbohydrate as a tetrose, pentose, or hexose.

a.

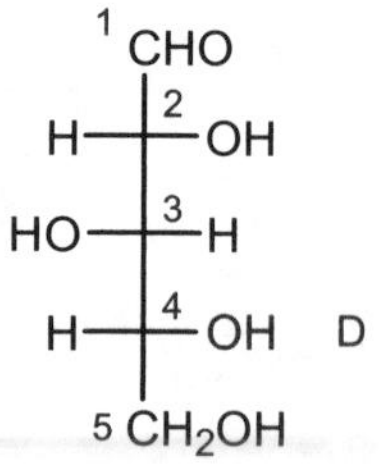

A D aldopentose

b.

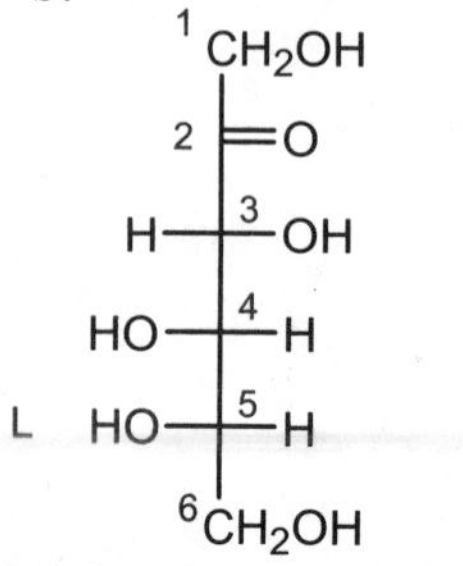

An L 2-ketohexose

c.

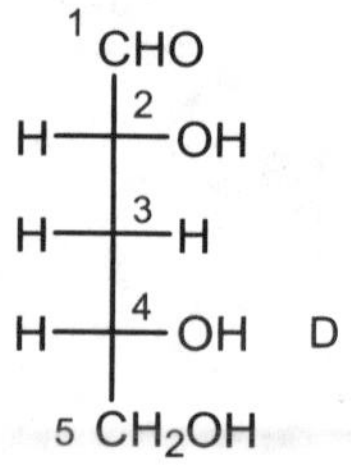

A D aldopentose

d.

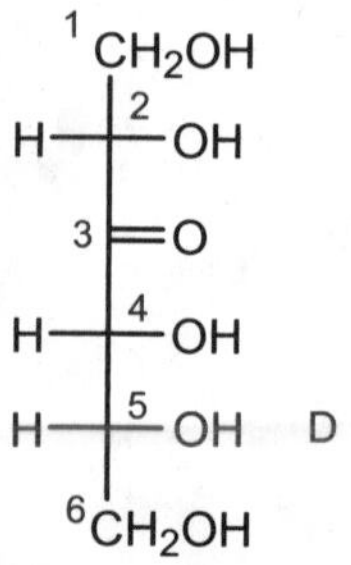

A D 3-ketohexose

19.23. Follow the procedures outlined in the solutions to Exercise 19.11 and 19.13.

a.   The β anomer has the OH group to the left in the Fischer projection. The furanose suffix means that the cyclic form has a five-membered ring. For an aldohexose, the furanose form has the chiral center at C5, which is an exocyclic center. Therefore, the absolute configuration must be assigned in order to make certain that the correct stereochemistry is drawn.

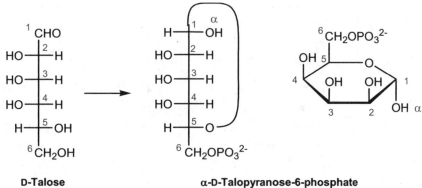

b.   The α anomer has the OH group to the right in the Fischer projection. The pyranoside suffix means that the cyclic form has a six-membered ring. The deoxy prefix means that the OH group has been replaced by H (at C2).

c.   The α anomer has the OH group to the right in the Fischer projection. The pyranoside suffix means that the cyclic form has a six-membered ring. The phosphate suffix means that the OH group has been replaced by the OPO_3^{2-} group (at C6).

d.   The β anomer has the OR group (R = methyl) to the left in the Fischer projection. The furanoside suffix means that the cyclic form has a five-membered ring. For an aldohexose, the furanose form has the chiral center at C5, which is an exocyclic center. Therefore, the absolute configuration must

19.23. (continued)

be assigned in order to make certain that the correct stereochemistry is drawn. Note that because the endocyclic oxygen atom (at C4) is on the left in the Fischer projection, C5 and C6 are below the plane. This molecule is a still a D-carbohydrate, so the anomeric carbon atom is at the right.

D-Galactose   **Methyl β-D-galactofuranoside**

19.24. The anomer of a carbohydrate is its C1 epimer (only the configuration of C1 changes). The enantiomer is the isomer in which every configuration has been inverted. The enantiomer is the non-superimposable mirror image of the original carbohydrate.

a.

b.

19.25. Follow the procedure outlined in the solution to Exercise 19.19. Note that the solution to this exercise asks about the non-meso compounds. The possible starting aldoses are D-arabinose and D-lyxose.

D-Ribose   **D-Arabinose**   **D-Xylose**   **D-Lyxose**

D-Ribitol   **D-Arabinitol**   **D-Xylitol**   **D-Lyxitol**

19.26. The aldoses that produce meso aldaric acids are the same ones that form meso alditols, as described in the solution to Exercise 19.17.

L-Allose → L-Allaric acid

L-Galactose → L-Galactaric acid

19.27. To predict the structures of the products in each of the following transformations, learn the details of each reaction type, which can be found in the reaction summary section of the chapter.

a. This transformation constitutes formation of the ethylene acetal derivative of an aldehyde. The product is achiral.

$HOCH_2CH_2OH$, TsOH, Δ   achiral

b. This transformation constitutes formation of the glycoside (acetal) derivative of a carbohydrate. The product mixture comprises a mixture of anomers; the remaining chiral centers are not affected.

CH_3CH_2OH, TsOH, Δ   OEt β + OEt α

c. This transformation constitutes oxidation of an aldose to an aldonic acid. The product can exist as a lactone. Except for the anomeric carbon atom, which is converted to an achiral carbonyl group, the configurations of the other chiral centers are not affected.

Br_2, H_2O, pH 5–6   or

d. This transformation is an example of acetal formation by addition of an alcohol to a vinyl ether. The product is achiral.

OCH_3 $(CH_3)_2CHOH$, TsOH, Δ OCH_3 $OCH(CH_3)_2$ achiral

e. An aldehyde is converted to a protected derivative when treated with a hydroxyl thiol. The product is formed as a racemic mixture.

CHO $HOCH_2CH_2CH_2SH$, TsOH, Δ   racemic

19.27. (continued)

f. The first step of this transformation is a substitution reaction. The product is a hemiacetal, which is unstable and loses a molecule of methanol to form the corresponding ketone. The product is achiral.

g. This transformation constitutes the oxidation of an aldose to an aldaric acid. (The product can also exist as a lactone.) The configurations of the chiral centers are not affected.

h. The starting compound is the hemiacetal derivative of a 2-ketose and is in equilibrium with the open form. The ketone carbonyl group is reduced to form an alcohol. Two epimers are formed, so the product exists as a mixture of diastereomers.

i. A ketone is converted to a thioacetal when treated with a dithiol and a Lewis acid. The thioacetal group is replaced by two hydrogen atoms when treated with Raney nickel. The product is meso.

j. The hydroxyl aldehyde exists as a cyclic hemiacetal, so its treatment with alcohol and strong acid leads to the formation of a diastereomeric mixture of a cyclic acetal.

19.28. Because the reagent needed to prepare the alcohol from the alkene also reacts with the carbonyl group, the latter must be protected. Formation of the acetal derivative is a satisfactory method for protecting aldehydes and ketones.

a. The retrosynthesis and synthesis schemes are straightforward, consisting of only protection, deprotection, and hydroboration reactions.

19.28. (continued)

Retrosynthesis

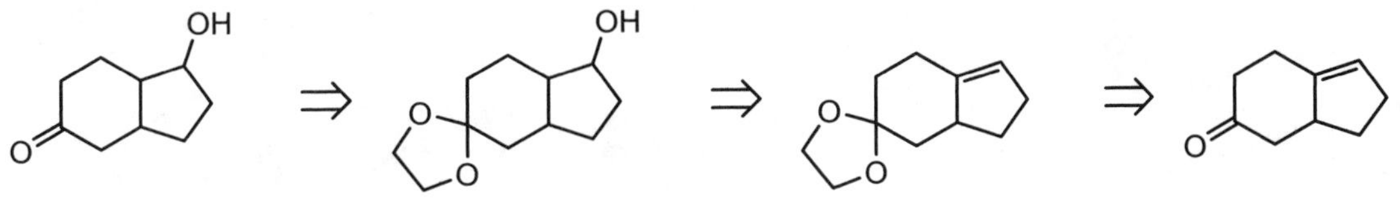

Synthesis

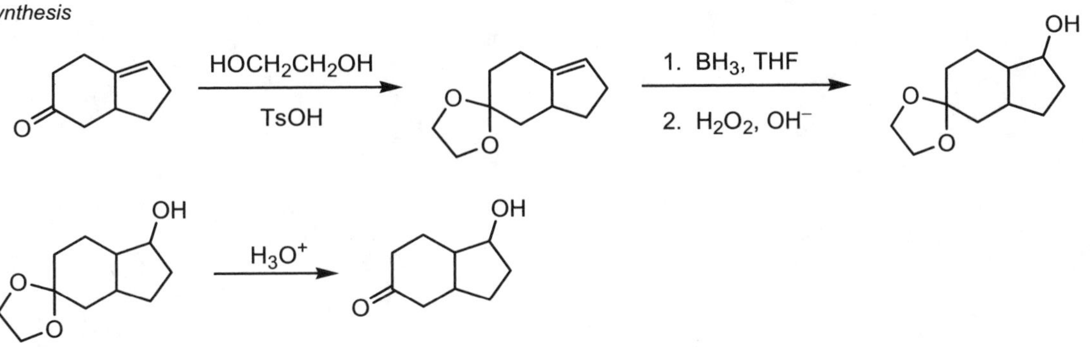

b.   The retrosynthesis of this molecule requires you to recognize that the product is the acetal derivative of a cyclic hemiacetal, which is equivalent to the corresponding hydroxyl aldehyde. The hydroxyl aldehyde molecule is made from the alkene after protecting the aldehyde group.

In the actual synthesis, the deprotection step also leads to formation of the cyclic hemiacetal. This molecule is converted to the acetal by treatment with methanol and a strong acid.

Retrosynthesis

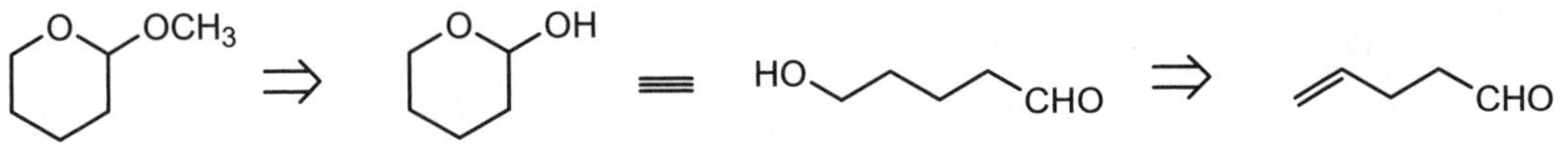

Synthesis

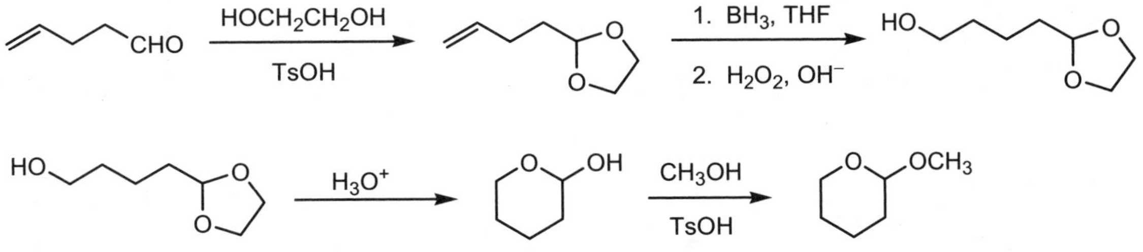

ADDITION-ELIMINATION REACTIONS
OF ALDEHYDES AND KETONES

20.1. The reaction between cyclohexanone and 2-amino-2-methylpropane (*tert*-butylamine) begins with addition of the nucleophilic nitrogen atom to the carbonyl group (step 1). Two proton transfer steps (steps 2 and 3) subsequently yield the neutral hemiaminal derivative. Protonation of the OH group (step 4) generates a good leaving group, and the electron pair on nitrogen displaces a molecule of water (step 5). Finally, the iminium ion is deprotonated to form the neutral imine (step 6).

20.2. The reaction between 3-pentanone and hydroxylamine begins with addition of the nucleophilic nitrogen atom to the carbonyl group (step 1). Two proton transfer steps (steps 2 and 3) subsequently yield a neutral derivative. Protonation of the OH group attached to nitrogen (step 4) generates a good leaving group.

20.2. (continued)

The electron pair on nitrogen displaces a molecule of water (step 5), and an acid-base reaction yields the neutral oxime (step 6).

20.3. The steps in this mechanism mirror those shown in the solution to Exercise 20.1. The process begins with protonation of the imine nitrogen atom (step 1). After a molecule of water adds to the carbon–nitrogen double bond (step 2), two proton transfer steps (steps 3 and 4) subsequently produce a protonated hemiaminal derivative. Deprotonation of the OH group (step 5) displaces the amine molecule, 2-amino-2-methylpropane, to yield cyclohexanone (step 5).

20.4. This reaction is an example of the Beckmann rearrangement. The leaving group is made by forming the tosylate derivative of the OH group (step 1), but the subsequent steps are the same as the acid-catalyzed process shown in the text.

Thus, heating the tosyl derivative in water leads to migration of the decalin ring from carbon to nitrogen with displacement of the tosylate group (step 2).

20.4. (continued)

The carbocation that is formed is intercepted by a molecule of water (step 3), and three proton transfer reactions (steps 4-6) occur to form the amide product.

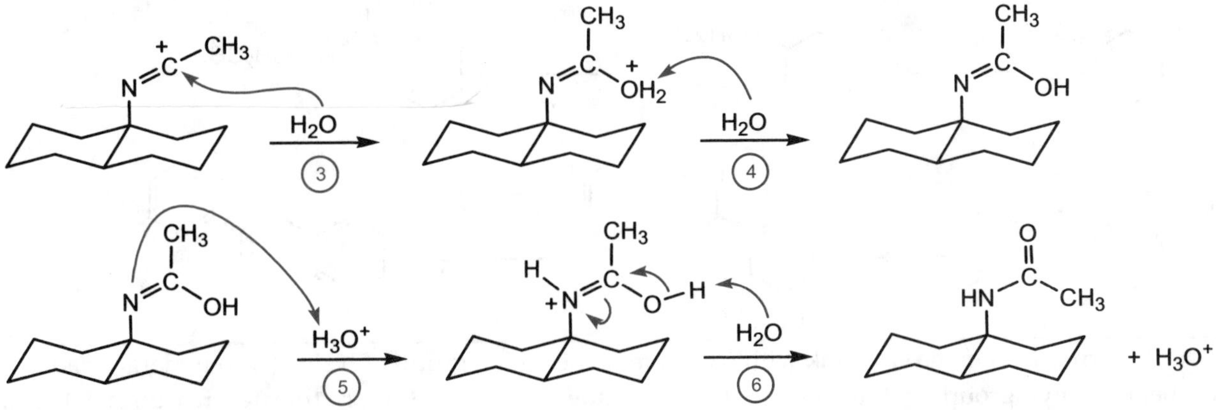

20.5. This transformation is an example of the classic Beckmann rearrangement. A good leaving group is formed after protonation of the oxime OH group (step 1). Migration of a carbon atom in the ring creates a seven-membered ring carbocation (step 2), which is intercepted by a molecule of water (step 3). Three proton transfer reactions (steps 4-6) occur to form the cyclic amide product, ε-caprolactam.

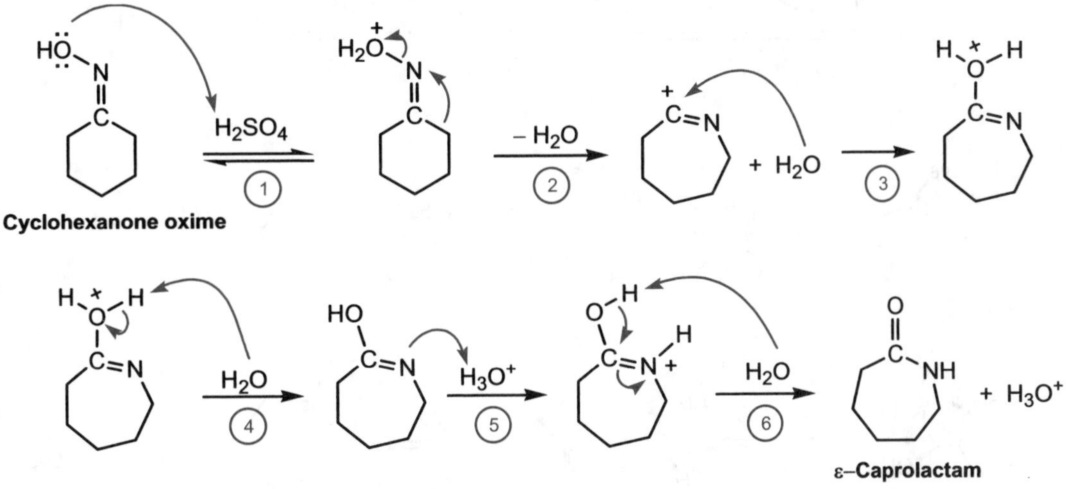

Cyclohexanone oxime

ε–Caprolactam

20.6. The first step of the reaction scheme in this exercise is the Wolff-Kishner reaction, which converts the ketone group to a methylene group via the hydrazone derivative (shown in brackets). Step 2 is the hydrolysis of an acetal, which unmasks the ketone group that had been protected. This transformation serves to underscore the fact that acetal protecting groups are stable toward bases and nucleophiles.

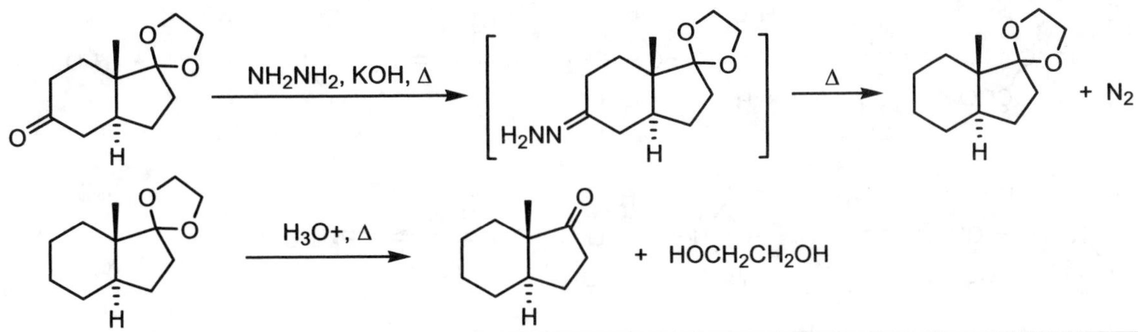

20.7. The first step in each of these reaction schemes is the formation of the imine derivative of the carbonyl compound. The combination of hydrogen and a metal catalyst reduces the carbon-nitrogen double bond to form the corresponding amine; sodium borohydride in aqueous ethanol accomplishes the same result.

a.

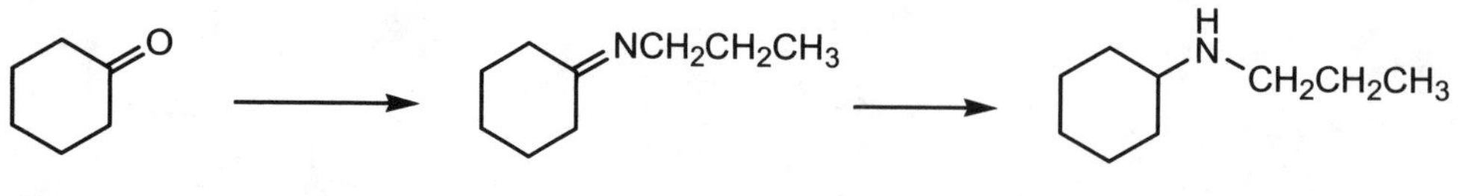

b.

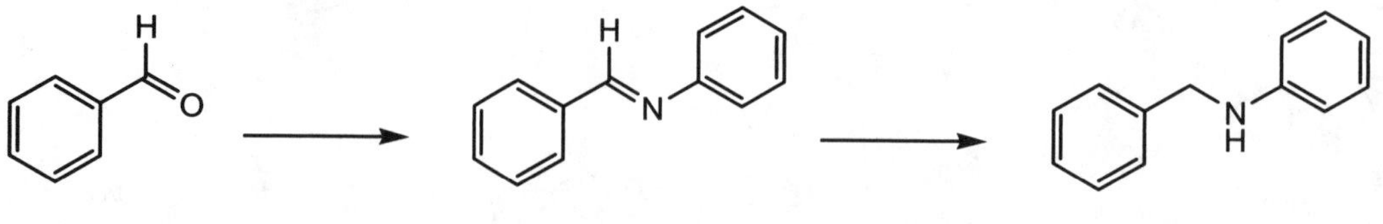

20.8. In forming an imine crosslink within a protein, the amino group of lysine (Lys) undergoes addition to the carbonyl group of the allysine (Als) residue (step 1). Proton transfer reactions yield the hemiaminal (steps 2 and 3). Protonation of the OH group generates a good leaving group (step 4), and the electron pair on nitrogen displaces a molecule of water to create the carbon-nitrogen double bond (step 5). Removal of the proton from the nitrogen atom yields the imine link (step 6).

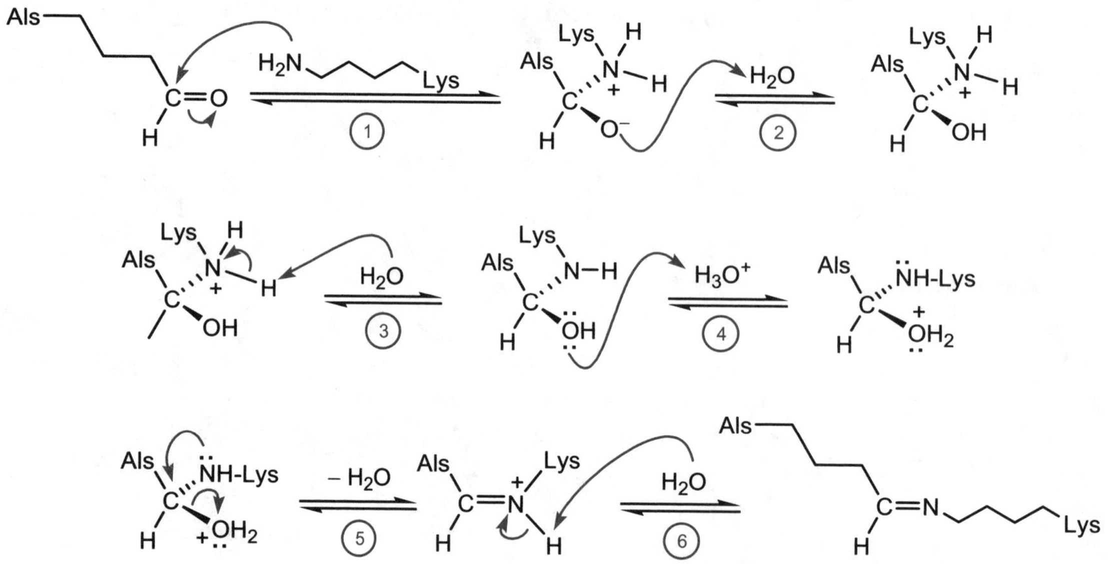

20.9. The mechanism for hydrolysis of the iminium C=N bond follows the same steps shown in the solution to Exercise 20.3. In the scheme shown below, Px is an abbreviation for the pyridine portion of pyridoxal phosphate.

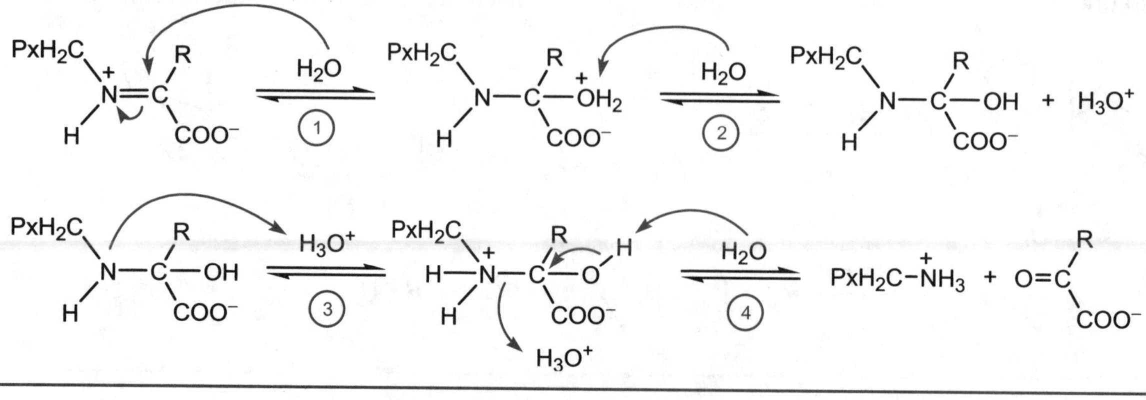

20.10. The carbon atom attached to the amino and carboxyl groups of an amino acid is converted during metabolism to a carbonyl group. The IUPAC prefix for this group is "oxo" and it appears at the 2-position. Thus, the products are either carboxylic acids or diacids, and to name each one, we look for the longest carbon chain that contains the acid group(s), and then specify the substituents and the positions to which they are attached.

Glutamic acid

2-Oxopentanedioic acid

Alanine

2-Oxopropanoic acid

Phenylalanine

2-Oxo-3-phenylpropanoic acid

Aspartic acid

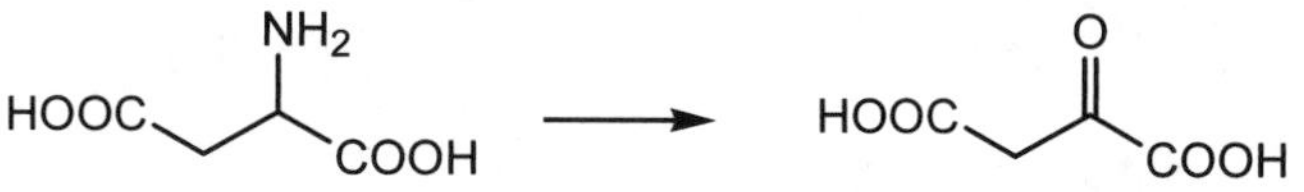

2-Oxobutanedioic acid

Valine

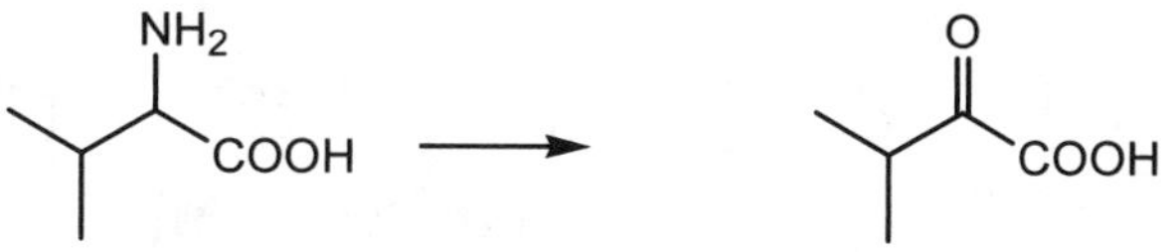

3-Methyl-2-oxobutanoic acid

Leucine

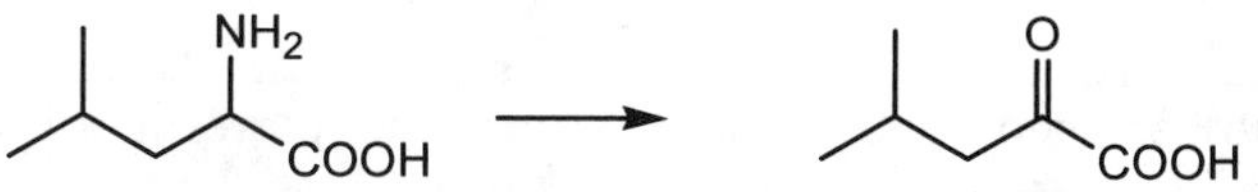

4-Methyl-2-oxopentanoic acid

20.11. A cyclic secondary amine reacts with a ketone to form an enamine. The reaction between cyclohexanone and pyrrolidine begins with nucleophilic addition of the nitrogen atom to the carbonyl group (step 1). Two proton transfer steps (steps 2 and 3) subsequently yield the neutral hemiaminal derivative. Protonation of the OH group (step 4) generates a good leaving group, and the electron pair on nitrogen displaces a molecule of water (step 5). Finally, the iminium ion is deprotonated at the carbon atom adjacent to the C=N bond, which forms the neutral enamine (step 6).

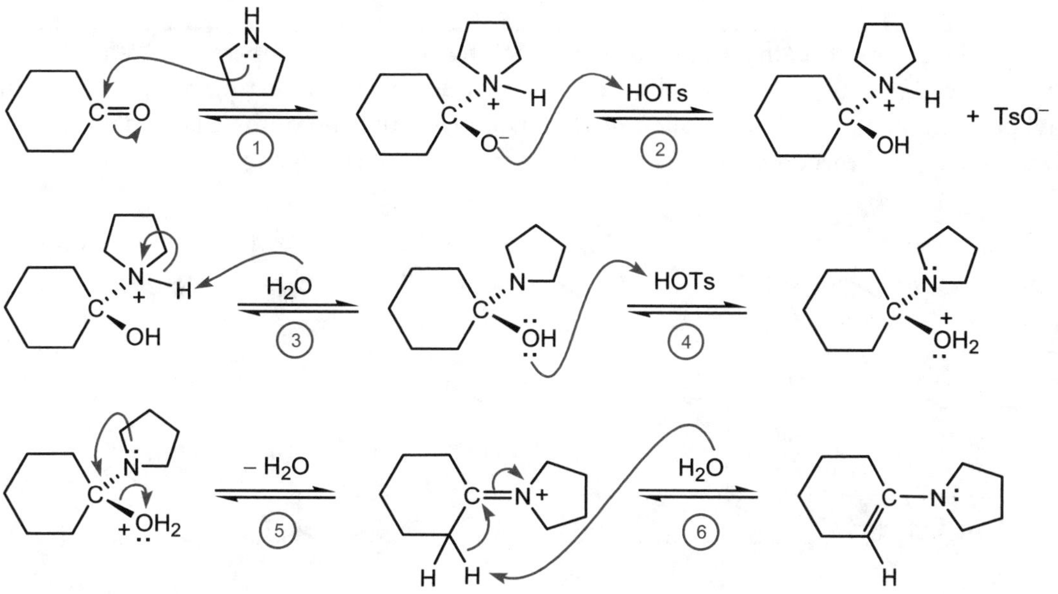

20.12. The mechanism for the first stage of this procedure was shown in the solution to Exercise 20.11. Stage 2 is the alkylation step, in which the enamine exerts its nucleophilic character to displace the iodide ion from methyl iodide. This is an S_N2 reaction.

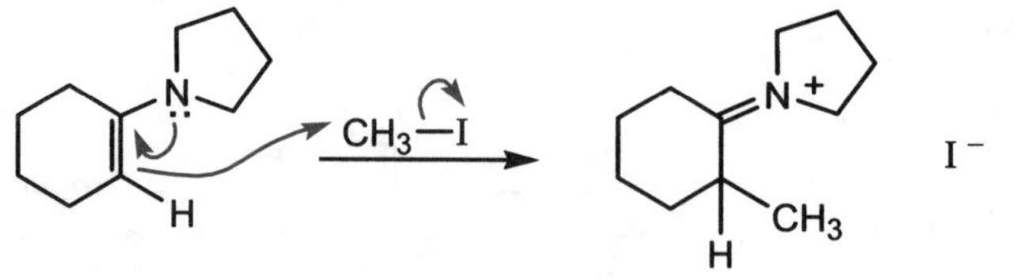

Stage 3 is a hydrolysis reaction. Water adds to the C=N bond (step 1), and two proton transfer reactions (steps 2 and 3) make the amino group into a good leaving group. Finally, a proton transfer reaction creates the carbonyl group with displacement of the amine molecule (step 4).

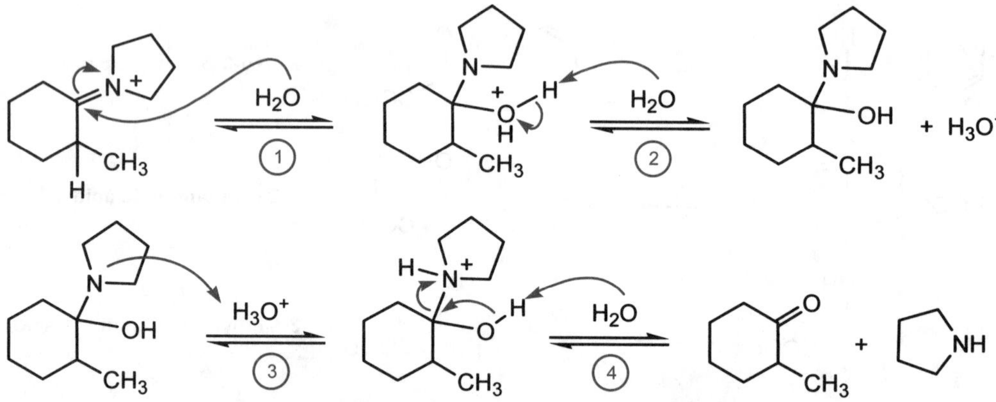

20.13. An alkyl halide reacts with triphenylphosphine to form the corresponding alkyltriphenyl-phosphonium salt. Treating this salt with strong base generates the ylide.

a.

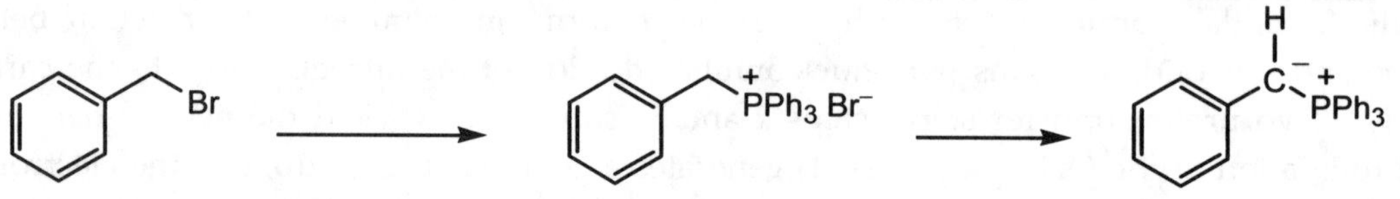

b.

20.14. The logical disconnection that makes use of the Wittig reaction to prepare an alkene is the double bond between the carbon atoms. If the double bond is exocyclic to a ring, the ring component normally is the carbonyl partner. In most cases, the ylide reactant comprises the end of the alkene double bond with the less highly-substituted carbon atom. The starting compounds are shown in color.

a.   *Retrosynthesis*

Synthesis

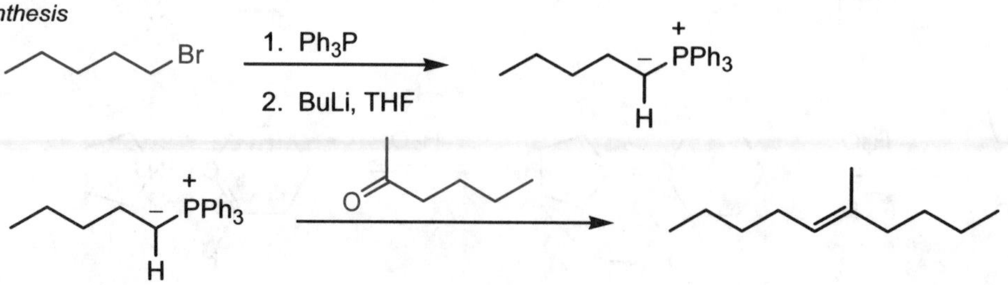

20.14. (continued)

b. *Retrosynthesis*

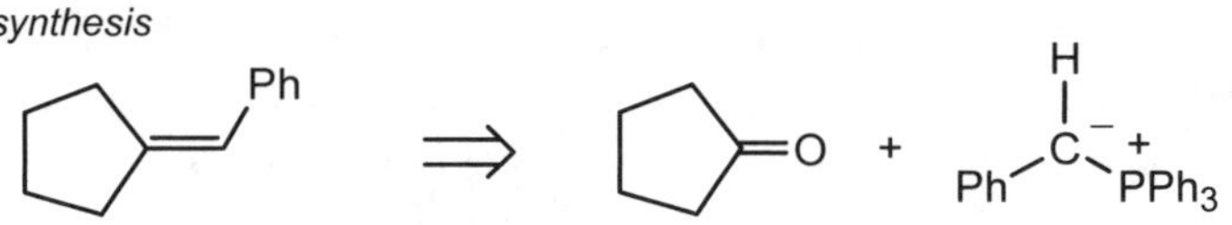

Synthesis

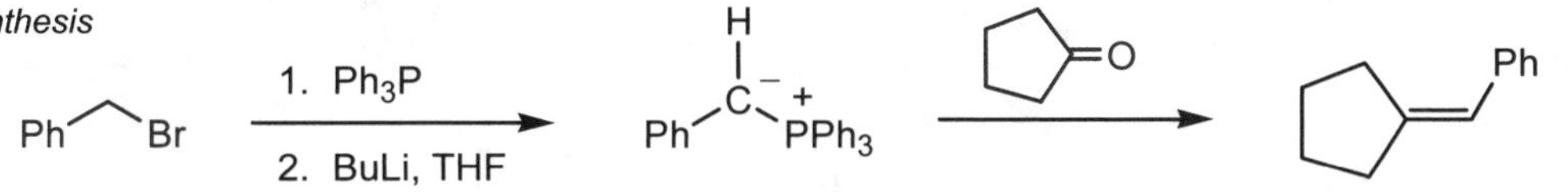

20.15. The Arbuzov reaction takes place between a reactive alkyl halide and trimethylphosphite. After the phosphorus atom reacts as a nucleophile with the alkyl halide (step 1), the halide ion reacts at one of the methyl groups to create the P=O bond (step 2).

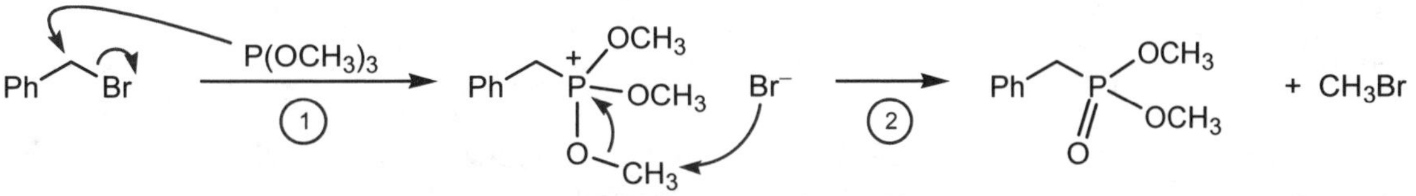

20.16. The logical disconnection that makes use of the Horner-Emmons or Wittig reaction to prepare an alkene is at the double bond between the carbon atoms. If the double bond is exocyclic to a ring, the ring component normally is the carbonyl partner. In most cases, the ylide reactant comprises the end of the alkene double bond with the less highly-substituted carbon atom. The needed phosphonate ylide is made using the Arbuzov reaction. The starting compounds are shown in color.

a. *Retrosynthesis*

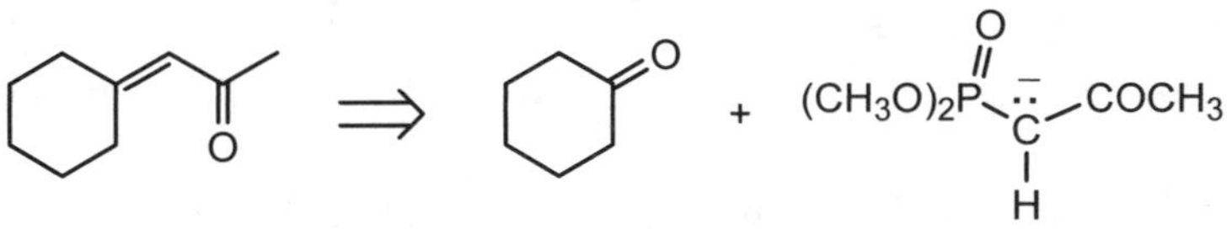

Synthesis

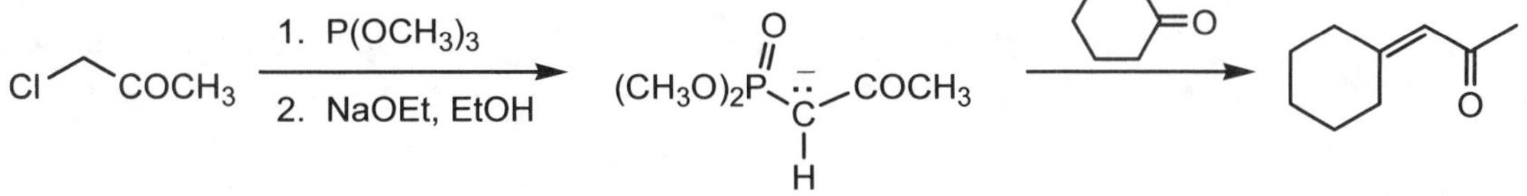

b. *Retrosynthesis*

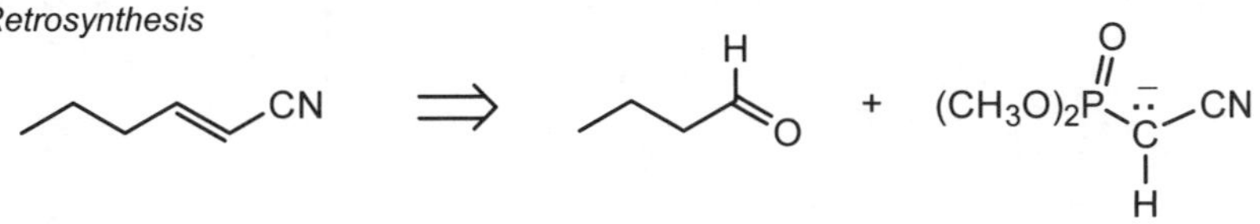

Synthesis

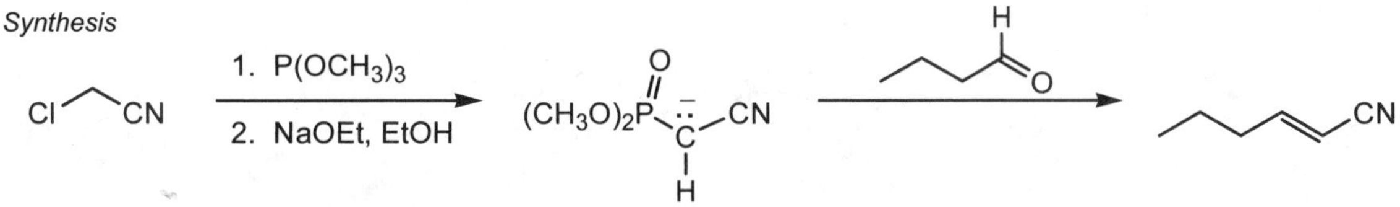

20.17. The cyclopropyl diphenylsulfonium ylide reacts with the carbonyl group of decanone to form epoxide **A**.

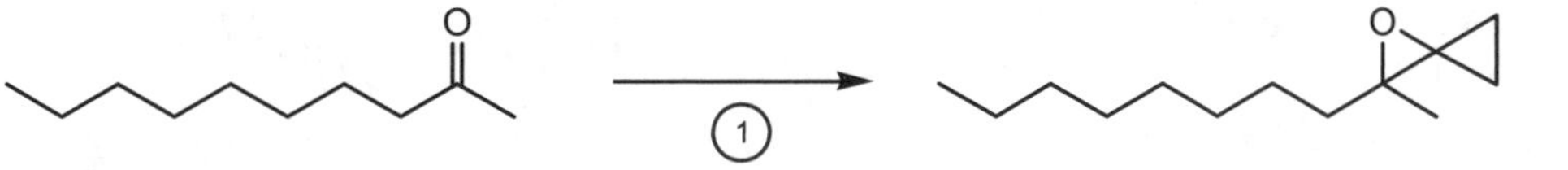

20.17. (continued)

This epoxide, when treated with acid, is first protonated at its oxygen atom (step 2). The strained three-membered ring opens to form a 3° carbocation (step 3), which rearranges to form the more stable cation (step 4). Deprotonation of the rearranged species yields the cyclobutanone derivative (step 5).

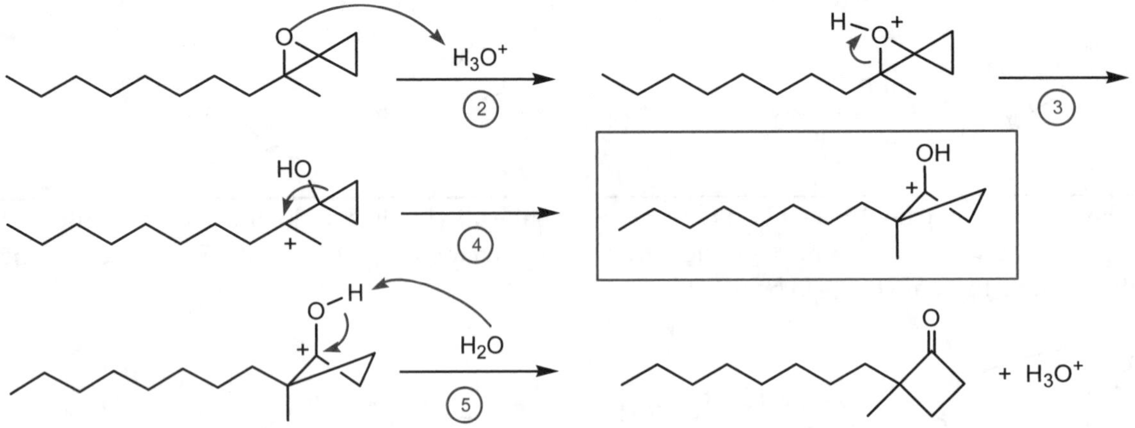

The second cation that is formed (shown in the colored box, above) is more stable than the tertiary carbocation because of resonance stabilization, as illustrated below. Formation of this more stable carbocation drives the rearrangement processes forward.

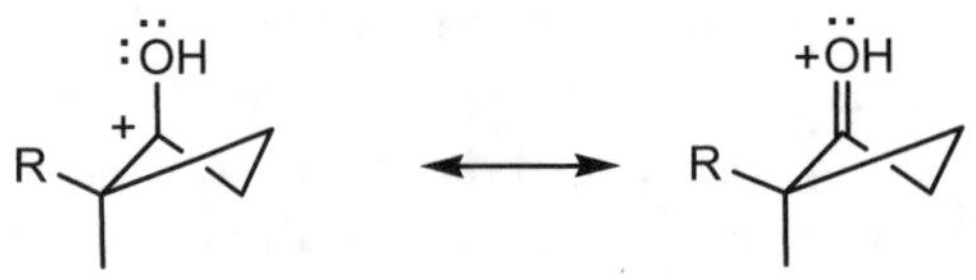

20.18. Semicarbazide is a hydrazine derivative with an aminocarboxy substituent. The NH₂ group at the hydrazine end of the molecule reacts in addition–elimination reactions with carbonyl groups (the other NH₂ group is attached directly to a carbonyl group, so it is more like an amide, which does not normally react with ketones and aldehydes to form C=N bonds).

The reaction begins with nucleophilic addition of the nitrogen atom to the carbonyl group (step 1). Two proton transfer steps (steps 2 and 3) subsequently yield a neutral derivative. Protonation of the OH group (step 4) generates a good leaving group, and the electron pair on nitrogen displaces a molecule of water (step 5). Finally, an acid-base reaction occurs to generate the product, a semicarbazone (step 6).

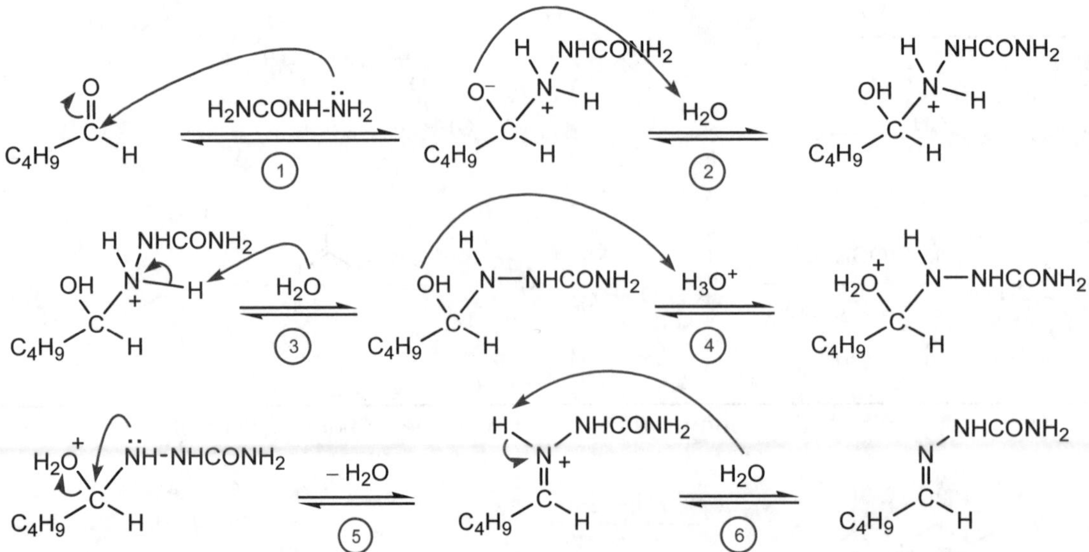

A semicarbazone

20.19. An imine is formed by the reaction between a primary amine and an aldehyde or ketone. The nitrogen atom replaces the oxygen atom in the carbonyl component, with loss of a molecule of water.

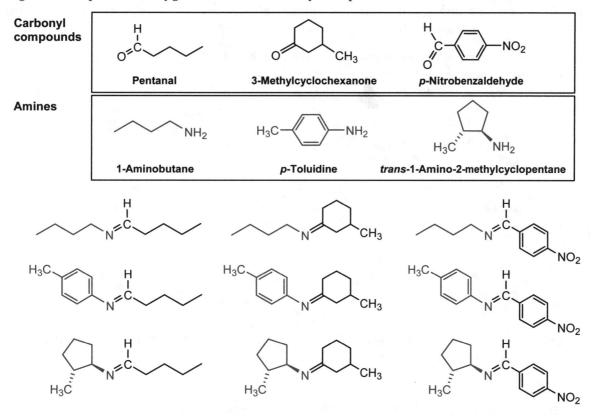

20.20. An enamine is formed by the reaction between a secondary amine and an aldehyde or ketone. The nitrogen atom replaces the oxygen atom in the carbonyl component, and a carbon-carbon double bond is formed between the original carbonyl carbon atom and one of the adjacent carbon atoms.

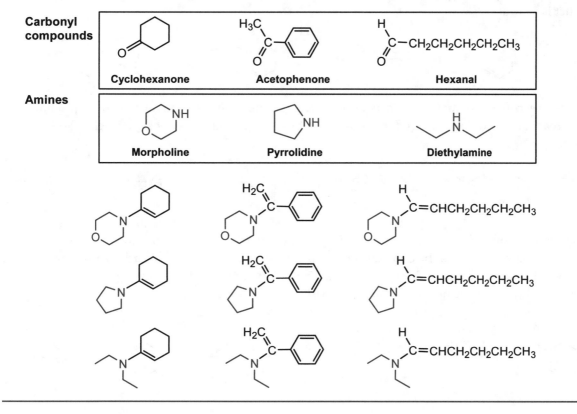

20.21. The compound 1-phenyl-1-propanone has two reactive units: the carbonyl group and the benzene ring. The ketone carbonyl group can undergo addition, addition-substitution, and addition-elimination reactions. The benzene ring can undergo electrophilic aromatic substitution reactions. All of these possibilities have to be assessed in predicting the major product that will be formed.

a.   Organolithium compounds and Grignard reagents react by addition of the "R" group to the carbonyl double bond. Aqueous acid workup gives the tertiary alcohol product. A new chiral center is formed, so the product exists as a racemic mixture.

b.   Ketones are reduced by hydride reagents to form secondary alcohols. A new chiral center is formed, so the product exists as a racemic mixture.

c.   The carbonyl group of a ketone is replaced by two hydrogen atoms when heated with hydrazine and base to high temperatures (the Wolff-Kishner reaction). The product is achiral.

d.   The benzene ring undergoes electrophilic bromination; the ketone group is a meta director. The product is achiral.

e.   Ketones are converted to the corresponding alkenes when treated with a phosphorus ylide (the Wittig reaction). Two isomers are formed because the Wittig reaction is not stereospecific.

f.   The carbonyl group of a ketone is converted to an epoxide when treated with the methylene ylide of dimethyl sulfide. In step 2, the epoxide ring is opened at its less substituted carbon atom by reaction with the Grignard reagent, and the resulting alkoxide ion is protonated to form the alcohol (step 3). A new chiral center is formed, so the product exists as a racemic mixture.

20.21. (continued)

g. Ketones with protons on the carbon atom adjacent to the carbonyl group are converted to enamines when treated with secondary amines and acid. The carbon–carbon double bond can exist in isomeric forms because it is trisubstituted with four different groups (including the H atom).

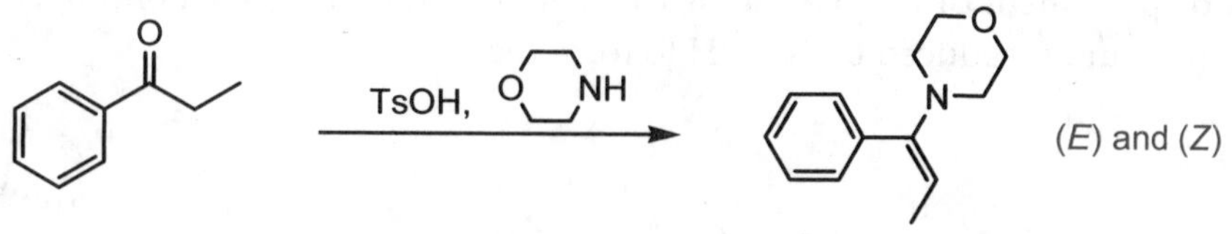

(E) and (Z)

h. Ketones are converted to oximes when treated with hydroxylamine and acid. The carbon–nitrogen double bond can exist in isomeric forms.

(Z) + (E)

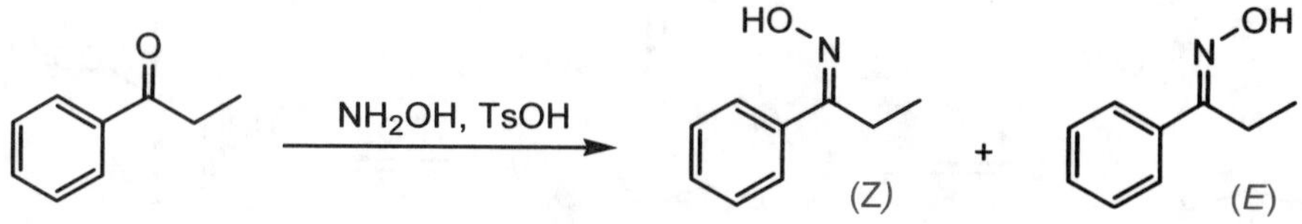

20.22. If the final step in the synthesis of Valium is the formation of the imine bond, then the immediate precursor must be a compound in which the carbon-nitrogen double bond has been disconnected and replaced by the elements of water. The carbon atom is the one attached to the =O group, and the nitrogen atom is attached to two atoms of H.

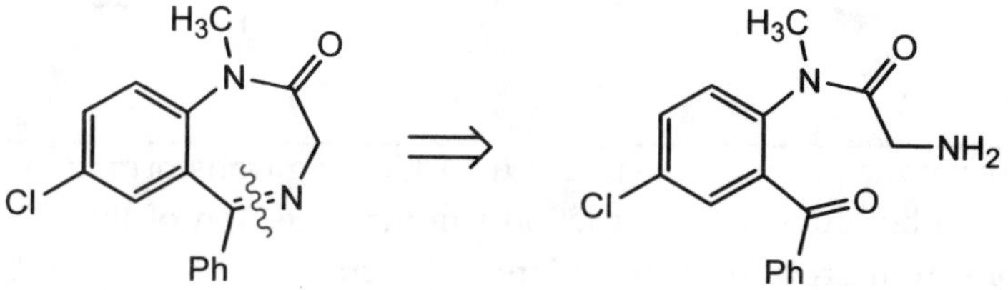

The mechanism for ring closure begins with addition of the amino group to the C=O double bond (step 1). A proton is transferred from the nitrogen to the oxygen atom (probably via H_3O^+ and H_2O in solution, but represented here simply as step 2), and then the OH group is protonated to form a good leaving group (step 3). Removing the proton from the nitrogen atom displaces a molecule of water and forms the C=N bond (step 4).

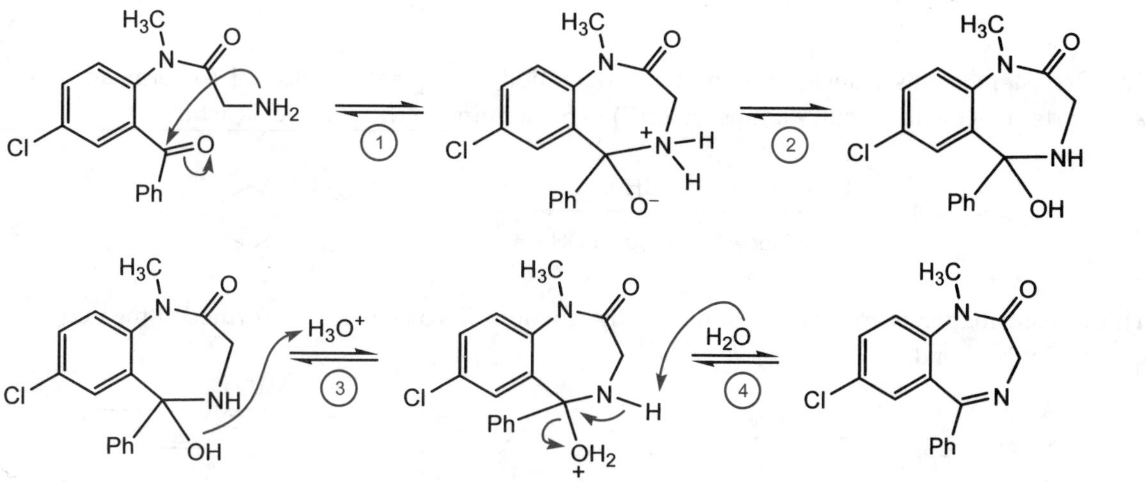

20.23. The two steps needed to convert glutamate-5-semialdehyde to proline involve cyclization, which creates the imine linkage, and reduction, which converts the imine to an amine.

The mechanism is straightforward. A proton is transferred from the ammonium ion to the carbonyl group (step 1), and then nucleophilic nitrogen atom reacts with the resulting carbocation (step 2). A proton is transferred from the N to the O atom (step 3) and the remaining proton is removed from the nitrogen atom with displacement of a molecule of water to generate the C=N bond (step 4). The imine group, like a carbonyl group, is reduced by NADH to form the 2° amine.

20.24. To predict the structures of the products in each of the following transformations, learn the details of each reaction type, which can be found in the reaction summary section of the chapter. In some of these transformations, reactions from previous chapters are included.

a. This transformation constitutes the formation of an enamine derivative of a ketone. The carbon–carbon double bond can exist in isomeric forms because it is trisubstituted with different groups.

b. The first step of this transformation is the oxidation of a primary alcohol to form an aldehyde. Aldehydes react with primary amines (step 2) to form imines. The product is achiral.

c. This transformation is an example of the Wittig reaction. Two isomers are formed at the new carbon–carbon double bond.

20.24. (continued)

d. The carbonyl group of a ketone is converted to an epoxide when treated with the methylene ylide of dimethyl sulfide. A new chiral center is formed, so the product exists as a racemic mixture.

racemic

e. The first step of this transformation converts the acetal to the corresponding aldehyde by hydrolysis. The amino group reacts with the aldehyde group to form an imine. The chiral center that exists in the starting compound is not affected by formation of the C=N bond, so the reaction proceeds with retention of stereochemistry.

retention

f. The first step of this transformation is a Friedel-Crafts acylation reaction to form acetophenone. The ketone reacts with a secondary amine in the presence of an acid catalyst to form an enamine. The double bond cannot exist as isomers, and the product is achiral.

achiral

g. The first step of this transformation is a hydroboration reaction of the alkyne, which yields an aldehyde. The second part is a Wittig reaction to form the terminal alkene.

$(CH_3)_2CHCH_2-C{\equiv}C-H$
1. catecholborane
2. H_2O_2, OH^-
$(CH_3)_2CHCH_2-CH_2-CHO$

$(CH_3)_2CHCH_2-CH_2-CHO$
$Ph_3P-\overset{+}{}\overset{..}{C}H_2$
$(CH_3)_2CHCH_2-CH_2-CH{=}CH_2$ achiral

h. The first step of this transformation is the formation of an oxime, which should exist mainly as the isomer shown in order to minimize the steric interaction between the OH and methyl groups. The second step constitutes the Beckmann rearrangement. The aryl group is anti to the OH group, so it migrates to the nitrogen atom to give the illustrated product.

achiral

20.25. When you perform a reaction in which you know what the product will be (or is expected to be), you want to focus on those spectroscopic features that are unique for the starting material and product. Although many subtle changes will undoubtedly occur in many parts of the spectrum, it helps to look for a limited number of specific changes.

For each of these transformations, the very intense band attributed to the carbonyl stretch vibration will be absent in the spectrum of the product. The weaker band for the C=C or C=N stretching vibration is likely to be obscured somewhat by the C=C stretching vibrations of the benzene ring, so it may or may not be readily observed.

20.25. (continued)

a.

$v \sim 1715$ cm^{-1}
very strong intensity

$v \sim 1650$ cm^{-1}
weak to medium intensity

b.

$v \sim 1725$ cm^{-1}
very strong intensity

$v \sim 1650$ cm^{-1}
medium intensity

20.26. As noted in the solution to Exercise 20.25, focus on those spectroscopic features that are unique to the starting material and product molecules. Many other subtle changes will undoubtedly occur in other parts of the spectrum as well

a. For this transformation, the ^{1}H NMR spectrum of the product will have a signal in the alkene region (δ 5-6) that is not present in the spectrum of the starting ketone. The spectra will be similar otherwise, with small changes in the chemical shift values.

^{1}H NMR spectrum

CH$_2$ — δ 5-6, 2H

—CH$_2$CH$_3$

δ 5-6, 2H

δ 7-8, 5H

δ 2.9, 2H, triplet

δ 7-8, 5H

δ 3.2 (estimated), 2H, triplet

The ^{1}H NMR spectrum of the starting ketone will have resonances in the aromatic region with a total integrated intensity value of 5, and three sets of signals in the aliphatic proton region: two triplets and a sextet.

Proton NMR spectrum

(5)

(2)

(2)

(3)

TMS

8   6   4   2   0

The ^{13}C NMR spectra of the starting material and product molecules will differ substantially because they have different types of carbon atoms (carbonyl versus alkene). The product has an additional carbon atom, too. These spectra will be similar with regard to the chemical shifts of the aryl and the propyl group carbon atoms.

^{13}C NMR spectrum

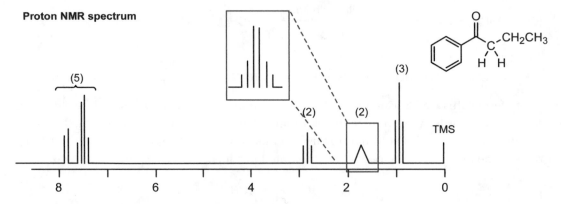

δ 210
0H (DEPT)

δ 110-150
2H (DEPT)

CH$_2$

δ 110-150
0H (DEPT)

δ 110-150 4 resonances { 3 with 1H (DEPT)
1 with 0H (DEPT)

δ 110-150 4 resonances { 3 with 1H (DEPT)
1 with 0H (DEPT)

20.26. (continued)

b. For this transformation, the proton NMR spectrum of the product will have signals for the isopropyl group in the aliphatic proton region (δ 1-3) that are not present in the spectrum of the starting material. The spectra will be similar otherwise, with small changes in the chemical shift values. The spectrum of the starting aldehyde will have the signals for an ortho-disubstituted benzene ring (two doublets and two triplets) in addition to the resonance for the aldehyde proton.

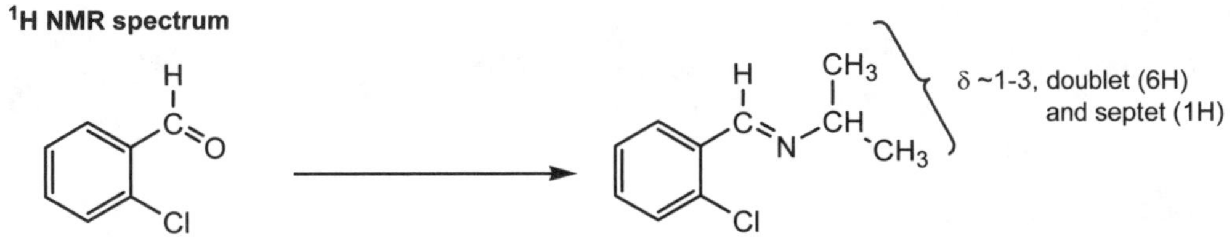

¹H NMR spectrum
δ ~1-3, doublet (6H)
and septet (1H)

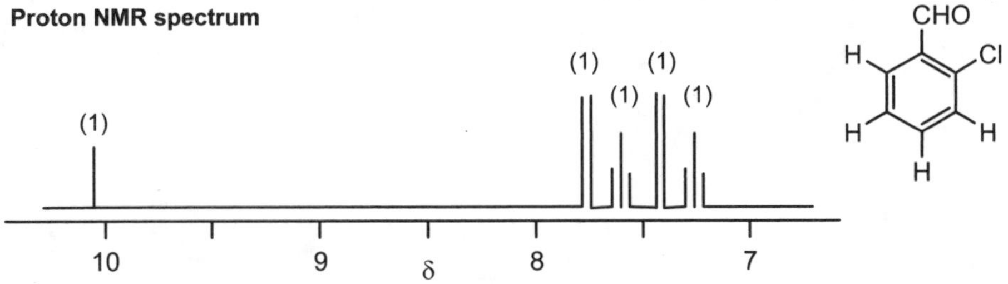

Proton NMR spectrum
(1)
(1) (1)
(1) (1)
10
9
δ
8
7
CHO
Cl
H

The ^{13}C NMR spectra of the starting material and product molecules will differ because the product has additional carbon atoms that produce signals in the aliphatic region (δ 10-50). These spectra will be similar with regard to the chemical shifts attributed to the *sp*²-hybridized carbon atoms.

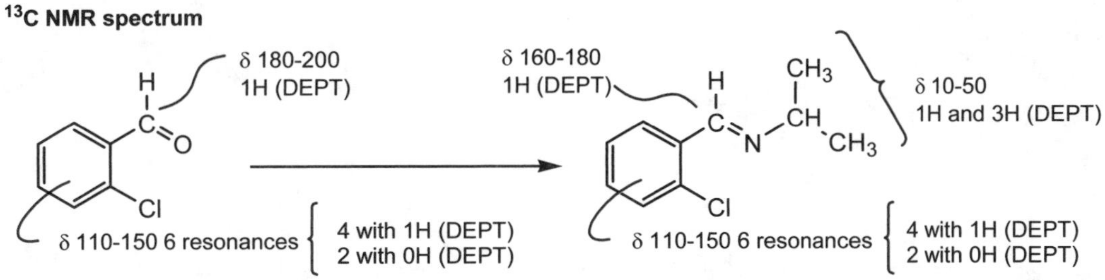

¹³C NMR spectrum
δ 180-200
1H (DEPT)
δ 160-180
1H (DEPT)
δ 10-50
1H and 3H (DEPT)
δ 110-150 6 resonances
4 with 1H (DEPT)
2 with 0H (DEPT)
δ 110-150 6 resonances
4 with 1H (DEPT)
2 with 0H (DEPT)

20.27. These reactions are standard examples of addition–elimination processes. In (a.), the ketone is converted to an alkene using an ylide (the Wittig reaction). In (b.), the aldehyde is converted to an imine by its reaction with a primary amine in the presence of an acid catalyst.

a.

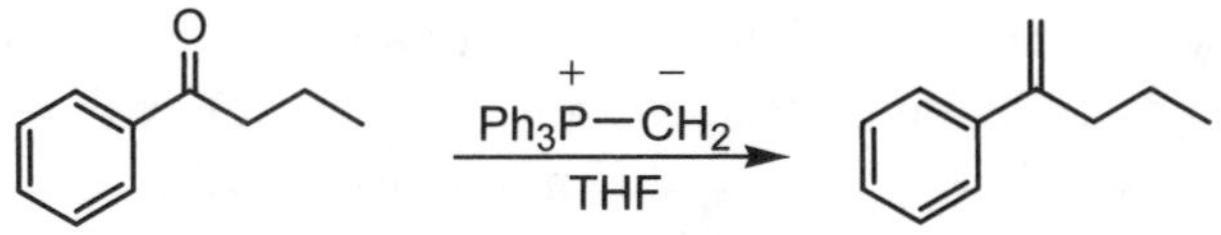

Ph3P—CH2
THF

b.

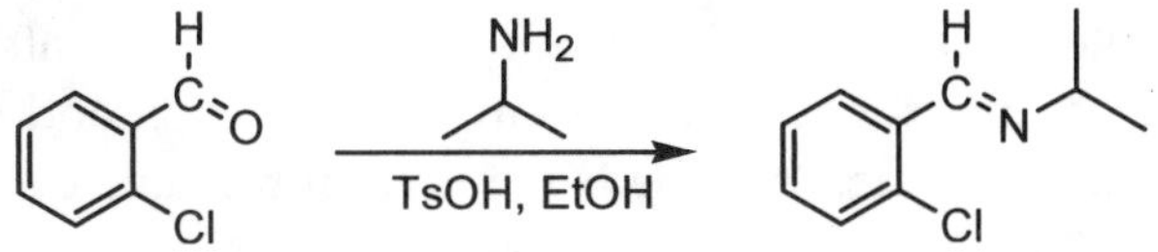

NH2
TsOH, EtOH

20.28. If a Wittig reaction is used to prepare an alkene, the retrosynthetic disconnection is at the double bond. One end of the alkene comes from the ylide reagent and the other from the carbonyl component (aldehyde or ketone). The starting materials in each of the following schemes are shown in color.

20.28. (continued)

a.   This Wittig reaction uses an ylide derived from toluene to react with acetone.

Retrosynthesis

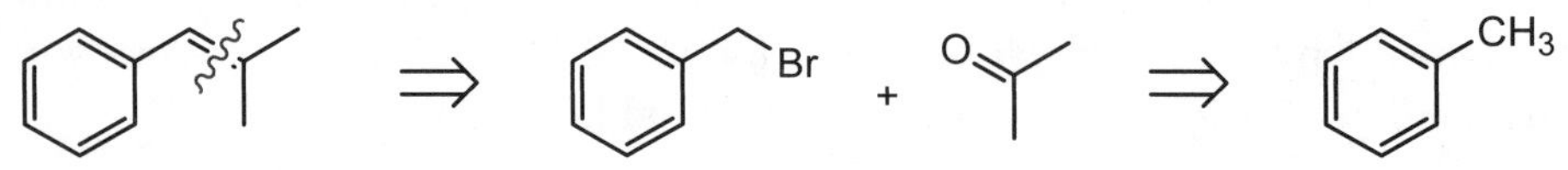

Synthesis

- - - - - - - - - - - - - - - - - - - -

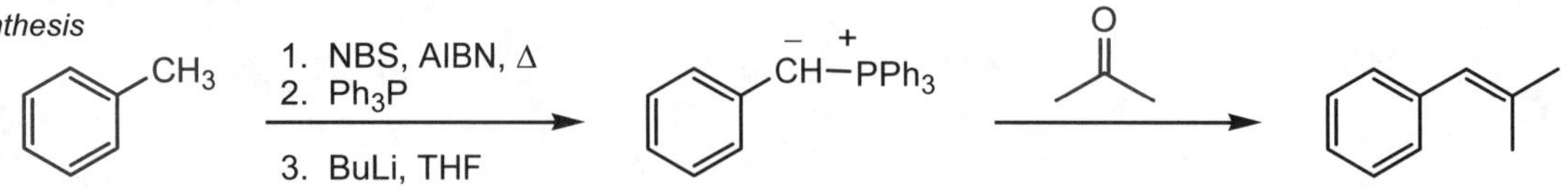

b.   This Wittig reaction uses an ylide derived from methyl iodide to react with 3-methylbutanal.

Retrosynthesis

Synthesis

- - - - - - - - - - - - - - - - - - - -

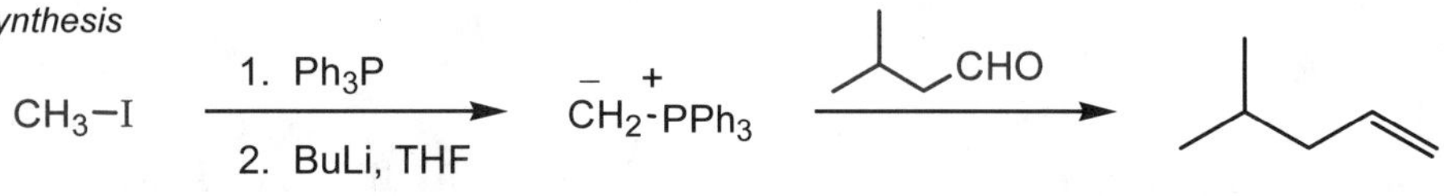

c.   This Wittig reaction uses an ylide derived from ethyl iodide to react with cyclohexanone.

Retrosynthesis

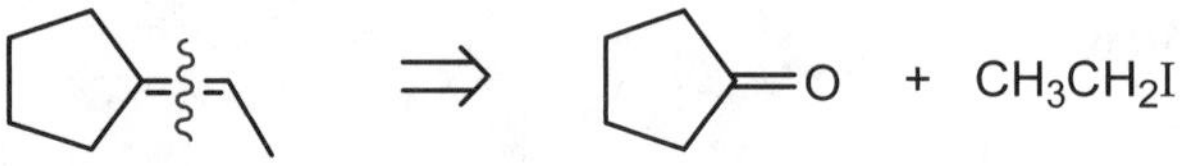

Synthesis

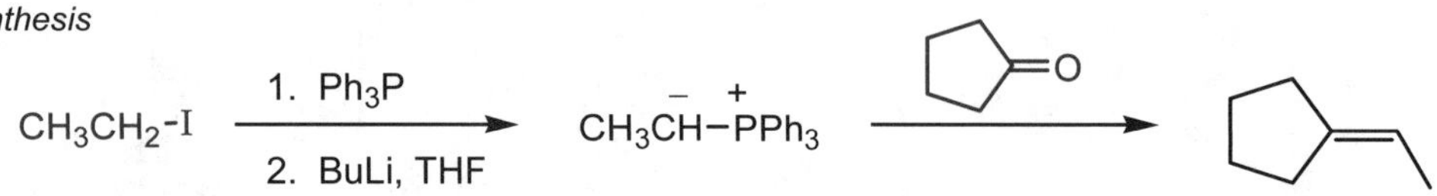

20.29. In this chapter, you were introduced to reactions that make compounds with C=C and C=N double bonds from aldehydes and ketones: Imines, enamines, and alkenes can be made via addition–elimination reactions. In the retrosynthetic analysis, look particularly for disconnections at double bonds. If the synthetic goal has a ring as part of the carbon skeleton, consider ways to include the ring as a starting material. There are several ways to prepare the molecules shown below. Only one route for each is presented. The molecules shown in color are the organic starting materials.

a.   The imine bond is a logical disconnection site. The aldehyde is made from benzene by attaching a two carbon fragment, which is done using the reaction between a Grignard reaction and ethylene oxide. The alcohol product is oxidized to the aldehyde. Aniline is made from benzene by nitration followed by reduction.

Retrosynthesis

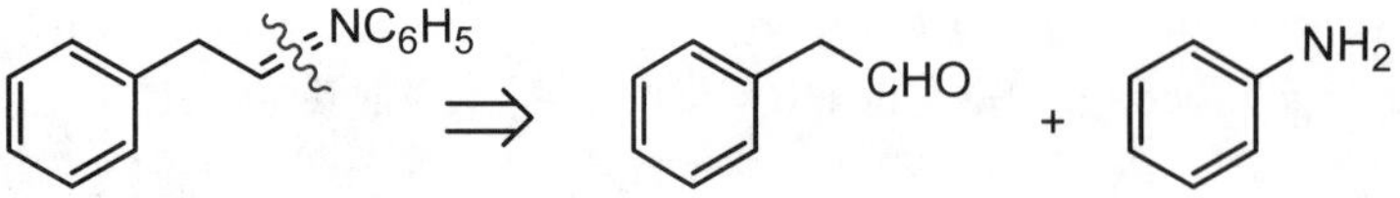

20.29. (continued)

Synthesis

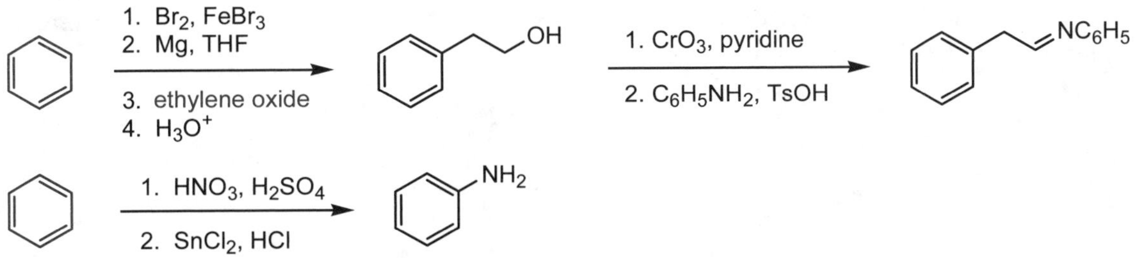

- - - - - - - - - - - - - - - - - - - -

b. When a desired product has no functional group, you have to insert one (or more) in a place at which it can be readily converted to a saturated hydrocarbon unit. A Wittig reaction can be used to make an alkene, and the double bond is readily reduced using catalytic hydrogenation. The symmetry of this molecule makes it logical to break the molecule into two, six-carbon fragments, which are made from the same four-carbon starting material by adding a two-carbon fragment.

Retrosynthesis

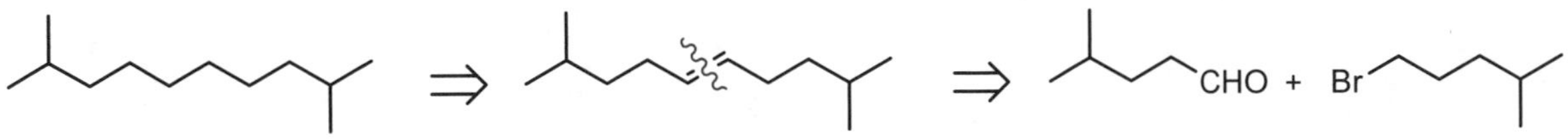

Synthesis

- - - - - - - - - - - - - - - - - - - -

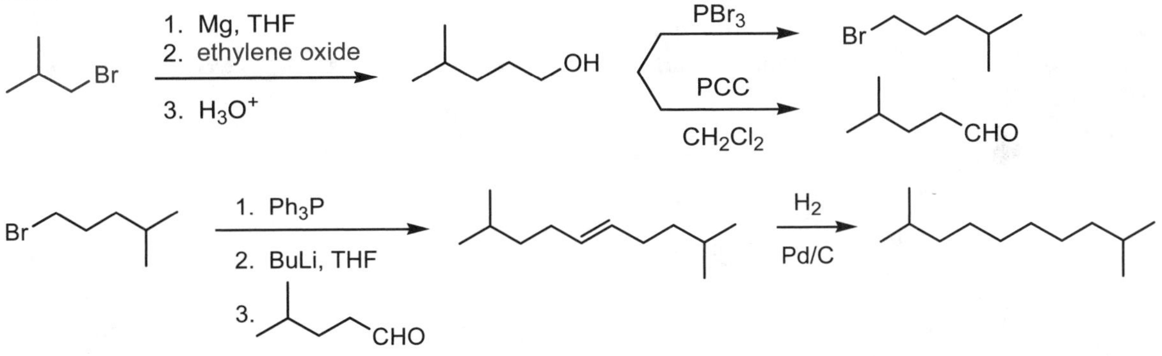

c. A Wittig reaction could be used to make this alkene, but the product has eight carbon atoms, so a disconnection to give two, four-carbon atom fragments would expedite the synthesis. The location and type of double bond in the product means that it can be made by an elimination reaction (E1).

Retrosynthesis

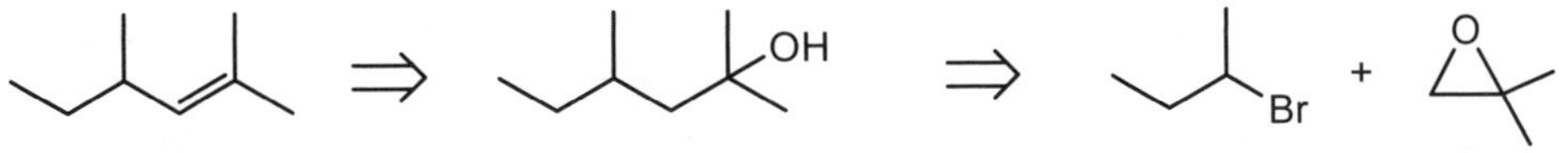

Synthesis

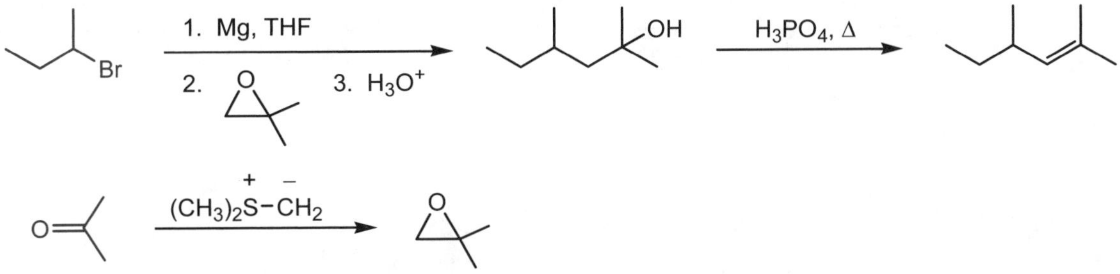

20.30. Deprotonation of an alkyltriphenylphosphonium salt generates an ylide (step 1), which has a nucleophilic carbon atom. Alkylation occurs by reaction between the nucleophilic carbon atom and the electrophilic carbon atom of benzyl bromide (step 2). Deprotonation with NaH (step 3) produces another ylide. Its reaction with benzaldehyde (step 4) forms the carbon-carbon double bond of the product by the Wittig reaction. Both geometric isomers are formed.

20.31. A six-membered ring is formed in the first step by the Diels-Alder reaction (Section 10.4b between a conjugated diene and the dienophile, vinyltriphenylphosphonium bromide. The resulting cyclohexenyl phosphonium ion is subsequently deprotonated by reaction with LDA (step 2), which forms an ylide.

The reaction of the ylide with propanal yields the product with an exocyclic carbon-carbon double bond (in addition to the endocyclic double bond from the Diels-Alder reaction). Two geometric isomers are formed in this final step.

20.32. Dimethylsulfonium methylide is a nucleophile, so it reacts with the alkyl halide to form a homologous sulfonium salt. The dimethylsulfonium group is a good leaving group (Section 7.3b) so reaction of the alkylated sulfonium salt with strong base produces the alkene by the E2 pathway.

20.33. First, summarize the reactions described in the accompanying paragraph.

20.33. (continued)

The identity of **Z** is straightforward to reckon because it is made by the Friedel-Crafts acylation reaction between toluene and 3-methylbutanoyl chloride. Compound **Z** is 1-(4-methylphenyl)-3-methyl-1-butanone.

Compound **Y** is 4-methylbenzaldehyde (*p*-tolualdehyde), the structure of which is determined from the given IR and proton NMR spectra.

Working backwards from the structure of compound **Z**, we conclude that dichromate ion was probably used to oxidize a 2° alcohol to make the carbonyl group in **Z**. The alcohol must have been made from an alkene by hydration.

To check this possibility, we ask whether **X** can be converted to **Y** (and a four-carbon aldehyde) by ozonolysis. The answer is yes.

To prepare **X** from **Y**, we employ the Wittig reaction between 4-methylbenzaldehyde and the four-carbon ylide derived from isobutyl bromide.

ADDITION–ELIMINATION REACTIONS OF CARBOXYLIC ACIDS AND DERIVATIVES

21.1. Follow the procedure outlined in Example 21.1.

tert-Butyl (E)-2-hexenoate

hex	6 carbon atoms
en	a carbon–carbon double bond, starting at C2
oate	the ester functional group; its carbonyl carbon atom defines C1
(E)	the higher priority substituents at the ends of the double bond are on opposite sides of the bond
tert-butyl	the C(CH₃)₃ group is attached to the non-carbonyl oxygen atom of the ester functional group

---

21.2. A carboxylic acid reacts with an alcohol in the presence of a mineral acid to form an ester. Dehydrating agents such as molecular sieves are sometimes added to shift the equilibrium toward formation of the ester product. To draw the structure of the ester formed from the reaction of a carboxylic acid with an alcohol, replace the H atom of the carboxylic acid group with the alkyl group of the alcohol.

a.

b.

21.3. If one of the alcohol groups of a diol is oxidized before the other one becomes oxidized, then cyclization can occur. The carbonyl group of the mono-carboxylic acid is protonated (step 1), and the alcohol OH group reacts with this carbocation intermediate (step 2). After protons are transferred to generate a good leaving group (steps 3 and 4), the carbonyl group of the lactone is regenerated (step 5).

21.4. A hydroxy carboxylic acid can undergo either intra- or intermolecular esterification. If there are three or four atoms separating the alcohol and carboxylic acid groups, a lactone will be formed. When the hydroxy group is attached to the carbon atom adjacent to the carboxylic acid group, a dimer is produced. This transformation proceeds by the same steps outlined in the solution to Exercise 21.3.

21.5. To prepare an ester from a carboxylic acid, you have the choice of treating a carboxylic acid with a primary alcohol in the presence of an acid catalyst, with diazomethane (methyl esters only), or with alcohol after converting the carboxylic acid to its acid chloride derivative (all types of alcohols). To prepare an ester with a tertiary alcohol as the reactant, only the last option is viable.

21.6. A carboxamide is made by the reaction between an amine and either an ester or an acid chloride. If an acid chloride is used to make an ester, a base like pyridine, triethylamine, or hydroxide ion is normally included in the reaction mixture to remove the HCl that is formed. An ester or lactone needs no added base because the leaving group is the alkoxide ion, which removes a proton attached to the nitrogen atom of the amine reactant.

a. To prepare this carboxamide, the acid is converted to the acid chloride and treated with aniline.

21.6. (continued)

b.   This carboxamide is made by converting the acid to its acid chloride derivative followed by reaction
with the amine.

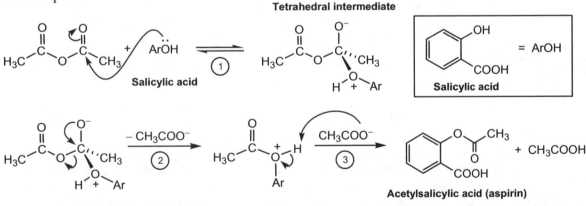

21.7. Hydrazine reacts with an acid chloride in the same way that an amine does. The second nitrogen
atom can act as a base, so aqueous OH⁻ has to be added at the end of the reaction to generate the neutral
hydrazide derivative.

a carboxylic acid hydrazide

21.8. The phenol oxygen atom is the nucleophile that adds to the carbonyl group of acetic anhydride to
form a tetrahedral intermediate (step 1). Regeneration of the carbon-oxygen double bond displaces
acetate ion (step 2), which removes a proton from the oxygen atom (step 3) to form the products, acetic
acid and aspirin.

Tetrahedral intermediate

Salicylic acid

Salicylic acid   = ArOH

Acetylsalicylic acid (aspirin)

21.9. A thiol reacts with an acid chloride by addition of the sulfur atom to the carbonyl group (step 1).
The tetrahedral intermediate collapses to regenerate the carbonyl group by expulsion of chloride ion
(step 2), and an acid-base reaction yields the product, a thioester (step 3).

21.10. In any acid-catalyzed hydrolysis reaction of a carboxylic acid derivative, the carbonyl group is first protonated, which activates it for the subsequent reaction with water (step 2). After a proton transfer reaction (step 3), the OPh group (in this transformation) is protonated to form a good leaving group (step 4) that is displaced when the C=O bond is regenerated (step 5).

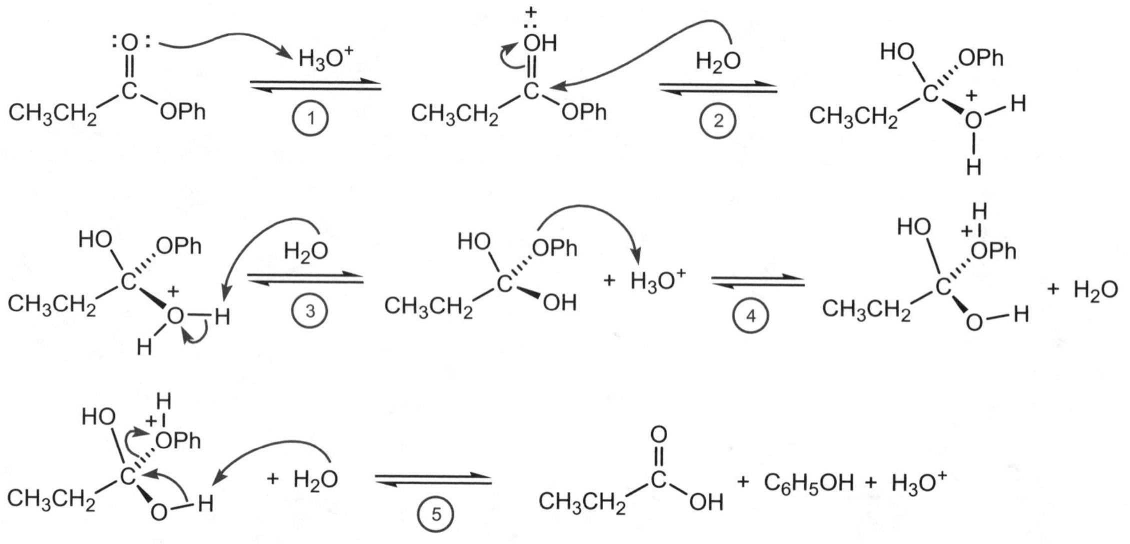

21.11. The base-promoted hydrolysis of a lactone follows the same mechanism as the saponification reaction of any ester. Hydroxide ion adds to the carbonyl group, forming a tetrahedral intermediate (step 1). That intermediate collapses to regenerate the carbon-oxygen double bond, which leads to opening of the lactone ring (step 2), shown here as assisted by a proton-transfer step from water. An acid-base reaction subsequently occurs with the hydroxide ion reactant to form the carboxylate salt (step 3). Acid workup, shown below as step 4, yields the hydroxy acid.

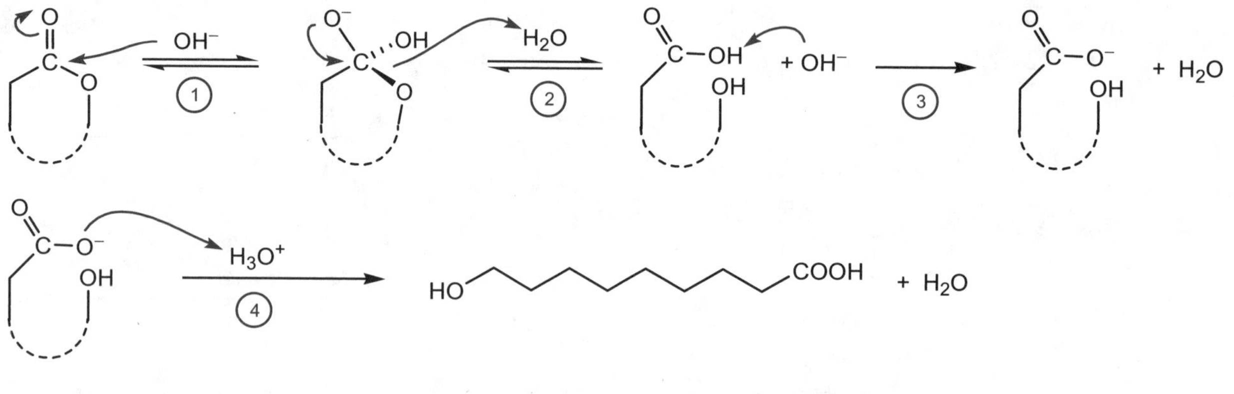

21.12. Transesterification occurs under basic conditions by nucleophilic addition of the alkoxide ion to the carbonyl group, which forms the tetrahedral intermediate. When the carbon-oxygen double bond is regenerated (step 2), the original alkoxy group is displaced to react with the alcohol solvent to generate more alkoxide ion. If a low-MW alcohol such as methanol is formed, its removal from the reaction mixture by distillation drives the equilibrium in the direction of the desired product.

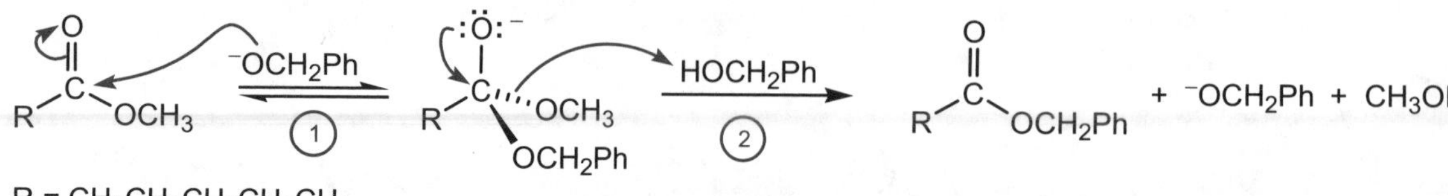

21.13. In the acid-catalyzed hydrolysis reaction of a lactam, the carbonyl group is first protonated, forming an activated intermediate that is trapped by reaction with water (step 2). After a proton transfer reaction (step 3), the nitrogen atom reacts with acid to generate a good leaving group (step 4), which is displaced when the carbon-oxygen double bond is regenerated (step 5). An acid-base reaction occurs between the amino group and the acidic solution (step 6).

21.14. The chymotrypsin-catalyzed hydrolysis of *p*-nitrophenyl acetate starts with binding of the aromatic molecule in the active site of the enzyme. In the first step, the serine hydroxyl group adds to the carbonyl group of the substrate, generating a tetrahedral intermediate. The generation of the alkoxide ion nucleophile is facilitated by the acid-base reaction in which the histidine side chain participates.

In the next step, the tetrahedral intermediate collapses with displacement of *p*-nitrophenol. The proton on the imidazole group of histidine is used to generate the neutral phenol leaving group.

21.14. (continued)

A molecule of water enters the enzyme active site and is used to form hydroxide ion, which adds to the carbonyl group of the acetyl serine derivative (step 3). This step also creates a tetrahedral intermediate.

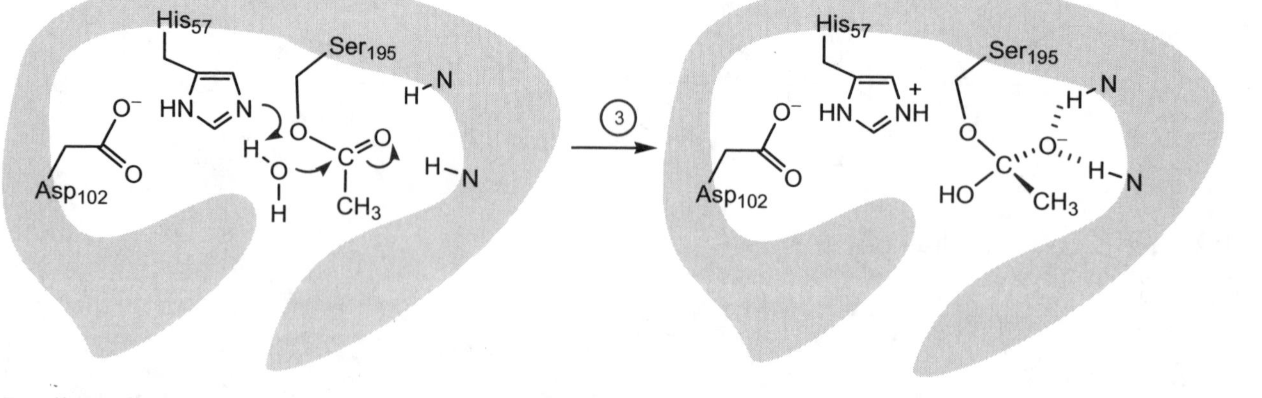

Finally, collapse of the tetrahedral intermediate yields acetic acid and regenerates the active site in the form needed to catalyze the next round of reaction.

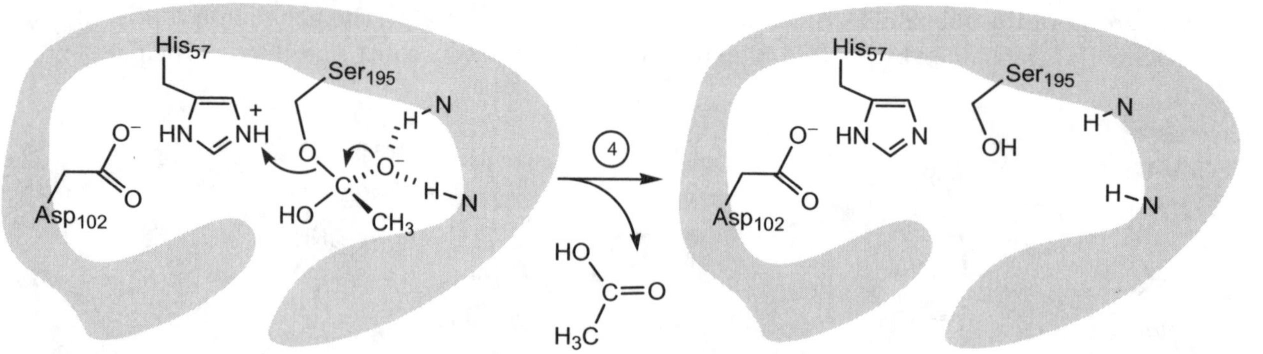

21.15. A nitrile reacts with HCl by protonation of its nitrogen atom (step 1). The cation so formed is intercepted by the nucleophilic oxygen atom of methanol (step 2). A proton transfer reaction between the oxygen and nitrogen atoms yields the product (step 3).

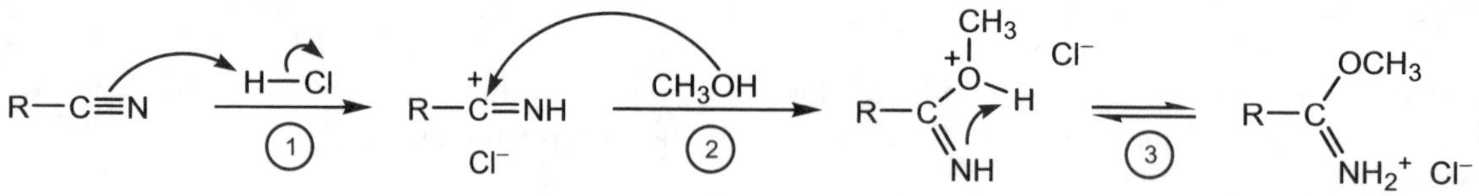

21.16. A ketone is readily prepared from a carboxylic acid derivative by the reaction between an acid chloride and an organocuprate reagent, which in turn is made from copper(I) iodide and an organolithium compound. (The reaction used to prepare the organocuprate reagent is not shown in the following schemes).

a.

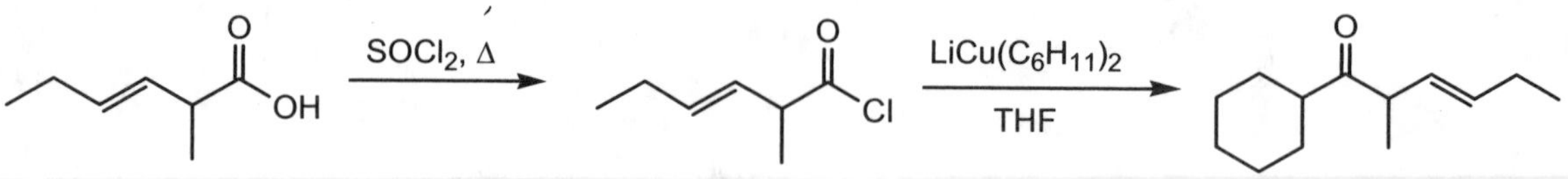

b.

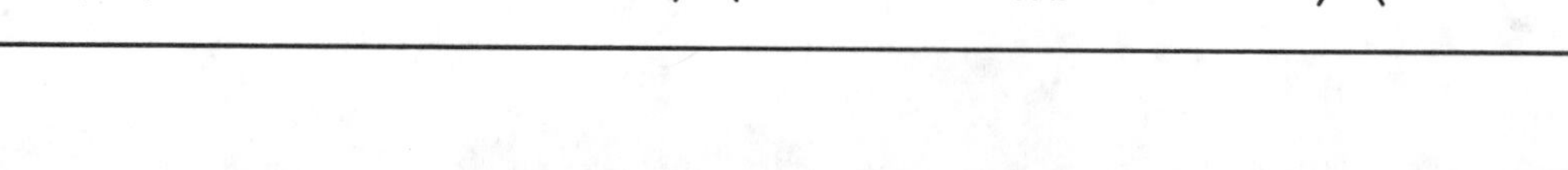

21.17. In the reaction between a Grignard reagent and a carboxamide, complex formation (see Section 15.2b) is followed by addition of the nucleophilic R group to the C=O bond (step 1), which creates a tetrahedral intermediate. Protonation during workup (step 2) generates the geminal amino alcohol (a hemiaminal) which is unstable and eliminates a molecule of the amine by protonation of the nitrogen atom (step 3) followed by formation of the carbon-oxygen double bond of the ketone (step 4).

21.18. In the reaction between a Grignard reagent and a formate ester, complex formation (Section 15.2b) is followed by addition of the nucleophilic R group to the C=O bond (step 1), which creates a tetrahedral intermediate. Regeneration of the carbonyl group (step 2) yields an aldehyde, which forms a second complex (step 3; Section 15.2b) and undergoes a second addition reaction (step 4). Acid workup (step 5) produces the secondary alcohol.

21.19. A lactone is an ester, so two equivalents of a Grignard reagent will react with a lactone to form a 3° alcohol group at the original carbonyl carbon atom. The alcohol portion of the lactone is cleaved from the carbonyl group by this reaction, but it remains part of the product molecule.

21.20. Follow the procedures outlined in Example 21.1 and in the solution to Exercise 1.30.

a.   **N,N-Diethylacetamide**

acetamide	the carboxamide having 2 carbon atoms (CH_3CONH_2)
N,N-diethyl	two CH_2CH_3 groups are attached to the nitrogen atom of the amide group

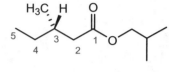

b.   **Isobutyl (S)-3-methylpentanoate**

pent	five carbon atoms
an	no carbon–carbon double or triple bonds
oate	ester functional group; its carbon atom defines C1
3-methyl	a CH_3 group is attached to C3
(S)	the configuration of the chiral carbon atom is (S)
isobutyl	the $CH_2CH(CH_3)_2$ group is attached to the oxygen atom of the ester functional group

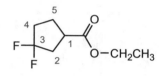

c.   **Ethyl 3,3-difluorocyclopentanecarboxylate**

cyclopentane	five-carbon ring with no double or triple bonds
carboxylate	ester functional group as a substituent on a ring; its point of attachment to the ring defines C1
3,3-difluoro	two fluorine atoms are attached to C3 of the ring
ethyl	the CH_2CH_3 group is attached to the ester oxygen atom

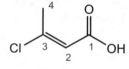

d.   **(E)-3-Chloro-2-butenoic acid**

but	four carbon atoms
en	a carbon–carbon double bond, starting at C2
oic acid	carboxylic acid functional group; its carbon atom is C1
(E)	the higher priority substituents at each end of the double bond are on opposites sides of the bond
3-chloro	a Cl atom is attached at C3

e.   **3-Bromo-2-nitrobenzoic acid**

benzoic acid	the carboxylic acid derivative of benzene
3-bromo	a Br atom is attached at C3
2-nitro	an NO_2 group is attached at C2

21.21. Follow the procedures outlined in the solution to Exercise 1.17.

a.   This compound has the nitrile functional group in a carbon chain with six carbon atoms and a double bond that starts at C4, so the core of its name is 4-hexenenitrile. A methyl group is also attached at C4, and the stereochemistry of the double bond is (E) because the higher priority groups at the ends of the C=C bond are on opposite sides. The name is **(E)-4-methyl-4-hexenenitrile**.

- - - - - - - - - - - - - - - - - - - -

b.   This compound has the carboxylic acid functional group attached to a benzene ring, so the root name is benzoic acid. The substituents include a methoxy group at C2, a chlorine atom at C3, and a nitro group at C6. (If we numbered the ring counterclockwise, the numbering would be 1, 2, 5, 6 instead of 1, 2, 3, 6, and we use the set with the lowest number at the first point of difference). We append the substituent names in alphabetical order to give the name: **3-chloro-2-methoxy-6-nitrobenzoic acid**.

21.21. (continued)

c. This compound has the acid chloride functional group in a carbon chain with five carbon atoms and a double bond that starts at C3, so the core of its name is 3-pentenoyl chloride. A methyl group is attached at C4 and no stereochemical designation is needed. The name is **4-methyl-3-pentenoyl chloride.**

- - - - - - - - - - - - - - - - - - -

d. This compound has the ester functional group in a carbon chain with seven carbon atoms (counting the carbonyl group) and no unsaturation, so the core of its name is heptanoate. The group attached to the ester oxygen atom is the allyl group. There are no other substituents, so the name is **allyl heptanoate.**

21.22. To create an equation, draw the structure of the reactant, and then write the names or structures of the solvents, reagents, and reaction conditions on the arrow, numbering the steps if more than one is required. To predict the structures of the products in each of the following transformations, learn the details of each reaction type, which can be found in the reaction summary section of the chapter. In some of these transformations, reactions from previous chapters are included. Pentanoic acid is a five-carbon, saturated carboxylic acid.

a. This reaction is an example of the direct (Fischer) esterification of a carboxylic acid.

$(CH_3)_2CHCH_2OH$, TsOH, Δ

b. This transformation shows the reduction of a carboxylic acid to form a primary alcohol.

1. $LiAlH_4$
2. H_3O^+

c. In this reaction, the carboxylic acid is first converted to its acid chloride derivative. An amide is formed by reaction of the acid chloride with an amine, and base is added to remove the HCl that is formed.

1. $SOCl_2$
2. $CH_3(CH_2)_3NH_2$
3. aq NaOH

d. The first two steps produce 1-pentanol [part (b.)], and PCC oxidizes a primary alcohol to the corresponding aldehyde.

1. $LiAlH_4$
2. H_3O^+
3. PCC, CH_2Cl_2

e. After formation of the acid chloride derivative in step 1, reduction with the hindered hydride reagent yields pentanal, the same product formed in part (d.).

1. $SOCl_2$
2. $LiAlH(OtBu)_3$
   low temperature

21.22. (continued)

f.   The first two steps are used to form the ester, methyl pentanoate. A Grignard reagent reacts with an ester to form a tertiary alcohol with two like groups (methyl in this case) attached to the alcohol carbon atom.

21.23. The rates among the addition-elimination reactions of carboxylic acid derivatives are affected by the electrophilic properties of the carbonyl carbon atom, by the steric effects of any substituents, and by the potential leaving group facility. Among acid derivatives, the order of reactivity is acid chloride > anhydride > thioester > ester > amide (3° > 2° > 1°). There is no anhydride in this series, so the order is:

most reactive — least reactive

21.24. The first step in this acid chloride forming reaction with oxalyl chloride is an acid-base reaction that produces a carboxylate ion. The resulting anion undergoes addition-elimination with oxalyl chloride (steps 2 and 3) to form a mixed anhydride. Next, chloride ion adds to the carbonyl group that was in the original carboxylic acid (step 4). Finally, that tetrahedral intermediate collapses to generate the acid chloride, displacing carbon dioxide, carbon monoxide, and chloride ion (step 5).

21.25. A cyclic anhydride reacts in the same way that an acyclic one does except that the leaving group remains part of the product molecule. Thus, the OH group of propanol reacts at one of the carbonyl groups of the anhydride (step 1) to form a tetrahedral intermediate.

Transfer of a proton (step 2) generates a protonated carboxyl group, which is a good leaving group; and collapse of the tetrahedral intermediate yields the half-acid half-ester product (step 3).

21.25. (continued)

21.26. When you perform a reaction in which you know what the product will be (or is expected to be), focus on those spectroscopic features that are unique for the groups in the starting material and product molecules. Although many subtle changes will undoubtedly occur in several parts of the spectrum, it helps to look for a limited number of specific changes.

a. For this transformation, the position of the very intense band attributed to the carbonyl stretching vibration will change significantly.   The value of $\nu C{=}O$ is near 1800 cm^{-1} for an acid chloride, and it appears at lower frequency for an ester. In addition, the ester will have a strong absorption at around 1200 cm^{-1} because of the C–O single bond stretching vibration.

b. For this transformation, the position of the intense band attributed to the carbonyl stretching vibration will change significantly. The value of $\nu C{=}O$ is expected to be around 1720 cm^{-1} for an ester that is conjugated with an aromatic ring, and it will appear at a lower frequency value for a carboxylic acid group attached to a benzene ring. In addition, the ester will have a strong absorption at around 1200 cm^{-1} because of the C–O single bond stretching vibration. The carboxylic acid will display a broad band between 3500 and 2500 cm^{-1} because of the presence of its OH group.

21.27. As noted in the solution to Exercise 21.26, focus on those spectroscopic features that are unique for the starting material and product molecules. Many other subtle changes will undoubtedly occur in other parts of the spectrum as well

a. For this transformation, the ^{1}H NMR spectra of both molecules will display similar patterns of signals at approximately the same chemical shift values. The only significant difference is that the product will have a singlet attributed to the presence of the methyl group attached to the ester oxygen atom. This signal will appear at about δ 3.7.

^{1}H NMR spectrum

21.27. (continued)

In the ^{13}C NMR spectra, the signals attributable to the carbonyl carbon atoms may have different chemical shift values, but the biggest difference will be a appearance of a signal in the aliphatic region of the product spectrum for the methyl group attached to the ester oxygen atom.

^{13}C NMR spectrum

b. For this transformation, the ^1H NMR spectra of both molecules will display similar patterns of signals in the aromatic region. The starting material will have the doublet/septet pattern associated with the isopropyl group (Figure 13.11), and the product will have a signal attributable to the acid proton at about δ 12. This signal will disappear when the sample is treated with D_2O.

    The carbon NMR spectra will be similar to each other in the carbonyl and aromatic regions. The biggest difference will be the signals for the starting compound in the aliphatic region that are associated with the isopropyl group. These signals will be absent in the spectrum of the product.

^1H NMR spectrum

^{13}C NMR spectrum

21.28. Both transformations in Exercise 21.26 represent interconversion reactions of carboxylic acid derivatives.

a. This transformation constitutes the conversion of an acid chloride to an ester.

b. This transformation comprises the hydrolysis of an ester to form a carboxylic acid.

21.29. To predict the structures of the products in each of the following transformations, learn the details of each reaction type, which can be found in the reaction summary section of the chapter. In some of these transformations, reactions from previous chapters are included.

a. An ester (a lactone in this instance) reacts with excess Grignard reagent to form a tertiary alcohol with two like groups bonded to the alcohol carbon atom, which is the original carbonyl carbon atom. The alkoxy portion of the original ester becomes an alcohol itself. For lactones, this "other" alcohol group remains part of the molecule. The product is achiral.

b. Aldehydes are not oxidized by chromium oxide in pyridine, so after step 1, the aldehyde is unchanged. An aldehyde reacts with a primary amine to form an imine (Section 20.1a).

c. Acid chlorides react with most nucleophiles to form carboxylic acid derivatives. A thiol reacts with an acid chloride to form a thioester.

d. Esters react with amines to form carboxamides with expulsion of a molecule of alcohol (methanol in this reaction).

e. A carboxylic acid reacts with thionyl chloride to form an acid chloride, which, in turn, reacts with the amine to form a carboxamide. The amide is reduced by lithium aluminum hydride to form the corresponding amine. Because the last step comprises acid workup, the ammonium ion is the actual product under these conditions. Addition of aqueous base would yield the amine itself.

f. Carboxylic acids react with primary alcohols in the presence of strong acid catalysts to form esters.

g. A carboxamide undergoes hydrolysis in aqueous acid to form the corresponding carboxylic acid and amine molecules. When acidic conditions are used, the ammonium ion is the actual nitrogen-containing product that is formed.

21.29. (continued)

achiral   H   achiral

h. Ozonolysis of an alkene followed by oxidative workup yields a carboxylic acid if the alkene carbon atom has a proton attached and a ketone if the carbon atom has no hydrogen atom substituent. A cyclic alkene yields a difunctional product. In this transformation, the product contains both ketone and carboxylic acid groups.

i. An acid chloride reacts with an organocuprate reagent to form a ketone. Acid chlorides are prepared from the corresponding carboxylic acids by treatment with thionyl chloride.

j. Aldoses react with aqueous bromine solution to form aldonic acids (Section 19.5b). These oxidized carbohydrate derivatives can exist as lactones.

k. Aldehydes react with hydroxylamine to form oximes, which in turn can be dehydrated to form the corresponding nitriles. The chiral center is not affected by the reactions at the aldehyde group.

l. A primary alcohol is oxidized in aqueous solution to form a carboxylic acid. Reaction of the carboxylic acid with alcohol and a strong acid catalyst yields the corresponding ester.

21.30. To deduce what reagents can be used to prepare a carboxylic acid from a given starting material, learn the details of each reaction type, which can be found in the reaction summary section of the chapter.

a. Oxidation of a primary alcohol with an aqueous solution of a metal-based oxidant yields a carboxylic acid (Section 11.4c).

21.30. (continued)

b. An alkyl chain attached to a benzene ring can be oxidized with potassium permanganate to form a benzoic acid derivative (Section 17.4a). If the oxidation step is done in basic solution, an acid workup step is needed to obtain the carboxylic acid itself.

c. A nitrile is hydrolyzed with aqueous acid or base to form the corresponding carboxylic acid.

d. Bromobenzene can be converted to the aryl Grignard reagent, which in turn reacts with carbon dioxide (Section 15.2b) to form benzoic acid after the aqueous acid workup step.

e. An amide undergoes hydrolysis with aqueous acid or base to form the corresponding carboxylic acid.

f. Ozonolysis of an alkene followed by oxidative workup yields a carboxylic acid if the alkene carbon atom has a proton attached (Section 11.3b).

21.31. Use the data listed in Table 14.5 to differentiate among the carboxylic acid derivatives by the positions and shapes of their IR absorption bands.

a. This spectrum has a strong C=O stretching vibration at 1720 cm^{-1} with an equally strong band at approximately 1230 cm^{-1}. The molecule producing this spectrum is an ester, RCOOR'.

b. This spectrum has a strong C=O stretching vibration at 1800 cm^{-1}. The molecule producing this spectrum is an acid chloride, RCOCl.

c. This spectrum has a strong C=O stretching vibration at 1690 cm^{-1} with a broad, medium intensity band between 3500 and 2500 cm^{-1}. Note that there are several peaks resulting from C–H stretching vibrations around 3000 cm^{-1}, which may distract you from seeing the broad peak under them, but if you compare this region with the same region in the other spectra, you can see that this one looks quite different. The molecule producing this spectrum is a carboxylic acid, RCOOH.

21.32. Use the data listed in Figure 13.23 and Table 13.4 to differentiate among the carboxylic acid derivatives by the chemical shift values of their carbonyl carbon atom resonances.

a. This spectrum has a weak signal at δ 168, which is in the range expected for the ester carbonyl resonance.

b. This spectrum has a medium intensity signal at δ 175, which is in the range expected for the amide carbonyl resonance.

c. This spectrum has a signal at δ 120, which is in the range expected for the cyano carbon atom resonance of a nitrile.

21.33. Hydride and organometallic reagents differ in their patterns of reactivity toward various types of carbonyl compounds. As with any reaction, learn how the reagents react with different functional groups as described in the reaction summary section of the chapter. In some of these transformations, reactions from previous chapters are included.

a. Sodium borohydride is capable of reducing the ketone functional group, but not the ester. Therefore, the 2° alcohol is produced. A new chiral center is formed, so the product is racemic.

b. Lithium aluminum hydride is capable of reducing all types of carbonyl groups. A ketone is converted to a secondary alcohol, and an ester is reduced to form an alcohol group. The alkoxy portion of the ester is converted to an alcohol molecule as well (methanol in this reaction).

c. Lithium aluminum hydride reduces an ester to form two molecules of alcohol. A lactone is therefore converted to a diol. Because an ester is reduced by breaking the single bond between the carbonyl carbon atom and the oxygen atom, the stereochemistry of the chiral center is not affected during this particular transformation.   The alkoxy portion of the lactone remains in the structure of the product, unlike the case in part (b.), in which the alkoxy portion becomes a separate product.

d. An ester (a lactone in this instance) reacts with excess Grignard reagent to form a tertiary alcohol with two like groups bonded to the alcohol carbon atom. The alkoxy portion of the original ester becomes an alcohol itself. For lactones, this "other" alcohol group remains part of the molecule. Because a lactone reacts by breaking the single bond between the carbonyl carbon atom and the oxygen atom, the stereochemistry of the chiral center is not affected during this transformation.

21.34. The Friedel-Crafts acylation reaction provides a straightforward way to make an aryl alkyl ketone such as 1-phenyl-1-butanone. Using organometallic reagents, you can exploit the low temperature reaction between a Grignard reagent and an acid chloride.   Reaction of an acid chloride with an organocuprate reagent is usually more reliable.

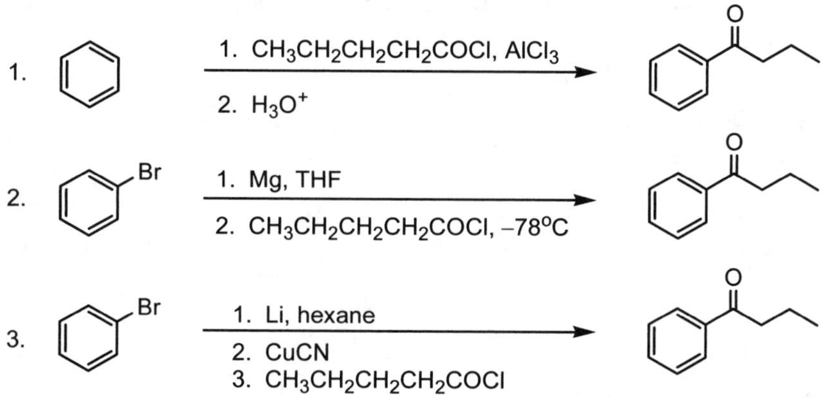

1.

$$\text{1. } CH_3CH_2CH_2CH_2COCl, AlCl_3$$
$$\text{2. } H_3O^+$$

2.

Br

$$\text{1. } Mg, THF$$
$$\text{2. } CH_3CH_2CH_2CH_2COCl, -78°C$$

3.

Br

$$\text{1. } Li, hexane$$
$$\text{2. } CuCN$$
$$\text{3. } CH_3CH_2CH_2CH_2COCl$$

21.35. In this chapter, you were introduced to reactions that interconvert carboxylic acid derivatives. In the retrosynthetic analysis of acid derivatives, look particularly for disconnections at the carbonyl group. If the synthetic goal has a ring as part of the carbon skeleton, consider ways to include the ring as a starting material. There are several ways to prepare the molecules shown below. Only one route for each is presented. The molecules shown in color are the organic starting materials.

a. The reaction of a Grignard reagent with carbon dioxide is one of the simplest ways to prepare a carboxylic acid.

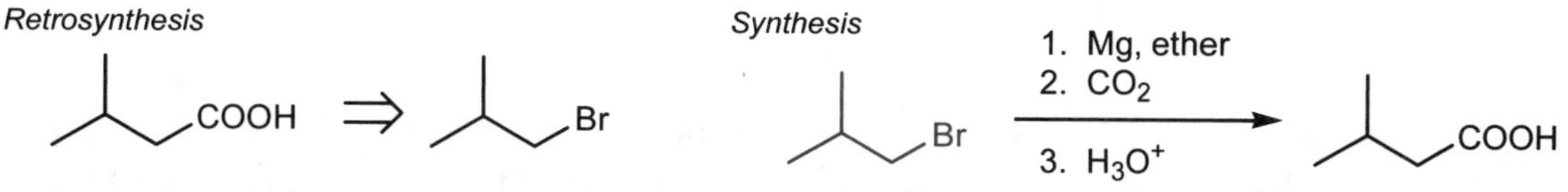

Retrosynthesis   *Synthesis*

COOH ⇒ Br

$$\text{1. } Mg, ether$$
$$\text{2. } CO_2$$
$$\text{3. } H_3O^+$$

Br → COOH

b. The carboxylic acid group is a meta director, so it can be used to introduce a bromine atom substituent that is subsequently be replaced by the methyl group using lithium dimethylcuprate. (Direct introduction of the methyl group by Friedel-Crafts alkylation is unlikely because a meta director deactivates the ring too much).   The carboxylic acid group will have to be protected as its ester derivative before the organocuprate reaction in order to avoid an acid-base reaction. Hydrolysis of the ester group in the last step will regenerate the carboxylic acid group.

Retrosynthesis

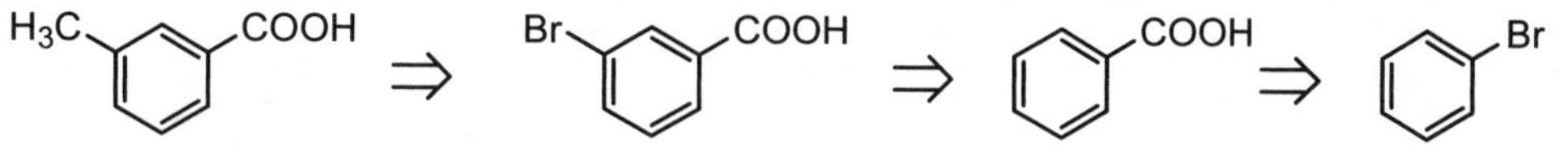

H_3C COOH ⇒ Br COOH ⇒ COOH ⇒ Br

Synthesis

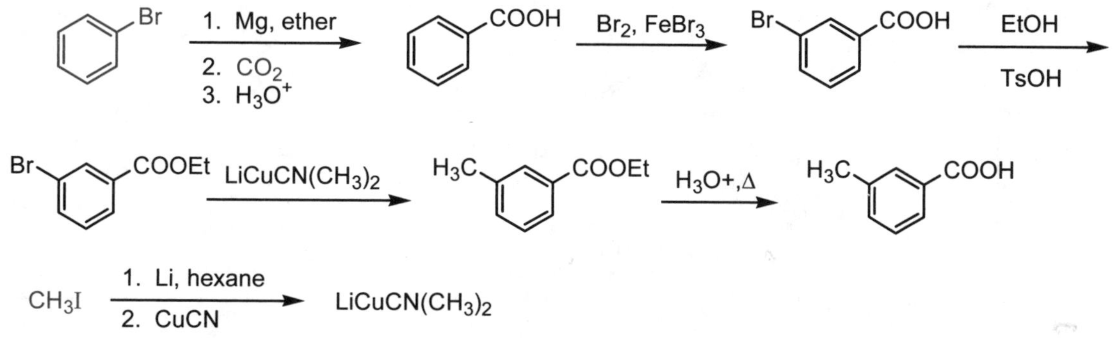

Br

$$\text{1. } Mg, ether$$
$$\text{2. } CO_2$$
$$\text{3. } H_3O^+$$

COOH $Br_2, FeBr_3$ Br COOH

EtOH
TsOH

Br COOEt $LiCuCN(CH_3)_2$ H_3C COOEt H_3O+,Δ H_3C COOH

CH_3I

$$\text{1. } Li, hexane$$
$$\text{2. } CuCN$$

$LiCuCN(CH_3)_2$

21.35. (continued)

c. The carboxylic acid group can be introduced by means of a Grignard reaction, but the aldehyde group must first be protected as its acetal derivative. After the Grignard reagent is made and treated with carbon dioxide, the acetal group is converted back to the aldehyde group using aqueous acid.

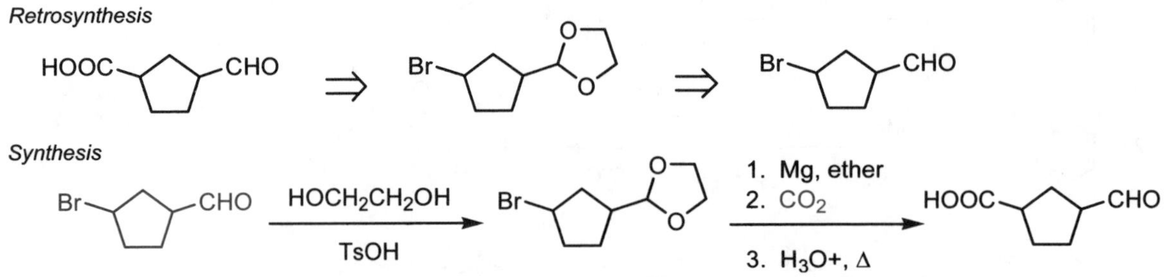

Retrosynthesis

Synthesis

21.36. The Wohl degradation converts an aldose to the aldose having one less carbon atom. Cleavage occurs from the aldehyde end. The first step is an addition-elimination reaction between the aldehyde group and hydroxylamine, which produces an oxime (steps 1-4).

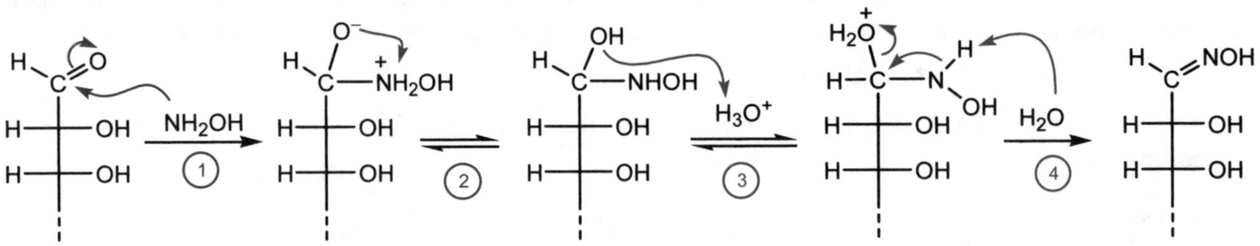

Next, the oxygen atom of the oxime reacts with acetic anhydride via an addition-elimination process (steps 5 and 6) to attach an acetyl group to the oxime oxygen atom. Ethoxide ion then reacts to in an E2 reaction (step 7), which produces a nitrile by loss of the elements of acetic acid. The product of step 7 is a cyanohydrin, which is not stable toward base (see section 18.2b), so HCN is eliminated to generate the aldehyde group (step 8). The carbon chain of the carbohydrate is thereby decreased in length by one carbon atom.

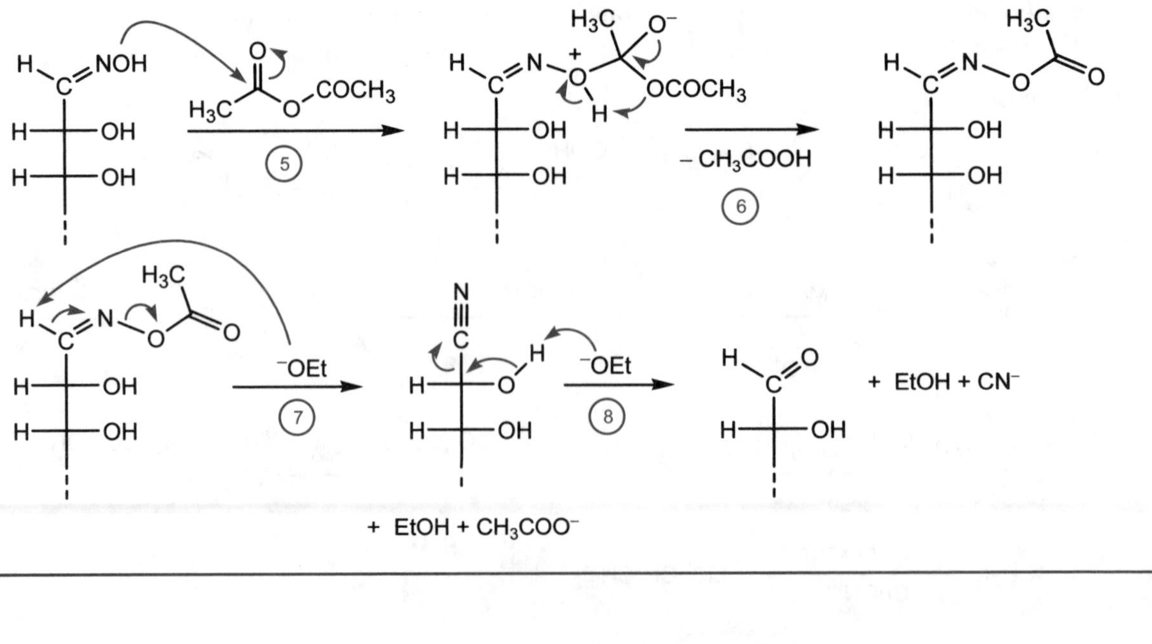

21.37. The Wohl degradation, described in Exercise 21.36, converts a carbohydrate to its next lower homolog. The configurations of the chiral centers that appear in the product are not affected.

a.

D-Mannose → D-Arabinose

b.

D-Ribose → D-Erythrose

c.

L-Arabinose → L-Erythrose

21.38. The Wohl degradation, described in Exercise 21.36, converts a carbohydrate to its next lower homolog. The configurations of any chiral centers that appear in the product are not affected. The starting carbohydrate used to prepare a particular sugar can have either configuration at the carbon atom that becomes the new aldehyde group (that is, C2), because the chirality of that center is lost during the transformation. Note that these processes are shown in the retrosynthetic format whereas the answers to Exercise 21.37 were given in the synthesis format.

a.

D-Glyceraldehyde ⇒ D-Erythrose or D-Threose

b.

L-Threose ⇒ L-Xylose or L-Lyxose

c.

D-Erythrose ⇒ D-Ribose or D-Arabinose

21.39. Follow the procedure outlined in the solution to Exercise 14.21.

a. From the given molecular formula, C_4H_6O, calculate the number of unsaturation sites: $[2(4) + 2 - 6]/2$ = 2 sites of unsaturation. This compound contains two double bonds, a triple bond, two rings, or a double bond and a ring.

   The IR spectrum displays bands that can be associated with the COOH group: $1690\ cm^{-1}$ (vs) and a broad, strong intensity band between 3500 and $2500\ cm^{-1}$.

   The data from the proton NMR spectrum are summarized in the following table:

Chemical shift (ppm)	Integrated intensity	Assignment	Multiplicity	No. adjacent protons (n)
12.1	1	COOH	singlet	0
6.20	1	=CH	singlet	0
5.75	1	=CH	singlet	0
1.95	3	CH$_3$	singlet	0

21.39. (continued)

Note the presence of the three general features deduced from the NMR spectrum: a COOH proton, two alkene protons, and three aliphatic protons. The spectrum only has singlets, so the hydrogen atoms are all separated by more than three bonds. The only structure that can produce such a spectrum is **2-methyl-2-propenoic acid**, which has two double bonds (C=O and C=C), hence two sites of unsaturation.

b.   From the given molecular formula, $C_6H_{11}ClO_2$, calculate the number of sites of unsaturation. Thus, $[2(6) + 2 - 11 - 1]/2 = 1$ site of unsaturation. This compound contains either a double bond or a ring.

The IR spectrum displays bands that can be associated with the ester group [1740 cm^{-1} (vs) and approximately 1200 cm^{-1} (vs)].

The data from the proton NMR spectrum are summarized in the table below. Only aliphatic protons are present, and the splitting patterns clearly show an ethyl group (triplet and quartet), and three other features that can be attributed to the presence of a 1,3-propylene group. Putting these fragments together along with the Cl atom and the ester functional group yields the structure of this molecule: **Ethyl 4-chlorobutanoate**. The methylene group that produces the signal farthest downfield has to be attached to the oxygen atom of the ester group, and it must be adjacent to the methyl group because its signal appears as a quartet.

Chemical shift (ppm)	Integrated intensity	Assignment	Multiplicity	No. adjacent protons (n)
4.2	2	CH$_2$	quartet	3
3.6	2	CH$_2$	triplet	2
2.5	2	CH$_2$	triplet	2
2.1	2	CH$_2$	quintet	4
1.3	3	CH$_3$	triplet	2

ethyl group

propylene group

$$Cl-CH_2-CH_2-CH_2-\overset{\displaystyle O-CH_2-CH_3}{\underset{\displaystyle O}{C}}$$

21.40. Friedel-Crafts acylation reactions occur through the participation of an acyl cation. An acyl cation can be created from an anhydride by complexation of a carbonyl group with aluminum chloride (step 1). The benzene ring intercepts this cation (step 2), and the aromatic ring is regenerated by removal of a hydrogen atom (step 3). Hydrolysis of the oxygen–aluminum bond completes the reaction (step 4).

Solutions to Exercises for CHAPTER 22

THE ACID-BASE CHEMISTRY OF CARBONYL COMPOUNDS

22.1. The dianion of a phosphoric acid derivative is stabilized by the delocalization of unshared and π electron pairs among the terminal oxygen atoms.

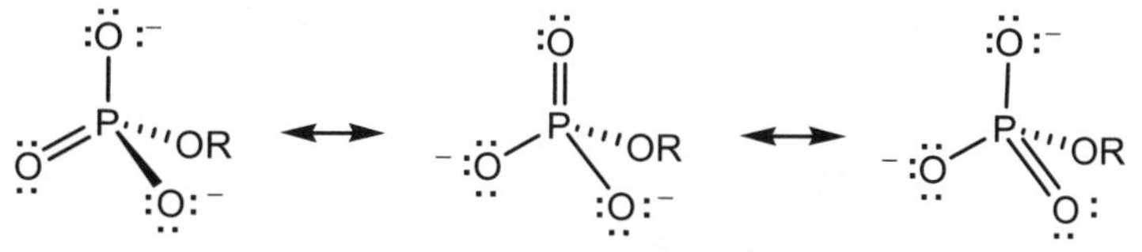

22.2. The dissociation of a proton from oxalic and malonic acids produces a monoanion that can form a hydrogen bond in a five- or six-membered ring. Acids that have additional carbon atoms between the carboxylic acid groups would have to form larger rings, which are entropically less favorable.

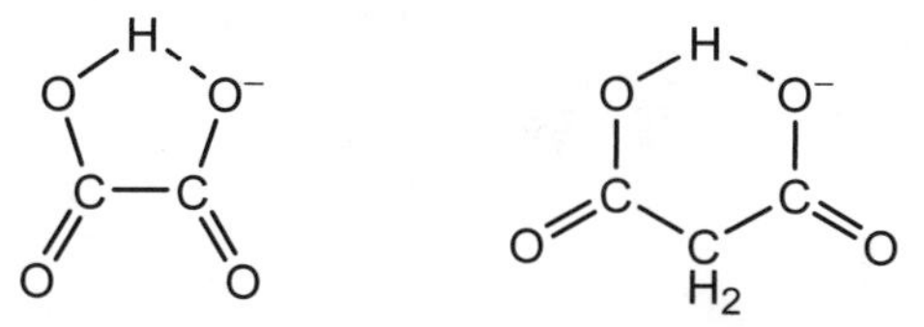

Oxalic acid, monoanion **Malonic acid, monoanion**

22.3. Two electrophilic centers are present in butanoamide: the protons bonded to nitrogen and the carbonyl carbon atom. In *N,N*-dimethylbutanoamide, the only electrophilic center is the carbon atom of the carbonyl group. Thus, butanoamide can either undergo an acid-base reaction to form a resonance-stabilized anion or it can undergo addition, which leads to hydrolysis (dashed arrow, subsequent steps not shown). On the other hand, the dimethyl derivative only undergoes nucleophilic addition to the carbonyl group, which leads to hydrolysis (dashed arrow, subsequent steps not shown).

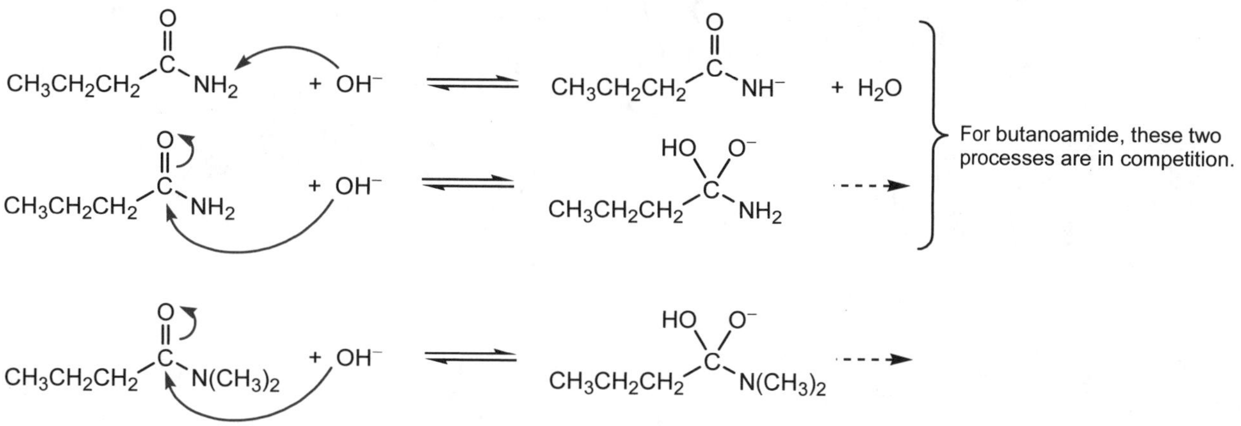
For butanoamide, these two processes are in competition.

22.4. The enol form of phenol is aromatic (6 π electrons). Recall from Section 17.1a that the resonance energy of a benzene derivative is substantial (> 30 kcal/mol), so it more than offsets the normal high stability of the C=O double bond.

22.5. In the conversion of glyceraldehyde-3-phosphate to dihydroxyacetone phosphate, protonation of the aldose carbonyl group (step 1) leads to formation of the enediol (step 2). Protonation of the enediol double bond (step 3) followed by deprotonation of the OH group at C2 yields the ketose (step 4).

22.6. The proton attached to the alpha carbon atom of phenylalanine ethyl ester is slightly acidic (pK_a ~ 25). This proton can be removed by base to form the enolate ion derivative. Protonation can occur at either face of the enolate double bond, which creates the enantiomeric forms.

22.7. Each round in the iodination of a methyl ketone consists of two steps: formation of the enolate ion (steps 1, 3, and 5) and reaction of the anion with I_2, displacing iodide ion (steps 2, 4, and 6).

22.8. At low temperatures (usually -78 °C), the kinetic enolate derivative of a ketone is formed by the removal of a proton from the less highly substituted alpha carbon atom. If this enolate ion is allowed to warm to room temperature with a small amount of ketone added to the solution, the thermodynamic enolate ion (the one with the more highly-substituted double bond) is formed.

a.

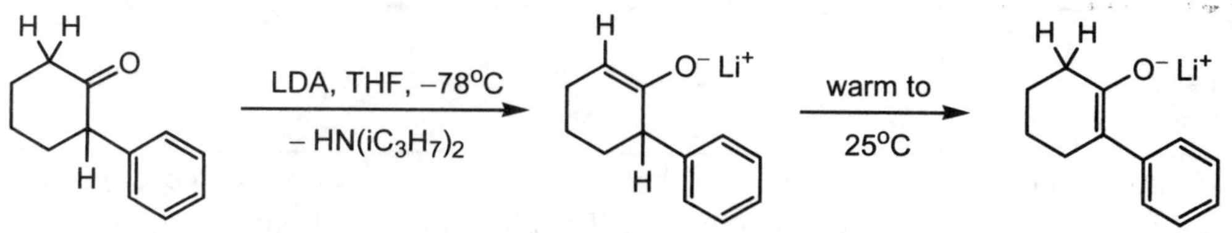

b.

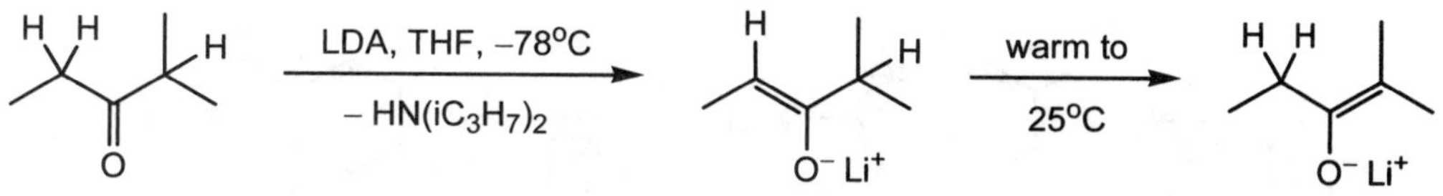

22.9. Both of the compounds in this exercise are prepared by alkylation reactions that occur between a ketone enolate ion and a reactive 1° alkyl halide [benzylic in (a.) and allylic in (b.)]. The starting compounds are shown in color.

a.   *Retrosynthesis*

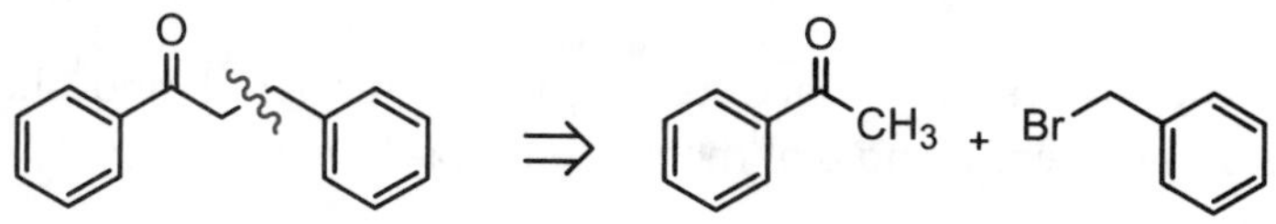

Synthesis

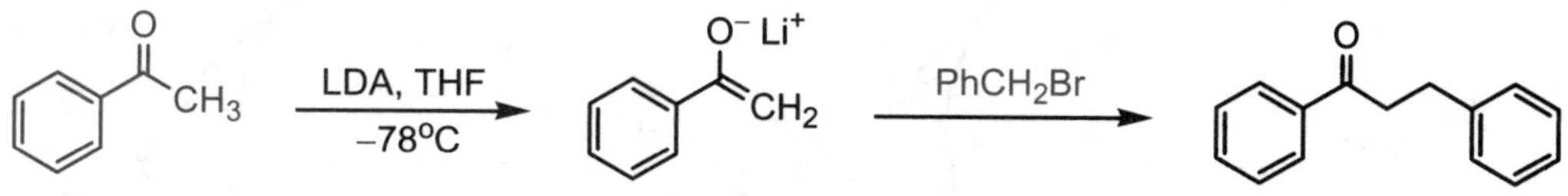

b.   *Retrosynthesis*

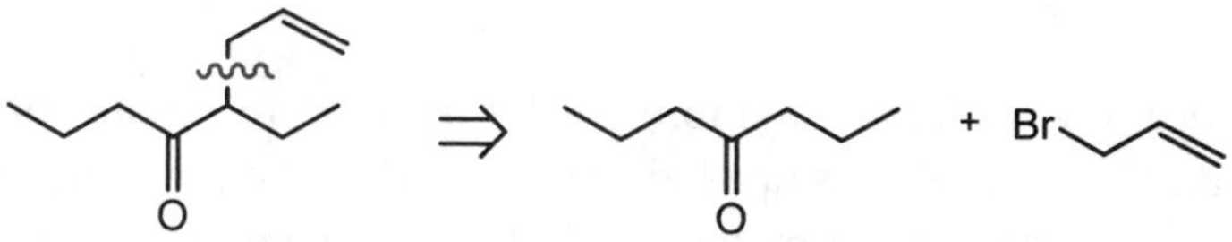

Synthesis

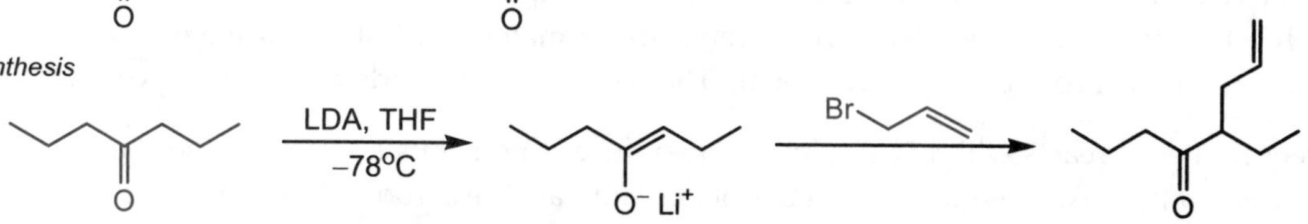

22.10. In the given transformation, the first step generates an enolate ion from the ketone molecule. Only one of the alpha carbon atoms has protons, so there is only one possible enolate ion that can form. The enolate ion reacts with PhSeCl to form the phenylselenium derivative. Oxidation with hydrogen peroxide in step 3 leads to formation of the selenoxide derivative, which then undergoes elimination of PhSeOH to form the unsaturated ketone product.

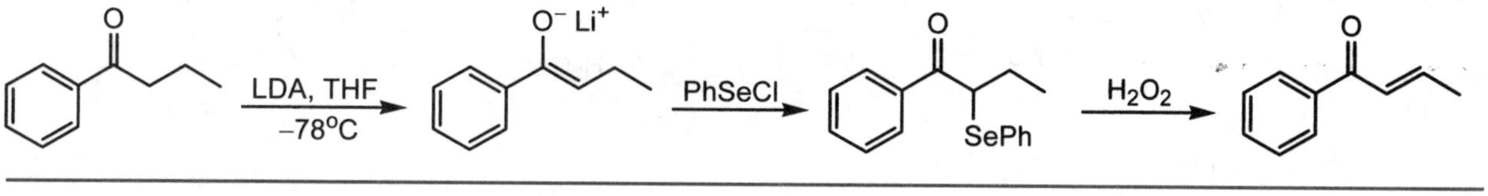

22.11. An enolate ion is formed by deprotonating the ketone at its alpha carbon atom. This nucleophile reacts with diphenyl disulfide at one sulfur atom, displacing the phenylthiolate ion as a leaving group.

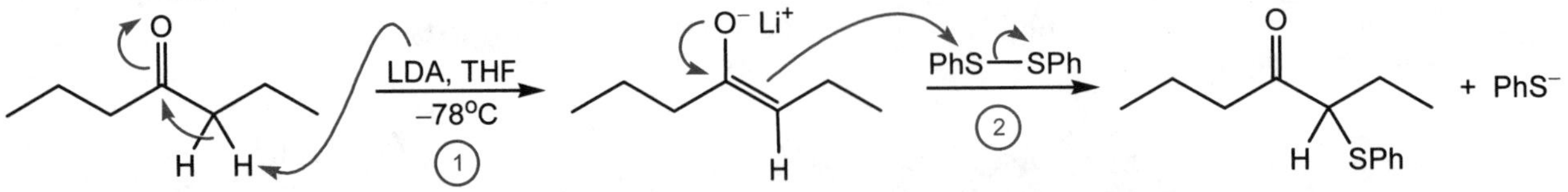

22.12. Lactones form enolate ions in the same way that ketones and esters do, by deprotonation at the position alpha to the carbonyl group (reactions 1 and 3). The alkylation reactions (reactions 2 and 4) are S_N2 processes in which the enolate ions serve as nucleophiles.

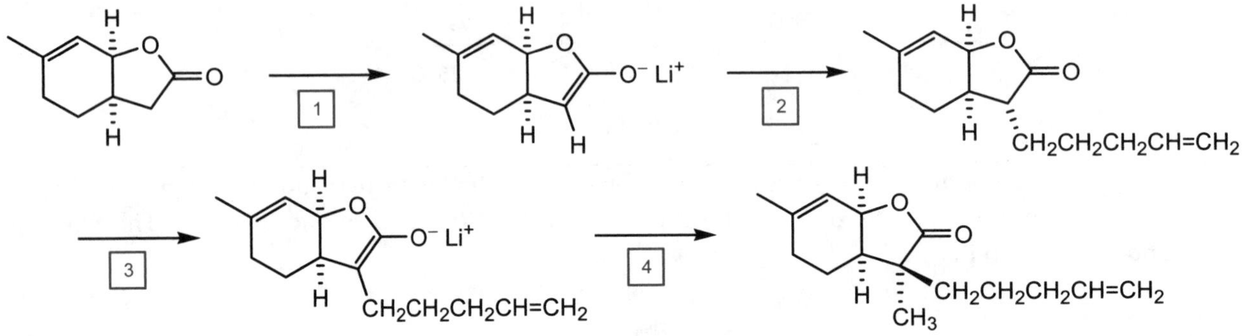

This molecule is saddle-shaped, so the bottom face of the lactone ring is less hindered, and alkylation occurs from below the ring (as drawn below) in steps 2 and 4 of the preceding scheme.

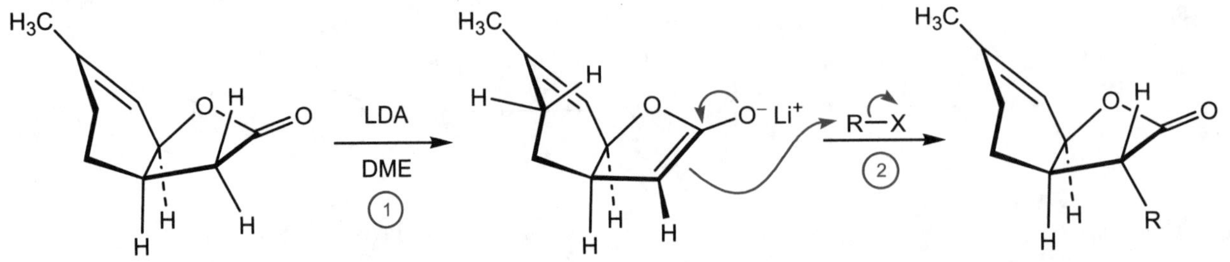

22.13. In any retrosynthesis that makes use of an ester enolate ion, work the target molecule back to an ester derivative and consider which primary alkyl halide is needed for the carbon–carbon bond forming reaction. If a two-carbon ester is identified as the starting material, *tert*-butyl acetate is used instead of ethyl acetate (to prevent condensation reactions). The starting compounds are shown in color.

a.   The required ester reacts with LDA at low temperatures to form its enolate ion derivative; alkylation is followed by hydrolysis with hot, aqueous acid to liberate the carboxylic acid product.

Retrosynthesis

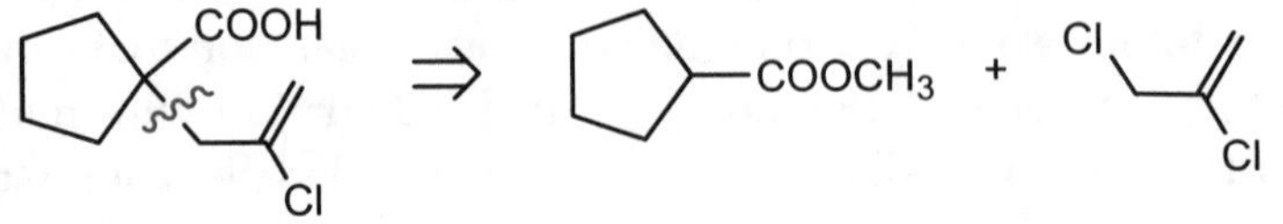

Synthesis

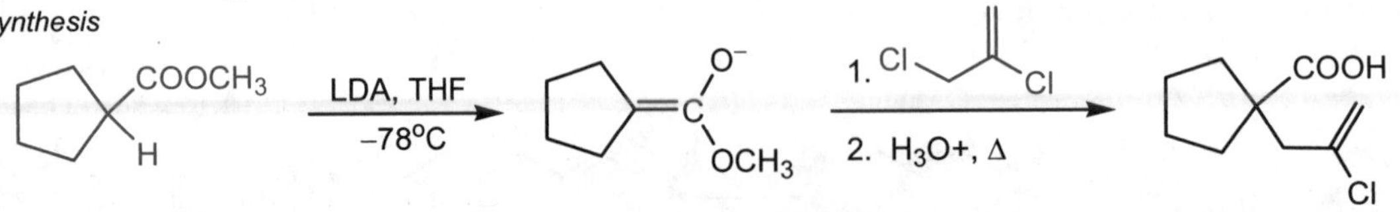

22.13. (continued)

b. The required ester reacts with LDA at low temperatures to form its enolate ion derivative; alkylation is followed by reduction to form the primary alcohol molecule.

Retrosynthesis

Synthesis

22.14. The resonance forms for the anions of diethyl malonate and ethyl cyanoacetate are like those that can be drawn for ethyl acetoacetate: The negative charge and electron pair are delocalized onto the appropriate atoms of the cyano and/or carbonyl groups.

Diethyl malonate

Ethyl cyanoacetate

22.15. An active methylene compound has the form Z–CH₂–Z′, where Z and Z′ are electron withdrawing substituents, often carbonyl groups. (A compound that has two or more "Z-groups" may be called an "active methylene compound," even if it is actually a methine derivative.) To use these types of compounds in synthesis, base is added to a solution of the β-dicarbonyl compound (step 1), and the resulting enolate ion is treated with the appropriate alkyl halide (step 2).

a.

b.

22.16. To plan the retrosynthesis of a carbonyl compound in which alkylation will be a key step, attach an ester group at the position alpha to the carbonyl group of the given product. If the product is a carboxylic acid, think of this acid group as its ester derivative. Then consider which alkylating agent is needed to prepare the carbon atom skeleton. The last step of the synthesis will comprise the hydrolysis/decarboxylation procedure, which hydrolyzes any ester groups present. Decarboxylation will occur if a carboxylic acid group has a carbonyl group at its beta position. The starting materials in the following procedures are shown in color.

22.16. (continued)

a. The given ketone is evaluated as its β-keto ester precursor; the cyclobutane ring is made by the double alkylation of ethyl acetoacetate with a dihaloalkane. Hydrolysis/decarboxylation is carried out by heating the diester in aqueous acid.

Retrosynthesis

Synthesis

1. NaOEt, EtOH

2. BrCH₂CH₂CH₂Cl
3. NaOEt, EtOH

H₃O+, Δ

b. This carboxylic acid is evaluated as its diester precursor. Hydrolysis/decarboxylation is carried out by heating the diester in aqueous acid.

Retrosynthesis

Synthesis

1. NaOEt, EtOH

2. BrCH₂CH₂CH₂CH₃

H₃O+, Δ

22.17. To prepare a primary amine from a primary alkyl halide having the same number of carbon atoms, use a substitution reaction in which the nucleophile can react with only a single equivalent of RX. In such cases, the azide ion is a best suited, and it can be converted to an amine by hydrogenation.

1. NaN₃, DMF

2. H₂, Pd/C

With 1-aminobutane in hand, we can prepare the dialkyl and trialkyl analogs by attaching alkyl groups, one at a time. An easy way to accomplish sequential alkylation of an amine is by use of acylation followed by reduction of the corresponding carboxamide. We first convert 1-bromobutane to butanoyl chloride via substitution, oxidation, and reaction with thionyl chloride.

1. NaOH, DMSO

2. CrO₃, H₂SO₄

SOCl₂, Δ

Reaction of the acid chloride with 1-aminobutane yields the corresponding carboxamide; reduction using lithium aluminum hydride generates the dialkyl amine.

NaOH, H₂O

1. LiAlH₄, ether

2. H₃O⁺
3. OH⁻

22.17. (continued)

Tributylamine can be made in similar fashion to that used for the synthesis of dibutylamine: Reaction of butanoyl chloride with dibutylamine is followed by reduction with lithium aluminum hydride.

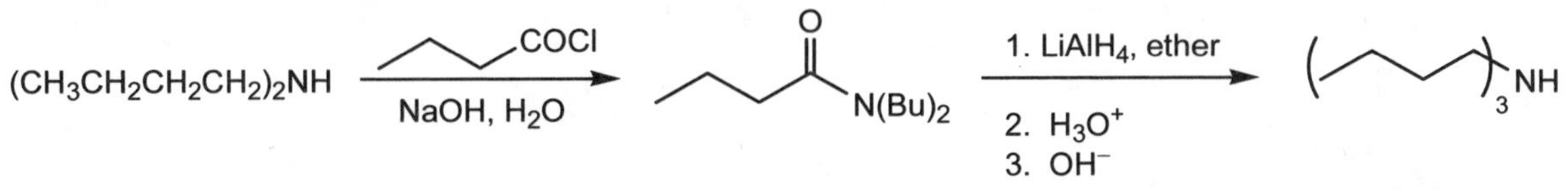

22.18. The steps of this transformation follow the same ones shown for the corresponding Hofmann rearrangement carried out in water. The only difference occurs at the end: A carbamic acid molecule is not formed, so decarboxylation does not take place. Instead, a carbamate ester is the final product.

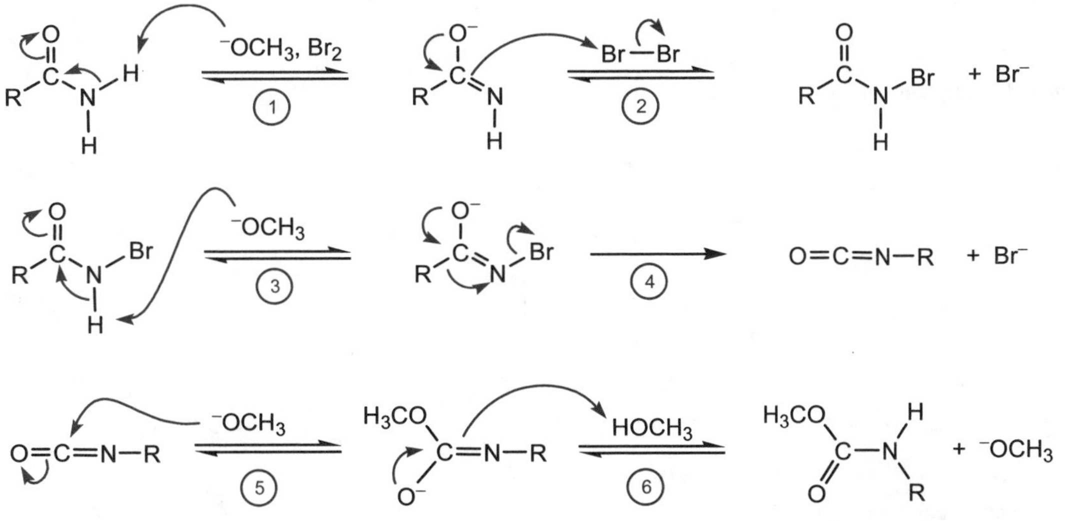

A carbamate ester

22.19. The reaction of hydrazine with a benzimide derivative starts with addition of the hydrazine nitrogen atom to one of the carbonyl groups (step 1). The tetrahedral intermediate can then collapse to regenerate the carbonyl group. At the same time, the imide nitrogen atom removes a proton from the hydrazine group to form the carboxamide/hydrazide derivative of phthalic acid (step 2). The same two steps occur again, except that the other nitrogen atom of the hydrazino group is the nucleophile.

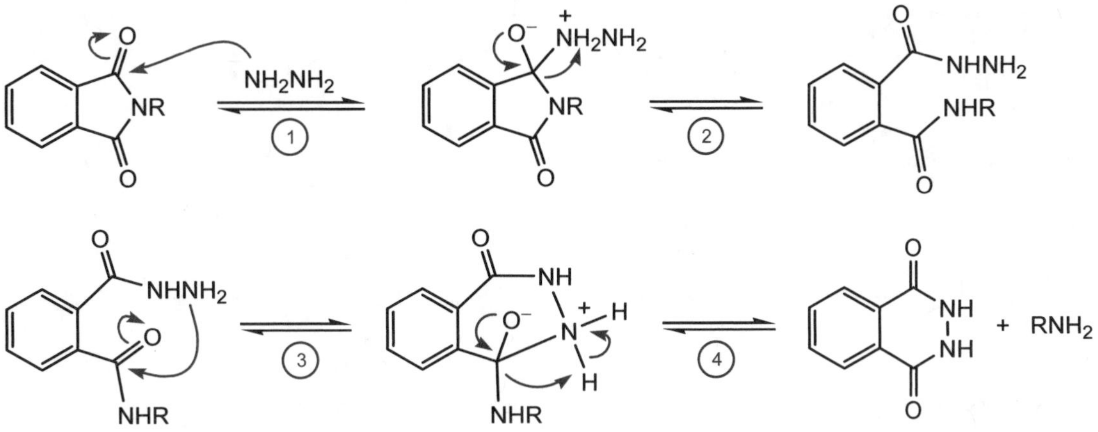

22.20. To draw the structure for the conjugate base of a carbonyl compound, first, interpret the name of the starting compound (the "acid"). Then identify the most acidic proton (shown in color in the following structures), which is normally the one alpha to a carbonyl, sulfonyl, nitrile, or nitro group.

22.20. (continued)

If two electron withdrawing groups are present, the most acidic proton is attached to the atom between them.

a.

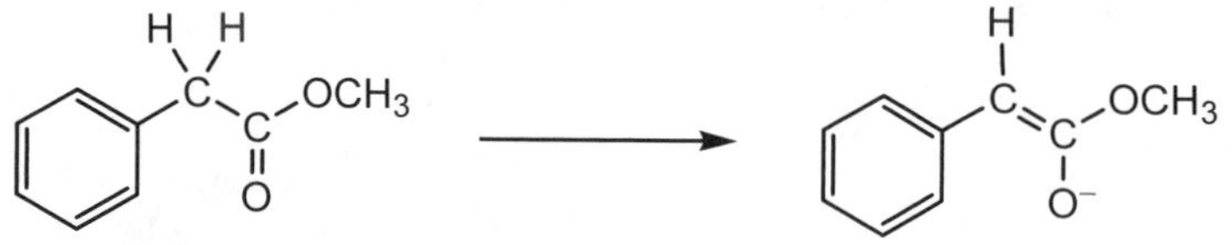

Methyl phenylacetate

b.

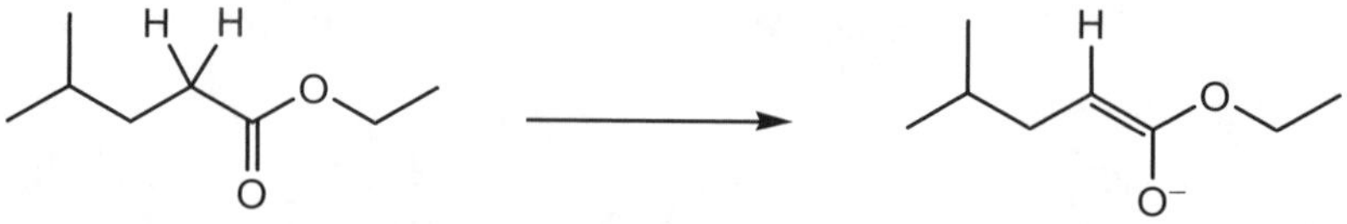

Ethyl 4-methylpentanoate

c.

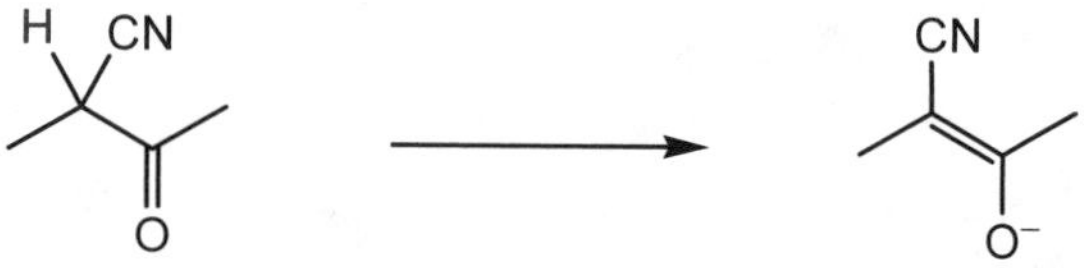

3-Cyano-2-butanone

d.

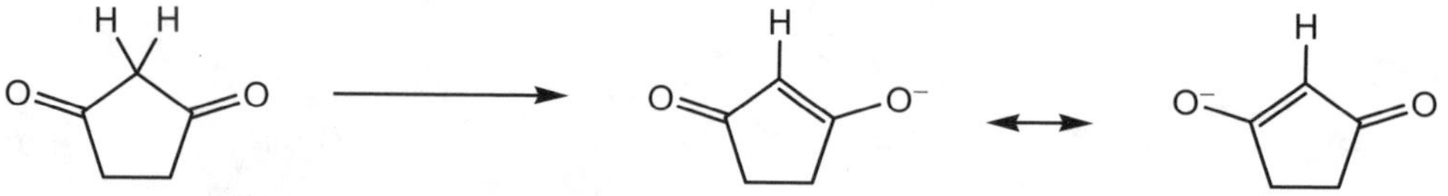

1,3-Cyclopentanedione

22.21. To draw the structure of the enol form of a carbonyl compound, first interpret the name of the starting compound, then change the C=O group to a C—OH group. Remove a proton from the carbon atom adjacent to the carbonyl group and insert a double bond between the original carbonyl carbon atom and the adjacent carbon atom. For a β–keto ester, the ketone carbonyl group forms the enol derivative more readily than the ester carbonyl group.

a.

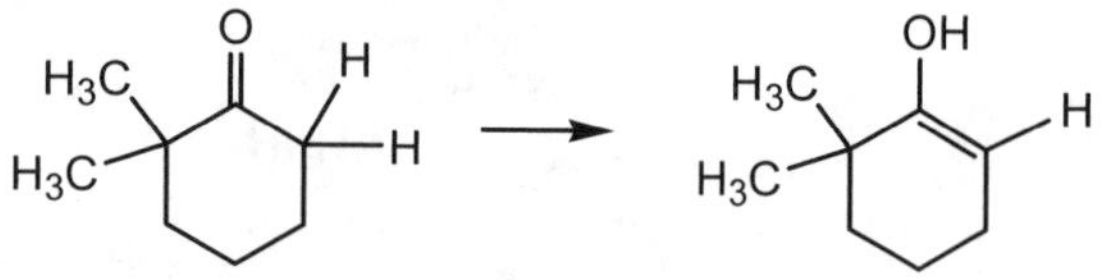

2,2-Dimethylcyclohexanone

b.

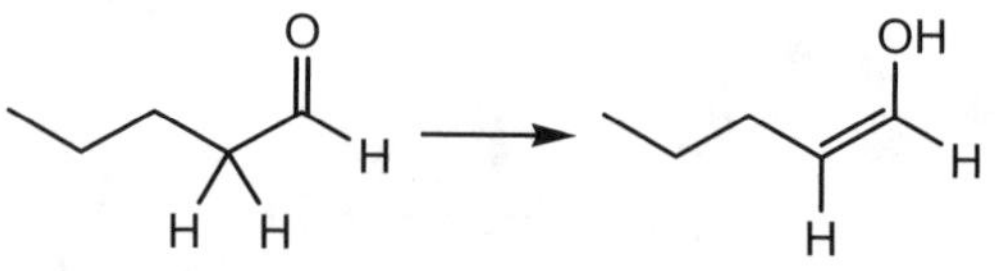

Pentanal

c.

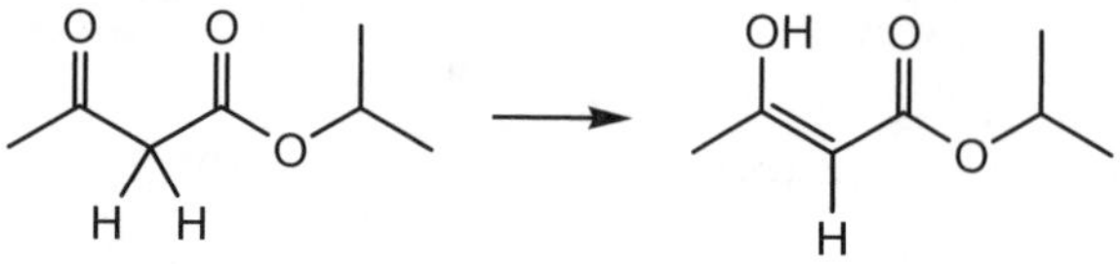

Isopropyl acetoacetate

d.

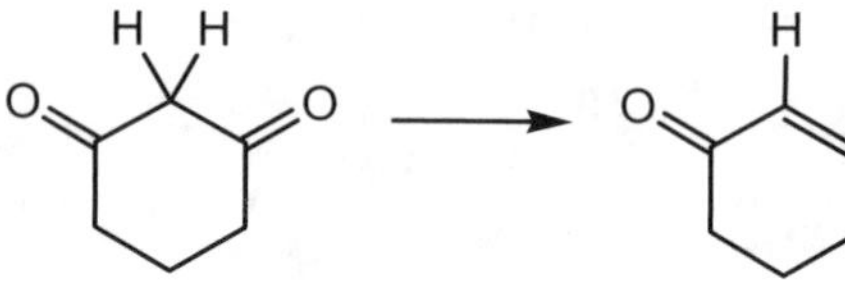

1,3-Cyclohexanedione

22.22. Follow the procedures outlined in the solutions to Exercises 2.2 and 2.4.

a.

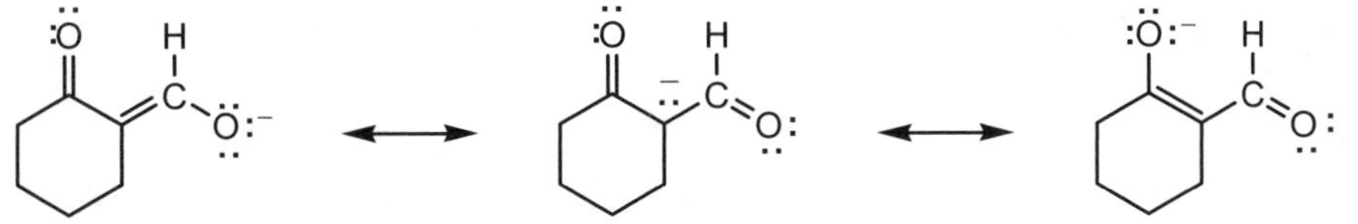

----

b.

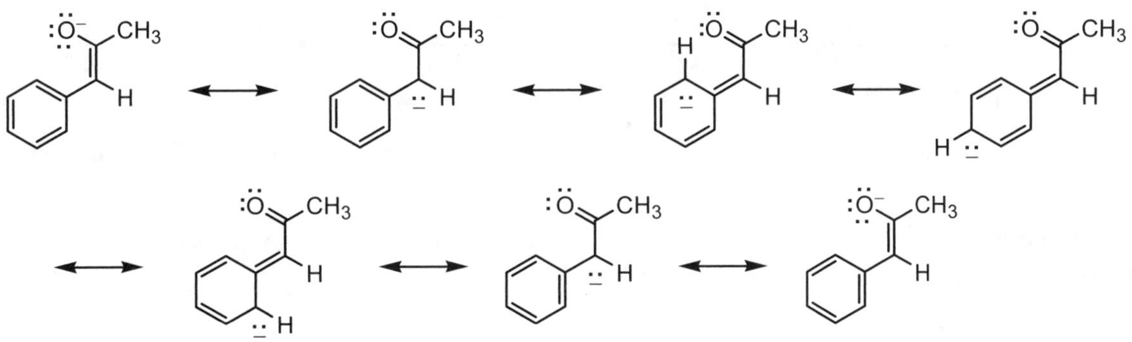

----

c.

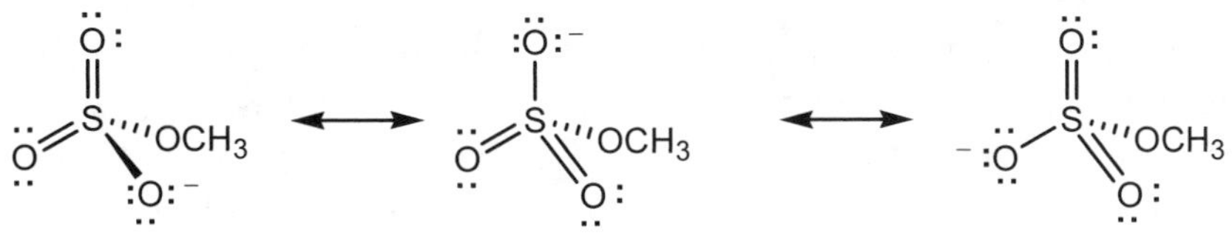

----

d.

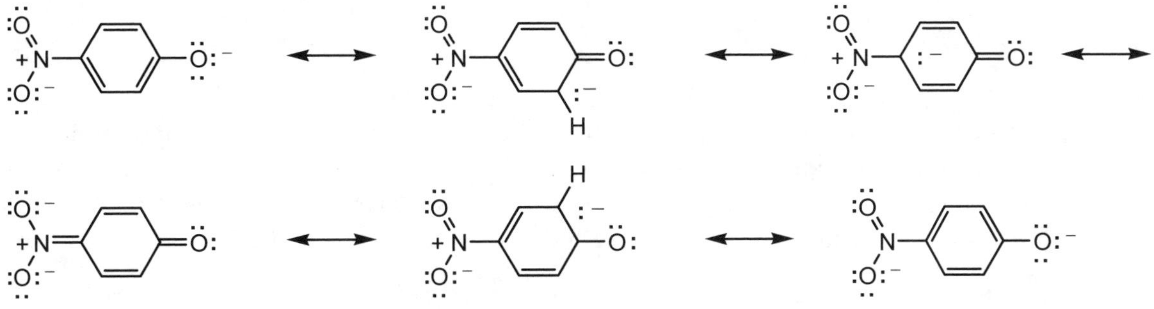

22.23. Use the pK_a values given in the table on the inside front cover to evaluate the acidity of each type of proton. Generally, a proton attached to an oxygen atom is relatively acidic. If unsaturation is present, then acidity increases (i.e., a phenol OH group is more acidic than an alcohol OH group). If no proton is attached to a heteroatom, the most acidic proton is the one bonded to a carbon atom adjacent to one or more carbonyl groups. A sulfur atom also increases the acidity of protons attached to the adjacent carbon atom.

a.

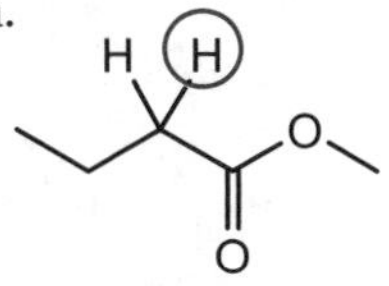

$pK_a \sim 20$

b.

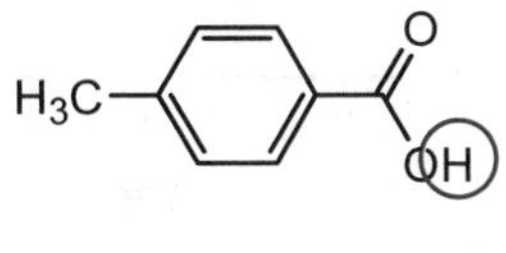

$pK_a \sim 5$

c. CH_2OH

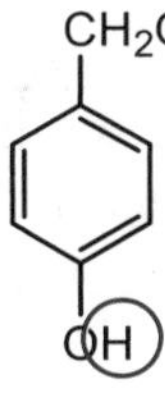

$pK_a \sim 10$

d.

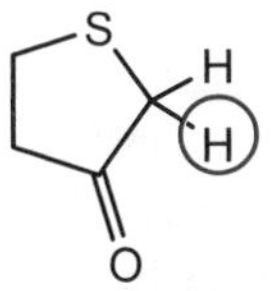

$pK_a < 20$

e.

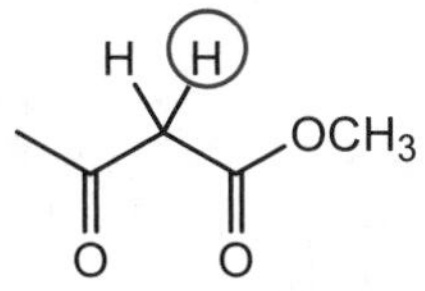

$pK_a \sim 10$

22.24. Ethyl acetoacetate has two carbonyl groups, so two carbanions can be stabilized by delocalization with the π bonds of the adjacent carbonyl groups. The protons attached to the methylene group have a pK_a ~12, and the protons attached to the terminal methyl group have a pK_a ~20.

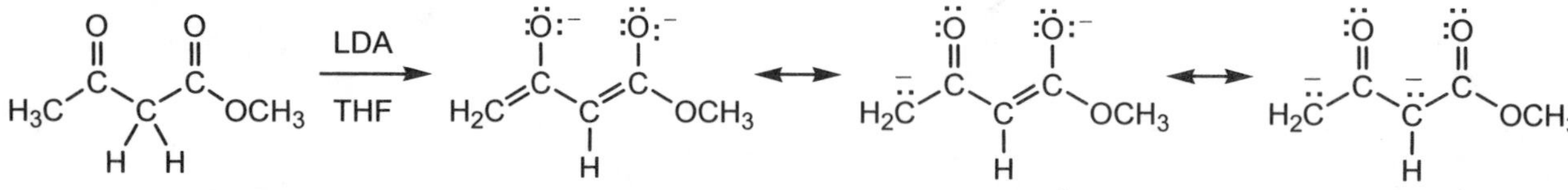

22.25. Assess the acidity of each compound by looking for structural differences that are based on the relative electronegativity values of any substituents, the identities of the atoms to which the acidic protons are attached, and the possibilities for delocalization.

a. Based on electronegativity influences alone (O > Cl > H), the methoxy group is the most electron-withdrawing substituent, so 2-methoxypentanoic acid should be the most acidic molecule shown below. Pentanoic acid itself is the least acidic.

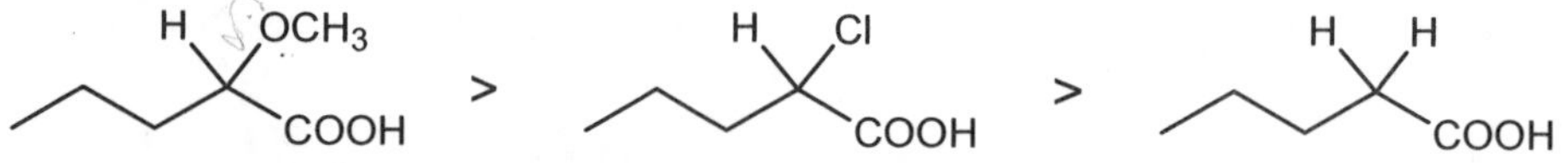

b. A proton attached to a sulfur atom is generally more acidic than one attached to oxygen. The nitro group on the benzene ring is an electron withdrawing substituent, so it makes the proton of the carboxylic acid group even more acidic. Even though a methyl group is an electron donating substituent, benzoic acid derivatives with a group in the *ortho* position are more acidic than benzoic acid itself.

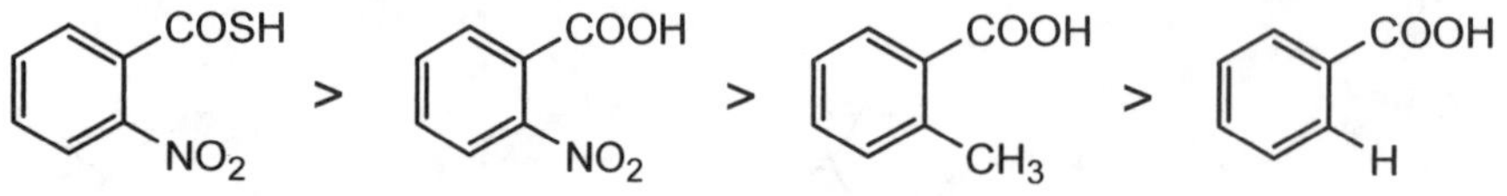

c. Protons adjacent to a carbonyl group are relatively acidic, but for esters and amides, the delocalization of electrons from the oxygen or nitrogen atom into the carbonyl group offsets some of the resonance stabilization that the carbonyl group provides to a neighboring carbanion. An amide is even more stabilized than an ester, so it is expected that its α protons will be less acidic than the α protons of an ester.

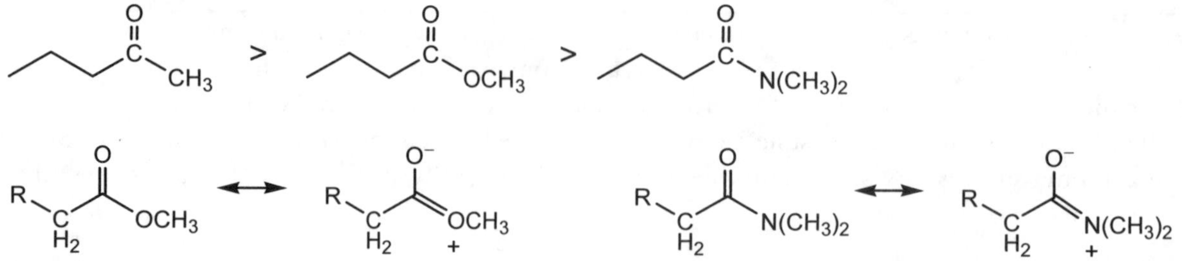

22.26. We can divide the compounds in this exercise into two groups. The first group of compounds have pKa values >20, which requires LDA in order to form their enolate ions. This group includes ketones, esters, and nitriles with α protons. The other compounds have two carbonyl groups (β-diesters, β-keto esters, and β-diketones), which can be deprotonated using NaOEt in ethanol.

22.26. (continued)

Starting material	pK$_a$ value	Base and solvent	Product from reaction with R"X
	20	LDA, THF, -78°C	
	25	LDA, THF, -78°C	
	25	LDA, THF, -78°C	

Starting material	pK$_a$ value	Base and solvent	Product from reaction with R"X
	11	NaOEt, EtOH	
	14	NaOEt, EtOH	

22.27. In acid solution, (E)-5-methyl-3-hexen-2-one is protonated at its carbonyl group to form a carbocation (step 1). The allylic nature of this cation leads to formation of a dienol when deprotonation occurs at the γ-position(step 2). Tautomerism regenerates the carbonyl compound in which the carbon-carbon double bond is no longer conjugated with the carbonyl group. This process occurs by protonation of the enol double bond (step 3) followed by deprotonation of the carbonyl group (step 4).

22.28. To predict the structures of the products in each of the following transformations, learn the details of each reaction type, which can be found in the reaction summary section of the chapter. In some of these transformations, reactions from previous chapters are included.

22.28. (continued)

a. The ketone in this exercise has protons on only side of the carbonyl group, so the enolate ion forms there. The second step attaches the phenylselenyl group to the αcarbon atom. Step 3 oxidizes the Se atom and causes elimination of PhSeOH, which makes the molecule achiral. This initial product is not stable and undergoes tautomerism to form the aromatic naphthalene derivative (see Section 25.1a).

b. A ketone undergoes halogenation at its alpha carbon atom when treated with one of the halogens (except F) and either acid or base. In acid solution, the reaction is readily stopped at the monosubstitution stage.

c. This amide has protons on the carbon atom alpha to the carbonyl group, so the first step generates the corresponding enolate ion. (If the nitrogen atom had a proton instead of the methyl group, deprotonation would occur there instead.) Alkylation occurs at the alpha carbon atom, which creates a new chiral center. Thus, the product is formed as a racemic mixture.

d. This ketone has protons on each side of the carbonyl group. Treating the ketone with LDA at low temperature generates the kinetic enolate; warming to room temperature yields the thermodynamic enolate, which is associated with the carbon atom that is more highly substituted. Alkylation occurs there, which generates a new chiral center. Because the starting material has a chiral center already (the configuration of which is retained), the product exists as a mixture of diastereomers.

e. Treating a primary amide with excess hydroxide ion and bromine yields the corresponding amine with one less carbon atom (the carbonyl group is effectively removed). This is an example of the Hofmann rearrangement.

f. The first step generates the enolate ion of the aryl methyl ketone; the second step leads to alkylation at the original methyl carbon atom.

22.28. (continued)

g. The methylene group between the carbonyl groups has the most acidic protons, and the resulting anion undergoes alkylation. The acid hydrolysis step generates the β–keto acid, which undergoes decarboxylation under the conditions of the third step.

h. This transformation is an example of the Gabriel reaction, which converts an alkyl halide to a primary aliphatic amine.

i. This is another example of enolate ion formation followed by alkylation. Only the methylene group can form an enolate ion because the molecule would be too strained if the proton were removed from the bridgehead carbon atom. The alkylation step creates a new chiral center, so the product is formed as a racemic mixture.

j. Treating a nitrile having alpha protons with base yields a carbanion that is stabilized by delocalization into the carbon–nitrogen triple bond. Alkylation occurs as with enolate ions.

22.29. When you perform a reaction in which you know what the product will be (or is expected to be), focus on those spectroscopic features that are unique for the groups in the reactant and product molecules. Although many subtle changes will undoubtedly occur in several parts of the spectrum, it helps to look for a limited number of specific changes.

a. This transformation converts a ketone to a carboxylic acid. Both have carbonyl groups, and their respective carbonyl stretching vibrations have similar energies. The carboxylic acid will display an additional strong, broad absorption because of the hydrogen-bonded OH stretching vibration.

22.29. (continued)

b. This transformation converts a saturated ketone to its unsaturated analog. Conjugation of a carbonyl group with an alkene double bond lowers the frequency of the C=O stretching frequency. In addition, the product will have a new band corresponding to the C=C stretching vibration. These two peaks may overlap, so they may appear as a broad, strong absorption band.

νCO ~ 1715 cm^{-1} (vs)

νCO ~ 1690 cm^{-1} (vs)

νC=C ~ 1600-1650 cm^{-1} (m)

22.30. As noted in the solution to Exercise 22.29, focus on those spectroscopic features that are unique for the starting material and product molecules. Many other subtle changes will undoubtedly occur in other parts of the spectrum as well

a. In both ^{1}H and ^{13}C NMR spectra, the majority of the resonances will be the same (the aromatic ring and three carbon atoms of the aliphatic portion). The starting material has an extra carbon (with associated protons), and the product has the carboxylic acid group. These structural differences will manifest themselves in the numbers and positions of their associated resonances.

^{1}H NMR spectrum

δ 2.2, 3H, singlet

δ 12, 1H
exchanges with D$_2$O

^{13}C NMR spectrum

δ 10-30
3H (DEPT)

δ >200
0H (DEPT)

δ 160-185
0H (DEPT)

b. In both ^{1}H and ^{13}C NMR spectra, the changes will be apparent in the regions associated with the alkene protons and carbon atoms, which are blank in the spectra of the reactant molecule. The chemical shifts of the carbonyl resonances will undoubtedly be different, but the change may be small.

^{1}H NMR spectrum

all resonances < δ 2.5

δ 5-6, 1H, doublet

δ 5-6, 1H, doublet

δ 5-6, 1H, doublet of doublets

^{13}C NMR spectrum

δ ~200
0H (DEPT)

δ ~200
0H (DEPT)

δ 120-150
2H (DEPT)

δ 120-150
1H (DEPT)

22.31. Deducing which reagents are needed to bring about a particular transformation requires you to be familiar with the reactions described in this chapter, which are summarized in the reaction summary section.

a. A methyl ketone is converted to the conjugate base of a carboxylic acid by the haloform reaction. An acid workup step is included to protonate the carboxylate ion formed in the first step.

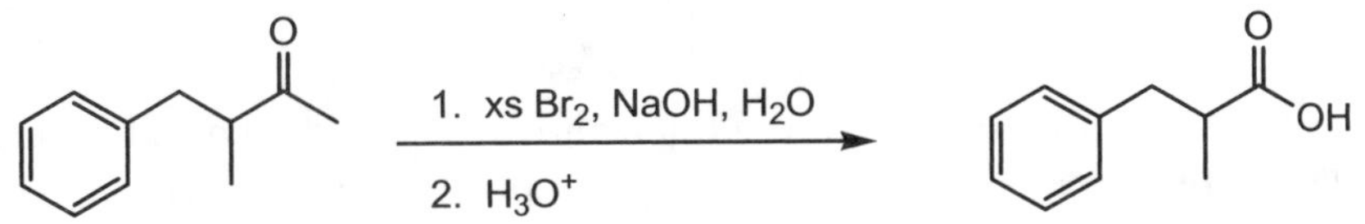

1. xs Br_2, NaOH, H_2O

2. H_3O^+

b. Unsaturation can be introduced into the carbon skeleton of a ketone by formation of its enolate ion followed by treatment with PhSeCl and then oxidation.

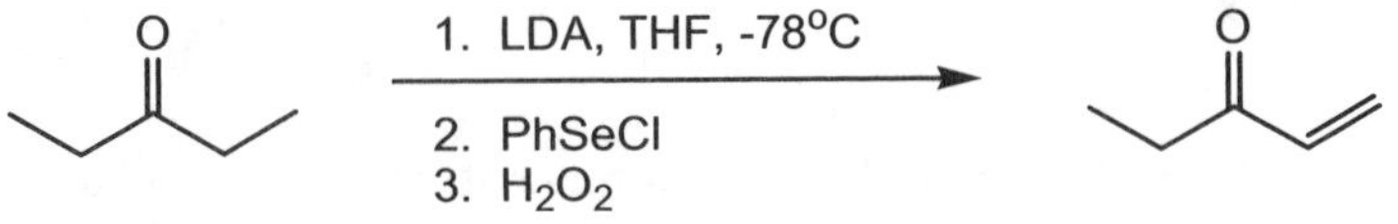

1. LDA, THF, -78°C

2. PhSeCl

3. H_2O_2

22.32. MCPBA converts a cyclic ketone to a lactone, an example of the Baeyer-Villiger reaction (Section 18.4b). If the oxidation step is done first, then enolate formation can occur only on the side of the lactone carbonyl group and alkylation follows in the normal way.

If the enolate ion is formed first, then the regiochemistry is also not an issue because the ketone is symmetrical. After alkylation, however, the ketone is no longer symmetrical, and it undergoes the Baeyer-Villiger reaction by insertion of an oxygen atom into the side that was alkylated because the more highly-substituted carbon atom is the one that migrates preferentially.

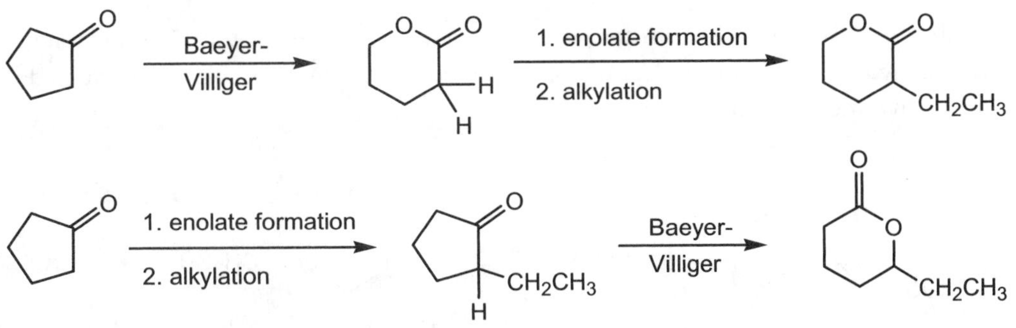

Baeyer-Villiger

1. enolate formation

2. alkylation

CH_2CH_3

1. enolate formation

2. alkylation

CH_2CH_3

Baeyer-Villiger

CH_2CH_3

22.33. We can choose reagents for a given reaction by paying attention to its type (substitution, addition, etc.) and by keeping track of the numbers of carbon atoms. If the product has more carbon atoms than the reactant, then we have to include a carbon–carbon bond-forming reaction in our planning as one of the steps. Otherwise, the transformation becomes a matter of functional group interconversion.

a. 2-Hexanone has more carbon atoms than 1-pentnaol, so 1-pentanol is first oxidized to pentanal, and then addition of the methyl Grignard reagent is used to add the needed carbon atom. The secondary alcohol product is oxidized to form the ketone.

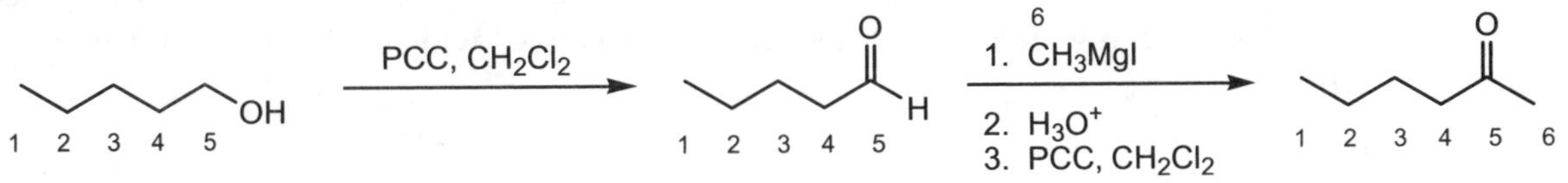

PCC, CH_2Cl_2

1. CH_3MgI

2. H_3O^+

3. PCC, CH_2Cl_2

22.33. (continued)

b. 2-Hexene has more carbon atoms than 1-pentnaol, so 1-pentanol is first oxidized to pentanal and the methylene Wittig reagent is used to form the terminal alkene with the addition of one carbon atom.

c. Converting 1-pentanol to pentanoic acid and then to the ester derivative yields the compound with the correct chain length. The double bond is introduced using the phenyl selenium chloride reagent and its subsequent oxidation.

d. The product has two more carbon atoms than 1-pentnaol, so 1-pentanol is first oxidized to pentanal and the appropriate Horner-Emmons reagent is used to prepare the unsaturated ester.

22.34. Follow the procedures outlined in the solution to Exercise 22.16. The needed starting materials are shown in color.

a. This carboxylic acid is evaluated as its diester precursor. Two alkyl groups (ethyl and benzyl) can be attached to the methylene group of diethyl malonate. Hydrolysis/decarboxylation is carried out by heating the diester in aqueous acid.

Retrosynthesis

Synthesis

b. An amine can be thought of as a reduced amide, which is a carboxylic acid derivative. The needed carboxylic acid is evaluated as its diester precursor. After alkylating diethyl malonate with the appropriate alkyl halide, hydrolysis/decarboxylation is carried out by heating the diester in aqueous acid. The resulting acid is converted to its amide derivative, which is reduced using lithium aluminum hydride.

Retrosynthesis

22.34. (continued)

Synthesis

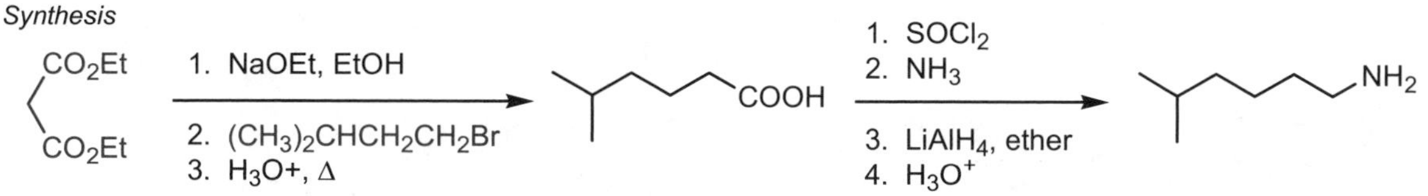

c. This ketone can be evaluated as its β-keto ester precursor. After alkylating ethyl acetoacetate with the appropriate alkyl halide, hydrolysis/decarboxylation is carried out by heating the keto ester in aqueous acid.

Retrosynthesis

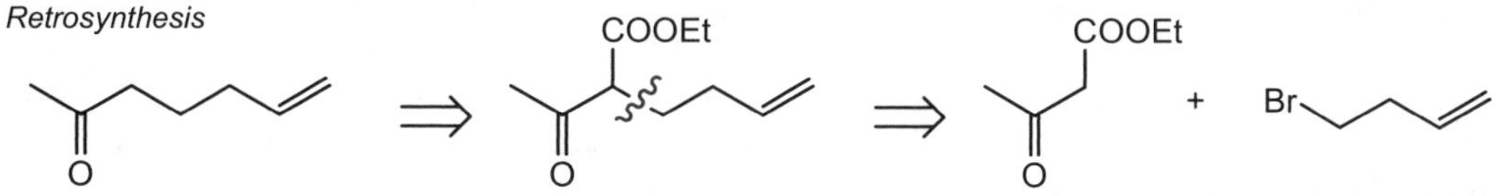

Synthesis

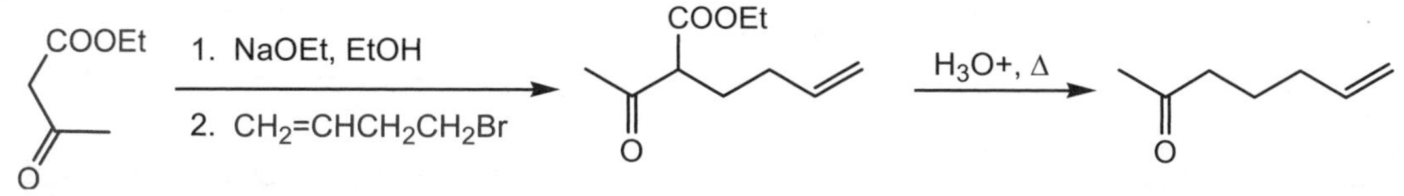

22.35. Because the starting material is specified (shown in color in the following schemes), you can treat this exercise as one that asks you which reagents are needed to go from one substance to another. Use the information in Chapter 16 to decide how the indicated enantiomer is to be formed.

a. In this synthetic scheme, the epoxide ring will be created as a single stereoisomer by making use of a chiral reagent (Section 16.4d). The conversion of bromocyclohexane to the ester will make use of a Grignard reaction with CO_2 followed by esterification. The double bond is introduced using PhSeCl followed by oxidation.

Retrosynthesis

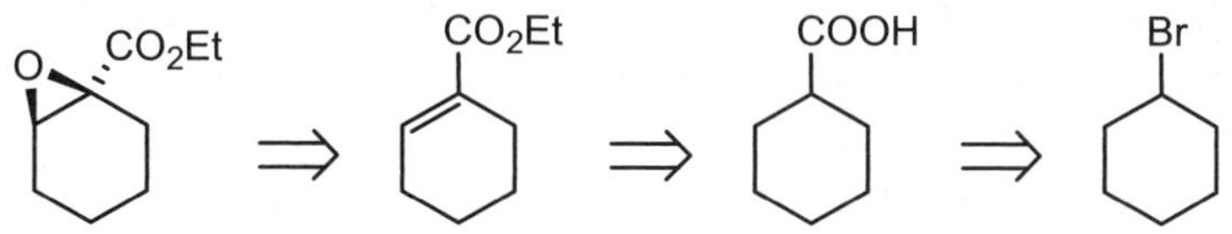

Synthesis

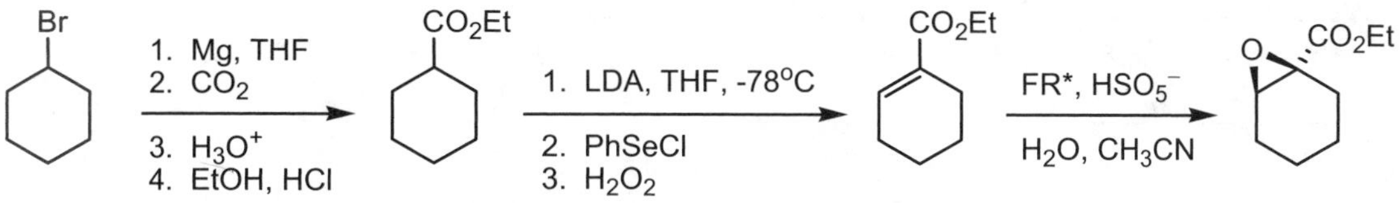

b. In this synthetic scheme, the chiral alcohol will be created as a single stereoisomer by making use of a chiral reducing agent acting on a ketone precursor (Section 18.3b). The conversion of benzene to the ketone is accomplished using the Friedel-Crafts acylation procedure; chlorination of the ketone is done under acidic conditions in order to stop the reaction with formation of the product with a single chlorine atom.

22.35. (continued)

Retrosynthesis

Synthesis

22.36. The Curtius rearrangement occurs with the initial loss of molecular nitrogen, which generates a acyl nitrene intermediate. Migration of the R group from C to N yields the isocyanate.

22.37. An acyl azide is made by a reaction that is typical of acid chlorides: addition of the nucleophile and expulsion of chloride ion as a leaving group to regenerate the C=O double bond.

22.38. Follow the procedure outlined in the solution to Exercise 14.21.

a.   From the given molecular formula, $C_{13}H_{16}O_3$, first calculate the number of sites of unsaturation: $[2(13) + 2 - 16]/2 = 6$ sites.

From the proton NMR spectrum, we conclude that the molecule has both aromatic and aliphatic portions. A benzene ring accounts for four sites of unsaturation, so there are two others.

No protons appear in the alkene region, so this molecule probably has carbonyl groups or rings. The data from the compound's proton NMR spectrum are summarized in the table below.

Chemical shift (ppm)	Integrated intensity	Assignment	Multiplicity	No. adjacent protons (n)
7.2	5	ArH	multiplet	–
4.1	2	CH$_2$	quartet	3
3.8	1	CH	triplet	2
3.2	1	CH$_2$	doublet	1
2.2	3	CH$_3$	singlet	0
1.2	3	CH$_3$	triplet	2

22.38. (continued)

From the splitting patterns, you can see that the molecule has an ethyl group; the chemical shift value of δ 4.1 is consistent with the presence of a methylene group bonded to the oxygen atom of an ester. With its splitting pattern, we conclude that the fragment $CH_3CH_2O-C=O$ is present. The remaining NMR signals indicate that the following fragments are also present: $CH-CH_2$, C_6H_5, and CH_3. Subtracting these elements from the molecular formula ($C_{13}H_{16}O_3 - CH_3CH_2O-C=O - CHCH_2 - C_6H_5 - CH_3$) leaves only C and O, which is another carbonyl group.

Assembling these groups to fit the chemical shift values of the signals, we conclude that this molecule is **ethyl benzylacetoacetate**.

b. From the given molecular formula, $C_6H_{10}O_6$, calculate the number of unsaturation sites: [2(6) + 2 – 10]/2 = 2 sites of unsaturation. This compound contains two double bonds, a triple bond, two rings, or a double bond and a ring.

All of the protons in this molecule are aliphatic ones; the data from the proton NMR spectrum are summarized in the table below.

Chemical shift (ppm)	Integrated intensity	Assignment	Multiplicity	No. adjacent protons (n)
3.8	3	CH$_3$	singlet	0
3.5	2	CH$_2$	singlet	0
2.6	2	CH$_2$	quartet	3
1.1	3	CH$_3$	triplet	2

ethyl group

From the splitting patterns, you can see that the molecule has an ethyl group; the chemical shift value of δ 2.6 is consistent with the presence of a methylene group bonded to the carbonyl group of a ketone. We conclude, therefore, that the fragment $CH_3CH_2-C=O$ is present. The chemical shift of the singlet at δ 3.8 is consistent with the presence of a methyl ester, so the fragment $CH_3O-C=O$ is also present. With the remaining methylene group, all of the elements are accounted for.

Assembling these groups to fit the chemical shift values of the signals, we conclude that this molecule is **methyl 3-oxopentanoate**.

22.39. When a bromoketone is treated with base, an enolate ion is formed (step 1) and it undergoes an intramolecular alkylation reaction to form intermediate **C** (step 2). The three-membered ring ketone is strained, so hydroxide ion adds readily to the carbon-oxygen double bond, producing a tetrahedral intermediate (step 3). Ring opening yields the carboxylate ion by transfer of a proton to the carbon–carbon bond that is broken (step 4).

22.40. First summarize the reactions that interconvert the lettered compounds.

The IR spectrum of compound **A**, along with its given formula, indicates that this molecule is a primary carboxamide. Its reaction with base and bromine proceeds via the Hofmann rearrangement pathway, which means compound **B** is a primary amine, $C_4H_{11}N$. A primary amine reacts with dimethyl-dioxirane to form a nitro compound (Section 11.5). Compound **C** is therefore $C_4H_9NO_2$, and the strong peaks at 1550 and 1390 cm^{-1} in its IR spectrum confirm the presence of the nitro group.

The proton NMR spectrum of **C** has a single peak, and the only isomer of C_4H_9 that has nine equivalent protons is the *tert*-butyl group. Compound **C** is therefore 2-methyl-2-nitropropane, and compounds **A** and **B** correspondingly have the *tert*-butyl group.

THE NUCLEOPHILIC ADDITION REACTIONS OF ENOLATE IONS

23.1. The two products from the aldol reaction of propanal are 3-hydroxy-2-methylpentanal, which is a β-hydroxy aldehyde, and the corresponding α,β-unsaturated aldehyde, (*E*)-2-methylpentenal. The mechanism follows the same steps illustrated in Sections 23.1a and 23.1b for the aldol reaction of acetaldehyde. These steps are as follows:

1) Formation of the enolate derivative of propanal
2) Addition of the enolate ion to the carbonyl group of propanal
3) Protonation of the alkoxide ion
4) Formation of the enolate derivative of 3-hydroxy-2-methylpentanal
5) Elimination of hydroxide ion to form (*E*)-2-methylpentenal.

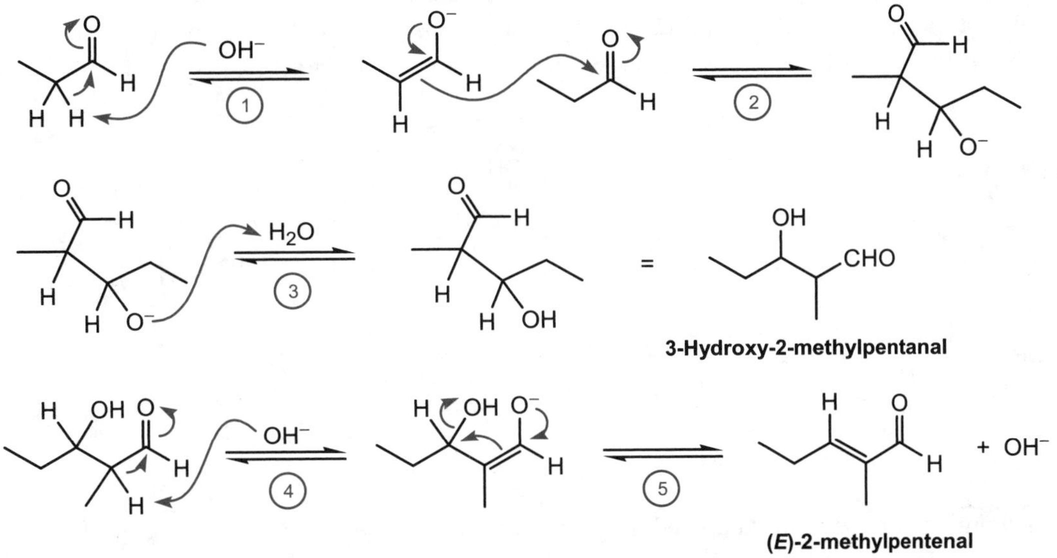

3-Hydroxy-2-methylpentanal

(*E*)-2-methylpentenal

23.2. The aldol reaction performed with acid catalysis requires initial formation of the enol derivative of the aldehyde:

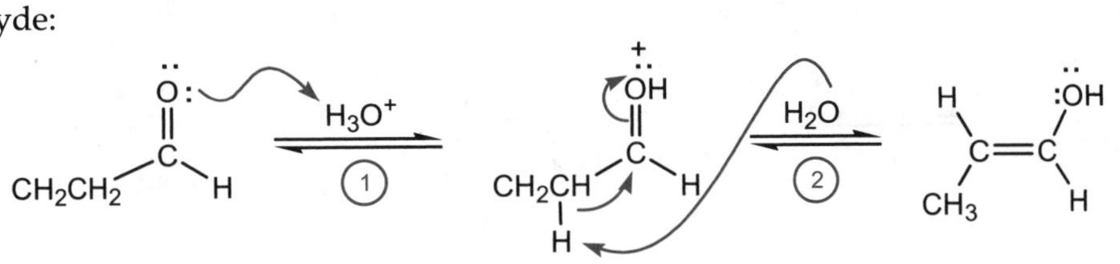

Next, proton activation of a second molecule of aldehyde occurs (step 3). The enol derivative that was formed in step 2 then reacts with the proton-activated aldehyde (step 4).

23.2. (continued)

The remaining steps comprise deprotonation of the initial aldol product (step 5) and dehydration via formation of a good leaving group (step 6) and an E2 reaction (step 7).

23.3. A retroaldol reaction that takes place under acidic conditions begins with protonation of the carbonyl group (step 1). The carbon–carbon bond is then broken (step 2), and the enol derivative that forms is converted to the aldehyde molecule by two proton transfer reactions (steps 3 and 4).

23.4. The mechanism for the acid-catalyzed aldol reaction of acetone is the same as that for the acid-catalyzed aldol reaction of propanal shown in the solution to Exercise 23.2. In this instance, the acid is H—(P), where (P) is the conjugate base of an acidic polymer. The enol form of acetone is generated first (steps 1 and 2).

This enol then reacts with the cation formed by protonation of acetone by the acidic polymer (step 3). A series of proton transfer steps leads to formation of the product.

23.4. (continued)

23.5. Elimination of water from the β-hydroxy ketone product of a crossed aldol reaction occurs in two steps. First, an enolate ion is formed by deprotonation at the position adjacent to the carbonyl group (step 1). Regeneration of the carbon-oxygen double bond occurs with the displacement of OH⁻ (step 2).

23.6. The base-catalyzed crossed-aldol reaction between an unsymmetrical ketone and an aldehyde with no alpha hydrogen atoms begins with formation of the ketone enolate ion (step 1). Addition to the aldehyde carbonyl group (step 2), protonation of the resulting alkoxide ion (step 3), and dehydration via the E1cb mechanism (steps 4 and 5) yields the unsaturated ketone.

23.7. The enolate derivative of a 3° carboxamide reacts as if it has a negative charge at the carbon atom adjacent to the carbonyl group. This nucleophilic center adds to a ketone carbonyl group, forming a new carbon-carbon bond. After workup with aqueous acid, the β-hydroxy carboxamide can be isolated.

23.8. An iminium ion has an electrophilic carbon atom, so it is susceptible to reaction with a nucleophile such as water (step 1). Two proton transfer reactions yield the protonated amine, a good leaving group. A base in the enzyme active site [represented by :B) in the following scheme] subsequently deprotonates the OH group attached at C2 of the substrate molecule, which leads to formation of the carbonyl group in the product.

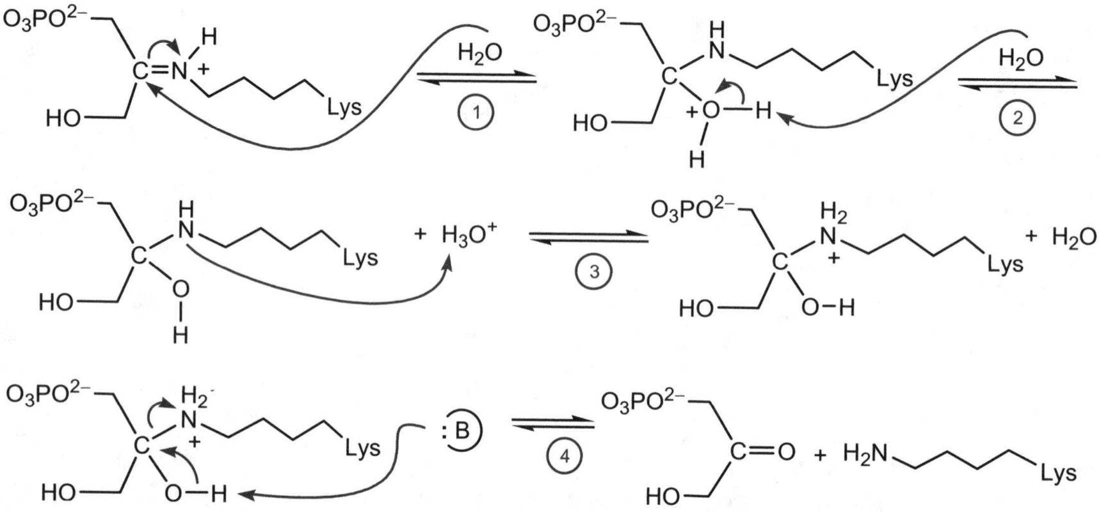

23.9. As with many transformations of carbohydrates in which carbon-carbon bonds are made or broken, an aldol (or retroaldol) reaction is involved. Biochemical aldol reactions often make use of Schiff-base catalysis, in this case from the reaction between the amino group of a lysine residue in the enzyme active site and the carbonyl group of 2-keto-3-deoxy-6 phosphogluconate (step 1).

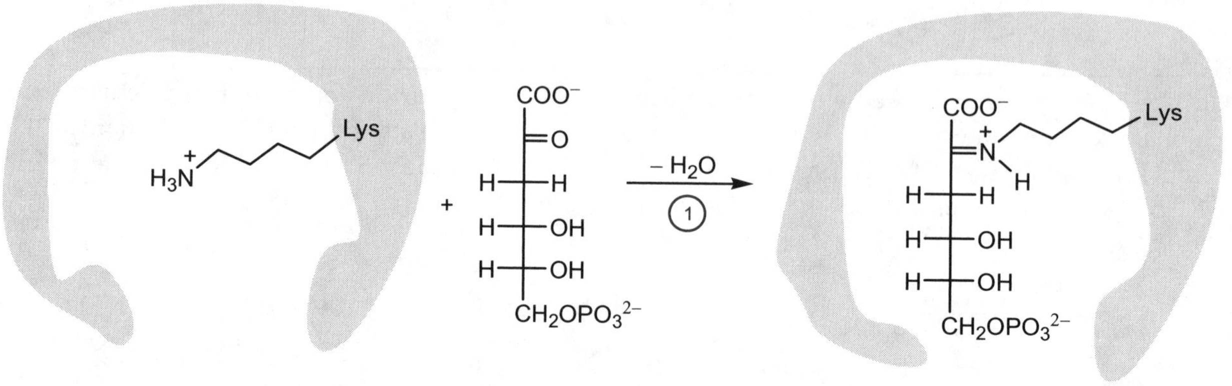

Next, a base removes the proton from the OH group attached to C4, cleaving the C3-C4 bond of the substrate molecule (step 2); D-glyceraldehyde-3-phosphate subsequently dissociates from the enzyme active site (step 3).

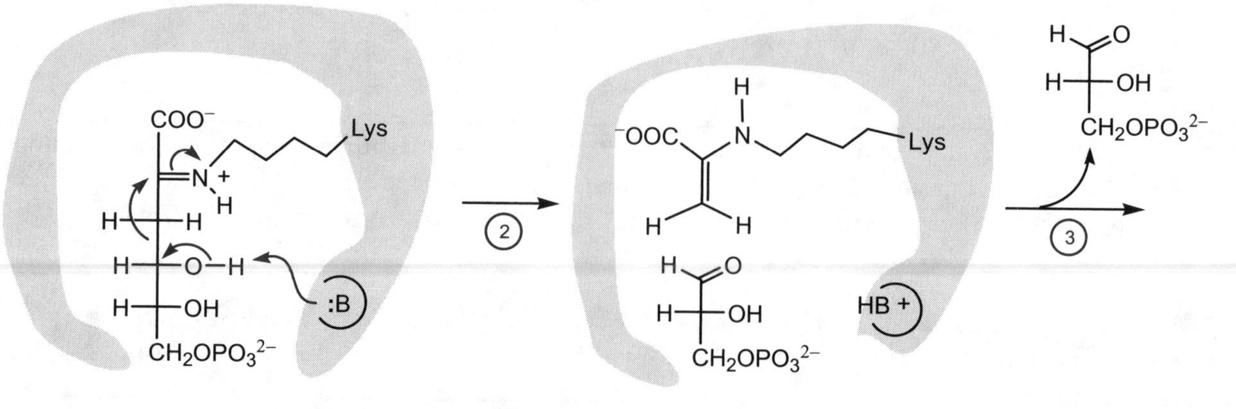

23.9. (continued)

The enamine can be protonated at the carbon atom adjacent to the imine group (step 4), and the Schiff-base is hydrolyzed (step 5) (see the solution to Exercise 23.8 for this mechanism) to form pyruvate.

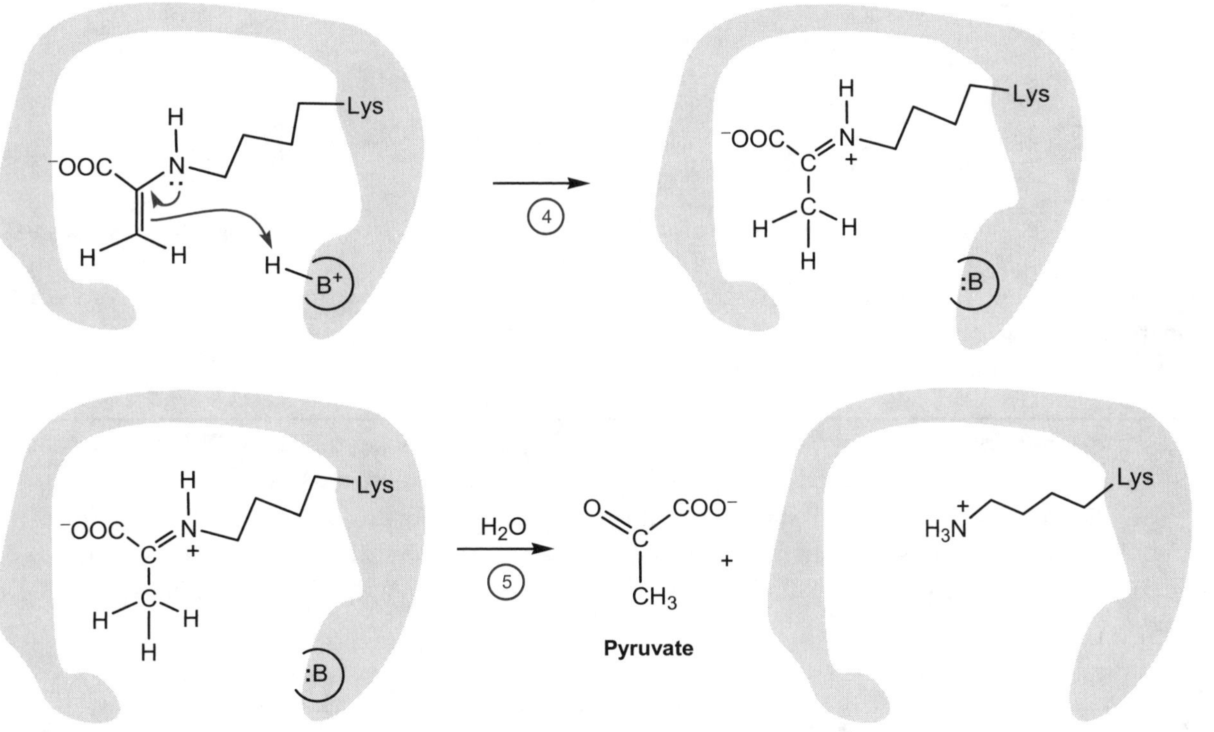

Pyruvate

23.10. If the carbon-carbon bond forming reaction catalyzed by *transaldolase* is non-stereospecific, then two new chiral centers are generated (at C3 and C4 of a hexose). Four diastereomers are therefore produced, the structures of which are shown below. The bond that is formed by this aldol reaction is shown in color. Notice that all of the possible 2-ketohexoses are formed. In the actual biochemical reaction, only fructose is generated. In that transformation, the reaction is stereospecific because it takes place within an enzyme active site.

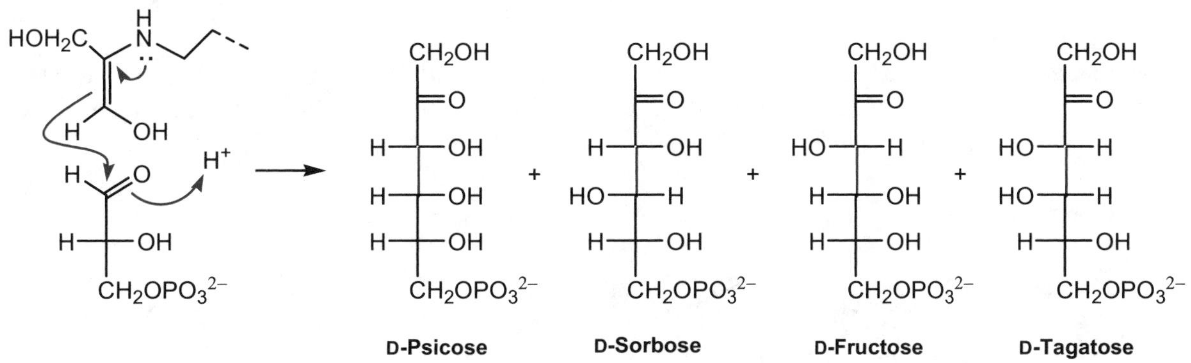

D-Psicose   **D-Sorbose**   **D-Fructose**   **D-Tagatose**

23.11. Ethyl acetoacetate can be alkylated twice at its methylene group under the conditions in the first step of the given reaction scheme. Dialkylation creates a quaternary center, so when this compound is treated with NaOEt in ethanol, the product undergoes the retro-Claisen reaction, generating two ester molecules. The mechanism occurs by addition of the ethoxide ion to the ketone carbonyl group (step 1), followed by the bond-breaking reaction (step 2). Tautomerism of the ester enol completes the transformation (step 3).

23.11. (continued)

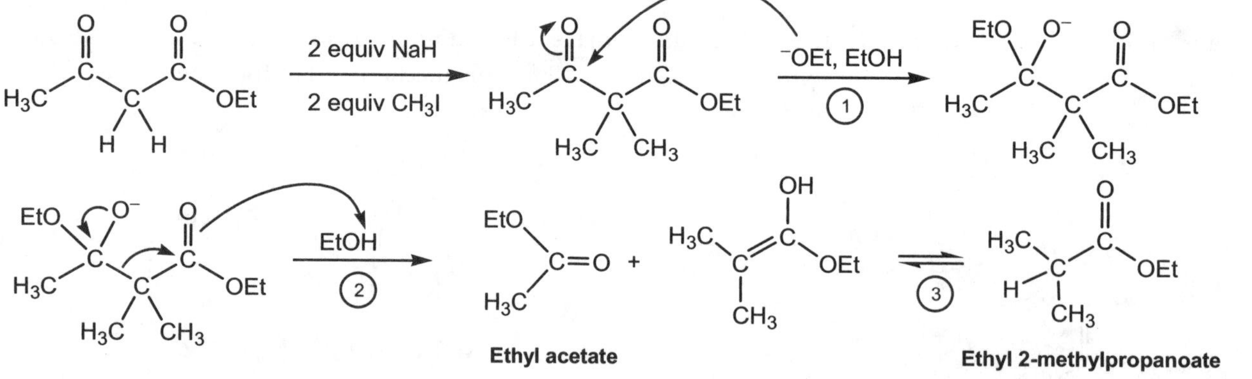

Ethyl acetate   **Ethyl 2-methylpropanoate**

The point of this exercise is to reinforce the idea that a retro-Claisen reaction is facile when the carbon atom between the carbonyl groups of a β-keto ester cannot be deprotonated upon treatment with a nucleophilic base such as ethoxide ion.

23.12. The transformation shown in this exercise comprises three stages: [1] a Dieckmann (Claisen) condensation reaction; [2] hydrolysis of an ester; and [3] decarboxylation of a β-keto acid.

The Dieckmann condensation starts with formation of an enolate ion (step 1). Addition of this nucleophile to the other ester carbonyl group (step 2) is followed by displacement of the ethoxide ion (step 3) and deprotonation of the initial keto ester product (step 4). Acid workup yields the β-keto ester (step 5).

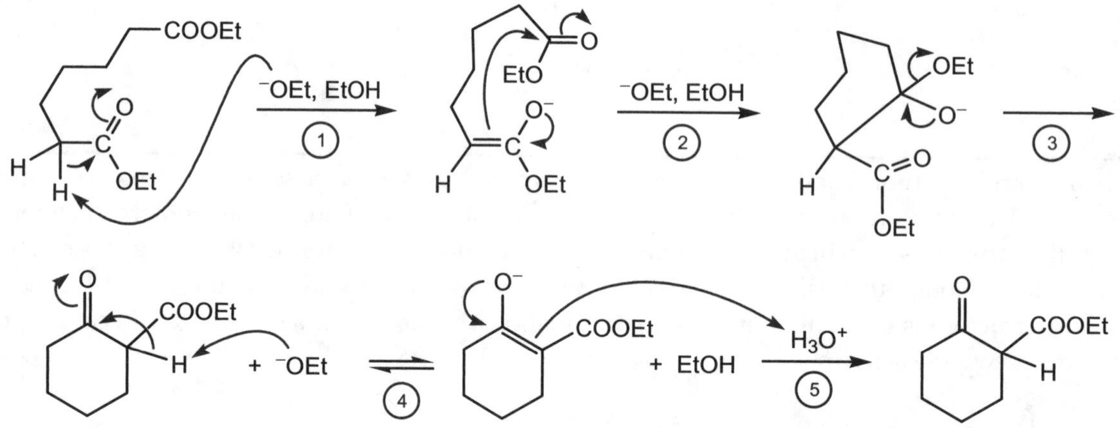

In the next stage, the β-keto ester undergoes hydrolysis by protonation of the ester carbonyl group (step 6), addition of a molecule of water to the double bond (step 7), a proton transfer reaction (step 8), and displacement of ethanol (step 9).

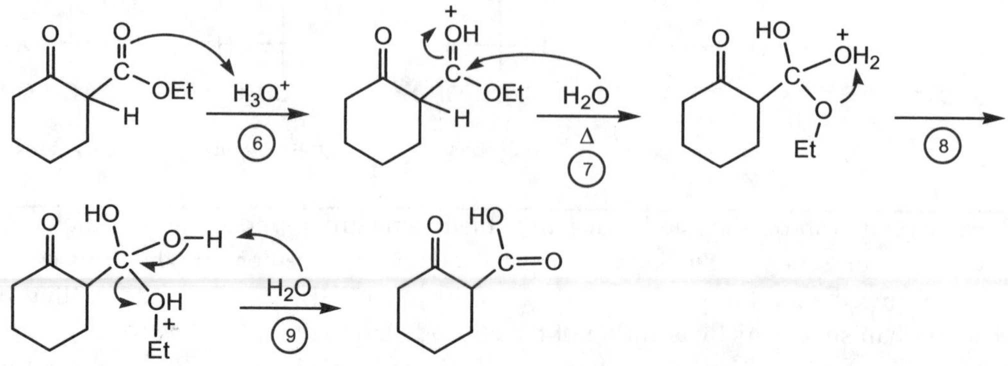

23.12. (continued)

Decarboxylation occurs via a cyclic transition state in step 10, followed by tautomerism (step 11) to yield the final product.

23.13. The transformation in this exercise constitutes the Dieckmann condensation of a diester in which only one ester group can be deprotonated at its alpha position. The first step creates an enolate ion that adds to the other ester carbonyl group (step 2). This addition step is followed by displacement of the ethoxide ion (step 3) and deprotonation (step 4). Acid workup yields the β-keto ester product (step 5).

23.14. When an enolate ion reacts with ethyl formate (HCOOEt), the product has an aldehyde group at the alpha position of the original carbonyl compound. This transformation is an example of a crossed Claisen condensation.

23.15. The reaction between diethyl carbonate, base, and a ketone leads to formation of a β-keto ester [reaction 1], which is readily alkylated upon treatment with base and a primary alkyl halide [reaction 2]. Hydrolysis and decarboxylation [reaction 3] forms the α-alkylated ketone.

This three-reaction procedure provides an excellent way to alkylate the position adjacent to a ketone carbonyl group without having problems with self-condensation of the ketone molecule.

23.16.   The biosynthesis of fatty acids proceeds as follows:

1) The acetyl thioester derivative of Acyl Carrier Protein (ACP) condenses with an enolate ion that has been generated by decarboxylation of malonyl~ACP (ACP~$SCOCH_2COO^-$) to form a β-keto thioester.

2) The ketone group of this β-keto thioester is then reduced to form the corresponding alcohol.

3) Loss of water (dehydration) yields an α,β-unsaturated thioester molecule.

4) The double bond of this unsaturated thioester is reduced to form the saturated thioester.

These same four steps are repeated to build the fatty acid two carbon atoms at a time.

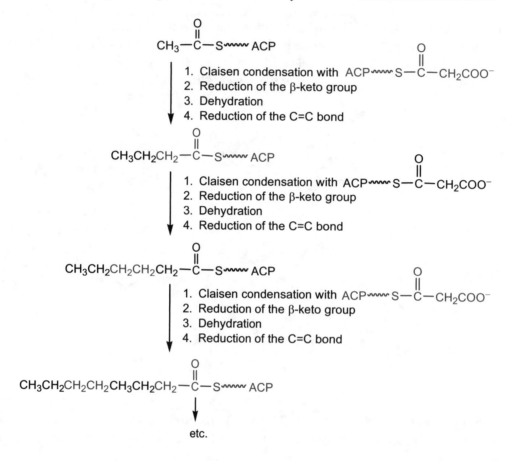

etc.

23.17. The enolate ion structure is derived by removing a proton from the alpha carbon atom of the parent ester molecule. The other molecules are named according to the procedures outlined in Chapter 1 (longest chain, principal functional group, and substituents).

a.

tert-Butyl acetate . . . . . . and its corresponding enolate ion.

b.

Methyl 3-keto-2-methylbutanoate

c.

(2*R*, 3*S*)-2,3-Dihydroxypentanoic acid

d.

3-^{14}C-Dihydroxyacetone phosphate

23.18   Aldehydes (and some ketones) react with enolate ions to form aldol products. If two reactants are present, decide which one(s) can form an enol or enolate ion, and then consider what product(s) will be formed if that enolate derivative reacts with the other carbonyl containing compound (or another molecule of the same carbonyl compound).

a.   In this crossed aldol reaction, only the acetophenone derivative can form an enolate ion. Addition of the enolate ion to formaldehyde yields the primary alcohol. Elimination of water will produce the α,β-unsaturated ketone.

b.   In this directed aldol reaction, the ester derivative is first converted to its enolate ion derivative with LDA. Addition of this enolate ion to the ketone molecule yields the β-hydroxy ester. Elimination of H_2O cannot occur because there is no proton adjacent to the carbonyl group.

c.   In this directed aldol reaction, a ketone enolate ion adds to an aldehyde carbonyl group to generate a β-hydroxy ketone.   Elimination of water will occur because dehydration creates an extensive conjugated π system that is more stable than the compound without the double bond adjacent to the carbonyl group.

d.   An aldehyde undergoes self-condensation to form a β-hydroxy aldehyde. Elimination does not occur because there is no proton alpha to the aldehyde carbonyl group.

e.   In this crossed aldol reaction, only the acetophenone derivative can form an enolate ion. Addition of the enolate ion to the aldehyde yields the β-hydroxy ketone. Elimination of water will occur because dehydration creates an extensive conjugated π system that is more stable than the compound without the double bond adjacent to the carbonyl group.

23.19. In the Henry reaction, a nitro compound is converted to a carbanion via an acid-base reaction (step 1). This nucleophile adds to the aldehyde carbonyl group, which is protonated on its oxygen atom (step 2). Elimination of water (E1cb mechanism) yields the unsaturated nitro compound.

23.20. In the Knoevenagel reaction, a β-keto ester reacts with a base to form the enolate ion derivative. This nucleophile adds to the aldehyde carbonyl group. Elimination of water via the E1cb mechanism (steps 3 and 4) produces the α,β-unsaturated keto ester.

23.21. Addition of an enolate ion to an ester (or acid chloride) carbonyl group yields products from the Claisen condensation. If two reactants are present, deduce which one(s) can form an enolate ion, and then consider what product(s) will be formed if that enolate derivative reacts with the other carbonyl-containing compound (or another molecule of the same carbonyl compound).

a. This reaction is a "standard" Claisen condensation—the enolate ion produced by deprotonation undergoes an addition–elimination reaction with another molecule of the same ester. Acid workup is required to obtain the neutral product.

b. In this directed Claisen reaction, the enolate ion is generated by deprotonation of the ester reactant, and it undergoes an addition–elimination reaction with the acid chloride molecule.

23.21. (continued)

c. This reaction is an example of the Dieckmann cyclization, the ring-forming variant of the Claisen condensation. The stereochemistry of each chiral center is not affected by the reaction, but one configuration appears to change because the priorities of the substituents are altered when the carbon–carbon bond is made. An additional *new* chiral center is formed, so a mixture of two diastereomers is created.

d. This is a crossed Claisen reaction between an enolizable ketone and a non-enolizable ester. The enolate forms (and reacts) on the side of the ketone that is less hindered. The product is a β-keto aldehyde, which normally exists in its hydrogen-bonded enol form, which is achiral.

23.22. In the Thorpe reaction, the first step is formation of a carbanion at the position alpha to the nitrile group. This nucleophile adds to the carbon-nitrogen triple bond of the other nitrile group (this exercise illustrates the intramolecular Thorpe reaction). Two proton transfer steps yield the enamine.

In the second stage of reaction, the nitrile group is hydrolyzed to form a carboxylic acid group (steps 5-10), and then the enamine portion undergoes hydrolysis to yield the ketone (steps 11-14).

23.22. (continued)

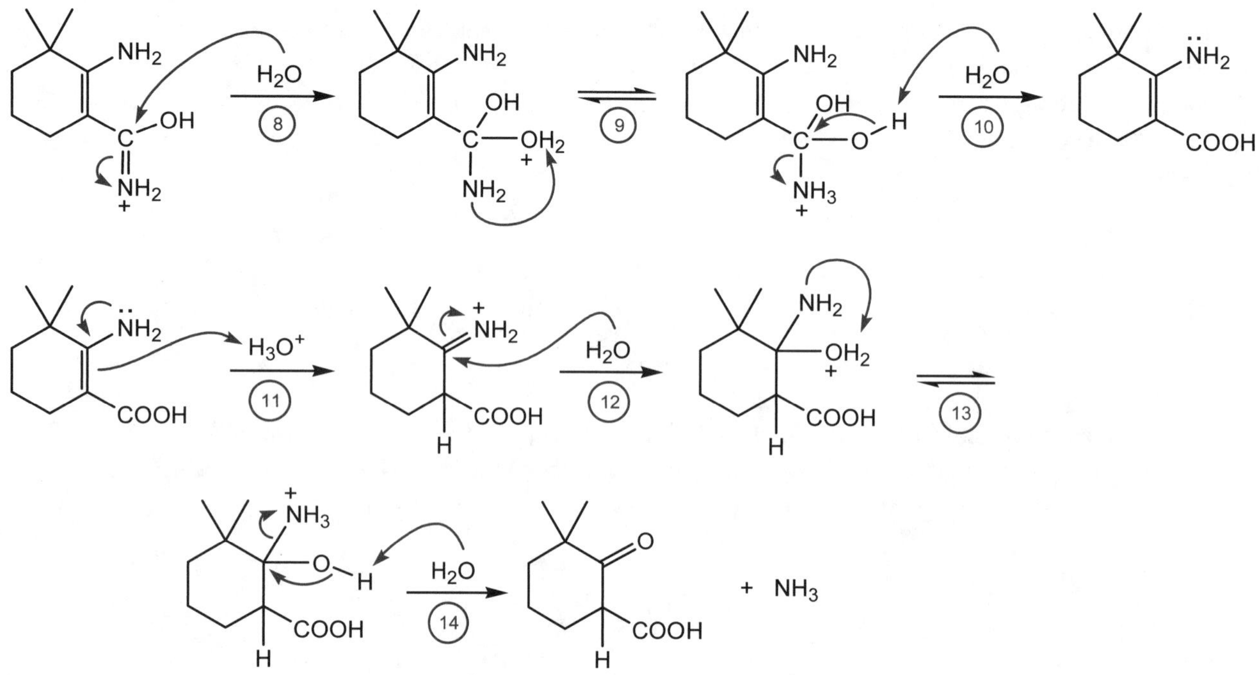

Finally, the keto ester undergoes decarboxylation (step 15) and tautomerism produces the ketone product (step 16).

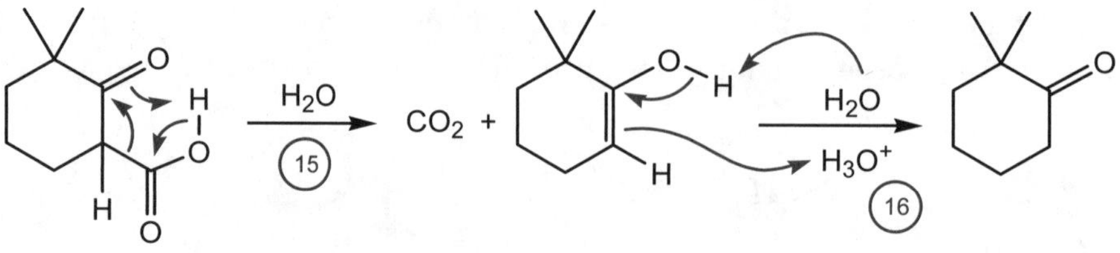

23.23. The Perkin reaction begins with the formation of an enolate ion from acetic anhydride by an acid-base reaction (step 1). This enolate ion next adds to the carbonyl group of the aldehyde (step 2).

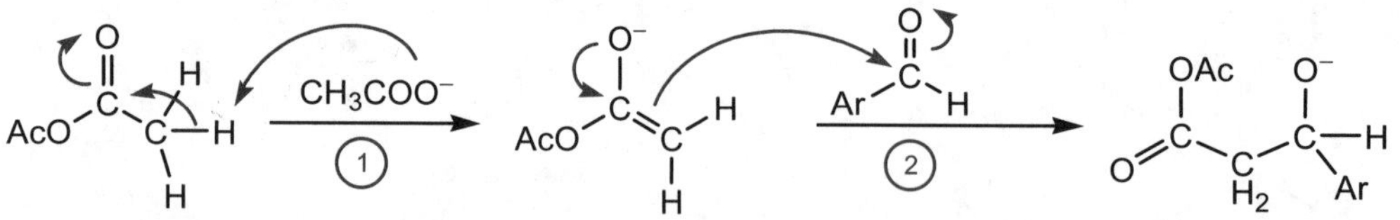

The acetyl group of the anhydride is then transferred to the oxyanion nucleophile via a tetrahedral intermediate (steps 3 and 4). With the addition of acid in the workup step, two proton transfer steps occur to generate the carboxylic acid group and to activate the acetyl group (steps 5 and 6).

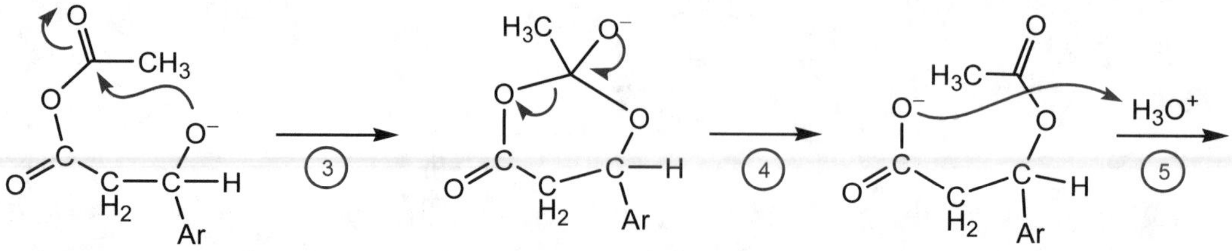

23.23. (continued)

The product is formed by an acid-base reaction that displaces acetate ion to generate the double bond (step 7).

23.24. The Darzens reaction begins with formation of the enolate ion from ethyl chloroacetate. This step is facilitated by the acidity of the methylene protons resulting from the presence of two strongly electron-withdrawing groups (the ester group and the chlorine atom). The resulting anion adds to the aldehyde carbonyl group (step 2), and the alkoxide ion that is generated participates in an intramolecular S$_N$2 reaction to produce the epoxide ring by displacing chloride ion.

23.25. When you perform a reaction in which you know what the product will be (or is expected to be), focus on those spectroscopic features that are unique for the groups in the starting material and product molecules. Although many subtle changes will undoubtedly occur in several parts of the spectrum, it helps to look for a limited number of specific changes.

a. This transformation converts a saturated ketone to an α,β-unsaturated ketone. The C=O stretch frequency will be lowered as a result of this change. In addition, new C=C stretching vibrations for the alkene and benzene groups will be observed.

b. This transformation converts an ester into a β-keto ester. The product will have two carbonyl stretching vibrations in place of the single absorption band observed for the starting ester molecule.

23.26. As noted in the solution to Exercise 23.25, focus on those spectroscopic features that are unique for the starting material and product molecules. Many other subtle changes will undoubtedly occur in other parts of the spectrum as well.

23.26. (continued)

a. In both ¹H and ¹³C NMR spectra, the majority of the resonances in the product spectra will appear mainly in regions that are blank in the spectrum of the starting material.

¹H NMR spectrum

all resonances < δ 2.5

δ 5-6, 1H, singlet

δ 7-8, 5H, multiplet

δ 2.4, 3H, singlet

¹³C NMR spectrum

δ ~200
0H (DEPT)

δ ~200
0H (DEPT)

δ 120-150
1H (DEPT)

δ 120-150
3 peaks: 1H (DEPT)
1 peak: 0H (DEPT)

δ 120-150
0H (DEPT)

δ 20-40
3H (DEPT)

- - - - - - - - - - - - - - - - - - - -

b. In both ¹H and ¹³C NMR spectra, the majority of the resonances in the product will be ones that do not appear in the starting ketone. The creation of a ketone group in the product (from an ester) will be readily apparent in the carbon NMR spectrum because the resonance for a ketone carbonyl group appears farther downfield than the analogous signal for an ester carbonyl group.

¹H NMR spectrum

δ 4.1, 2H, quartet

δ 1.2, 3H, triplet

δ 3.8, 1H, quartet

δ 1.5, 9 H, singlet

δ 1-2, 3H, doublet

¹³C NMR spectrum

Two signals in the aliphatic region

δ 160-185
0H (DEPT)

Four signals in the aliphatic region, one with 0H (DEPT)

δ >200
0H (DEPT)

δ 160-185
0H (DEPT)

23.27. The two reactions shown in Exercise 23.25 constitute examples of the crossed (or directed) aldol reaction and the crossed (or directed) Claisen condensation, respectively.

a. A crossed aldol reaction between the ketone and benzaldehyde will yield the given product.

NaOH, H₂O, EtOH

23.27. (continued)

b. A directed Claisen condensation between the ester and *tert*-butyl acetate will yield the given product.

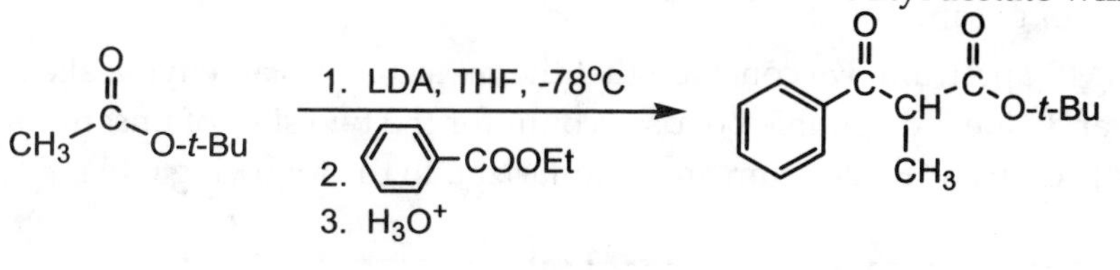

23.28. The Dieckmann cyclization procedure is used to prepare cyclic β-keto esters or β-diketones from the appropriate diesters.

a. This product is a cyclic β-keto ester, so the Dieckmann cyclization will yield the product directly. The starting diester has more than six carbon atoms, so the diacid (six carbon atoms) is first converted to its diester derivative. The starting compounds are shown in color.

Retrosynthesis

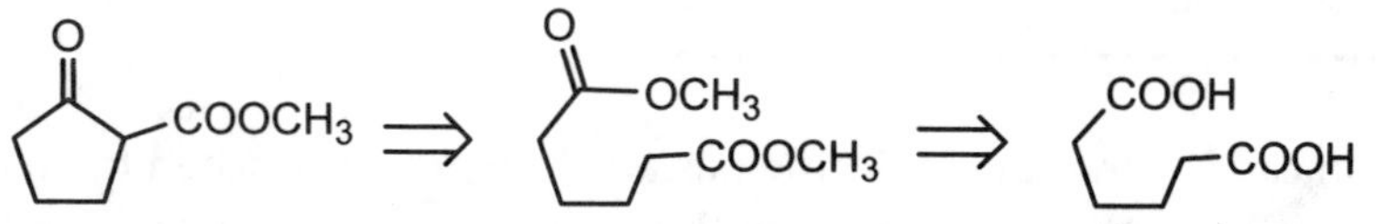

Synthesis

- -

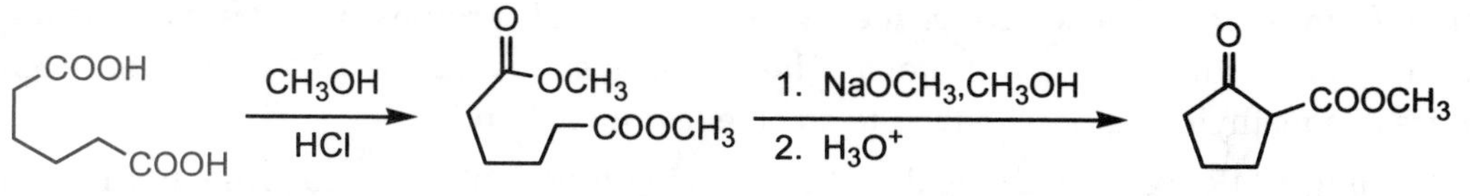

b. The desired product is an unsaturated cyclic ketone, which can be made from a cyclic ketone, itself formed by decarboxylation of a Dieckmann cyclization product. The starting diester has seven carbon atoms in its chain (not counting the alkoxy groups of the diester), so 1,5-dibromopentane is first converted to the needed diacid by adding two carbon atoms (two cyanide groups), which can be hydrolyzed and subsequently converted to the diester.

Retrosynthesis

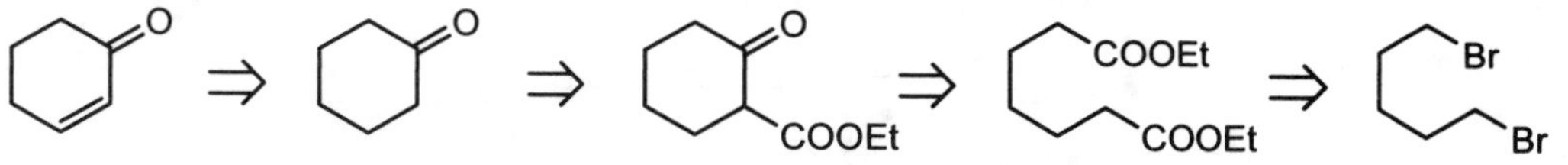

Synthesis

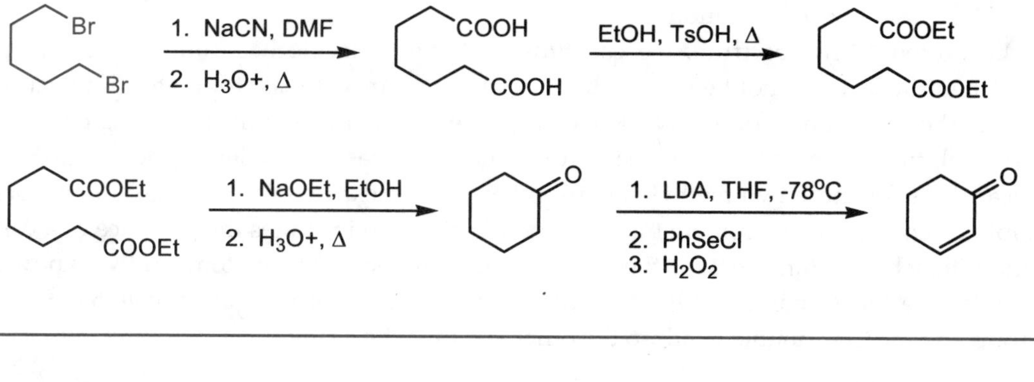

23.29. Follow the procedure outlined in the solution to Exercise 14.21.

a. From the given molecular formula, $C_6H_{10}O$, calculate the number of unsaturation sites:
$[2(6) + 2 - 16]/2 = 2$ sites of unsaturation.

From the proton NMR spectrum, we conclude that the molecule has aldehyde, alkene, and alkane portions. The alkene and aldehyde double bonds account for the two sites of unsaturation. The data from the compound's proton NMR spectrum are summarized in the following table.

Chemical shift (ppm)	Integrated intensity	Assignment	Multiplicity	No. adjacent protons (n)
9.4	1	CHO	singlet	0
6.5	1	=CH	triplet	2
2.4	2	CH$_2$	qunitet	5
1.8	3	CH$_3$	singlet	0
1.1	3	CH$_3$	triplet	2

From the splitting patterns, you can see that the molecule has an ethyl group attached to an alkene carbon atom, with additional coupling occurring between the alkene proton and the methylene group (see the substructure to the right of the table, above). If the alkene proton were more than three bonds from the methylene group, the signal for the CH$_2$ protons would appear as a quartet from coupling only with the methyl protons. The chemical shift value of δ 2.4 is also consistent with the presence of a methylene group bonded to an alkene double bond.

The remaining NMR signals indicate that the following fragments are also present: CHO and CH$_3$ (attached to the alkene double bond as well). Assembling these groups to fit the chemical shift values of the signals, we conclude that this molecule is **(E)-2-methyl-2-pentenal**. This molecule can be formed by applying the aldol reaction to propanal, followed by dehydration (see Exercise 23.1). The (E) stereochemistry is the natural consequence of having a double bond conjugated with a carbonyl group. The (E) isomer is much more stable than the (Z) isomer.

b. From the given molecular formula, $C_8H_{12}O$, calculate the number of unsaturation sites:
$[2(8) + 2 - 12]/2 = 3$ sites of unsaturation.

From the proton NMR spectrum, we conclude that the molecule has both alkene and alkane portions. A benzene ring cannot be present because that requires at least four sites of unsaturation. The data from the compound's proton NMR spectrum are summarized in the following table.

From the splitting patterns, you can see that the molecule has an ethylene group. Furthermore, the chemical shifts for these two methylene groups puts one next to a carbonyl group and one in a strictly aliphatic environment. You are told that the IR spectrum indicates the presence of a carbonyl group, and it must be a ketone because the molecule has only one oxygen atom, and we know that it is not an aldehyde (no signal at δ 10). Two of the sites of unsaturation are accounted for by the ketone group and the C=C double bond, so the other site must be a ring.

23.29. (continued)

Chemical shift (ppm)	Integrated intensity	Assignment	Multiplicity	No. adjacent protons (n)	
6.7	1	=CH	doublet	1	
5.8	1	=CH	doublet	1	
2.5	2	CH_2	triplet	2	
1.9	2	CH_2	triplet	2	
1.2	6	CH_3	singlet	0	

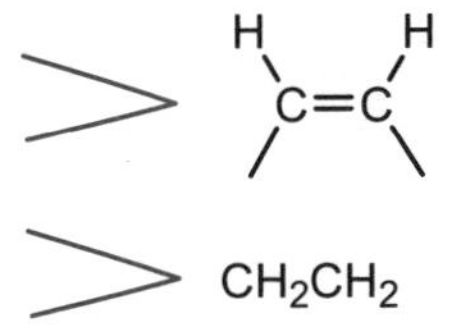

The remaining NMR signal indicates the presence of two methyl groups that are isolated from interactions with other protons. Therefore, we know that the following groups are present, which account for all of the atoms:

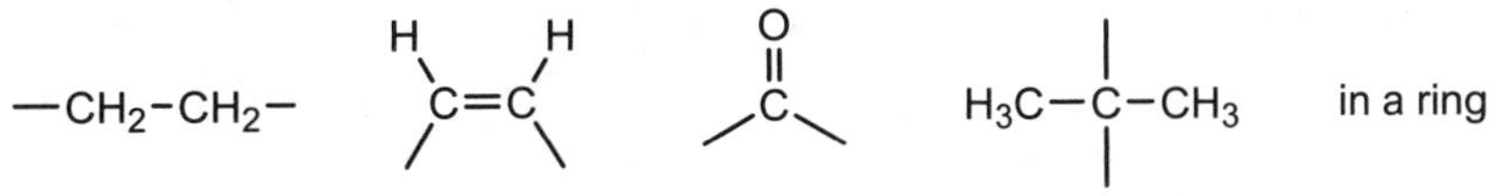

Assembling these groups to fit the chemical shift values of the signals, we conclude that this molecule is **4,4-dimethyl-2-cyclohexenone**.

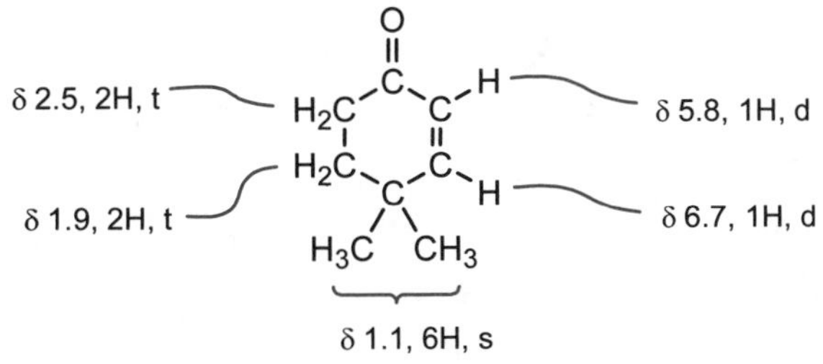

23.30. From the information presented in Chapters 22 and 23, we can draw a structure for each of the intermediates of the glycolysis scheme presented in this exercise. These structures are shown in the following scheme along with the important intermediates involved in each transformation.

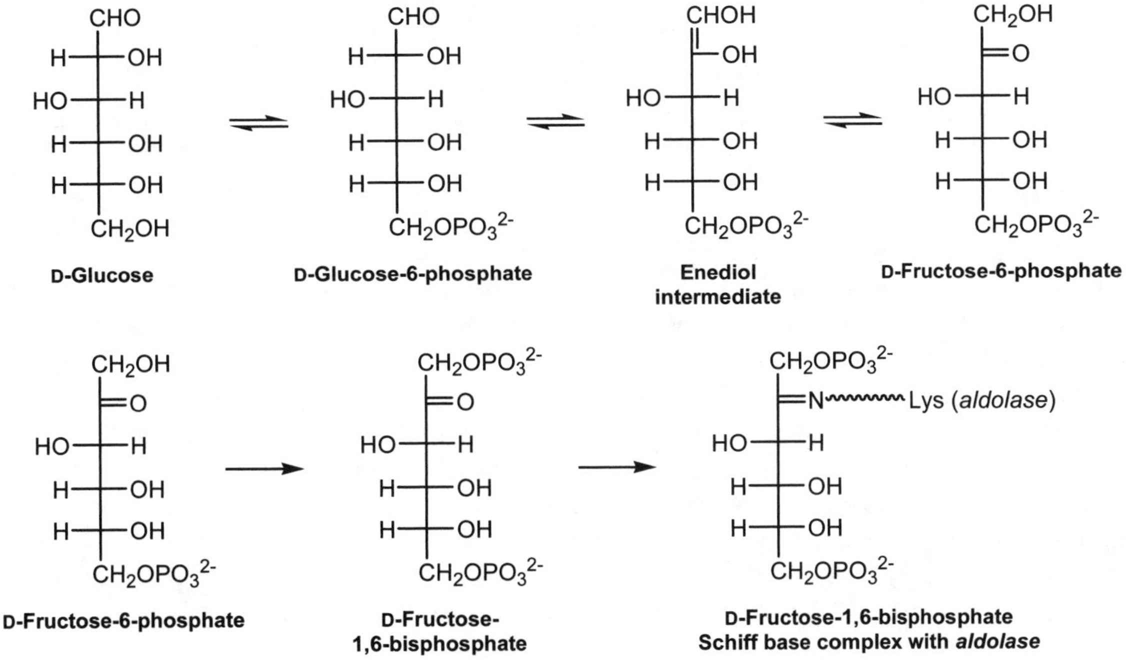

23.30. (continued)

After the Schiff base intermediate is formed between the fructose substrate molecule and the enzyme *aldolase*, a retroaldol reaction occurs to break the hexose into two trioses, as shown in the following the scheme. The trioses can be interconverted as well, and those transformations are shown as the schemes in the boxes, below.

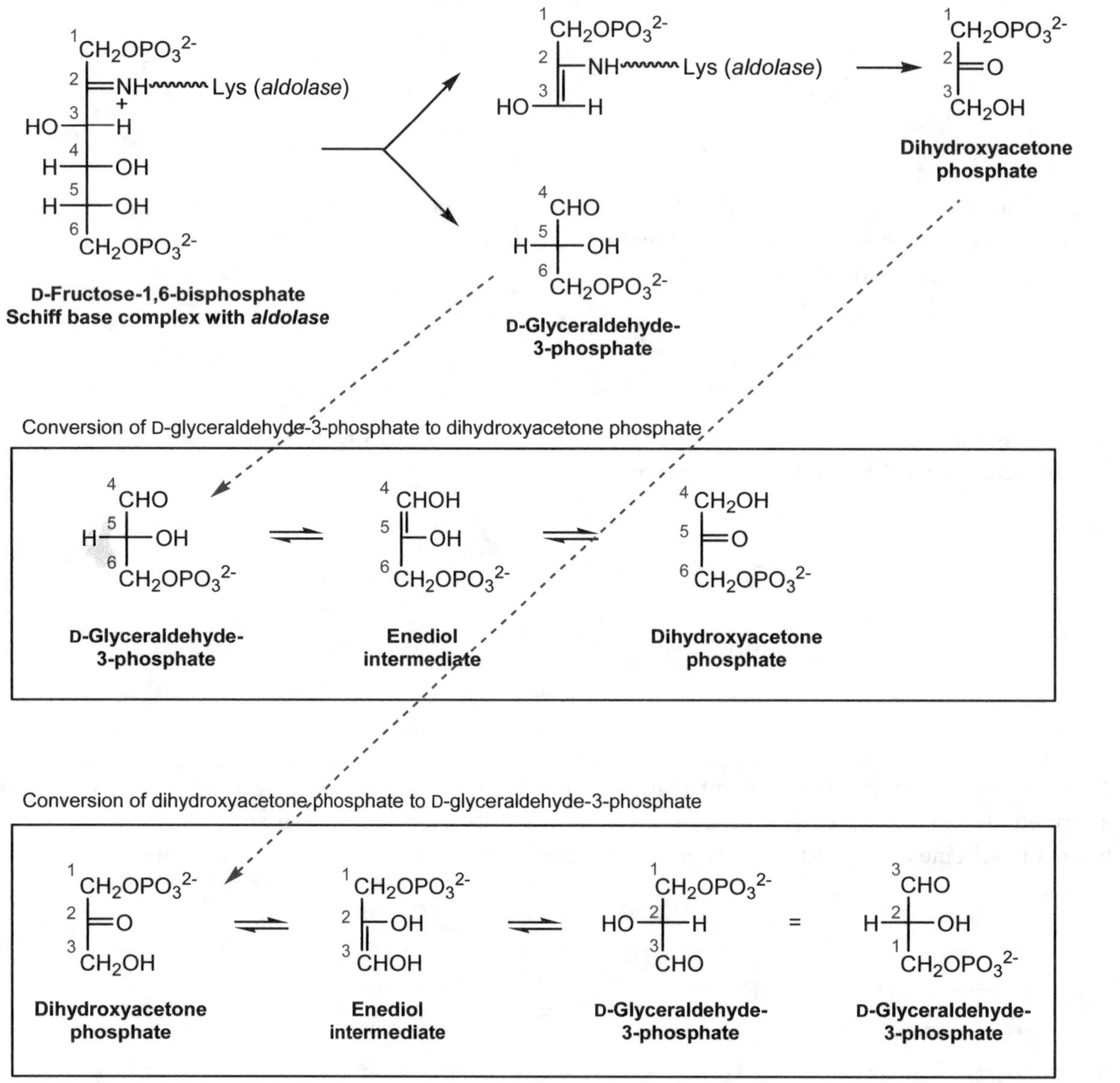

D-Fructose-1,6-bisphosphate Schiff base complex with *aldolase*

D-Glyceraldehyde-3-phosphate

Dihydroxyacetone phosphate

Conversion of D-glyceraldehyde-3-phosphate to dihydroxyacetone phosphate

D-Glyceraldehyde-3-phosphate   **Enediol intermediate**   **Dihydroxyacetone phosphate**

Conversion of dihydroxyacetone phosphate to D-glyceraldehyde-3-phosphate

Dihydroxyacetone phosphate   **Enediol intermediate**   **D-Glyceraldehyde-3-phosphate**   **D-Glyceraldehyde-3-phosphate**

The numbering of the molecules in the preceding schemes has been included to keep track of where each carbon atom goes in these metabolic reactions. *It is not necessarily the numbering that we would use for nomenclature purposes.*

a. According to the foregoing scheme, C1 of D-glucose becomes C1 of dihydroxyacetone phosphate, which then becomes the carbon atom attached to the phosphate group of D-glyceraldehyde-3-phosphate. The isotopically-labeled carbon atom is shown in color in the following scheme.

23.30. (continued)

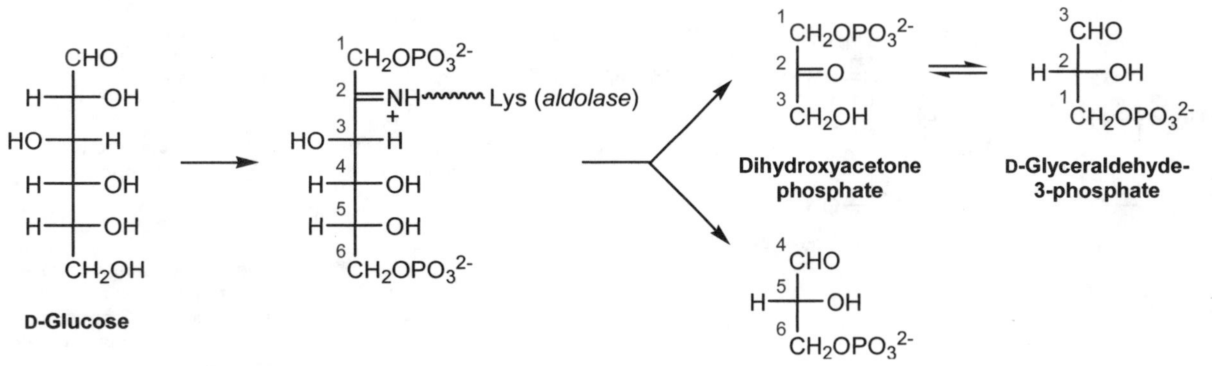

- - - - - - - - - - - - - - - - - - -

b. According to the foregoing schemes for glycolysis, C2 of D-glucose becomes C2 of dihydroxyacetone phosphate, which then becomes the middle carbon atom of D-glyceraldehyde-3-phosphate. The isotopically-labeled carbon atom is shown in color in the following scheme.

- - - - - - - - - - - - - - - - - - -

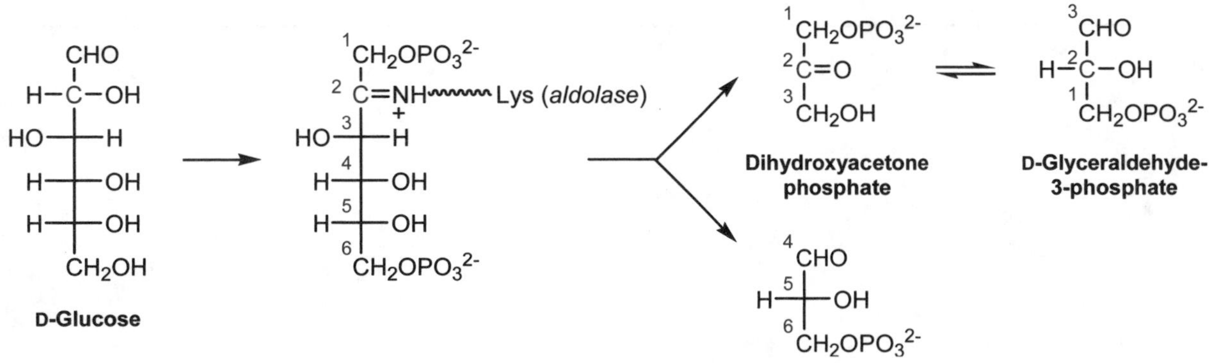

c. According to the foregoing schemes for glycolysis, the oxygen atom at C2 of D-glucose becomes the carbonyl oxygen atom of D-fructose. The *aldolase*-catalyzed cleavage of D-fructose-1,6-bisphosphate occurs by formation of the Schiff base between the D-fructose-1,6-bisphosphate carbonyl group and the lysine residue in the enzyme active site. The oxygen atom (shown in color in the following scheme) is lost as a molecule of water at this step, so it does not appear in either of the three-carbon sugars.

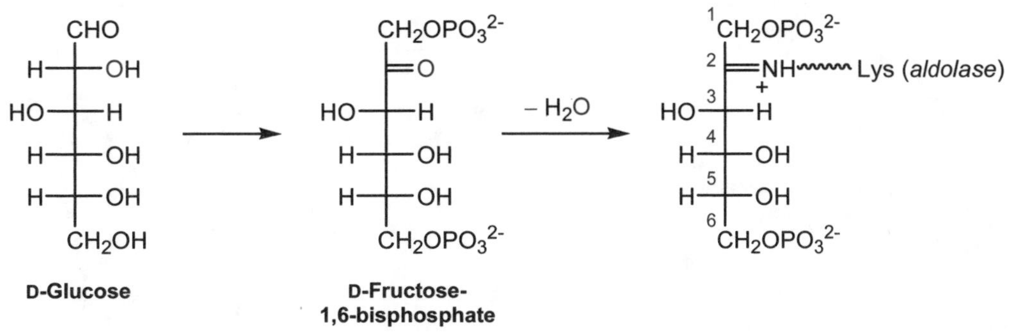

23.31. If the enzyme that interconverts D-glyceraldehyde-3-phosphate and dihydroxyacetone phos-phate is missing, then the labeled carbon atoms in D-glucose (Exercise 23.30a and 23.30b) will be found only in the dihydroxyacetone phosphate product (shown below as an example for the label at C2 of D-glucose). The carbon atoms in D-glyceraldehyde-3-phosphate derive from C4, C5, and C6 of D-glucose.

23.31. (continued)

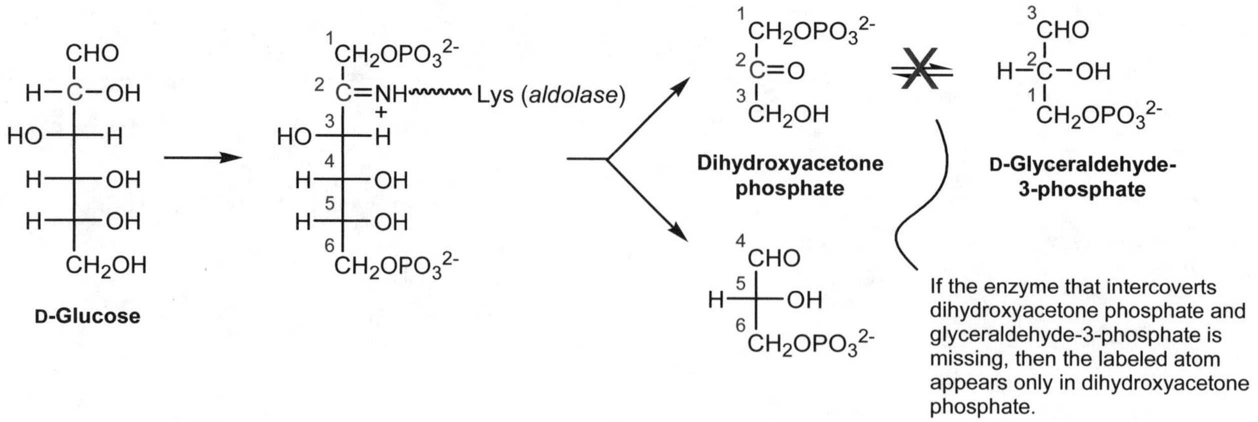

CHO
H−C−OH
HO−H
H−OH
H−OH
CH2OH
D-Glucose
CH2OPO3 2-
C=NH Lys (aldolase)
Dihydroxyacetone phosphate
D-Glyceraldehyde-3-phosphate
If the enzyme that intercoverts dihydroxyacetone phosphate and glyceraldehyde-3-phosphate is missing, then the labeled atom appears only in dihydroxyacetone phosphate.

The answer to Exercise 21.31c is the same as the answer to Exercise 23.30c: The oxygen atom is lost as a molecule of water during Schiff base formation between D-fructose-1,6-bisphosphate and the lysine residue of aldolase.

CONJUGATE ADDITION REACTIONS OF UNSATURATED CARBONYL COMPOUNDS

24.1. The most general way to prepare an α,β-unsaturated carbonyl compound from its saturated analog makes use of the corresponding enolate ion to attach the PhSe group adjacent to the carbonyl group. Oxidation of the Se atom with hydrogen peroxide and gentle warming leads to elimination of PhSeOH and production of the double bond. When preparing an α,β-unsaturated carbonyl compound in which the carbon chain must be lengthened, make use of the aldol or crossed-aldol reaction to attach the additional carbon atoms and to form the double bond in conjugation with the carbonyl group.

a. Because the carbon skeleton of the starting ketone is the same as that for the product, PhSeCl is used to introduce the carbon–carbon double bond.

b. This carbon skeleton can be made by using the crossed aldol reaction between an ester enolate ion and acetaldehyde. Conditions are chosen to maximize formation of the E1cb product after the new carbon–carbon bond is formed.

24.2. A secondary amine reacts with formaldehyde by nucleophilic addition of the amine nitrogen atom to the carbonyl group (step 1), which is followed by the transfer of a proton from nitrogen to oxygen (step 2). (In this scheme, the proton transfer is shown as an intramolecular reaction, but it more likely occurs by interactions with the aqueous solvent, which can supply and remove protons.)

24.3. An enolate ion reacts with Eschenmosher's salt to form the corresponding Mannich base. The enolate ion is formed by treating the ketone with LDA in THF at a low temperature. The enolate ion adds to the carbon–nitrogen double bond of the iminium ion in the second step.

24.3. (continued)

24.4. Hydrolysis of the β-enol ether derivative of a ketone starts with addition of water to the double bond, an acid-catalyzed process (step 1) that serves to activate the carbonyl group and make the β position more electrophilic. In step 2, a molecule of water intercepts the carbocation that was formed, and then a proton is transferred from one oxygen atom to the other (step 3). The carbonyl group is regenerated to form the enol derivative of the 1,3-diketone. Tautomerism can yield the β-diketone product (step 5), but the β-hydroxy enone is actually the more stable form.

24.5. The regeneration of citrate from aconitate proceeds by conjugate addition of water to the alkene double bond of the aconitate molecule. The process starts with the conjugate addition reaction of hydroxide ion, generated from water by reaction with a base within the enzyme active site (step 1); transfer of a proton yields citrate (step 2).

24.6. The retrosynthesis for a ketone with a heteroatom-containing substituent makes use of conjugate addition if the substituent is attached beta to the carbonyl group, and either alkylation or the Mannich reaction if the substituent is attached at the alpha position.

a. An amino group attached to the carbon atom beta to a carbonyl group requires the conjugate addition of an amine to the corresponding α,β-unsaturated compound, which is made according to the scheme shown in the solution to Exercise 24.1.

b. An aminomethyl group attached to the carbon atom beta to a carbonyl group requires conjugate addition of cyanide ion to the corresponding α,β-unsaturated compound followed by a reduction reaction. (If the ketone were the desired product, the alcohol group could be reoxidized or the ketone group could be protected as an acetal before the reduction step).

c. Attaching a dimethylaminomethyl group to the carbon atom alpha to a carbonyl group can be accomplished using the Mannich reaction. Hydroxide ion is added in a workup step to generate the Mannich base.

24.7. The base-catalyzed addition of a thiol molecule to an α,β-unsaturated ester takes place in three steps. The base first reacts with the thiol molecule to generate a small amount of the thiolate ion (step 1). That nucleophile subsequently adds to the terminus of the conjugated system (step 2), producing an enolate ion. This enolate ion deprotonates another equivalent of the thiol reactant, which forms the product (step 3) and regenerates the thiolate ion for further reaction.

24.8. An epoxide ring that spans the alpha and beta carbon atoms of a ketone can be made using the hydroperoxide ion. Isolated double bonds do not react with this reagent. To prepare the needed α,β-unsaturated ketone, start with the saturated ketone and treat its corresponding enolate ion with PhSeCl and then hydrogen peroxide. The saturated ketone can be prepared via the Diels-Alder reaction between 1,3-butadiene and cyclohexanone. The starting compounds are shown in color.

24.8. (continued)

Retrosynthesis

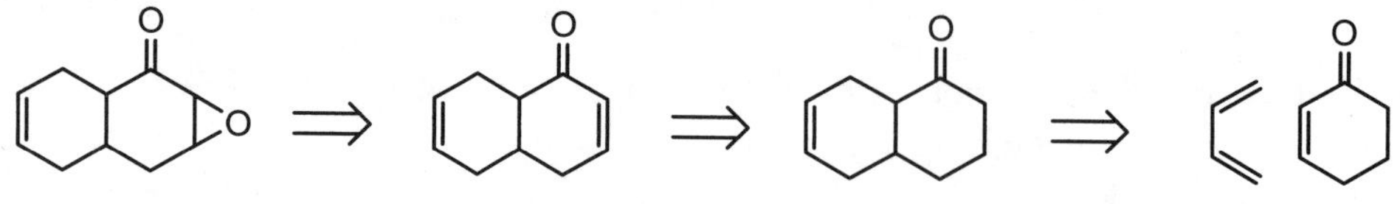

Synthesis

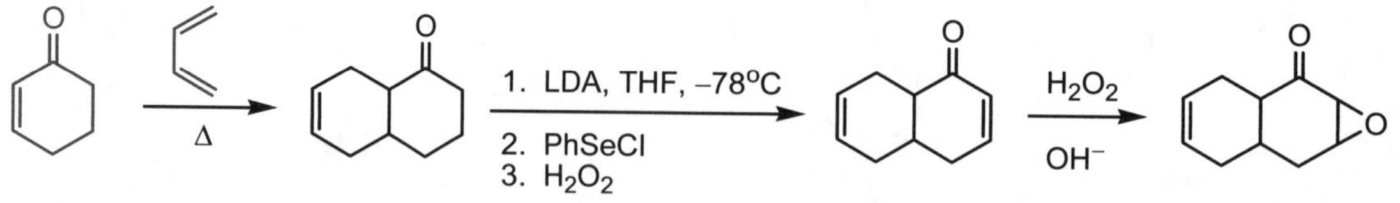
Δ
1. LDA, THF, –78°C
2. PhSeCl
3. H2O2
H2O2
OH–

24.9. In planning a synthesis that makes use of the Michael reaction, consider which stabilized enolate ion is needed to attach the desired fragment beta to the carbonyl group of an α,β-unsaturated substrate molecule. The starting compounds are shown in color.

a. The $-CH_2COOH$ group needs to be added to an α,β-unsaturated ketone to make the desired product, so diethyl malonate is used as the enolate ion precursor. Hydrolysis of the ester under conditions that cause decarboxylation leads to formation of the product.

Retrosynthesis

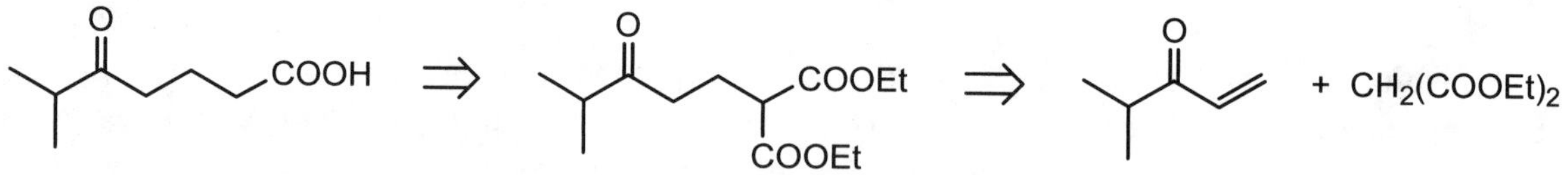
COOH
COOEt
COOEt
+ CH2(COOEt)2

Synthesis

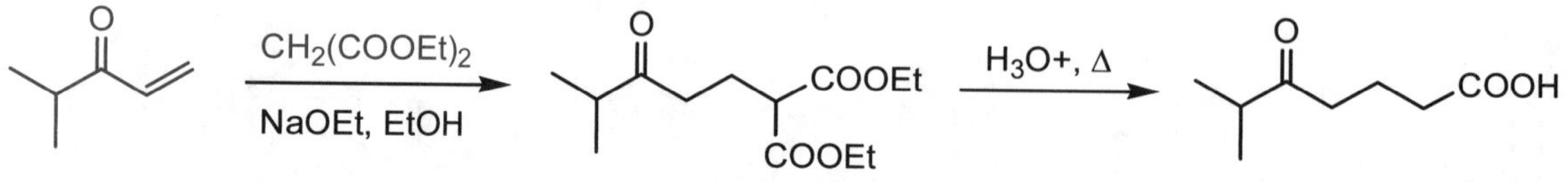
CH2(COOEt)2
NaOEt, EtOH
COOEt
COOEt
H3O+, Δ
COOH

b. The $-CH_2CH_2OH$ group is the reduced form of the $-CH_2COOH$ group, which comes from diethyl malonate. After the conjugate addition step, hydrolysis of the ester under conditions that cause decarboxylation yields the needed keto acid. Reduction of both carbonyl groups (ketone and carboxylic acid) generates the diol product.

Retrosynthesis

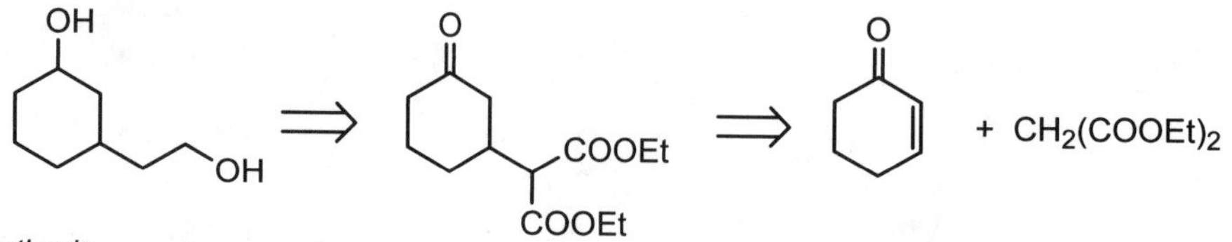
OH
OH
COOEt
COOEt
+ CH2(COOEt)2

Synthesis

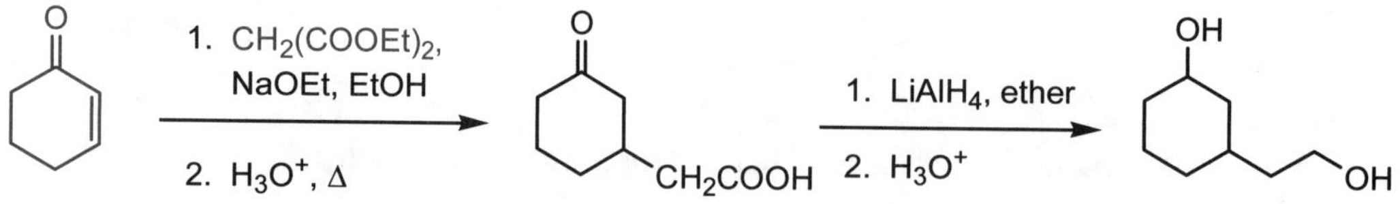
1. CH2(COOEt)2,
NaOEt, EtOH
2. H3O+, Δ
CH2COOH
1. LiAlH4, ether
2. H3O+
OH
OH

24.10. The interconversion between the two enolate ions in the Robinson annulation occurs by acid-base reactions in which the solvent (ethanol) functions as the proton donor.   These acid-base steps are equilibria, but the equations have been written below to show how the electrons move to produce the enolate ion needed to form the six-membered ring.

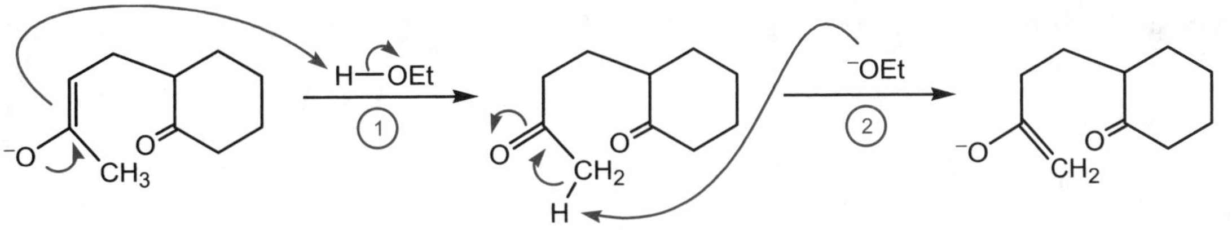

24.11. The E1cb reaction occurs by formation of the ketone's enolate derivative. When the carbonyl group is regenerated, hydroxide ion is expelled from the β-carbon atom to form the unsaturated ketone.

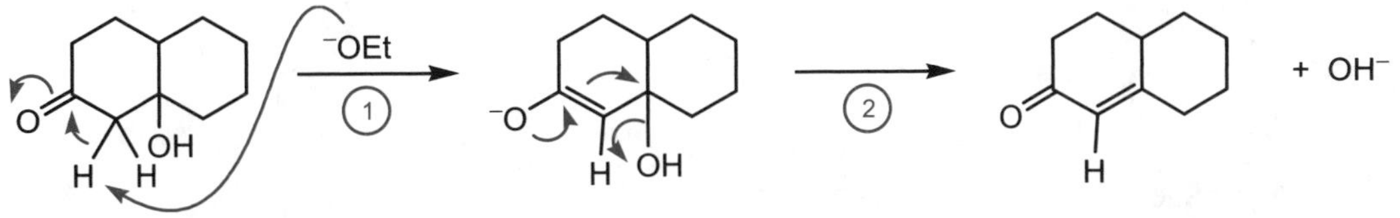

24.12. In planning a synthesis that makes use of conjugate addition of a hydrocarbon group from an organocopper reagent, identify the fragment attached beta to the carbonyl group. The organocuprate reagent have that group is then prepared and allowed to react with the ketone. Acid workup yields the product. The starting compounds are shown in color.

a. A Robinson annulation procedure is used to construct the bicyclic ring system. Conjugate addition of the vinyl group makes use of LiCu(CHCH₂)₂ or Li₂Cu(CN)(CHCH₂)₂ (shown).

Retrosynthesis

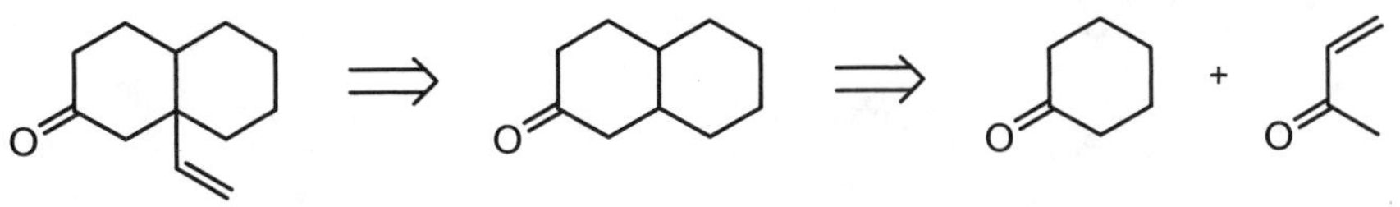

Synthesis

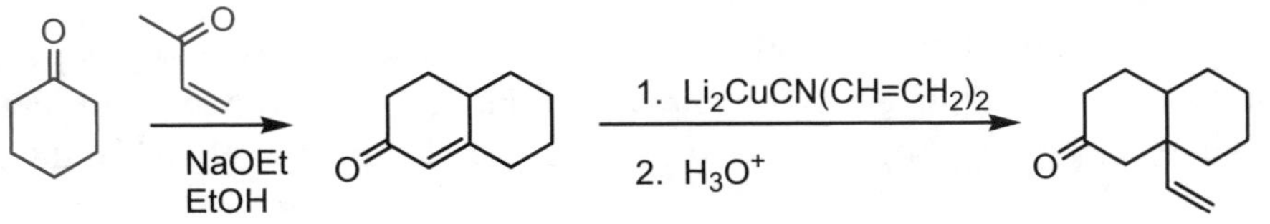

b. A seven-carbon ketone is an allowed starting material, so LiCu(C₃H₇)₂ (shown) or Li₂Cu(CN)(C₃H₇) ₂ is used to introduce the propyl group.

Synthesis

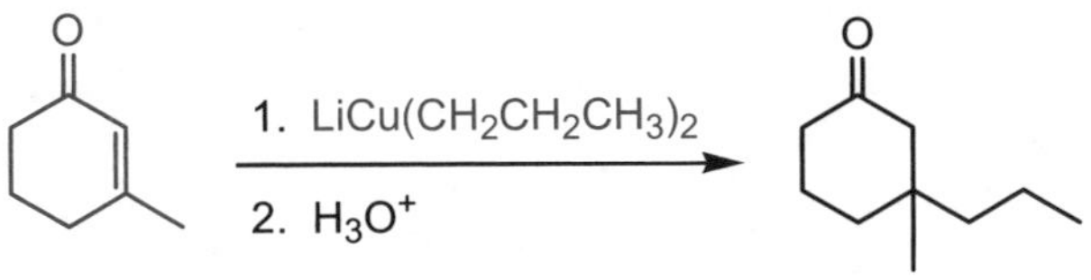

24.13. In planning a synthesis that makes use of conjugate addition of a hydrocarbon group from an organocopper reagent, identify the fragment attached beta to the carbonyl group. If an alkyl group is also attached at the alpha position and on the same side of the carbonyl group, then use the tandem addition/alkylation procedure.

a. Conjugate addition of the ethyl group from LiCu(C₂H₅)₂ is followed by a reaction of the resulting enolate ion with methyl iodide.

b. Conjugate addition of the ethyl group from LiCu(C₂H₅)₂ is followed by trapping the enolate ion with the PhSe group. Oxidation causes elimination and regeneration of the double bond. The methyl group is subsequently added by means of a second conjugate addition reaction.

24.14. An allylic alcohol is routinely made by reducing the carbonyl group of an enone. Many enones can be made by the route shown in the solution to Exercise 24.13b, which consists of conjugate addition of an alkyl group from an organocuprate reagent, trapping the resulting enolate ion with phenylselenyl chloride, and oxidation/elimination of PhSeOH after treatment with hydrogen peroxide.

a. After formation of the carbon skeleton by means of a conjugate addition and elimination reaction, the ketone carbonyl group is reduced using the combination of NaBH₄ and CeCl₃.

b. Conjugate addition of the methyl group from Li₂Cu(CN)(CH₃)₂ is followed by trapping the enolate ion with phenylselenyl chloride. Alkylation at the alpha position is accomplished by deprotonation with LDA followed by treatment with 1-bromobutane. Elimination of the PhSe group creates the double bond, and then the carbonyl group is reduced using the combination of NaBH₄ and CeCl₃.

24.15. Besides the combination of NaBH₄ and CeCl₃, reduction an enone can be accomplished using a ruthenium catalyst and hydrogen. The advantage of using this latter method is the possibility of carrying out enantioselective reduction reactions with a chiral catalyst. Making the needed enone follows the methods seen in the previous exercises.

a. The enone needed in this synthesis can be made using the crossed aldol reaction. The needed ketone and aldehyde molecules are made by reactions presented in previous chapters. Once the carbon skeleton has been made, the chiral Ru catalyst is used to make the chiral allylic alcohol.

Retrosynthesis

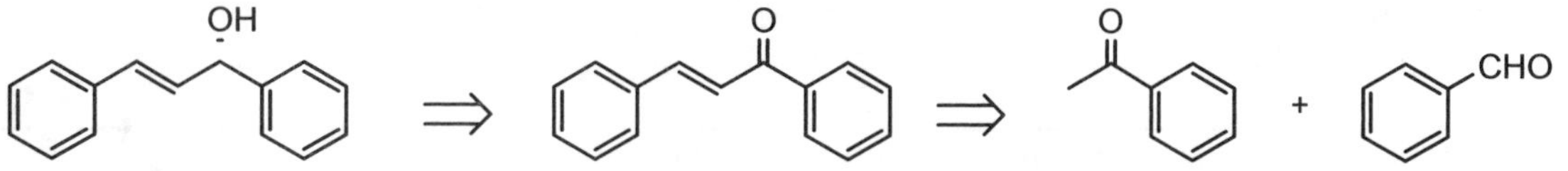

Synthesis

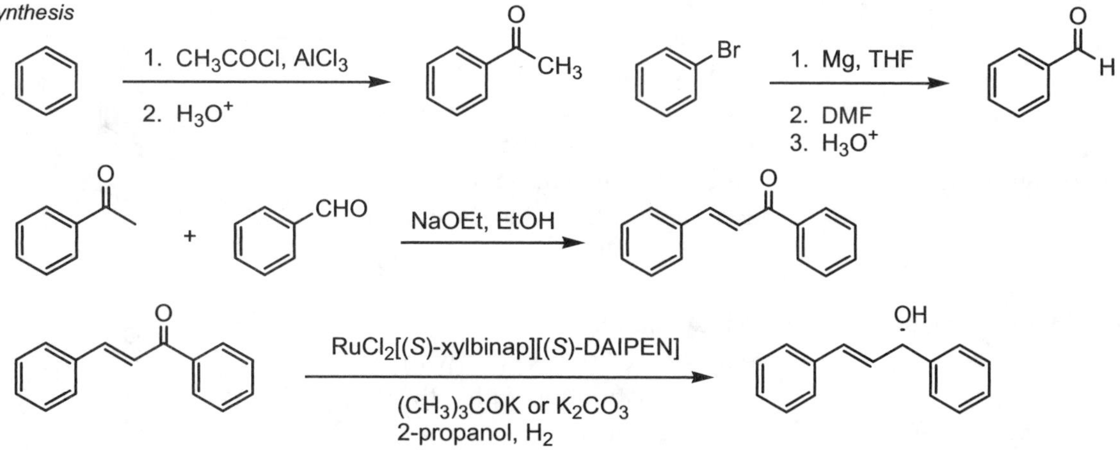

b. The enone needed for this synthesis can be made by oxidizing a racemic allylic alcohol formed as the product of a Grignard reaction. Once the carbon skeleton has been constructed, the chiral Ru catalyst is used to make the desired chiral allylic alcohol.

Retrosynthesis

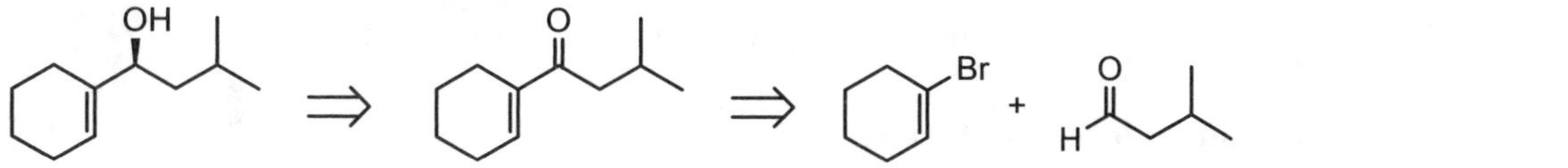

Synthesis

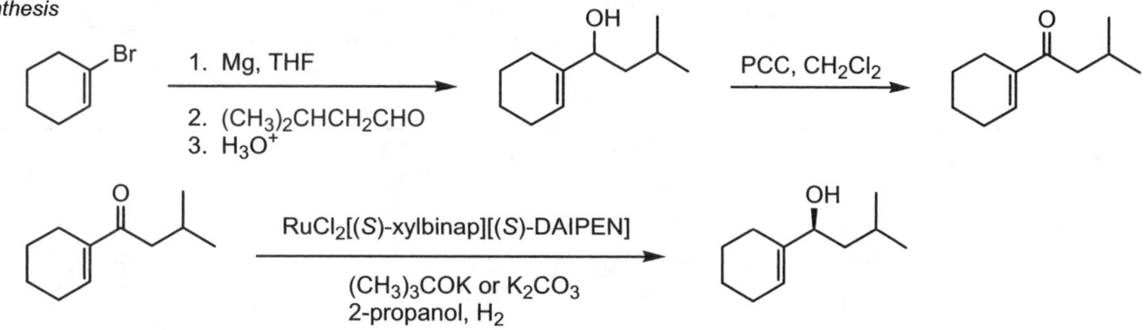

24.16. The dissolving metal reduction of a conjugated ketone leads to formation of a specific enolate ion, which is then alkylated with the given alkyl halide. The trans isomer is often the major product if alkyl substituents are attached at adjacent carbon atoms.

24.16. (continued)

a.

1. Li/NH$_3$, Et$_2$O, 1 eq t-BuOH

2. CH$_3$I

b.

1. Li/NH$_3$, Et$_2$O, 1 eq t-BuOH

2. Br—_/=/ Br

24.17. Follow the procedure outlined in the solution to Exercise 1.34.

a. **Methyl 2-phenylselenylbutanoate**

but	4 carbon atoms
an	no double or triple bonds
oate	ester functional group; the C=O group defines C1 of the carbon chain
2-phenylselenyl	a PhSe group is attached at C2
Methyl	a CH$_3$ group is attached to the oxygen atom of the ester functional group

b. **(S)-3-Methyl-2-isopropylidenecyclopentanone**

cyclopentanone	a five-membered ring ketone with no other sites of unsaturation within the ring
2-isopropylidene	a =C(CH$_3$)$_2$ group atom is attached at C2
3-Methyl	a CH$_3$ group is attached at C3
(S)	the configuration of C3 is (S)

c. **2,3-Epoxycyclohexanone**

cyclopentanone	a six-membered ring ketone with no other sites of unsaturation within the ring
2,3-Epoxy	an oxygen atom is bonded to C2 and C3 of the six-membered ring to form the epoxide ring

d. **2-Dimethylaminomethylcycloheptanone**

cycloheptanone	a seven-membered ring ketone with no other sites of unsaturation within the ring
2-Dimethylaminomethyl	the (CH$_3$)$_2$NCH$_2$– group is attached at C2

24.18. Follow the procedure outlined in the solution to Exercise 1.32.

a. This compound is a five-carbon, cyclic ketone with no other sites of unsaturation within the ring: **cyclopentanone**; the position of the carbonyl group defines C1 of the ring.

　　The substituents include a cyano group at C3 and a methyl group at C2, and their relationship is trans. The name is ***trans*-3-cyano-2-methylcyclopentanone**.

24.18. (continued)

b. This compound is a ketone with five carbon atoms in the longest chain, and the C=O group is positioned at C3: **3-pentanone**.

   The substituents include the dimethylamino group at C1, and the methyl group at C4. The name is **1-dimethylamino-4-methyl-3-pentanone**.

- - - - - - - - - - - - - - - - - - - -

c. This compound is an aldehyde with five carbon atoms in the chain, which has no double or triple bond: pentanal.

   The aldehyde functional group defines C1. Two methyl groups are attached at C2, and a nitro group is attached at C5. The name is **2,2-dimethyl-5-nitropentanal**.

- - - - - - - - - - - - - - - - - - -

d. This compound is a ketone with four carbon atoms in the chain, which has no double or triple bond: but/an/one = butanone.

   The principal functional group (ketone) is at C1, which can only occur with alkyl aryl ketones. A phenyl group is also attached at C1, and a methylthio group is attached at C3. The name is **3-methylthio-1-phenyl-1-butanone**.

24.19. Conjugate addition of water to the β-chloro-α,β-unsaturated ester shown in this exercise creates a situation in which two good leaving groups are attached to same carbon atom. Addition occurs by proton activation of the carbonyl group (step 1), which is followed by the conjugate addition step. Proton transfer reactions (steps 3 and 4) generate the carbonyl group of the ketone.

The next part comprises tautomerism of the enol form of the ester carbonyl group to its carbonyl form (step 5), the subsequent hydrolysis process (steps 6-9), and decarboxylation (step 10). The product at the end of this sequence is the enol form of the ketone.

24.19. (continued)

Finally, a tautomerism reaction forms the ketone product.

24.20. In performing the reaction in which our chemist friend adds water and base to the compound 2-isopropylidenecyclohexanone, conjugate addition of the hydroxide ion occurs as expected, forming compound **A**. This substance is a β-hydroxy ketone, however, so it undergoes a retroaldol reaction under the basic conditions being used (Section 23.1d). This retroaldol reaction generates acetone and cyclohexanone (compound **B**) as products.

24.21. The enolate ion of diethyl malonate cannot be alkylated using *tert*-butyl bromide because the latter would undergo elimination under the basic reaction conditions.

Instead, diethyl malonate is treated with acetone under conditions of the Knoevenagel reaction (see Exercise 23.20), which produces the isopropylidene malonate derivative. The desired diethyl *tert*-butylmalonate is subsequently made by conjugate addition of a methyl group.

24.21. (continued)

Even though you would probably choose lithium dimethylcuprate for this conjugate addition step, the simple Grignard reagent, CH$_3$MgI, actually suffices. The unsaturated diester is obviously a potent reactant toward conjugate addition, so the presence of copper is not required.

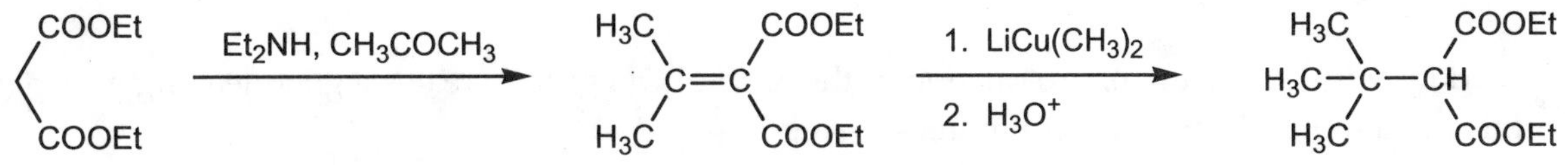

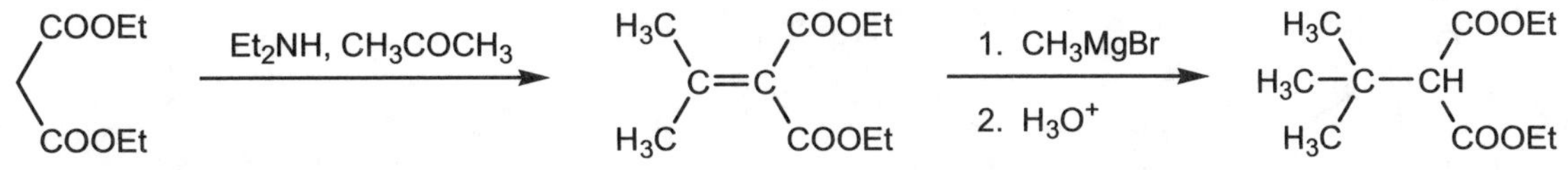

24.22. Follow the procedure outlined in the solution to Exercise 24.9. The starting compounds are shown in color.

a. The –CH$_2$COOH group needs to be added to an α,β-unsaturated ketone to make the desired product, so diethyl malonate is used as the enolate ion precursor. Hydrolysis of the diester under conditions that cause decarboxylation yields the product.

Retrosynthesis

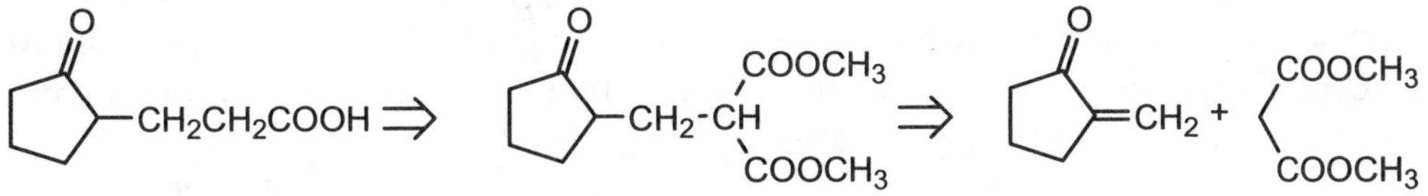

Synthesis

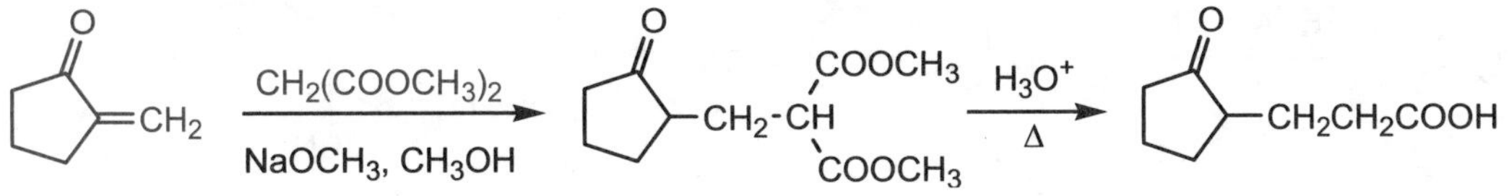

b. This Michael reaction makes use of acrylonitrile as the α,β-unsaturated component and ethyl acetoacetate as the enolate precursor.

Retrosynthesis

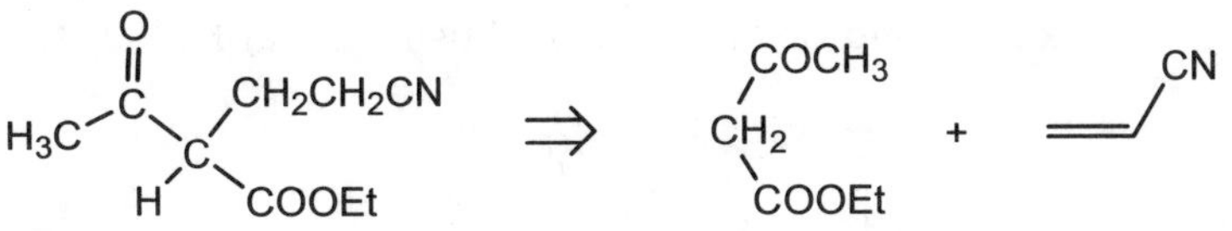

Synthesis

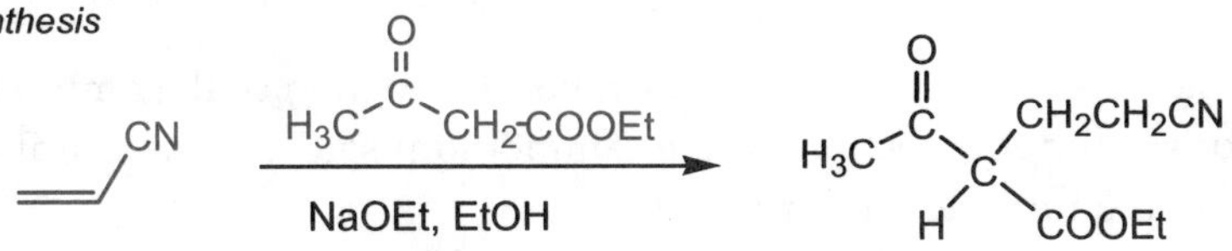

24.23. In this example of the Baylis-Hillman reaction, methoxide ion undergoes 1,4 addition to the unsaturated ester, which generates an enolate ion. This nucleophilic carbanion adds to the aldehyde carbonyl group in a crossed aldol reaction (step 2).

24.23. (continued)

Upon workup with acid (step 3), protonation of the alcohol OH group creates a good leaving group (step 4), and finally a molecule of water is eliminated (step 5).

24.24. The conjugate addition reaction of cyanide ion from diethylaluminum cyanide occurs by complexation of the diethylaluminum group to the carbonyl oxygen atom (step 1). This Lewis acid-Lewis base reaction activates the system for conjugate addition, and cyanide ion subsequently adds to the β carbon atom to form the diethylaluminum enolate (step 2). Workup replaces the aluminum atom with a proton, which tautomerizes to form the ketone product.

24.25. To predict the structures of the products in each of the following transformations, learn the details of each reaction type, which can be found in the reaction summary section of the chapter. In some of these transformations, reactions from previous chapters have been included.

a.   An α,β-unsaturated ester is reduced to the corresponding allylic alcohol using Dibal-H.

b.   An enolate ion reacts with Eschenmoser's salt to form a Mannich base. Reaction of the amino group of the Mannich base with ethyl iodide produces the quaternary ammonium salt. A new chiral center is formed, so the product is obtained as a racemic mixture.

24.25. (continued)

c. An α,β-unsaturated ester undergoes conjugate addition reactions with nucleophiles.   In this transformation, cyanide ion adds to the beta carbon atom.

d   Hydrogen peroxide and base react with α,β-unsaturated ketones to form the corresponding epoxide derivatives. Isolated double bonds do not react with this reagent combination.

e. An active methylene compound as its enolate derivative undergoes the Michael reaction with α,β-unsaturated ketones, esters, and nitriles. Two new chiral centers are formed, but they are made independently of the other, so four stereoisomers are generated.

f. An organocuprate reagent delivers its alkyl group to the beta carbon atom of an α,β-unsaturated ketone and generates an enolate ion. If an alkyl halide is added in the second step, its alkyl group becomes attached to the carbon atom alpha to the carbonyl group.

g. An α,β-unsaturated ketone is reduced to the corresponding allylic alcohol using sodium borohydride in the presence of cerium(III) chloride.

h. An α,β-unsaturated ester undergoes conjugate addition when treated with an amine.

24.26. To assign the prochiral configurations of the different groups, follow the procedures outlined in the solution to Exercise 16.3.

The identical carboxymethyl groups in the citrate molecule have priorities 3 and 4 in the Cahn-Ingold-Prelog system. If one of the methylene groups undergoes reaction (that is, if one of the methylene protons is replaced by an OH group), then these two groups no longer have the same priorities and C3 becomes chiral. The carboxymethyl group can be designated as pro-(R) or pro-(S).

24.26. (continued)

(*S*) as drawn, but the priority 4
group is coming forward, so
the actual configuration is (*R*).

Replacing a hydrogen atom of each carboxymethyl group makes the methylene carbon atom chiral, which in turn means that each hydrogen atom is either pro-(*R*) or pro-(*S*).

24.27. Draw the structures of the compounds (these can be found throughout the text), and then consider the type of reaction that is occurring. Use what you have learned about each reaction type to propose a reasonable mechanism.

1. **Citrate → Aconitate:** In this first step of the Krebs cycle, dehydration of citrate occurs via an E1cb mechanism. The enolate derivative is formed by deprotonation by a base, **B** (step 1a), and then hydroxide ion is expelled as a molecule of water when the carbonyl group is regenerated (step 1b).

24.27. (continued)

2. **Aconitate → Isocitrate:** Hydration occurs by reverse of the dehydration mechanism.

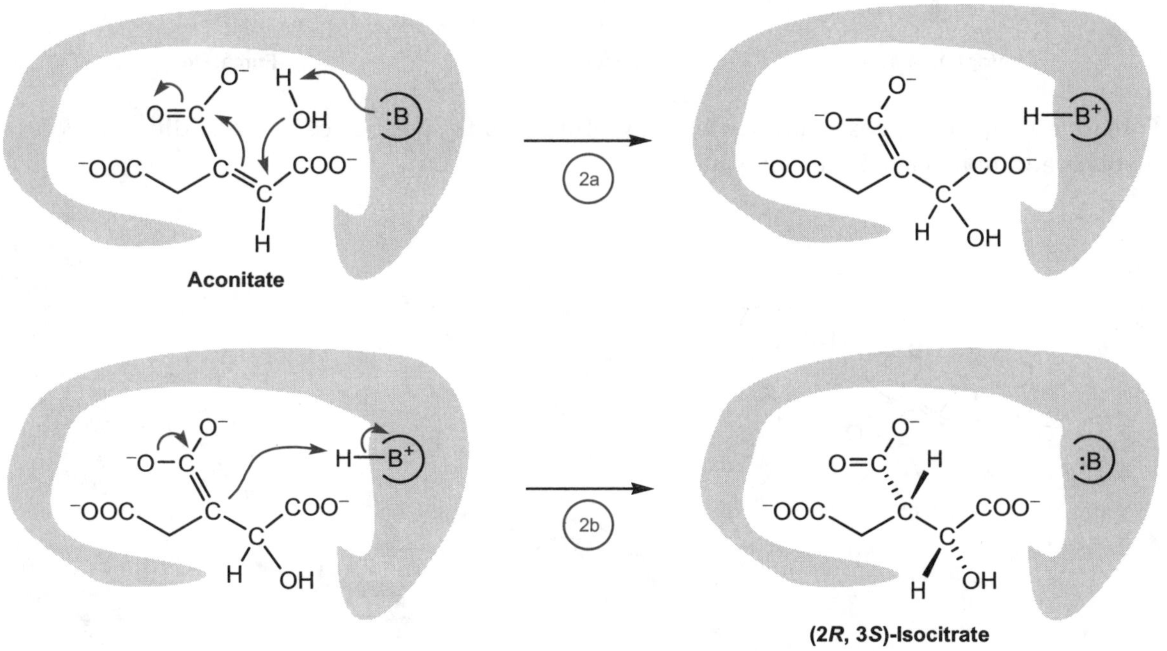

Aconitate

(2R, 3S)-Isocitrate

3. **Isocitrate → Ketoglutarate:** Oxidation of the alcohol to form the ketone group proceeds with involvement of NAD⁺ as a coenzyme (step 3a). Spontaneous decarboxylation occurs (step 3b) because the intermediate is a β-keto acid. This enol form tautomerizes to yield α-ketoglutarate (step 3c).

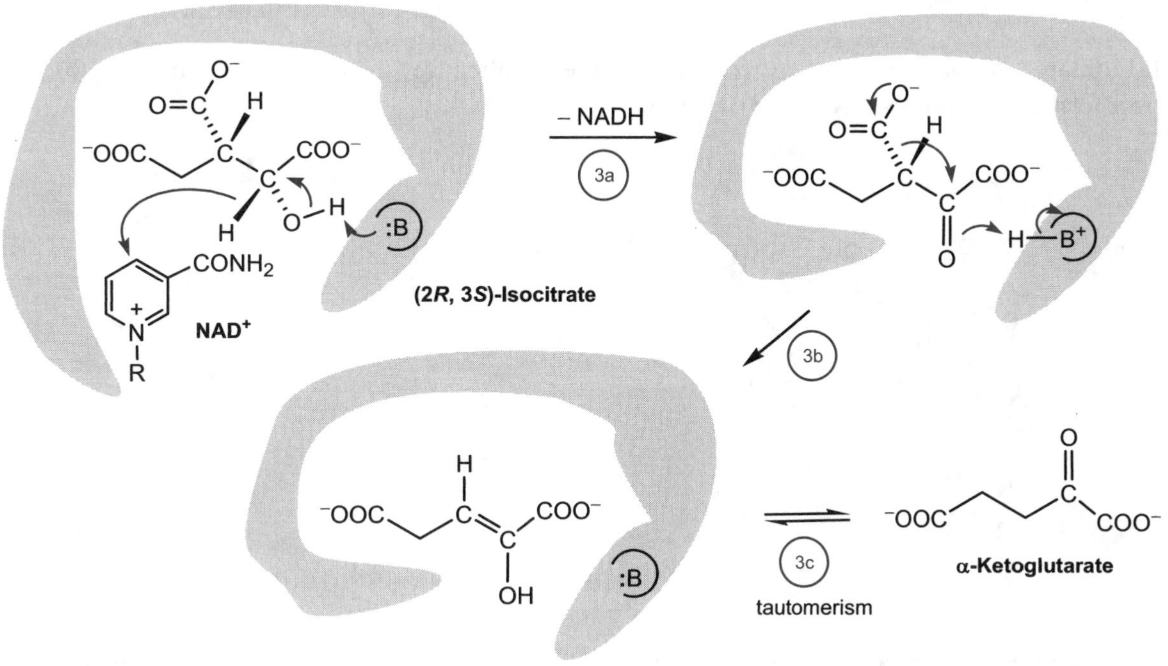

(2R, 3S)-Isocitrate

α-Ketoglutarate

tautomerism

4. **Ketoglutarate → Succinate → Fumarate → (S)-Malate:** After succinate is formed by another decarboxylation process (discussed later in the text), dehydrogenation produces fumarate.

24.27. (continued)

α-Ketoglutarate — $-CO_2$ → Succinate — $-2H$ → Fumarate

Fumarate then undergoes hydration by way of an enantiospecific conjugate addition process that involves addition of a molecule of water.

5. **(S)-Malate → Oxaloacetate:** Oxidation of the alcohol, (S)-malate, occurs in step (5). This transformation requires NAD⁺ as a coenzyme.

6. **Oxaloacetate → Citrate:** To complete the Krebs cycle, a crossed-aldol condensation takes place between oxaloacetate and acetyl coenzyme A (step 6a—see Section 23.2e). The resulting thioester undergoes hydrolysis via a tetrahedral intermediate to form citrate and coenzyme A.

24.27. (continued)

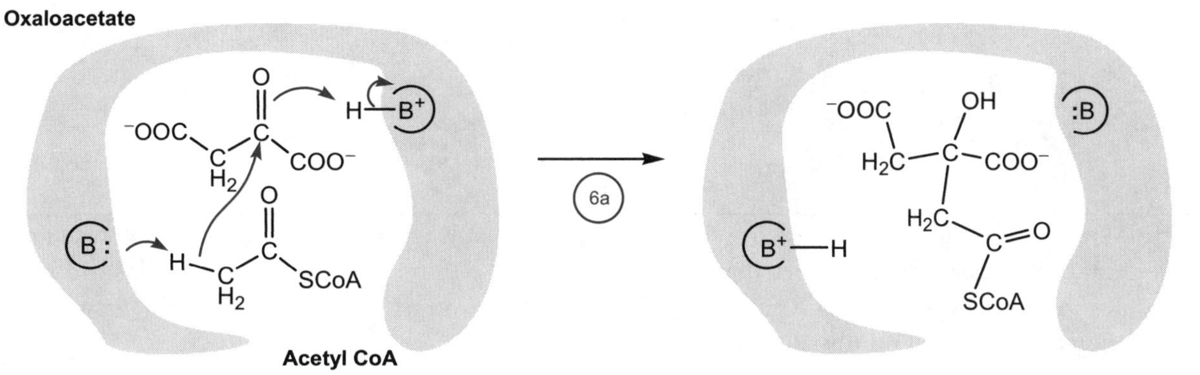

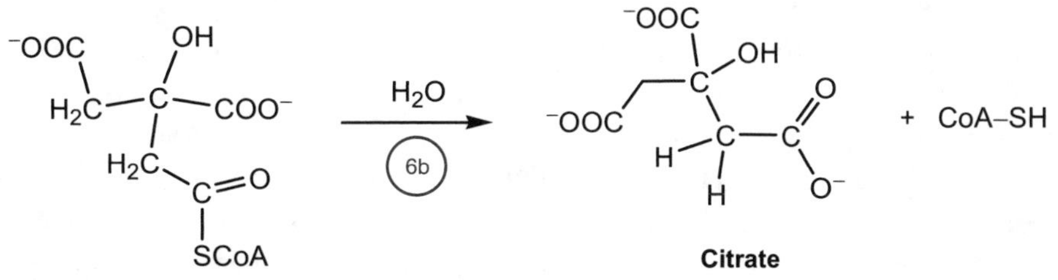

Oxaloacetate

Acetyl CoA

Citrate

24.28. A retrosynthetic analysis based on the Robinson annulation of methyl vinyl ketone breaks the bonds that remove the four-carbon ketone fragment from the product. If the six-membered ring has other functional groups, they are included as part of the ketone starting material. The starting compounds are shown in color.

a. Work backwards from the saturated ketone to the unsaturated ketone, and then disconnect the four-carbon unit.

Retrosynthesis

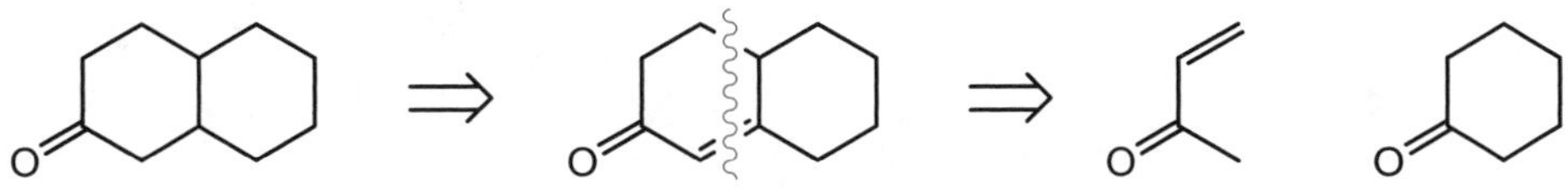

Synthesis

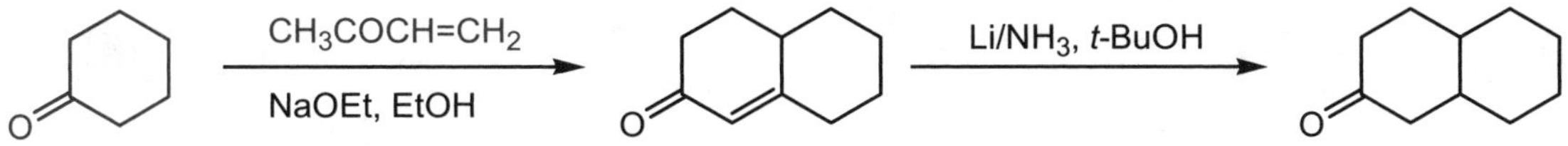

b. An α,β-unsaturated ketone group is already present, so disconnection of the four-carbon unit reveals the other needed starting material, 1,3-cyclohexeneone.

Retrosynthesis   *Synthesis*

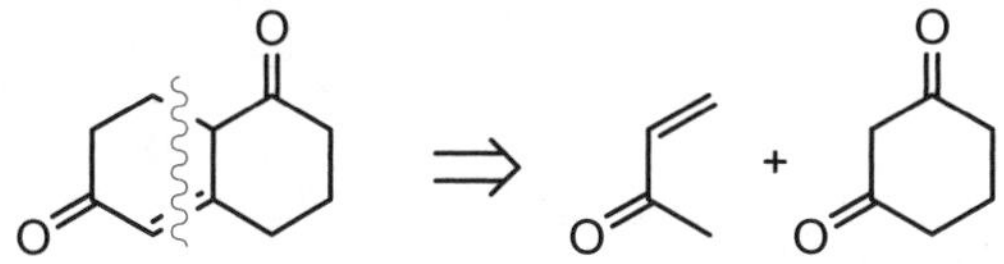

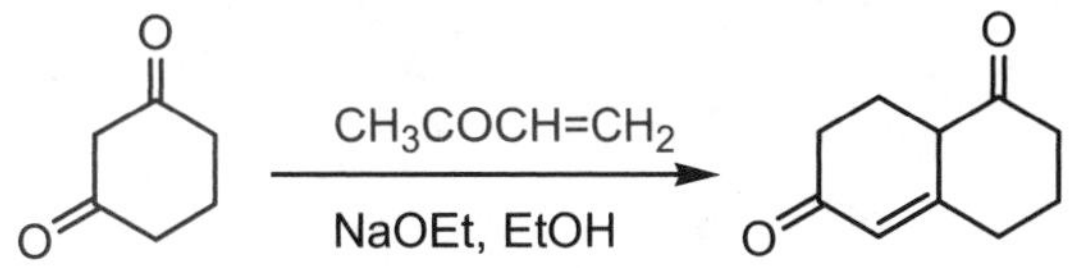

c. The presence of an alkyl group beta to the carbonyl group suggests the use of a conjugate addition process with an organocuprate reagent. The needed starting material is the product shown in part (a.) of this exercise.

24.28. (continued)

Retrosynthesis   *Synthesis*

24.29. When you perform a reaction in which you know what the product will be (or is expected to be), focus on those spectroscopic features that are unique for the groups in the starting material and product molecules. Although many subtle changes will undoubtedly occur in several parts of the spectrum, it helps to look for a limited number of specific changes.

a. For this transformation, the ¹H NMR spectra of both molecules will display similar patterns of signals in the aromatic region. The product will display aliphatic proton resonances (chemical shifts between δ 0 and 5), which are absent in the reactant, and the starting compound will have resonances in the alkene region (δ 5-7, more downfield than "normal" because of the strong electron-withdrawing cyano group) that do not appear in the spectrum of the product.

The carbon NMR spectra will differ in similar fashion: The product will have aliphatic carbon resonances, and the starting compound will have resonances in the alkene region. Both will have similar patterns of signals in the aromatic region.

¹H NMR spectrum

¹³C NMR spectrum

b. For this transformation, the ¹H NMR spectrum of the product will have significantly more signals: new signals will appear in the aliphatic and aromatic proton regions, and the signals for the starting material in the alkene region will disappear.

¹H NMR spectrum

24.29. (continued)

The carbon NMR spectra will also have more signals observed for the product molecule. The main difference will be the appearance of two additional peaks in the region δ 110-150.

¹³C NMR spectrum

δ 110-150
2 signals, 1H each (DEPT)

δ 110-150   4 signals
DEPT: three with 1H; one with 0H

24.30. To predict the structures of the products in each of the following transformations, learn the details of each reaction type, which can be found in the reaction summary section of the chapter.

a. An active methylene compound (as its enolate derivative) undergoes the Michael reaction with α,β-unsaturated ketones, esters, and nitriles. Hydrolysis in hot, aqueous acid leads to hydrolysis of all three ester groups and decarboxylation of the β–diacid unit.

b. An enolate ion undergoes the Michael reaction with α,β-unsaturated ketones, esters, and nitriles. Reduction occurs at both the ester and nitrile functional groups, forming a primary alcohol and a primary amine, respectively.

24.31. The retrosynthesis for ketones that have carbon-containing substituents makes use of conjugate addition if a substituent is attached beta to the carbonyl group, and either alkylation or the Mannich reaction if the substituent is in the alpha position. Compounds with rings are made via the Diels-Alder, Dieckmann, or Robinson reactions. The starting compounds are shown in color.

a. This product has a methyl group beta to the carbonyl group, so the use of an organocuprate reagent is called for. The unsaturated ketone required for the conjugate addition reaction is made using the PhSe elimination methodology. The double bond in the product is also generated by elimination of the PhSe group after the conjugate addition step.

Retrosynthesis

24.31. (continued)

Synthesis

----

b. The Robinson annulation provides the easiest route to prepare this bicyclic product.

Retrosynthesis   *Synthesis*

----

c. This product has a methylene group alpha to the carbonyl group, so the Mannich reaction is used to introduce that substituent. Formation of the enolate ion from cycloheptanone is followed by its reaction with Eschenmoser's salt. Alkylation of the nitrogen atom and elimination of trimethyl-amine yields the product.

Retrosynthesis

Synthesis

24.32. First, summarize the data with a set of equations.

The fact that compound **A** undergoes the Michael reaction suggests that it is an α,β-unsaturated ketone or aldehyde (only one oxygen atom is present). Addition of diethyl malonate followed by hydrolysis (and decarboxylation) adds a hydrogen atom and the –CH₂COOH group, which means that compound **B** has the formula C₇H₁₂O₃. Compound **C** has the same number of carbon and hydrogen atoms as compound **B**, so ozonolysis must be cleaving a ring to form a ketone and carboxylic acid (an aldehyde cannot be made under conditions that employ oxidative workup). If compound **B** has ketone and acid groups and was made by adding the –CH₂COOH group to **A**, then compound **A** must be an α,β-unsaturated ketone. Possible structure for **A** (C₅H₈O) are as follows.

24.32. (continued)

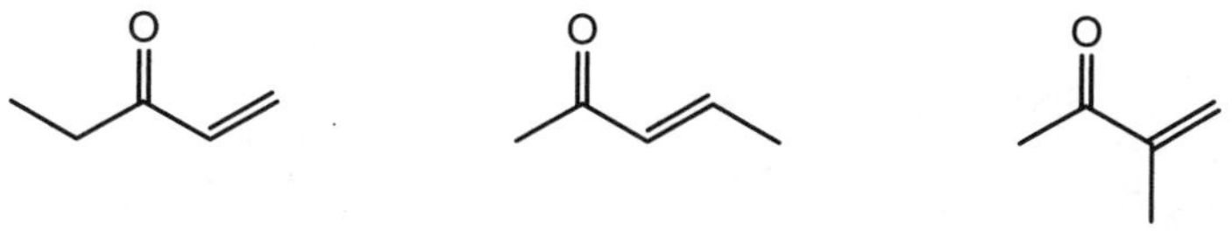

Compound **A** cannot be a methyl ketone, however, because the proton NMR spectrum of compound **B** has no feature that is a singlet (besides the resonance for the carboxylic acid proton), and the structures predicted for compound **B** would be those shown below. Two of these compounds would display singlets in their proton NMR spectra for the methyl groups shown in color.

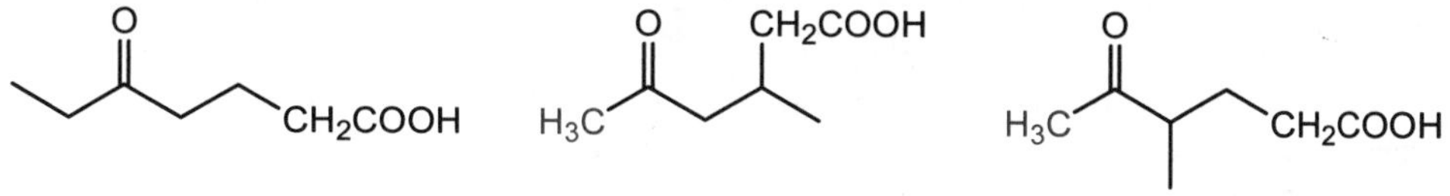

Therefore, compound **A** is **1-pentene-3-one**, and the other structures are as follows:

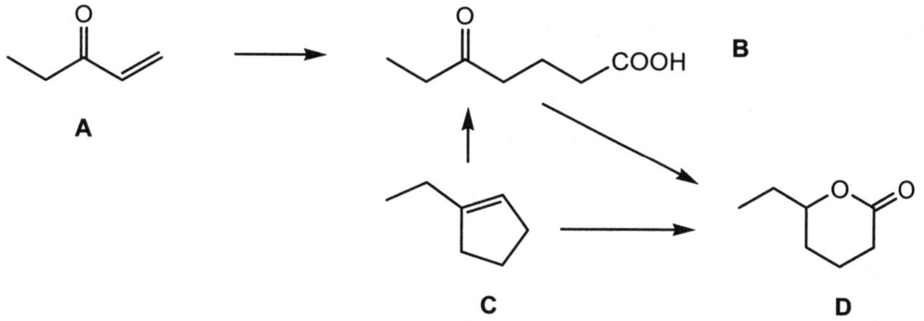

24.33. The conjugate addition reaction of a radical is similar to that of a nucleophile except that the electrons move individually rather than in pairs. Initiation takes place as you have seen before (Section 12.2d) by reaction of the radical derived from AIBN with tributyltin hydride.

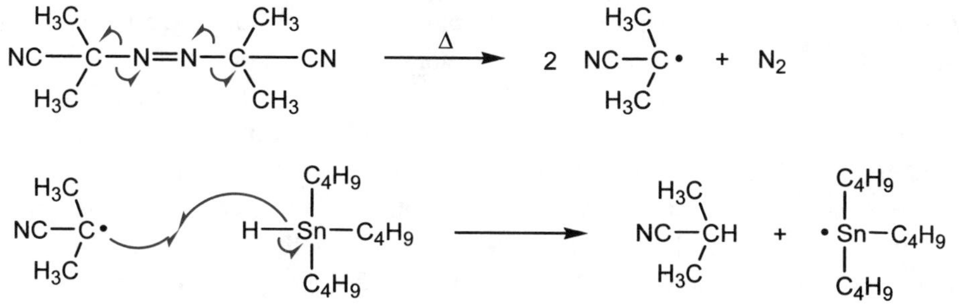

In the first step for the organic substrate molecule, the bromine atom is abstracted by the tributyltin radical. Addition of the radical to the double bond involves the carbonyl π electrons (step 2), which is what makes this process a conjugate addition process.

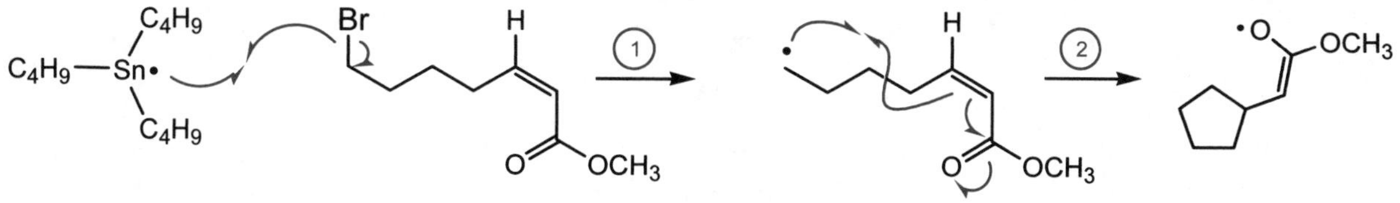

The resulting oxygen-centered radical reacts with tributyltin hydride (step 3), forming the cyclized ester product and regenerating the tributyltin radical, which can then react with more bromoalkyl substrate (step 1, above).

24.33. (continued)

24.34. Follow the procedure outlined in the solution to Exercise 14.21.

a. From the given molecular formula, C_4H_6O, we calculate $[2(4) + 2 - 6]/2 = 2$ sites of unsaturation. This compound contains two double bonds, a triple bond, two rings, or a double bond and a ring.

    You are told that the IR spectrum displays a strong band for the carbonyl group, which accounts for one double bond. The other site of unsaturation is either a C=C bond or a ring.

    The data from the proton NMR spectrum are summarized in the table below. There are two general features: three alkene protons and three aliphatic protons. The presence of the alkene resonances accounts for the other site of unsaturation, a C=C bond. The fact that there are three such protons indicates a monosubstituted double bond. The carbonyl group must be a ketone (one oxygen atom and no aldehyde resonance). The singlet at δ 2.3 is attributed to the resonance of a methyl group adjacent to the ketone carbonyl group. The structure that produces this spectrum is **methyl vinyl ketone (3-butene-2-one)**.

Chemical shift (ppm)	Integrated intensity	Assignment	Multiplicity	No. adjacent protons (n)
6.3	2	=CH₂	multiplet	-
5.9	1	=CH	doublet	1
2.3	3	CH₃	singlet	0

b. From the given molecular formula, $C_5H_8O_2$, we calculate $[2(5) + 2 - 8]/2 = 2$ sites of unsaturation. This compound contains two double bonds, a triple bond, two rings, or a double bond and a ring.

    You are told that the IR spectrum displays a strong band for the carbonyl group, which accounts for one double bond. The other site of unsaturation is either a C=C bond or a ring.

    The data from the proton NMR spectrum are summarized in the table below. There are three general features: a carboxylic acid proton, an alkene proton, and six aliphatic protons. The presence of the alkene resonance accounts for the other site of unsaturation, a C=C bond. The doublets at δ 2.18 and 1.93 (each with 3 protons), must correspond to methyl groups, each coupled with a single proton. The structure that can produce such a spectrum is **3-methyl-2-butenoic acid**. The two methyl groups are nonequivalent, and each is coupled with the alkene proton, which as a result appears as a septet (6 + 1 = 7 peaks).

Chemical shift (ppm)	Integrated intensity	Assignment	Multiplicity	No. adjacent protons (n)
12.2	1	COOH	singlet	0
5.71	1	=CH	septet	6
2.18	3	CH₃	doublet	1
1.93	3	CH₃	doublet	1

Solutions to Exercises for CHAPTER 25

THE CHEMISTRY OF POLYCYCLIC AND HETEROCYCLIC ARENES

25.1. The resonance forms for anthracene and phenanthrene are generated by moving the π bonds around the ring. In phenanthrene, C9 and C10 are connected by a double bond in four of the five resonance forms, so its length will be nearly the same as one in a typical alkene, which means that it will be shorter than the other carbon–carbon bonds in the rest of the molecule.

Anthracene

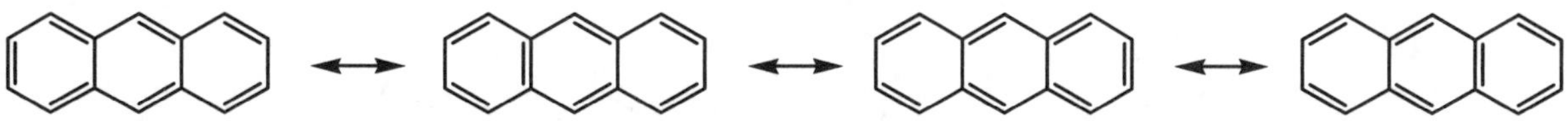

Phenanthrene

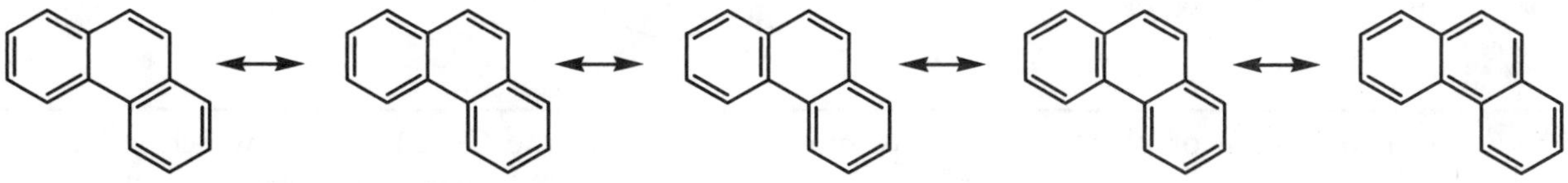

25.2. Six resonance structures can be drawn for the carbocation that is formed when an electrophile reacts at C2 of naphthalene.

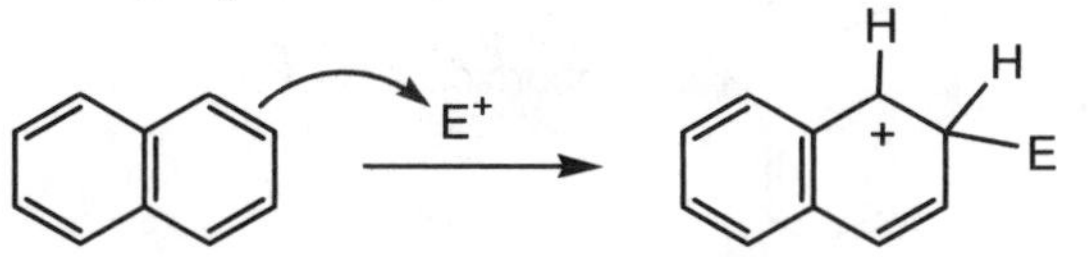

Only two of these have an intact benzene ring (shown in color, below).

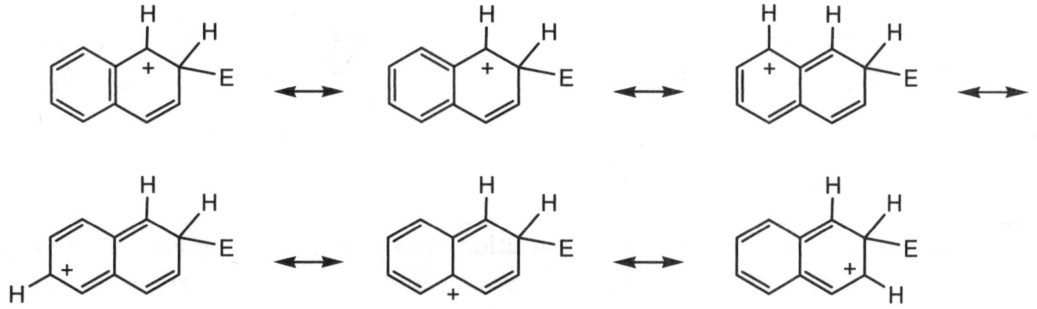

25.3. The methoxy group is a powerful ortho/para director and activator and the position para to it is blocked, but the products with a bromine atom ortho to the methoxy group are expected to form without trouble.

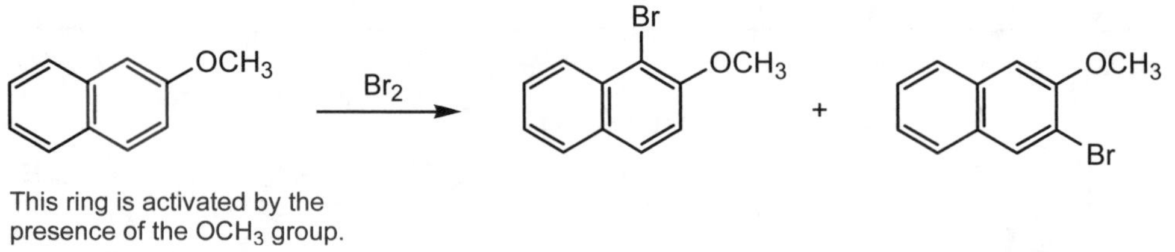

This ring is activated by the presence of the OCH₃ group.

The 1-position is generally the more reactive for naphthalene derivatives, but C1 is hindered in this starting material, so the 3-bromo compound may actually be the major product.

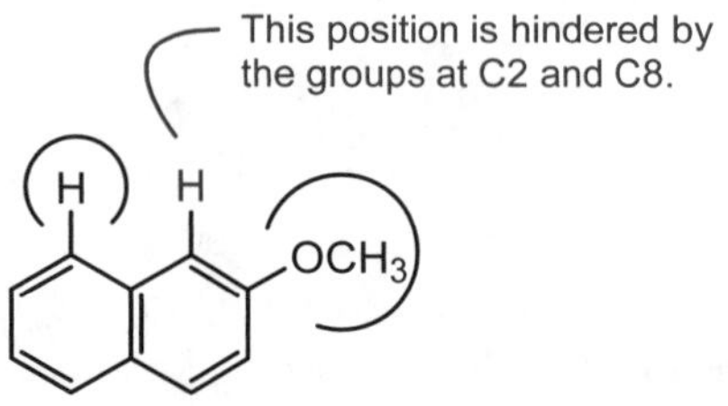

This position is hindered by the groups at C2 and C8.

25.4. The C9-C10 double bond is the most reactive in phenanthrene because a reaction there leaves two benzene rings intact. Oxidation produces a quinone in which the oxygen atoms are adjacent to each other.

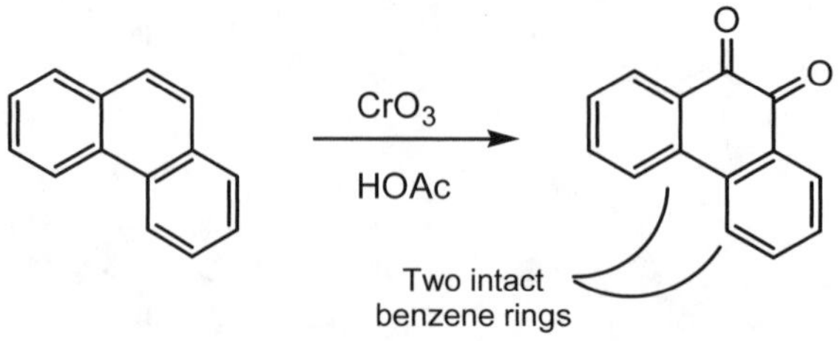

Two intact benzene rings

25.5. As with oxidation of phenanthrene, bromination occurs at the C9–C10 bond, which leaves two intact benzene rings in the product. The bromine atoms will likely add trans, as they do with alkenes.

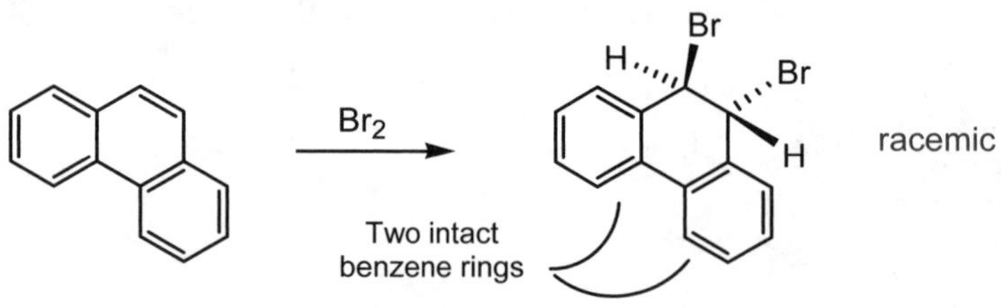

Two intact benzene rings

racemic

25.6. Follow the procedure shown in Example 25.1.

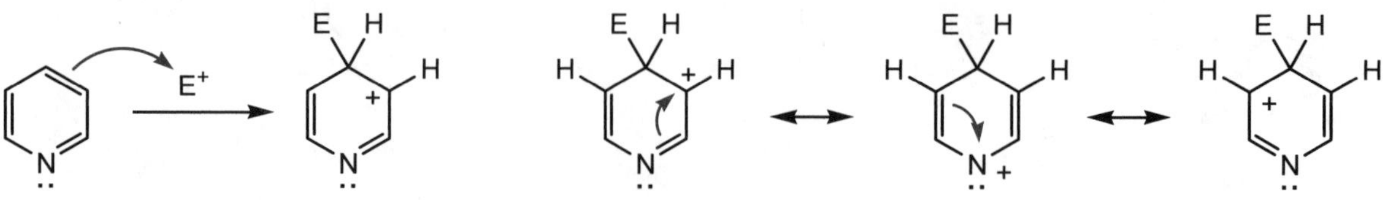

25.7. Pyridine derivatives undergo electrophilic substitution reactions slowly, but an activating group attached to the ring makes the reactivity toward substitution more like that of benzene. An activating group directs substitution to the positions *ortho* and *para* to itself.

25.7. (continued)

a. The amido group is an activating ortho/para director, so its influence directs the course of the reaction between this pyridine derivative and the electrophile NO_2^+.

b. The amino group is an activating ortho/para director, so its influence directs the course of the reaction between the pyridine derivative and the electrophile HSO_3^+.

25.8. Pyridine has approximately the same reactivity as nitrobenzene in the sense that the ring is electron deficient compared with benzene itself. Recall that a halogen atom ortho or para to a nitro group is readily substituted by nucleophiles (Section 17.4e). A halogen atom at C2 or C4 in a pyridine derivative reacts in the same fashion.

One resonance form of 4-chloropyridine places a positive charge at C4 (structure in box, above), which makes C4 even more attractive toward nucleophiles. The electronegativity of the Cl atom makes C4 electron deficient, too.

25.9. The negative charge on the oxygen atom in pyridine-*N*-oxide reacts with the positive metal ion of the Grignard reagent (step 1), and the phenyl group adds to the adjacent double bond, which has been activated by the presence of the nitrogen atom with its positive charge (step 2). When the reaction product of step 2 is treated with water during workup, the *N*-hydroxy compound is formed (step 3). Elimination of water takes place to regenerate the aromatic pyridine ring (step 4).

25.10. 2-Vinylpyridine reacts readily with nucleophiles, so the Michael reaction can occur just as it does with α,β-unsaturated ketones. A stabilized carbanion is formed by the acid-base reaction between diethyl malonate and ethoxide ion, and this enolate ion adds to the double bond conjugated with the C=N bond of the pyridine ring (step 1). Regeneration of the aromatic system occurs with the transfer of a proton from the solvent (step 2). The ethoxide ion that is formed subsequently reacts with diethyl malonate, so only a catalytic amount of base is required to make the overall transformation successful.

$$CH_2(COOEt)_2 + NaOEt \rightleftharpoons \bar{:}CH(COOEt)_2 + EtOH$$

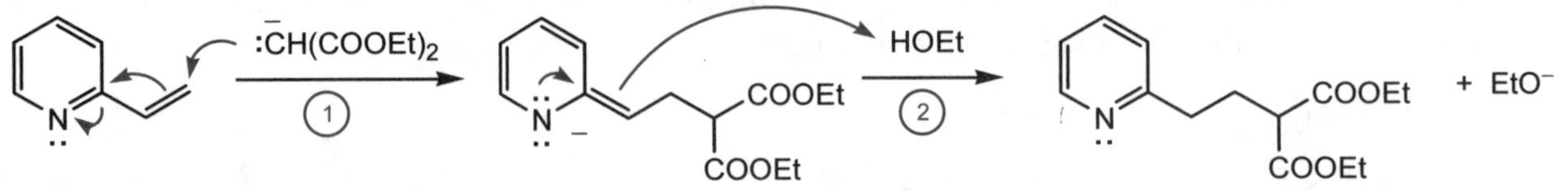

25.11. This reaction is an example of the crossed aldol reaction. A carbanion is formed by deprotonation of the methyl group of 2-methylpyridine-*N*-oxide (step 1), and this nucleophile adds to the carbonyl group of the aldehyde (step 2). A proton transfer occurs (step 3), and a molecule of water is lost via the E1cb pathway (steps 4 and 5) to form the unsaturated product.

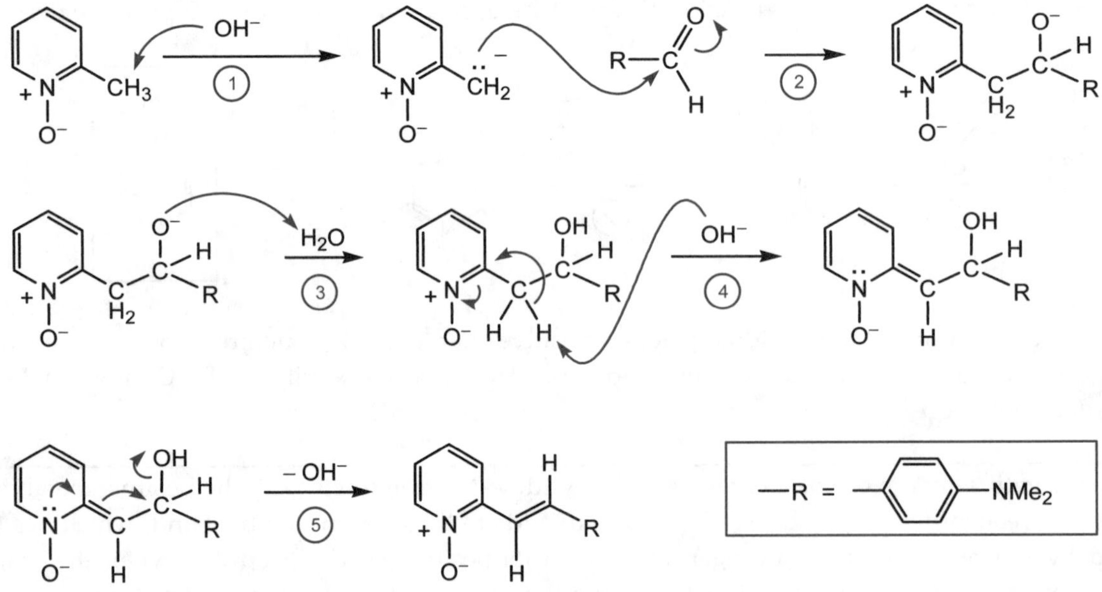

The following resonance structures for the conjugate base of 2-methylpyridine-*N*-oxide account for the relatively high acidity of the protons of the methyl group:

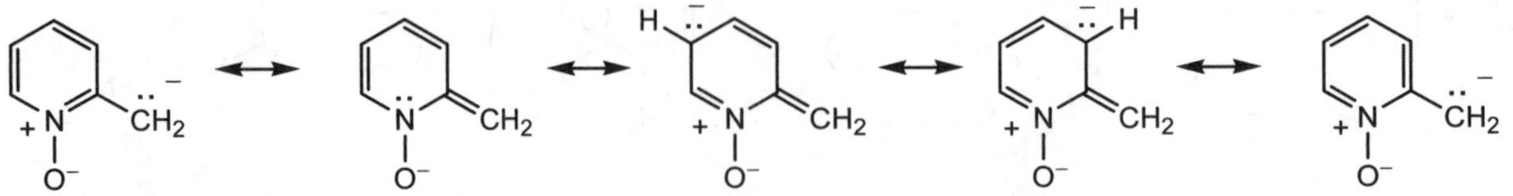

25.12. The reaction shown in this exercise is a variant of the crossed Claisen condensation. The heterocycle's methyl group is deprotonated by the strongly basic amide ion (step 1), and the resulting carbanion adds to the carbonyl group of the ester (step 2). Regeneration of the carbonyl group displaces the ethoxide ion (step 3). As in any Claisen condensation, deprotonation of the product occurs (step 4), but aqueous workup regenerates the neutral product (step 5).

25.12. (continued)

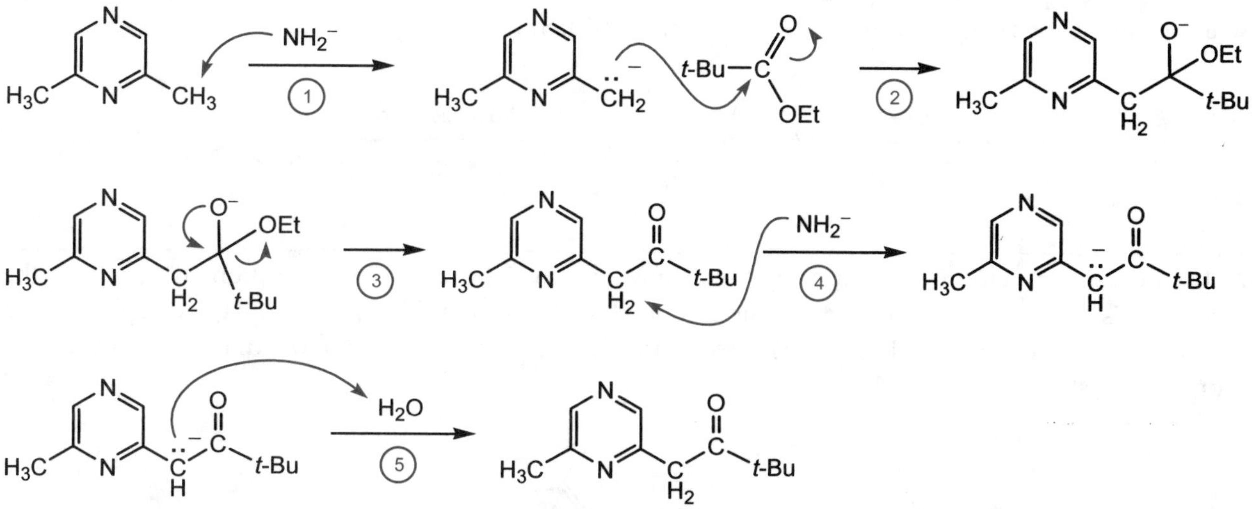

25.13. The five main steps in the pyrrole synthesis are given in the text, and the details of step 1 are given in Example 25.2. In step 2 of the overall process, the nitrogen atom—a nucleophile—adds to the ketone carbonyl group, which is then protonated (step 3). A good leaving group is formed by a second protonation of the OH group, and elimination of water creates the second double bond. What amounts to a tautomerism step completes the reaction sequence that produces the N-alkylpyrrole molecule (step 5).

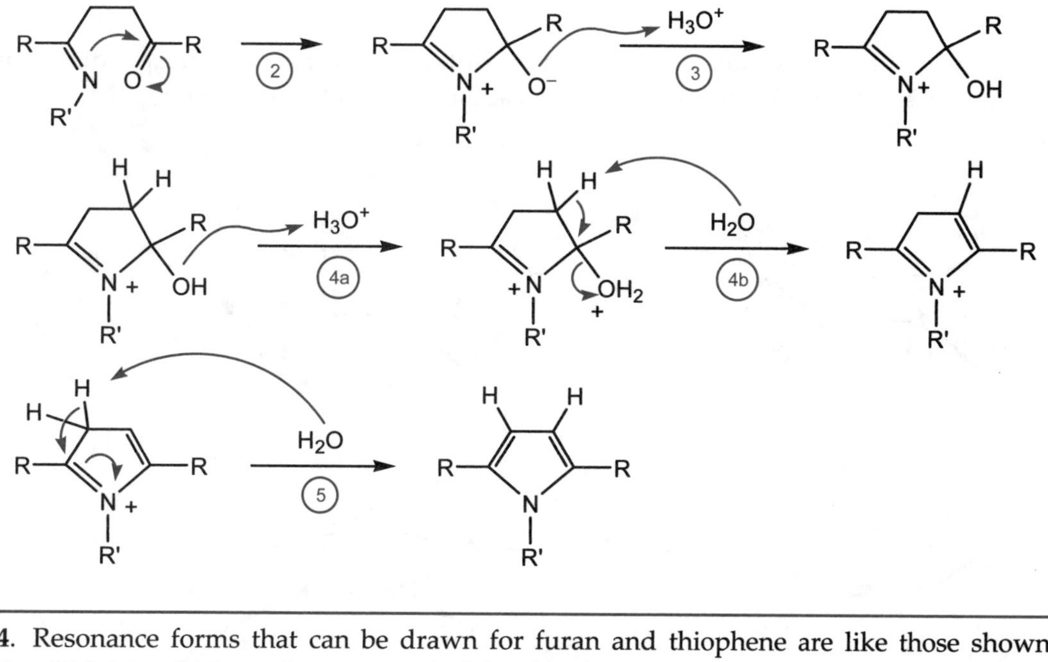

25.14. Resonance forms that can be drawn for furan and thiophene are like those shown for pyrrole (Section 25.3a) in which an electron pair is delocalized onto every atom of the ring.

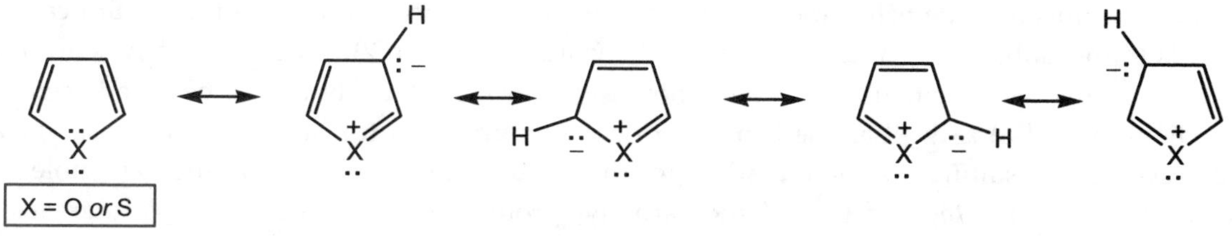

X = O or S

25.15. The first step in the given sequence is a Friedel-Crafts acylation reaction, and substitution takes place at the C2 position, as expected. The second step is the Wolff-Kishner reduction (Section 20.1e), which converts the acyl group to the hydrocarbon substituent.

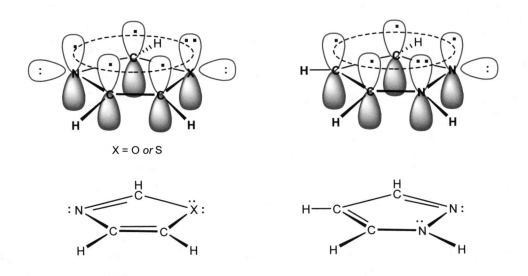

25.16. Recall that electrophilic aromatic substitution involves two steps (Section 17.2a). In the biosynthesis of tryptophan, the electrophilic carbon atom of the side chain precursor is intercepted by the π electrons of the heterocycle (step 1). Deprotonation of the cationic intermediate regenerates the aromatic π system (step 2).

25.17. Valence bond representations for the heterocycles oxazole, thiazole, and pyrazole are like the one shown for imidazole in Section 25.4a. In the thiazole and oxazole molecules, one unshared electron pair on the S or O atom is part of the π system. The other electron pair on the O or S atom is perpendicular to the π bonds.

For pyrazole, the orbital representation for each atom is the same as for the corresponding atom in imidazole; the difference is that the nitrogen atoms are adjacent to each other in pyrazole.

25.18. The mechanism of the benzoin condensation begins with formation of the thiazolium carbanion, which adds to the carbon-oxygen double bond of benzaldehyde (step 2). Intermediate **A** is an unusual species because the benzylic hydrogen atom (shown in color) is relatively acidic, being adjacent to the heterocyclic ring (THM) as well as the benzene ring. This proton is transferred to the basic oxyanion (step 2), and the resulting carbanion adds to the carbon-oxygen atom of another molecule of benzaldehyde (step 3). Regeneration of the carbonyl group occurs by the intramolecular acid-base reaction that displaces the thiazolium anion as a leaving group.

25.18. (continued)

25.19. The mechanism of the cyanide-catalyzed benzoin condensation begins with formation of the cyanohydrin anion, and then it follows the same series of steps shown in the solution to Exercise 25.18. Cyanide, like the thiazolium ion, has the feature that it makes the benzylic hydrogen atom of intermediate **B** relatively acidic.

25.20. To interpret the name of heterocyclic compounds, first deduce the structure of the parent heterocycle. Then number the atoms of the ring (the heteroatom is normally assigned as the 1-position), and indicate the positions of the substituents with the appropriate numbers.

a.   **2-Chloro-5-methylpyridine**

pyridine	parent heterocycle; the nitrogen atom is the 1-position
2-chloro	a Cl atom is attached to C2
5-methyl	a CH₃ group is attached to C5

b.   **(S)-3-methyltetrahydrothiophene**

thiophene	parent heterocycle; the sulfur atom is the 1-position
tetrahydro	the thiophene double bonds have been reduced
3-methyl	a CH₃ group is attached to C3
(S)	the chiral carbon atom has the (S) configuration

25.20. (continued)

c. **3-Thiacyclobutanone**

    cyclobutanone   cyclic four-carbon saturated ring with a ketone group;
                          the carbonyl carbon atom is C1.
      3-thia   a sulfur atom replaces a carbon atom at the 3-position

d. **3-Hydroxypyridine-*N*-oxide**

    pyridine   parent heterocycle; the nitrogen atom is the 1-position
    3-hydroxy   the OH group is attached at C3
    N-oxide   the nitrogen atom has an oxygen atom attached

e. **Methyl 3-pyrrolecarboxylate**

    pyrrole   parent heterocycle; the nitrogen atom is the 1-position
    3-carboxylate the carboxylic ester group is attached at C3
    methyl   a CH₃ group is attached to the oxygen atom of the ester

f. **2-Mercapto-4-nitroimidazole**

    imidazole   parent heterocycle; the nitrogen atom that defines the
                     the 1-position has a proton attached
    2-mercapto   the SH group is attached at C2
    4-nitro   the NO₂ group is attached at C4

g. **3-Acetyl-5-*tert*-butyl-1-methylpyrazole**

    pyrazole   parent heterocycle; one of the nitrogen atoms
                    is the 1-position
    3-acetyl   the CH₃CO group is attached to C3
    5-*tert*-butyl the (CH₃)₃C group is attached to C5
    1-methyl   a CH₃ group is attached to the nitrogen atom that
                    defines the 1-position

25.21. Follow the procedures outlined in the solution to Exercise 1.17. Identify the parent heterocycle, and then specify the substituents and their attachment points in the normal way.

a. This compound is a derivative of pyridine. An amino group
    is attached at C2 and a chlorine atom is attached at C6. The
    name is **2-amino-6-chloropyridine**.

– – – – – – – – – – – – – – – – – – –

b. This compound is a derivative of oxazole. A methoxy group
    is attached at the 2-position, so the name is **2-
    methoxyoxazole**.

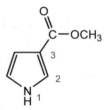

25.21. (continued)

c. This compound is a derivative of imidazole. An isopropyl group is attached at C2, and the carboxymethyl group is attached at the 1-position. The name is **1-carboxymethyl-2-isopropylimidazole**.

- -

d. This compound is a derivative of pyridine. A cyano group is attached at C3 and an oxygen atom is attached to the nitrogen atom. The name is **3-cyanopyridine-*N*-oxide**.

- -

e. This compound is a derivative of furan. A fluorine atom is attached at C3 and a methyl group is attached at C4. The name is **3-fluoro-4-methylfuran**.

- -

f. This compound is a derivative of pyrrole. An aldehyde group is attached at C2, so the name is **2-pyrrolecarbaldehyde**.

25.22. The nitrogen heterocycles that are present in these drug molecules are identified by removing all of their substituents.

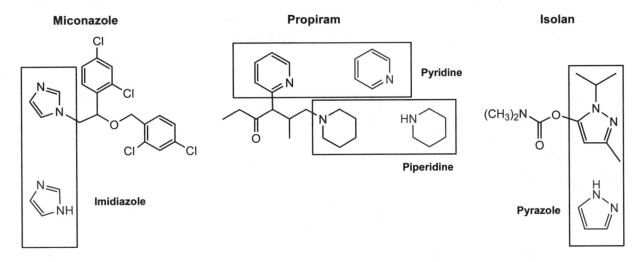

25.23. The tautomer of 2-pyridone is a lactam, which has the carbonyl group, hence the suffix "one."

When this molecule is deprotonated, the negative charge can be delocalized throughout the ring as well as on the oxygen atom.

25.23. (continued)

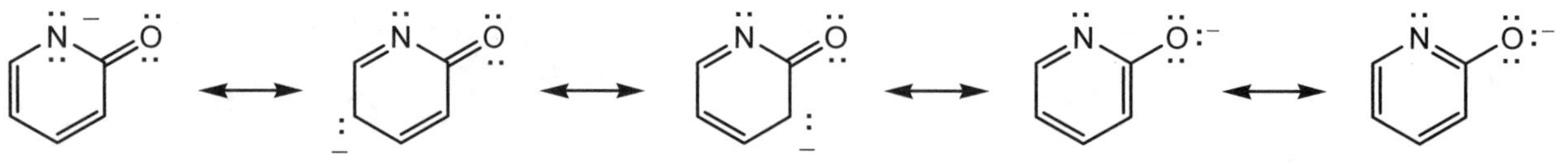

Thus, when this anion is treated with methyl iodide, alkylation can occur on either the nitrogen or oxygen atom.

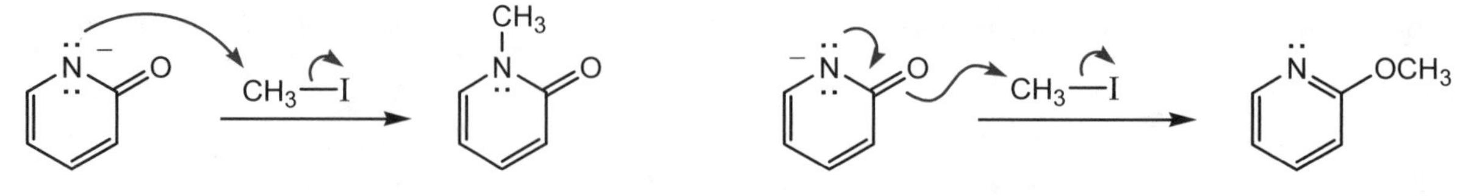

25.24. The resonance structures that can be drawn for pyridine-*N*-oxide place a negative charge at C2 and C4, which explains why electrophiles react at those positions.

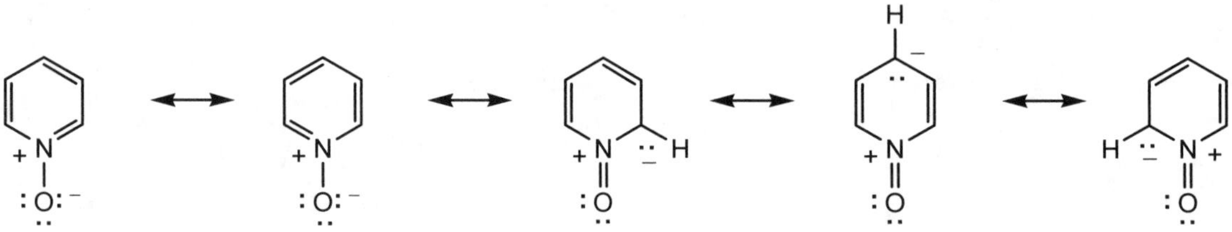

For the intermediates that are formed after pyridine-*N*-oxide reacts with an electrophile at the 2-, 3-, or 4-position, we see that only when reaction occurs at C2 and C4 are there contributors that have a single charge (shown in color below). Otherwise, the resonance forms will have three charges (two positive and one negative). As to why reactions occur at C4 rather than C2, steric effects between the incoming electrophile and the oxygen atom attached to nitrogen probably play a role.

reaction at C2:

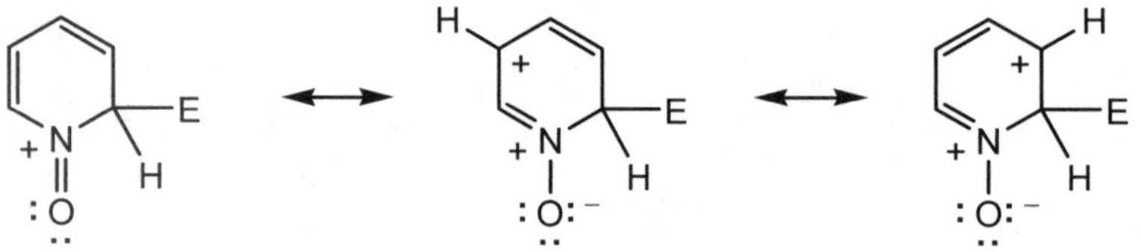

reaction at C3:

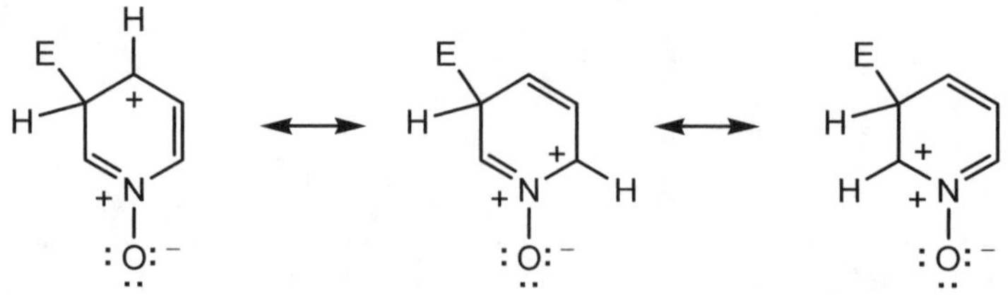

reaction at C4:

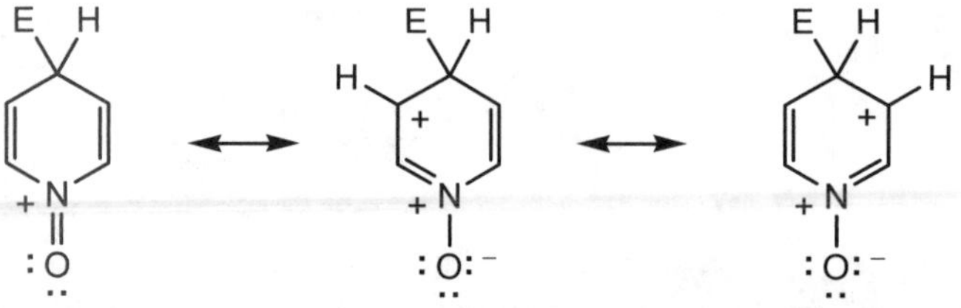

25.25. Pyridine–N–oxide undergoes alkylation at its oxygen atom by an S_N2 process (step 1). A nucleophile (cyanide ion in this exercise) subsequently adds to the π bond(s) at C4 or C2 (step2), and electrons flow toward the positively-charged nitrogen atom, the electrophilic center. Methanol is subsequently eliminated to regenerate the aromatic π system of the pyridine ring (step 3).

25.26. To assign chemical shifts to the resonances of the protons in pyridine, consider the resonance forms of the molecule as well as the splitting patterns. A proton attached to a carbon atom with a positive charge in an important resonance form tends to be deshielded, and its NMR signal appears farther downfield. A proton attached to a carbon atom with a negative charge in a resonance form tends to be more shielded, so its peak appears farther upfield.

The signal for H_c should be the farthest downfield because its positive charge is closer to the negative charge on the nitrogen atom, which will make these forms contribute more to the resonance hybrid. The signal for H_a should be farther downfield than the one for H_b.

The splitting patterns reflect the proximity of the different protons to each other, although the magnitudes of coupling constants are often small in heterocycles. The integrated intensity can also be used to make the correct assignments. The following assignments are the expected results:

1H, triplet δ 7.65

2H, triplet δ 7.25

2H, doublet δ 8.60

25.27. 3-Methylthiophene undergoes electrophilic substitution reactions at C2 and C5. The methyl group will hinder C2 somewhat, so reactions will occur preferentially at C5. Hydrogenation reactions are normally poisoned by the presence of a sulfur atom in organic compounds, and aromatic rings are not susceptible to catalytic hydrogenation at room temperature anyway.

25.27. (continued)

a. The electrophile in this reaction is Br⁺.

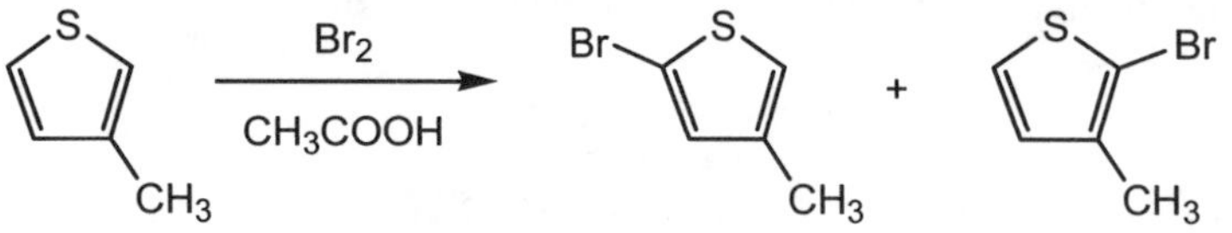

b. The electrophile in this Friedel-Crafts acylation reaction is PhCO⁺.

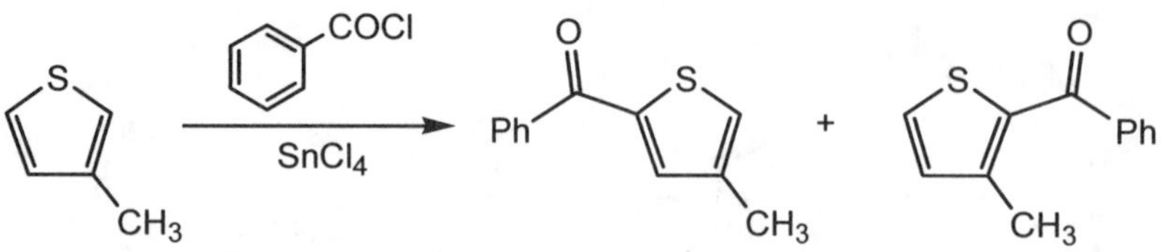

c. No reaction occurs with these reagents under these mild conditions (heat and high pressures of hydrogen would be required).

d. The electrophile in this reaction is HSO₃⁺.

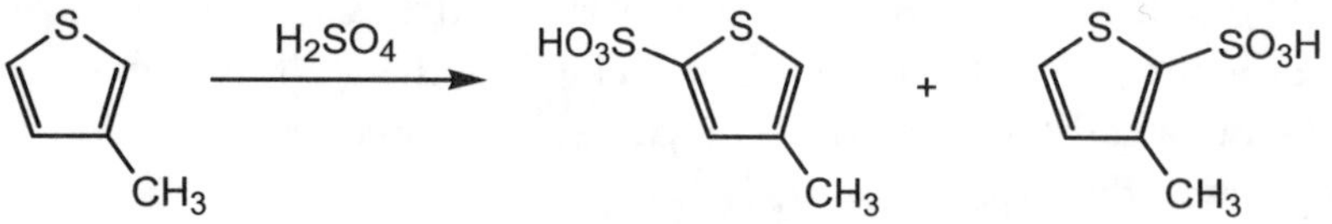

e. The electrophile in this reaction is NO₂⁺.

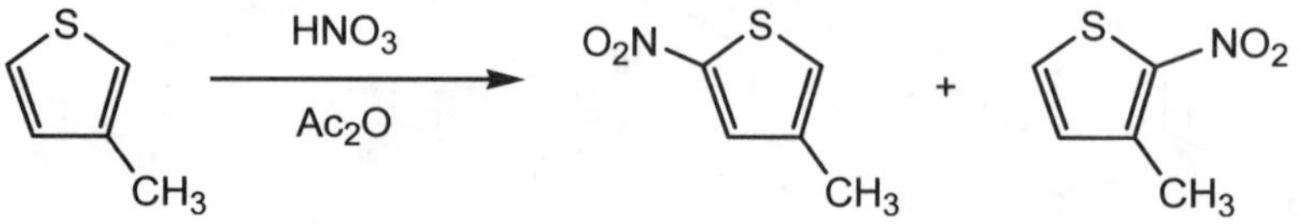

25.28. To predict the structures of the products in each of the following transformations, learn the details of each reaction type, which can be found in the reaction summary section of the chapter. In some of these transformations, reactions from previous chapters are also included.

a. Indole is a reactive heterocycle in electrophilic substitution reactions, and the incoming electrophile reacts at C3. The purpose of the hydroxide ion is to make sure that no acid is present, which in some cases will protonate a nitrogen-containing heterocycle.

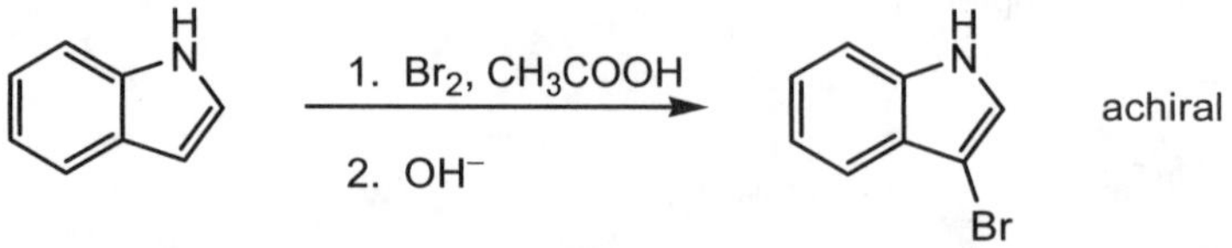

b. A strong base such as LDA deprotonates many heterocycles, especially if a proton is attached to the carbon atom between two heteroatoms. The resulting carbanion adds to the carbonyl group of the aldehyde in a crossed aldol reaction. Acid workup yields the racemic alcohol product.

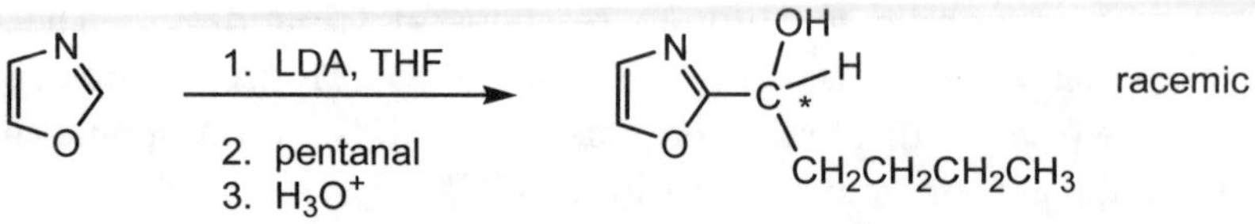

25.28. (continued)

c. Pyridine will undergo electrophilic substitution reactions, and the presence of a methoxy group will enhance this process. The electrophile reacts at C3, which is ortho to the strongly activating methoxy group. The purpose of the hydroxide ion is to make sure that no acid is present, which would protonate the nitrogen atom of pyridine.

$$OCH_3 \qquad \xrightarrow[\text{2. OH}^-]{\text{1. Br}_2,\ \text{CH}_3\text{COOH}} \qquad OCH_3,\ Br \qquad \text{achiral}$$

d. Pyridine can act as a nucleophile with reactive alkyl halides.. A salt is obtained as the product.

$$\xrightarrow{\text{BrCH}_2\text{COOEt}} \qquad Br^- \qquad \text{achiral} \qquad \overset{+}{N}\text{—CH}_2\text{COOEt}$$

e. The protons of a methyl group attached adjacent to the nitrogen atom of an aromatic heterocycle are often acidic, so a carbanion derivative can be formed in step 1. Addition of this carbanion to the aldehyde carbonyl group followed by hydrolysis yields the benzylic alcohol as the product. Under acidic conditions, elimination occurs to produce the alkene. The aqueous base in the last step neutralizes the acid used in step 3.

$$\xrightarrow[\text{2. PhCHO}]{\text{1. LDA, THF}} \qquad \xrightarrow[\text{4. OH}^-]{\text{3. H}_3\text{O}^+} \qquad \text{achiral}$$

f. An imidazole ring is alkylated at one of its nitrogen atoms when its conjugate base is treated with an alkyl halide. Subsequent treatment with LDA and an alkyl halide leads to alkylation at the 2-position. Hydrolysis of the methoxymethyl group, which is equivalent to the acetal functional group, yields the product 2-methylimidazole. The aqueous base in the last step neutralizes the acid used in step 3.

$$\xrightarrow[\text{2. ClCH}_2\text{OCH}_3]{\text{1. NaH, DMF}} \qquad \underset{\text{CH}_2\text{OCH}_3}{} \qquad \xrightarrow[\substack{\text{3. H}_3\text{O}^+,\ \Delta \\ \text{4. OH}^-}]{\substack{\text{1. LDA, THF} \\ \text{2. CH}_3\text{I}}} \qquad \text{—CH}_3 \quad \text{achiral}$$

25.29. Benzimidazole uses the same orbitals as those used to construct the imidazole ring.

a. The benzene portion has p orbitals that overlap with the π bonds of the heterocyclic ring. The NH nitrogen atom provides two electrons to the π system; the N atom has its unshared pair perpendicular to the π system.

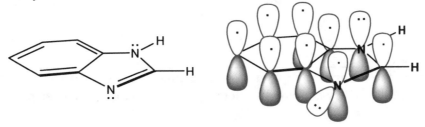

25.29. (continued)

b. Like imidazole itself, the NH group is acidic, so it can be converted to a nucleophile using a strong base. Alkylation then takes place via the S$_N$2 mechanism.

c. An amino group of *o*-phenylenediamine can participate in an acid-base reaction with the carboxylic acid, but it can also add to the carbonyl group of the carboxylic acid (step 1). A proton is transferred to produce the amino *gem*-diol (step 2).

Transfer of a proton from one of the OH groups to the amino group accompanies carbonyl group formation (step 3), and protonation of the oxygen atom yields a carbocation (step 4).

The amino group intercepts the carbocation (step 5), and after a proton transfer step (step 6), dehydration yields the benzimidazole derivative (step 7).

25.30. In forming the oxazole ring, the oxygen atom of formamide displaces chloride ion by an S$_N$2 process (step 1), which is followed by a proton transfer step to create a carbocation (step 2).

25.30. (continued)

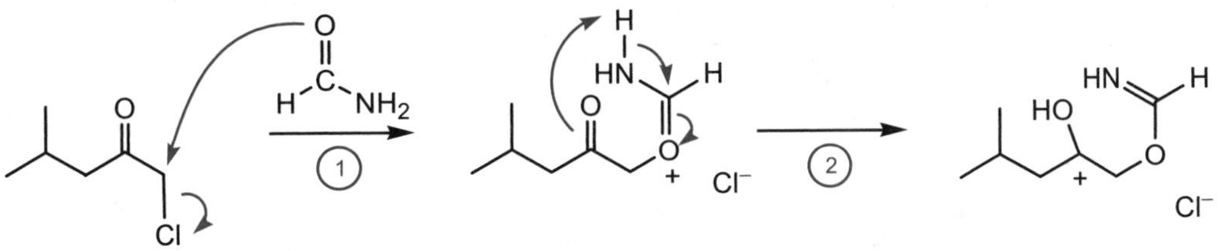

Next, the NH group intercepts this carbocation (step 3), which is followed by another proton transfer reaction (step 4).

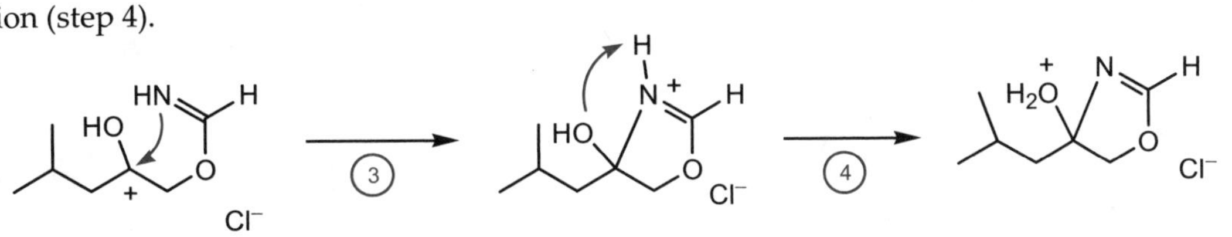

Finally, dehydration produces the oxazole ring.

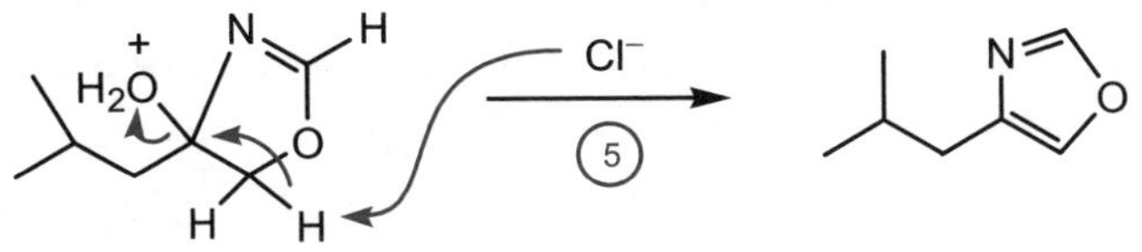

25.31. Many synthetic pathways that are used to make heterocycles rely on making a bond to a heteroatom (often nitrogen), which can be done using S$_N$2 or Michael reactions. With the constraint that the starting material must have six or fewer carbon atoms, many heterocyclic rings will not need to be prepared from acyclic precursors. The starting compounds are shown in color.

a. The imidazole ring has fewer than six carbon atoms, so it is one of the starting materials. A Michael reaction is used to create the C–N bond.

Retrosynthesis

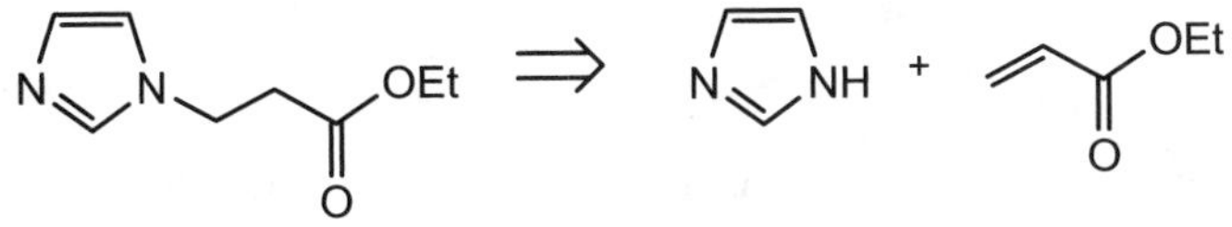

Synthesis

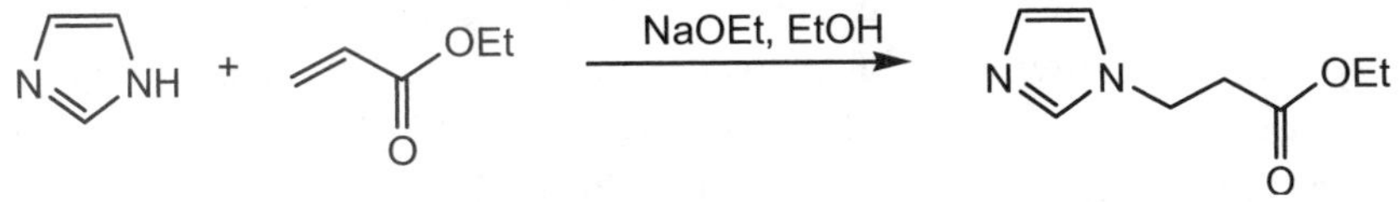

b. Both pyridine and piperidine have fewer than six carbon atoms, so these cyclic compounds constitute the needed starting materials. Substitution of a chloride ion occurs after converting the amine to its conjugate base using butyllithium.

Retrosynthesis   *Synthesis*

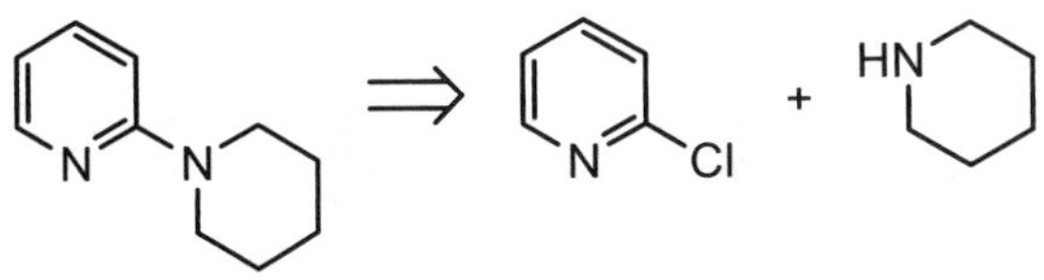

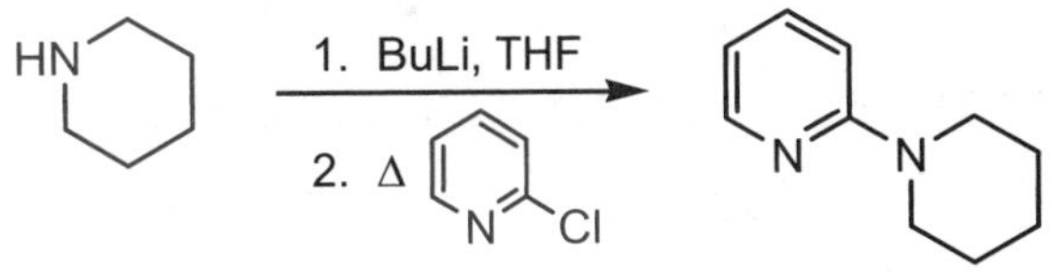

25.31. (continued)

c.   Pyridine has fewer than six carbon atoms, so it constitutes one starting material. The acetyl group is attached via the Friedel Crafts acylation reaction, and then a Horner-Emmons reaction is used to make the carbon–carbon double bond.

Retrosynthesis

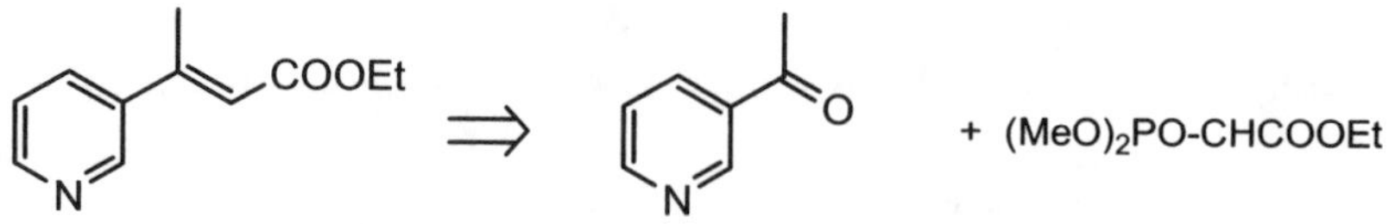

Synthesis

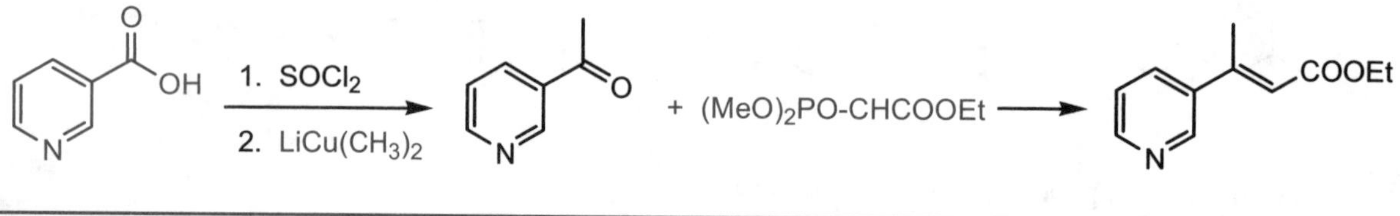

25.32. The NMR spectrum of **A** reveals the presence of an aldehyde group (signal at δ 9.7) and there are three protons in the aromatic region. This compound cannot be a benzene derivative because it contains only five carbon atoms, so it is most likely an aromatic heterocycle. The given reactions can be summarized as follows:

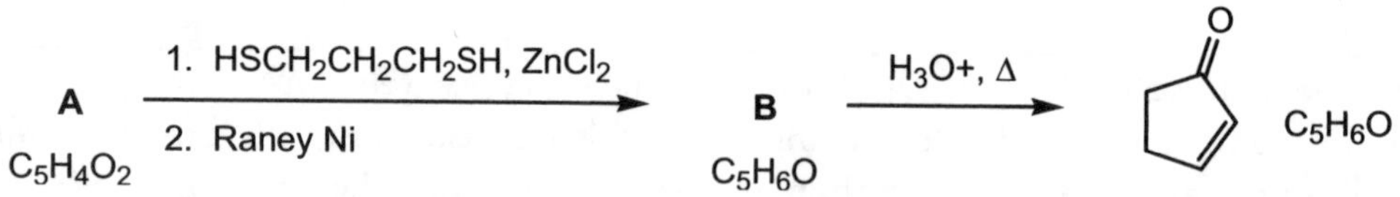

The formula suggests that compound **A** is a furan/aldehyde, so two possible structures are **I** and **II**, below.

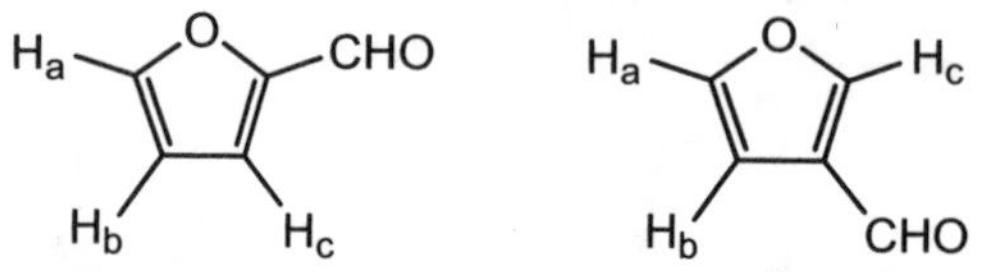

If structure **I** is correct, then the first series of reactions leads to removal of the carbonyl group via thioacetal formation and desulfurization. Hydrolysis yields the keto aldehyde, which can then undergo an acid-catalyzed crossed aldol reaction to form the unsaturated ketone. The splitting patterns in the proton NMR spectrum do not permit assignment of the correct structure because the coupling constants are much smaller than those observed for derivatives of benzene and the triplet and two doublets patterns *expected* for compound **I** appear instead as a doublet and two singlets.

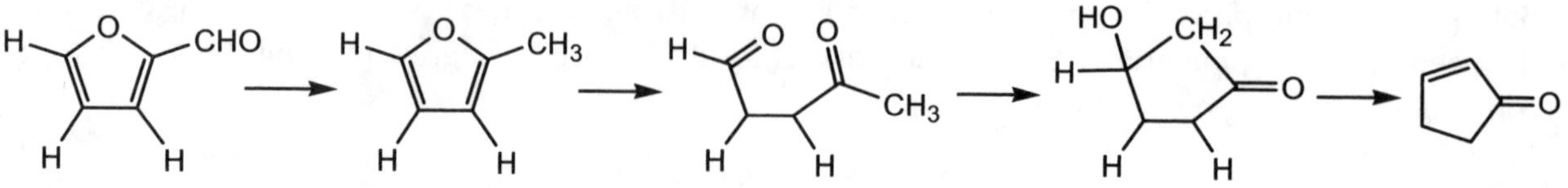

POLYMERS AND POLYMERIZATION

26.1 The monomer used to make each polymer can be deduced by looking for the repeating unit within the polymer structure. For an addition polymer (one made from an alkene and consisting of a carbon chain), the name of the polymer has the prefix "poly" in front of the monomer's name.

For condensation polymers (those with functional groups or heteroatoms within the chain), consider how that particular functional group can be made and adjust the structure of the repeating unit accordingly.

a. The name of this polymer is **poly(1,2-difluoroethene)**.

1,2-Difluoroethene

b. The name of this poly**ester** is **poly(3-hydroxybutanoate)**. An ester is made from the carboxylic acid and alcohol functional groups by removing a molecule of water, so we have to add the elements of water to the repeating unit to identify the monomer's structure. Notice that the name must also reflect the functional group that is present in the polymer. Even though the monomer's name is 3-hydroxybutanoic acid, the polymer's name ends in "oate" to denote the presence of the ester functional groups.

3-Hydroxybutanoic acid

26.2 The configuration of each stereogenic center is specified in the usual way after assigning the priority to each group attached to the stereogenic carbon atom (two specific carbon atoms are labeled to show how these assignments are made). Configurations of the atoms in the isotactic polymer occur in two blocks, (*R*) at one end of the chain, (*S*) at the other. At the midway point of the chain, the priorities of the two groups will switch, so the configurations will be opposite one another at equivalent points of the chain.

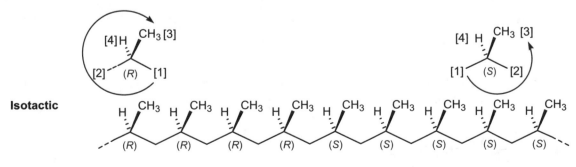

26.2 (continued)

Configurations in a syndiotactic polymer alternate, except at the very middle of the chain where the priorities switch, which results in two adjacent carbon atoms having the same configuration.

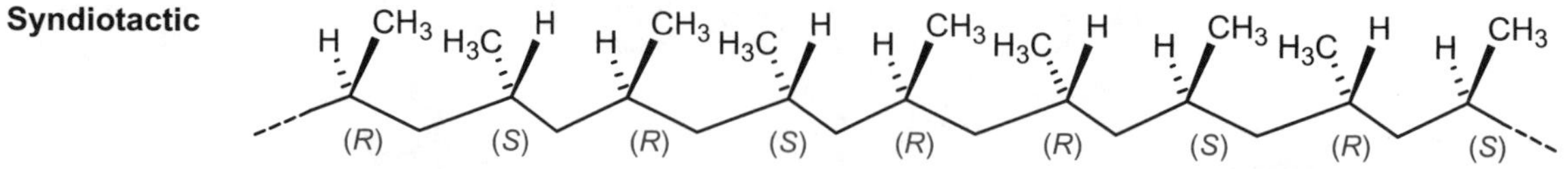

Syndiotactic

In a very long chain, the overall effect of chiral centers is negligible because an approximately equal number of each configuration exists. Even though the carbon atoms toward the middle of a chain are chiral (four different groups), the effect of this chirality on the rotation of plane polarized light is probably minimal. For instance in C1000–CH(CH₃)–C1001, the two chains are close enough in length as to be essentially the same.

26.3 Isobutylene reacts with H[BF₃OH] to form the *tert*-butyl carbocation (step 1). Another molecule of isobutylene usually intercepts this carbocation in the growing polymer chain (step 2).

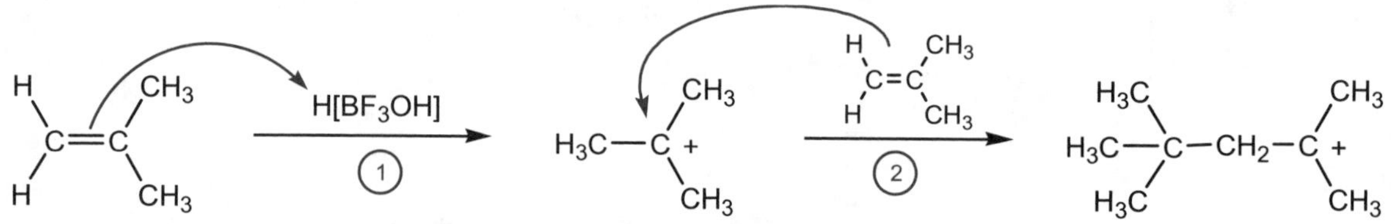

When isoprene reacts with the carbocation instead, the more highly substituted double bond reacts preferentially because it is the more nucleophilic of the two double bonds (step 3). Moreover, a tertiary carbocation is formed in this step.

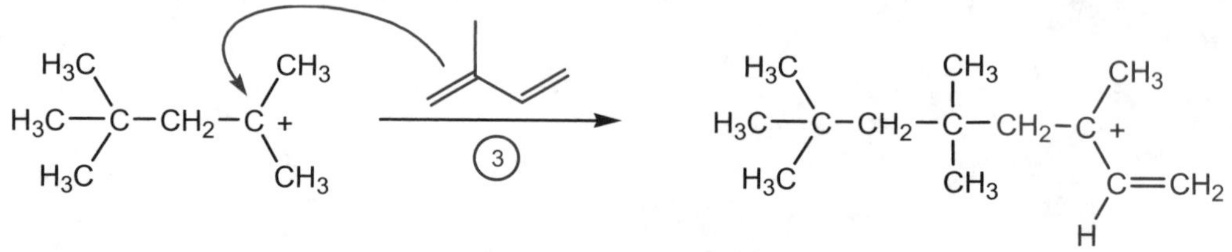

The carbocation formed in step 3 is also stabilized by resonance, as shown below. The cation that reacts in the next step is the one that leaves the more highly substituted double bond in place. Hence the primary allylic carbocation is the one that is intercepted by the next molecule of alkene (step 4).

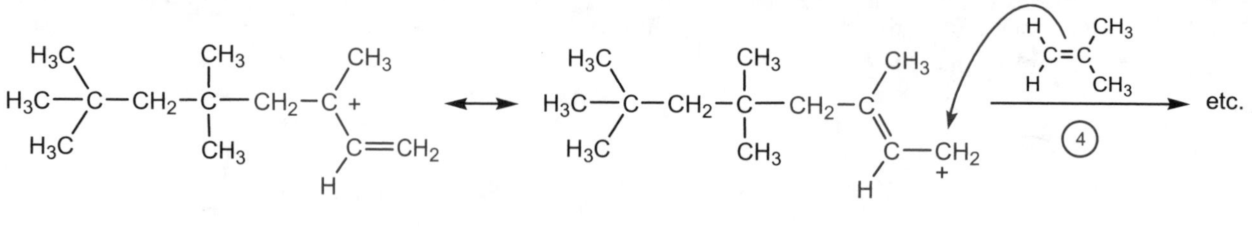

26.4 The free-radical polymerization of methyl methacrylate starts with formation of the isobutyronitrile radical from AIBN:

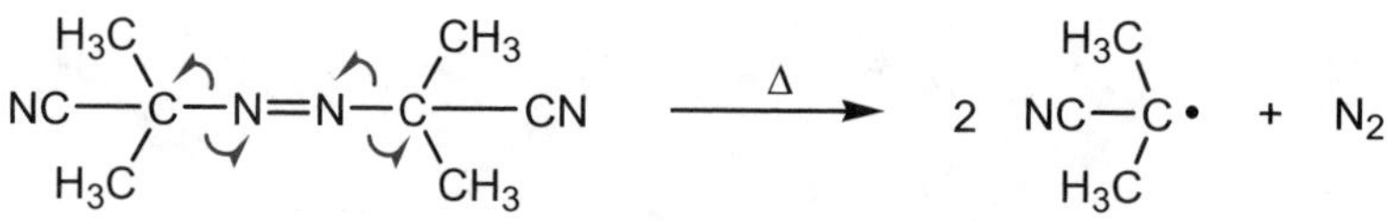

26.4 (continued)

The isobutyronitrile radical adds to the double bond of methyl methacrylate (step 1), and the resulting radical is intercepted by another molecule of methyl methacrylate (step 2). This addition reaction is repeated many times to form the polymer product [Init = $(CH_3)_2C(CN)-$].

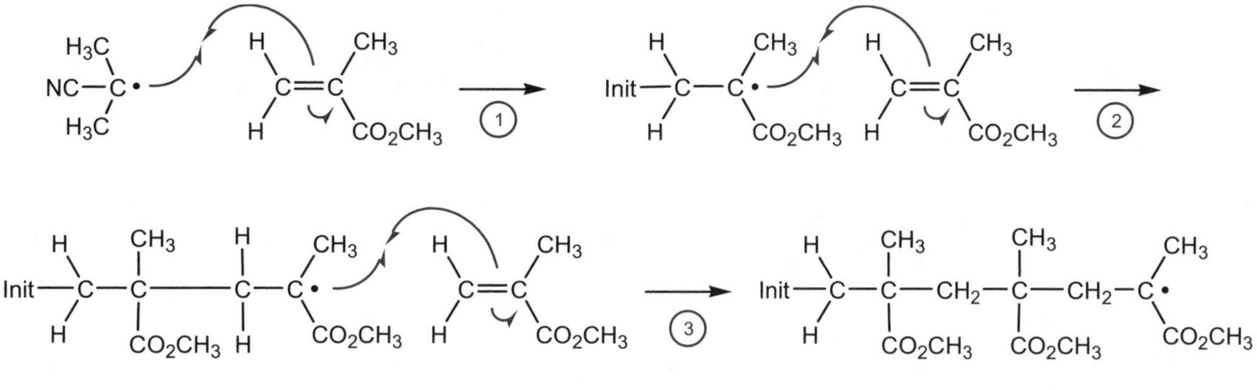

26.5 Any amino acid with a nucleophilic group in the side chain that can readily lose a proton is a candidate for initiating the polymerization reaction of methyl cyanoacrylate. The list includes amino acids with the alcohol, thiol, carboxylic acid, phenol, amine, and imidazole group in their side chains.

Serine	$-CH_2-OH$	**Threonine**	$-CH(-OH)CH_3$
Cysteine	$-CH_2-SH$	**Tyrosine**	$-C_6H_4OH$
Aspartic acid	$-CH_2COOH$	**Lysine**	$-CH_2CH_2CH_2CH_2NH_2$
Glutamic acid	$-CH_2CH_2COOH$	**Histidine**	$-CH_2(C_3H_3N_2)$

For example, the cysteine thiolate group can add as shown below (step 1). The resulting carbanion, which is resonance-stabilized by the cyano and ester carbonyl groups, adds to another molecule of cyanoacrylate (step 2), and this latter process continues with formation of the polymer.

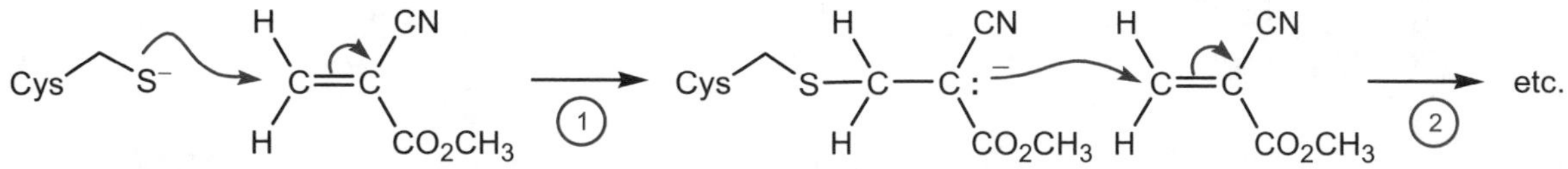

26.6 The monomer needed to make the given polyester can either be the hydroxy ester $HO(CH_2)_7COOCH_3$ or the eight-membered ring lactone shown below. With the addition of water, the lactone will be converted to the hydroxy acid (steps 1 and 2). The alcohol group adds to another molecule of the lactone to form the next ester group (steps 3 and 4), and this process continues.

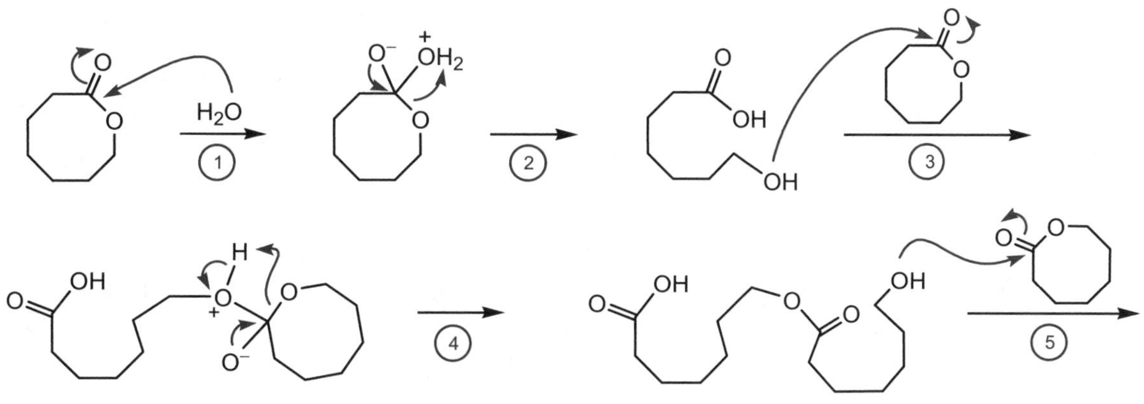

26.6 (continued)

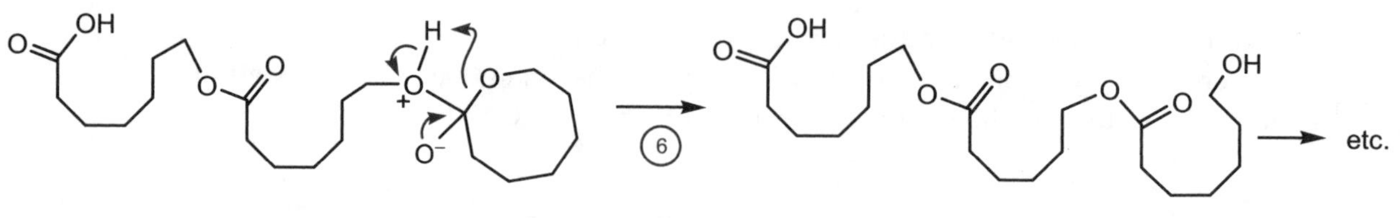

etc.

26.7 The preparation of dimethyl carbonate from phosgene and methanol takes place by successive sets of addition-elimination steps.

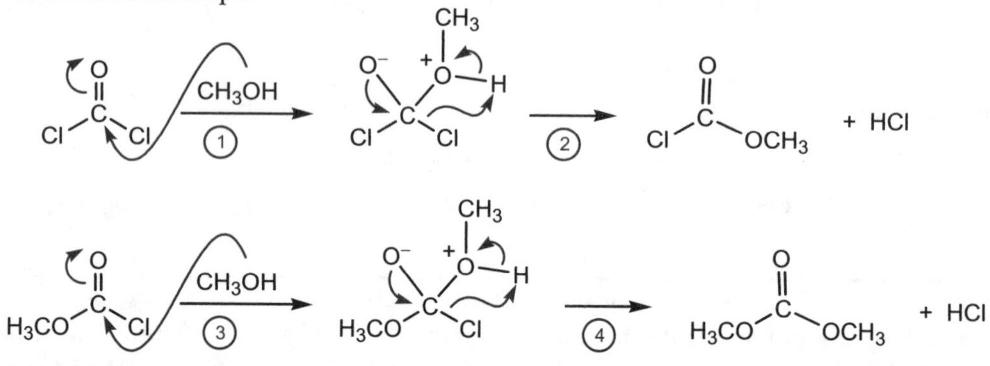

+ HCl

+ HCl

26.8 Epichlorhydrin is a difunctional molecule that reacts with two equivalents of alcohol to form a bis(ether). The epoxide ring is first opened by a molecule of alcohol (step 1), and the resulting alkoxide ion displaces the chloride ion to regenerate the epoxide ring (step 2). An acid-base reaction yields the methoxy epoxide. A second ring opening process subsequently forms the bis(ether) product.

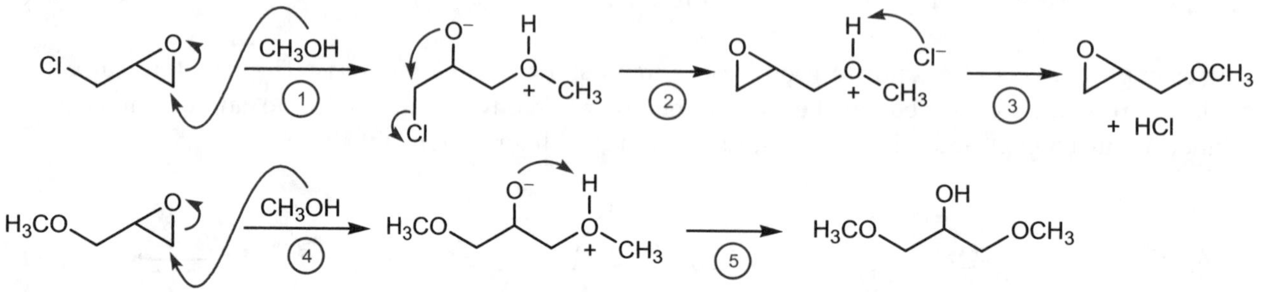

+ HCl

When epichlorhydrin reacts with 1,3-dihydroxypropane, the polymer that is formed has two ether groups in addition to the secondary alcohol group.

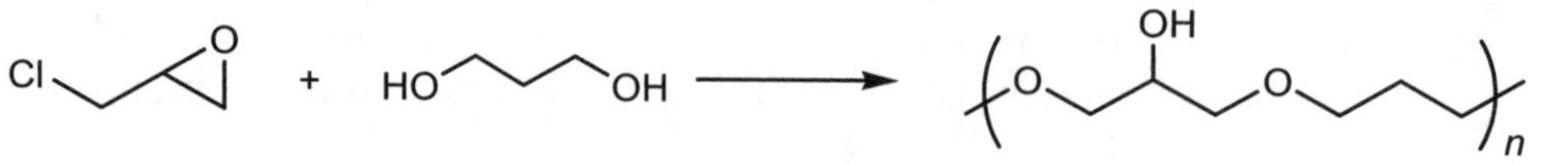

26.9. In the reaction between 1,2-diaminoethane and the aromatic bis(isocyanate), one amino group of the diamine (represented below as RNH$_2$) adds to the carbonyl portion of one isocyanate group. Regeneration of the carbonyl group accompanies proton transfer to form one urea group (step 2).

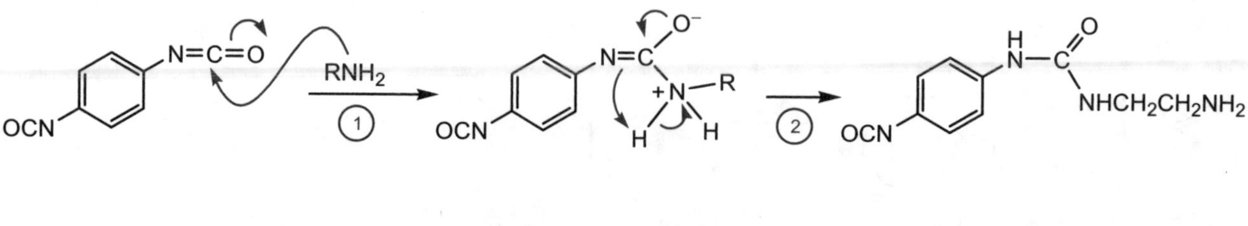

26.9. (continued)

In the subsequent polymerization process, these same two reactions (addition and proton transfer) between the amino and isocyanate groups are repeated over and over.

26.10. To prepare a polymer-supported reagent that can be used to convert carbonyl groups to epoxide rings, the thiophenol polymer is first methylated to form the dimethylsulfonium substituent. Butyllithium is then added to generate the sulfur ylide, which is attached to the polymer backbone.

To prepare an epoxide from a ketone, the polymer is simply treated with the carbonyl compound. The polymer can be recovered simply by filtration, and upon treatment with methyl iodide and butyllithium, the ylide reagent can be regenerated.

26.11. The hydrolysis reaction of the illustrated imine yields the primary amine and benzaldehyde.

The aminomethylpolystryrene reagent can act as a scavenger toward the aldehyde byproduct by formation of its imine derivative with the insoluble polymer. Filtration separates the polymer from the desired amine product.

26.12. To draw the structure of a random copolymer, link the given monomers together in any order you choose. The substituent side chains or functional groups are normally separated by an odd number of carbon atoms.

a. **Viton**: hexafluoropropene and 1,2-difluoroethylene

b. **Nitrile rubber**: 1,3-butadiene and acrylonitrile

c. **SBR**: styrene and 1,3-butadiene

26.13. When a polymer is formed using an acid catalyst, a carbocation is formed at each stage. This carbocation must be highly stabilized in order for the process to succeed.

a. When styrene undergoes the acid-catalyzed polymerization reaction, protonation of the double bond creates a benzylic cation (step 1), and another molecule of styrene intercepts this carbocation to form a second benzylic cation (step 2)

26.13. (continued)

Reaction of the carbocation with an additional molecule of styrene continues over and over, each step producing another carbocation.

b. When methyl vinyl ether undergoes acid-catalyzed polymerization, protonation of the double bond creates a cation that is stabilized by the neighboring oxygen atom (step 1), and another molecule of the alkene intercepts this carbocation to form a second such cation (step 2). Reaction of the carbocation with additional molecules of the alkene continues over and over, each step producing another stabilized carbocation.

26.14. The free-radical polymerization of tetrafluoroethylene to form Teflon begins with the addition of an initiating radical to the alkene double bond (step 1). The resulting carbon-centered radical is then intercepted by another molecule of tetrafluoroethylene (step 2).

26.14. (continued)

This addition process continues until the chain is terminated by the combination of two radicals.

26.15. Sodium hydroxide reacts with β-propiolactone by adding to the carbonyl group (step 1). Regeneration of the carbonyl group creates the 3-hydroxpropanoate ion (step 2), which can add to another molecule of the lactone (step 3). The carbonyl group is regenerated with formation of another alcohol group and carboxylate ion (step 4), which can react again and again.

The IR spectrum of this polymer will be dominated by the bands expected for the ester functional group. The proton NMR spectrum will have two signals for the protons of the different methylene groups. (NMR spectra of polymer samples often are broadened because the molecules cannot tumble freely in solution, so the given result assumes that good quality spectra can be obtained.)

νC=O ~ 1735 cm^{-1}

νC–O~ 1200 cm^{-1}

δ ~ 4.1, 2H, triplet

δ ~ 2.2, 2H, triplet

26.16. Bisphenol A can be prepared from phenol and acetone in the presence of strong acid. The transformation starts with the activation of the carbonyl group by protonation. The phenol ring intercepts this cation (step 2), and rearomatization yields the benzylic alcohol (step 3). These two steps amount to a Friedel-Crafts alkylation reaction. Protonation of the benzylic alcohol group of this initial product (step 4) is followed by dissociation of a molecule of water (step 5). The resulting carbocation is intercepted by another molecule of phenol (step 6), and rearomatization yields the product (step 7). This second stage is also an example of the Friedel-Crafts alkylation reaction.

26.16. (continued)

26.17. The transesterification reaction between diphenyl carbonate and bisphenol A to make Lexan starts with addition of one of the OH groups of bisphenol A (abbreviated below as ArOH) to the carbonyl group of diphenyl carbonate (step 1). The tetrahedral intermediate collapses and displaces a molecule of phenol (step 2). These two steps are repeated with a second molecule of bisphenol A (ArOH in steps 3 and 4). Continuation of these addition-elimination steps creates Lexan.

26.18. The self-condensation reaction between three molecules of urea to form melamine occurs with the removal of three molecules of water by a series of addition-elimination reactions.

26.20. (continued)

A series of possible steps in this overall transformation are shown in the following scheme, although other sequences are also feasible.

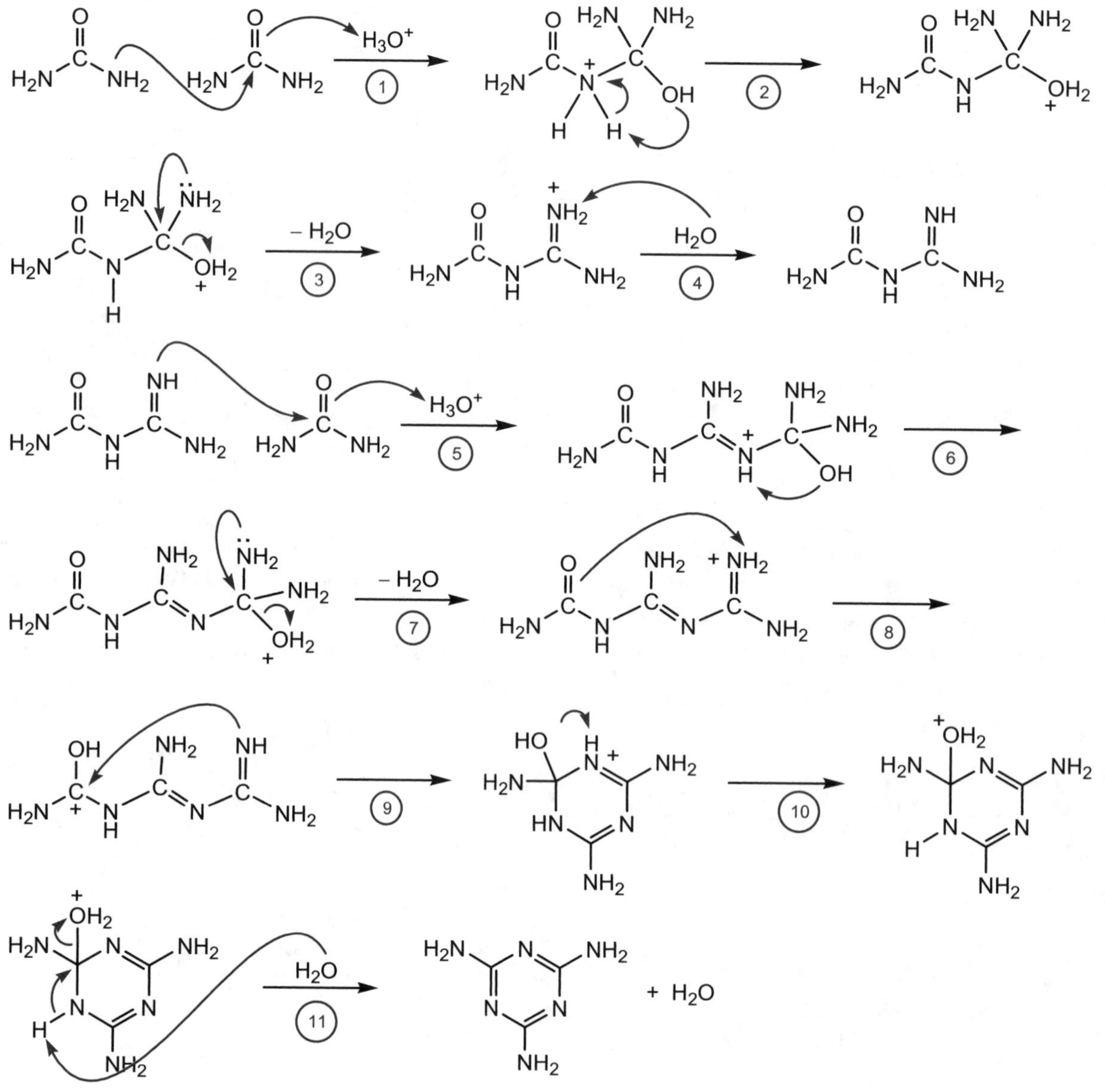

When melamine is treated with formaldehyde, the amino groups add to the carbonyl double bond (step 1), and a proton transfer step generates the hydroxymethyl derivatives.

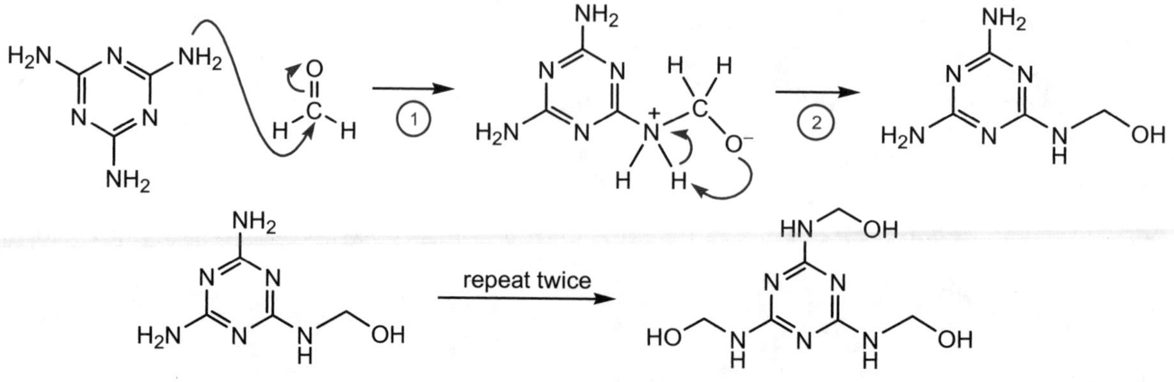

26.20. (continued)

Once the tris(hydroxymethyl)melamine has been formed, an acid catalyst reacts with the OH groups to form molecules of water, and the amino groups from neighboring molecules add to these methylene imine groups to form the crosslinks. There are several ways for the individual steps to occur; the scheme shown below is one possibility.

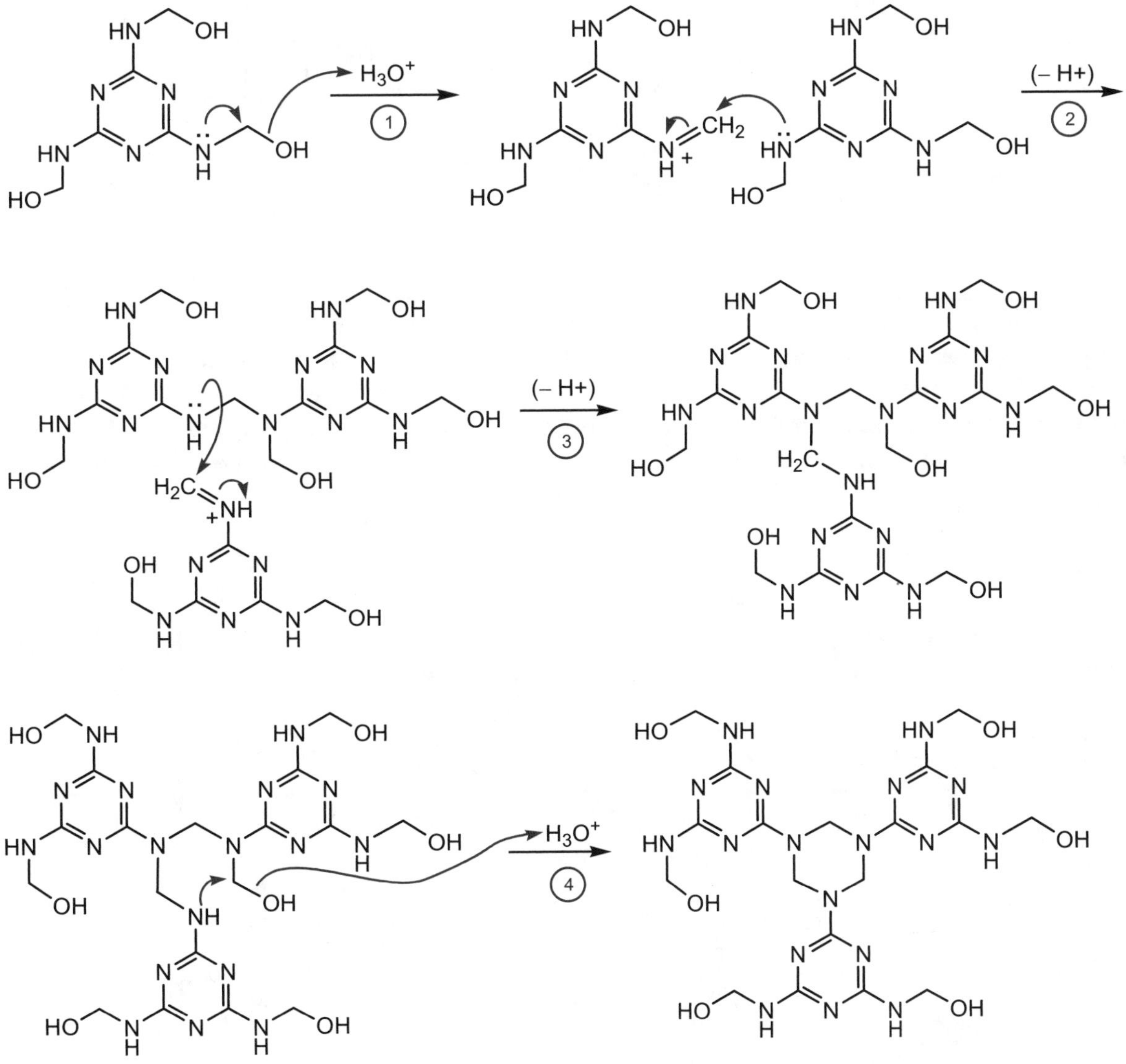

26.19. Qiana has amide groups that are made by the condensation polymerization reactions between a diacid and a diamine.

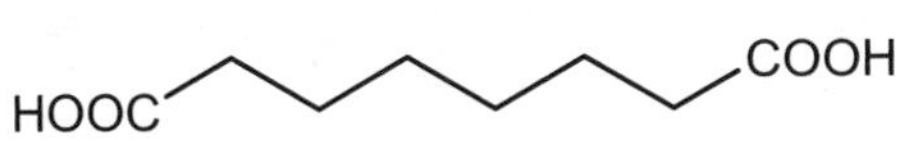

Octanedioic acid 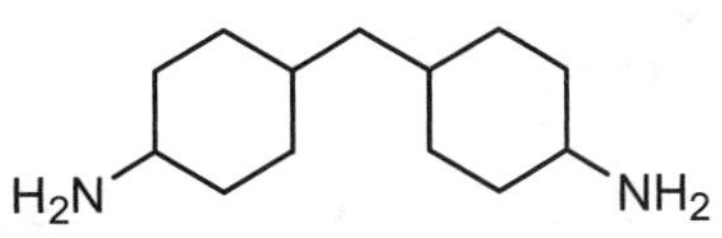   Bis(4-aminocyclohexyl)methane

Hydrogen bonds can readily form between the amide NH and carbonyl groups of neighboring polymer chains.

26.19. (continued)

26.20. The urea/formaldehyde polymer shown in this exercise can be formed by addition of a urea amino group to the carbonyl group of formaldehyde (step 1), which forms *N*-hyroxymethylurea after transfer of a proton (step 2). Under the influence of an acid catalyst (step 3), a C=N double bond is formed (step 4), and the amino group of another molecule of *N*-hyroxymethylurea adds to this C=N bond with help from proton activation (step 5). These addition steps can occur over and over to form the polymer.

26.21. The structure of poly(vinyl alcohol) reflects the structure of the monomer from which its name derives.

26.21. (continued)

This polymer cannot be made by polymerization of vinyl alcohol because the latter is unstable, existing as its more stable tautomer, acetaldehyde.

If *vinyl acetate* is made to polymerize, then poly(vinyl acetate) can be formed. Hydrolysis of the ester groups yields the alcohol groups of poly(vinyl alcohol).

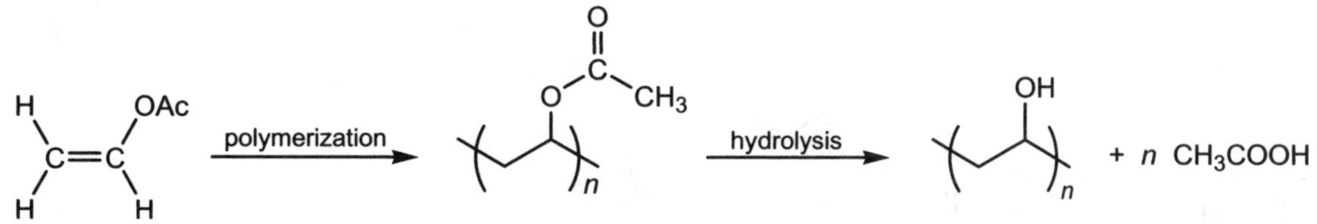

The mechanism of the polymerization process follows the same steps shown in the solution to Exercise 26.14.

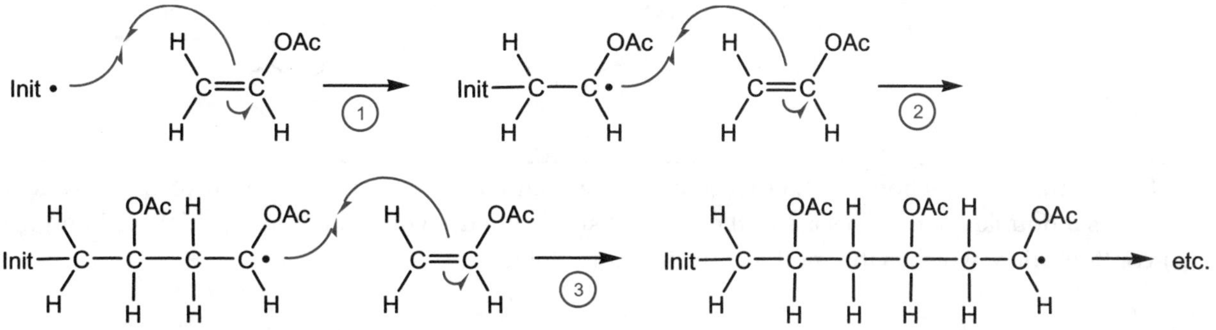

26.22. Natural rubber has the structure shown in the following scheme. Because it has double bonds, we would expect allylic radicals to form readily because they will be stabilized by resonance delocalization of the unpaired electron.

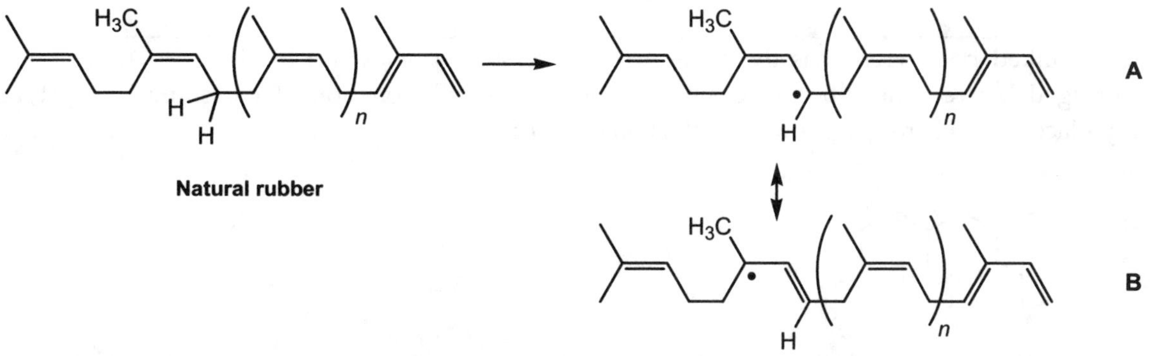

Natural rubber

26.23. The nucleophilic aromatic substitution reactions used to make polycarbonates rely on the presence of halogen atoms that are ortho or para to potent electron-withdrawing groups. In this transformation, the carbonyl group is apparently sufficient to activate the fluoride ion for displacement reactions. In each case, the nucleophilic addition-elimination mechanism operates. Once two ketone molecules have been linked by the carbonate group, additional substitution reactions occur to replace the fluorine atoms and form the polymer.

26.23. (continued)

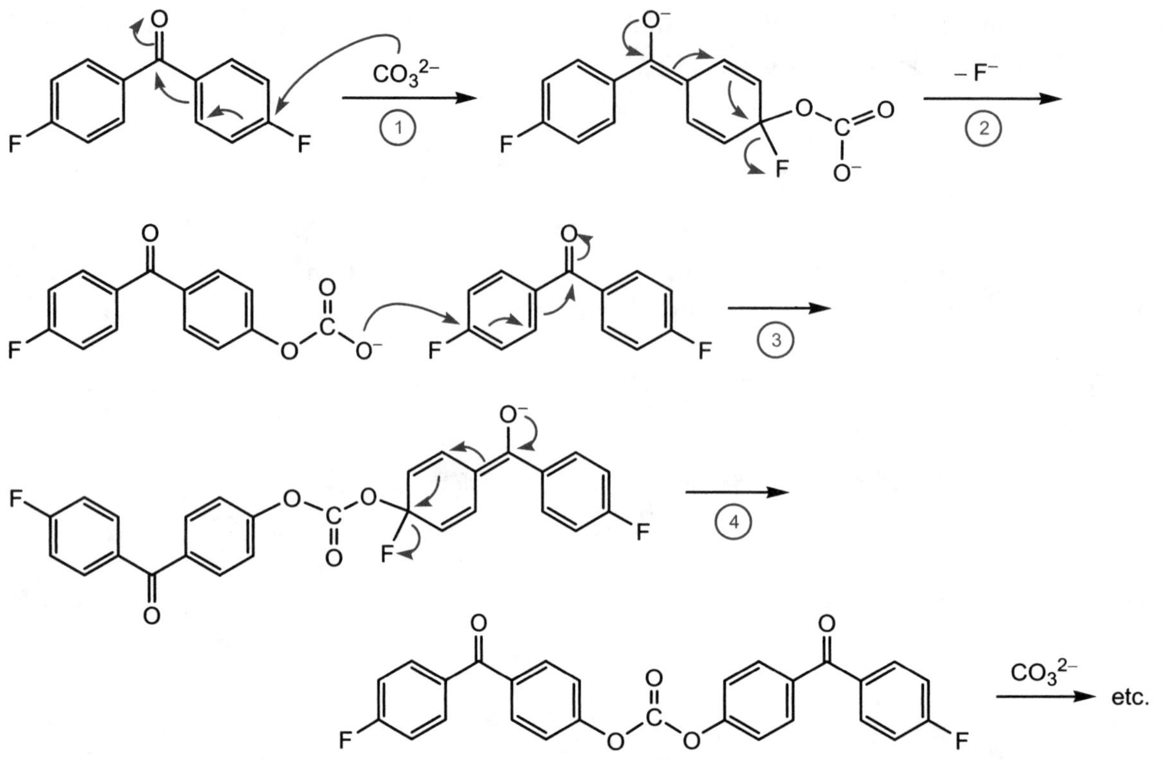

26.24. Polymerization reactions of 1,3-dienes occur by coupling between the termini of the π system, which leaves a double bond at the site of the original single bond. For Neoprene, this double bond has the (Z)-configuration, which is thermodynamically more stable.

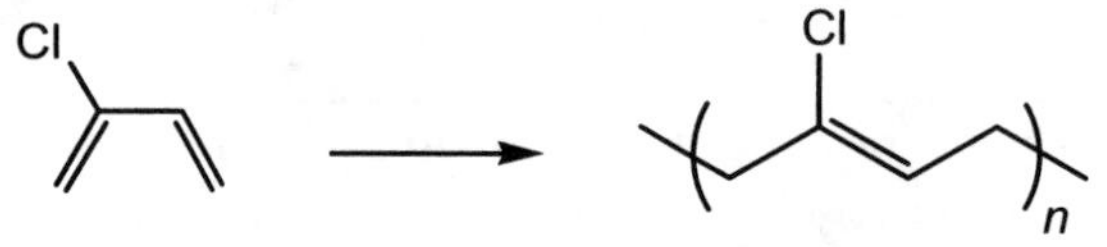

2-Chloro-1,3-butadiene   **Neoprene**

26.25. Crosslinked polystyrene has the general structure shown below (in black), with 1,4-phenylene groups bridged between neighboring chains of polystyrene. Sulfonation of crosslinked polystyrene randomly places –SO₃H groups (shown in color) on some of the rings.

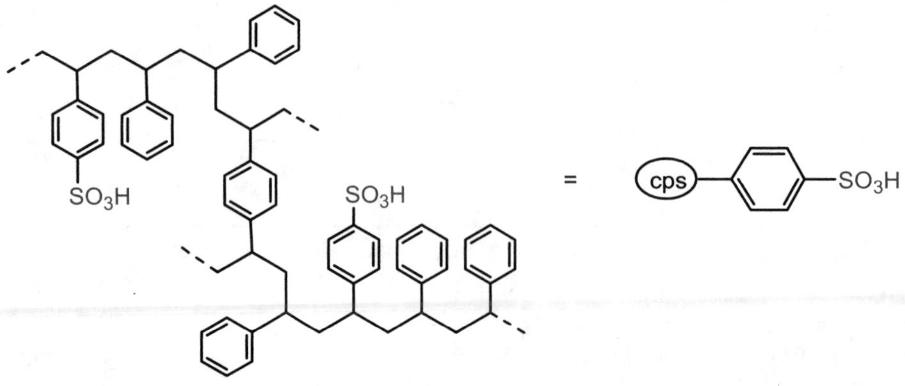

26.25. (continued)

Any reaction that requires the use of a strong acid for catalysis should be catalyzed by this acidic polymer. Some examples include the following:

Acetal formation

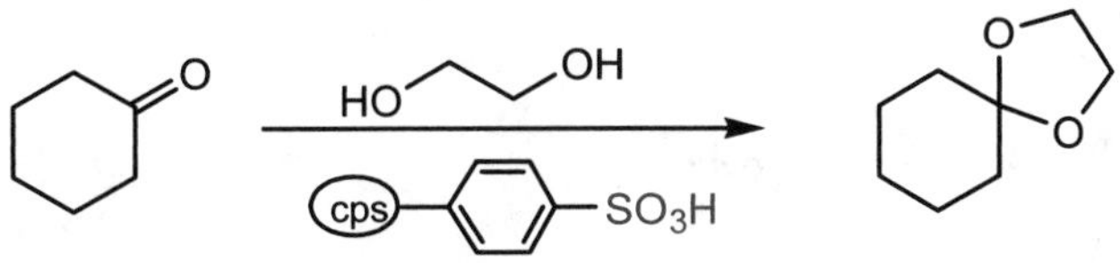

Esterification

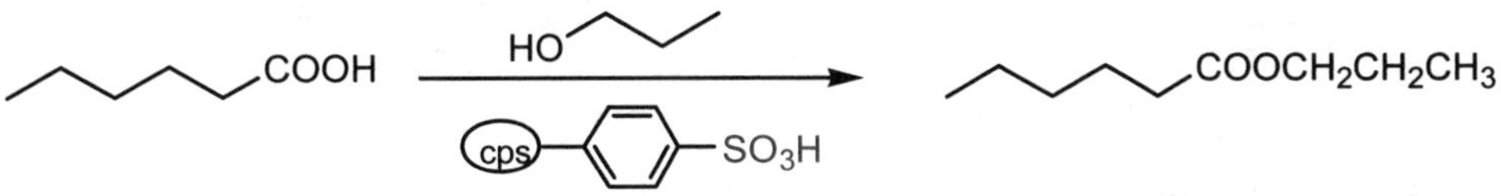

Aldol reaction

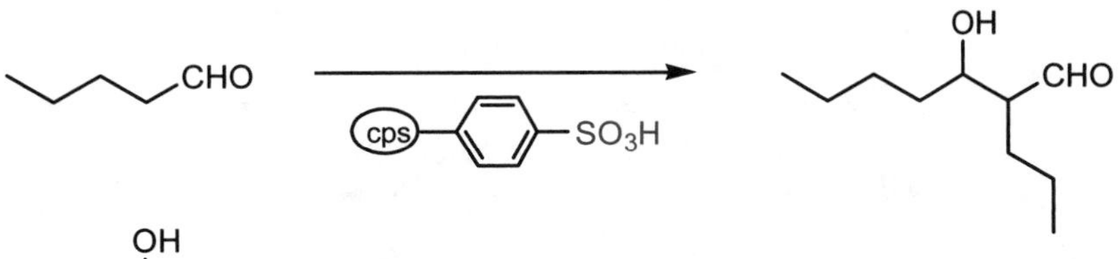

E1 dehydration

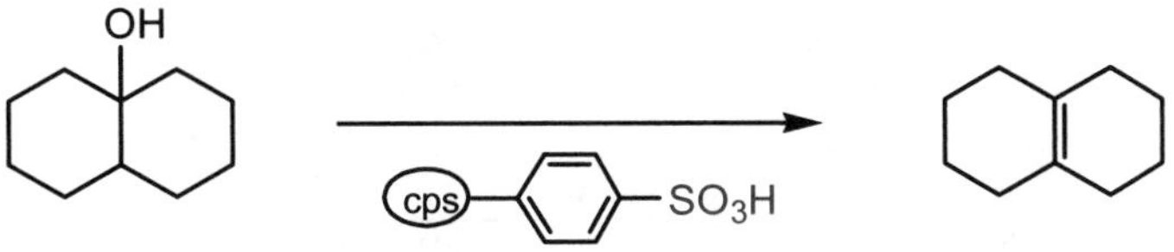

26.26. Wilkinson's catalyst has the formula RhCl(Ph₃P)₃. One or more of the phosphine groups in this catalyst could be provided by the polymeric phosphine ligand, and structures such as the ones shown below are possible. If the polymer is flexible enough, then two phosphine groups could bind to the metal ion.

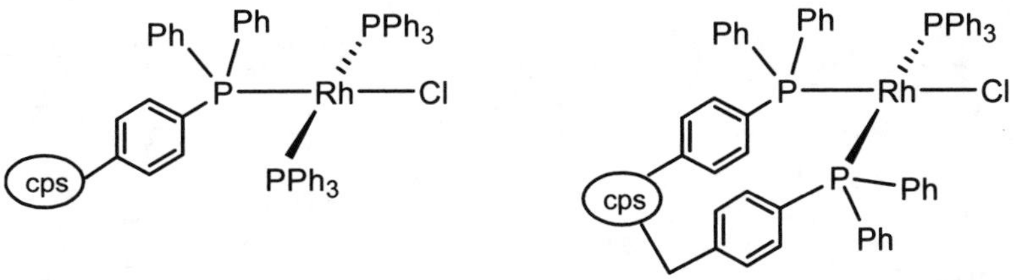

26.27. A crosslinked copolymer between styrene, divinylbenzene, and 4-vinylpyridine would have the following general structure.

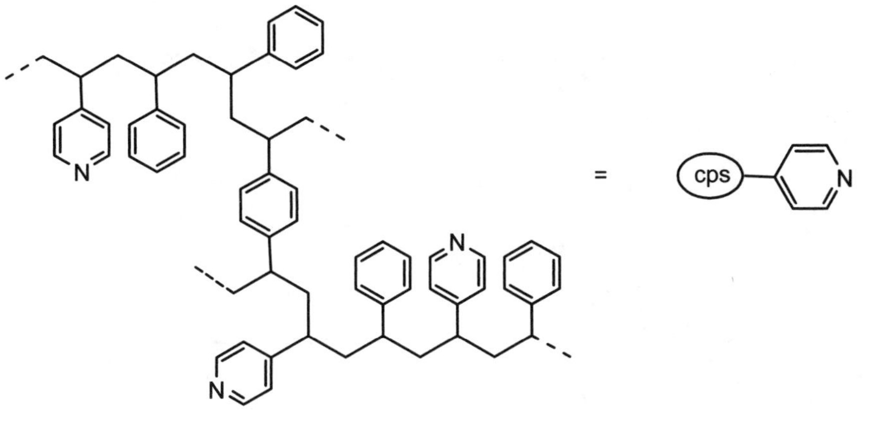

26.27. (continued)

A variety of chromium(VI) oxide-based reagents can be attached to or associated with such a polymer, as shown in the following equations. Recall that chromium oxide–pyridine and PCC are two oxidizing agents that find common use in synthesis (Section 11.4c).

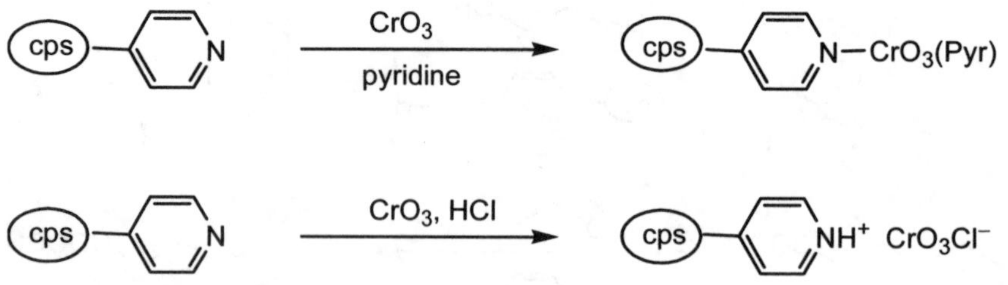

The use of these reagents for oxidation reactions that convert a primary alcohol to an aldehyde are shown in the following equations:

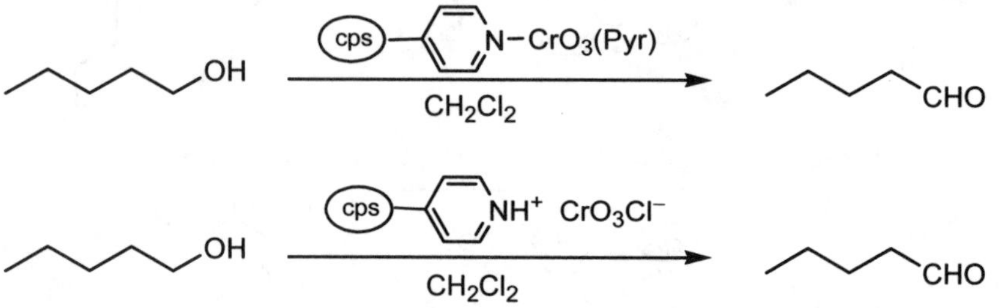

26.28. The Swern oxidation makes use of dimethylsulfoxide and oxalyl chloride to oxidize alcohols to carbonyl compounds (Section 11.4b). Sulfoxide polymers can be made as shown in the following scheme. A thioether is prepared, and then $NaIO_4$ oxidizes the sulfide to the sulfoxide group.

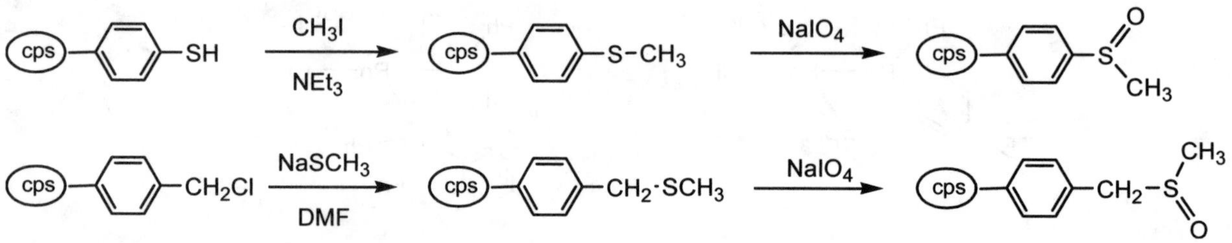

The Swern oxidation of aldehydes and ketones is then performed in the normal fashion with these reagents.

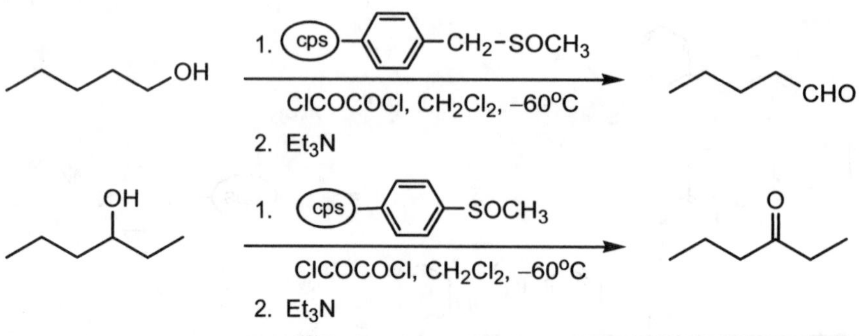

AMINO ACIDS, PEPTIDES, AND PROTEINS

27.1. The amino acids that have two chiral centers are isoleucine and threonine. Their structures (L-isomer), along with their IUPAC names, are shown below:

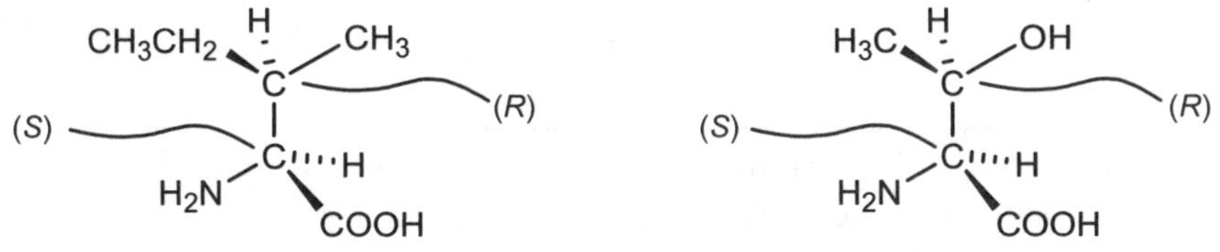

L-Isoleucine

(2S,3R)-2-amino-3-methylpentanoic acid

L-Threonine

(2S,3R)-2-amino-3-hydroxybutanoic acid

27.2. The structure of an amino acid at a particular pH value reflects the predominant form of each ionizable group. Use the procedure outlined in the solution to Example 27.1 to calculate whether each group (—COOH, —NH₂, —side chain) exists in its protonated (or deprotonated) form to the extent of >50%. If the percentage is greater than half, render the group in that predominate form.

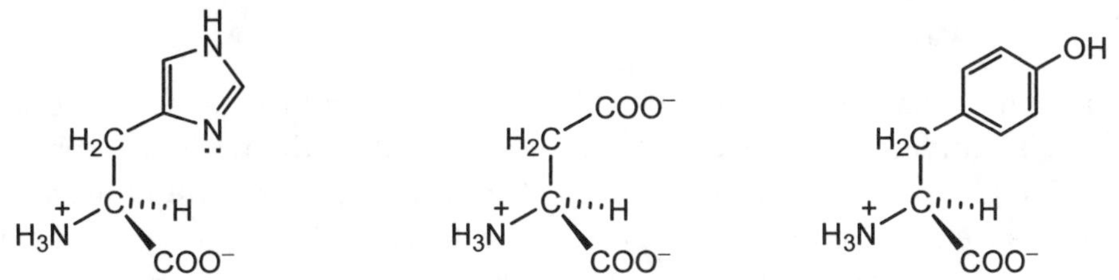

L-Histidine **L-Aspartic acid** **L-Tyrosine**

a. **Histidine.** At pH 7, the zwitterion form (–NH₃⁺ and –COO⁻) predominates and the imidazole group is not protonated (at pH 6 the imidazole ring is protonated to the extent of 50%).

c. **Aspartic acid.** At pH 5, the zwitterion form (–NH₃⁺ and –COO⁻) predominates and the side chain carboxylic acid group is in its conjugate base form (deprotonated).

c. **Tyrosine.** At pH 8, the zwitterion form (–NH₃⁺ and –COO⁻) predominates and the phenolic OH group remains protonated.

27.3. The electron pair on the nitrogen atom of tryptophan is part of the ten π electrons that make the indole ring aromatic. This electron pair, therefore, is not available to engage in hydrogen bonding as an acceptor, because doing so would disrupt the aromaticity of the heterocycle.

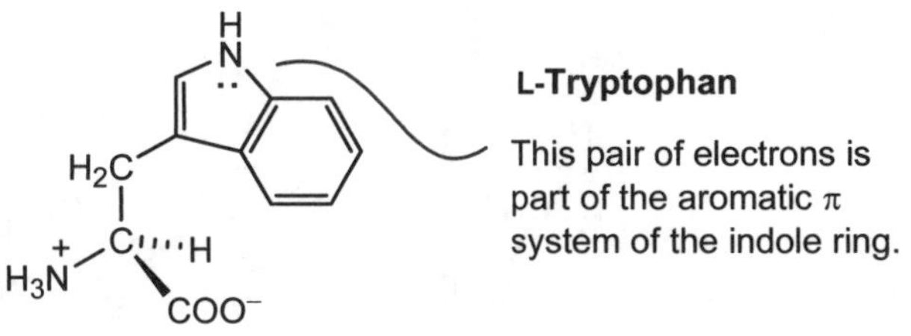

L-Tryptophan

This pair of electrons is part of the aromatic π system of the indole ring.

27.4. The classification of amino acids by type reflects the natures of the side chain groups that are present. An OH group makes a side chain polar, and an amino group is basic. An aldehyde group is only slightly polar and because most of allysine's side chain is hydrocarbon in nature, allysine is likely nonpolar. The zwitterion form of each amino acid is shown below; 5-hydroxylysine will exist with its side-chain amino group protonated at most pH values.

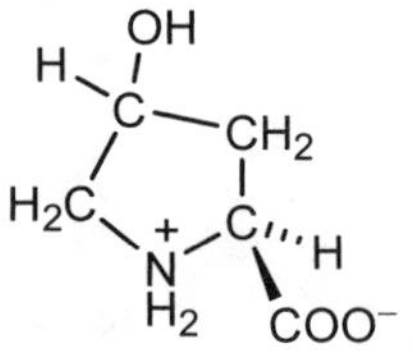

L-4-Hydroxyproline
Cyclic
Polar

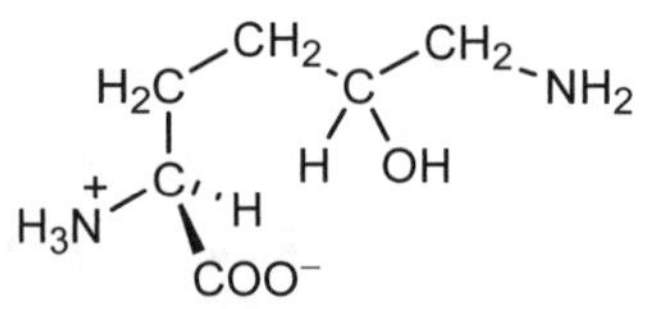

L-5-Hydroxlysine
Basic
Polar

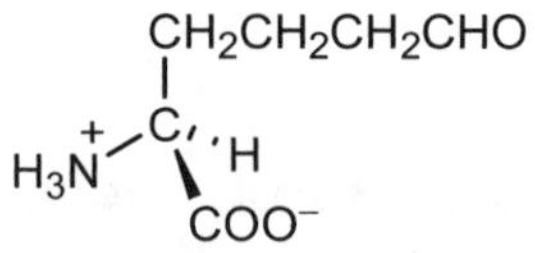

L-Allysine
Nonpolar

27.5. The aldehyde needed to make an amino acid by the Strecker synthesis is determined by considering first what nitrile will be formed. The carbon atom alpha to the cyano group is the aldehyde carbon atom in the retrosynthesis.

a. The aldehyde needed to make valine is a stable molecule and should present no problem in the Strecker reaction.

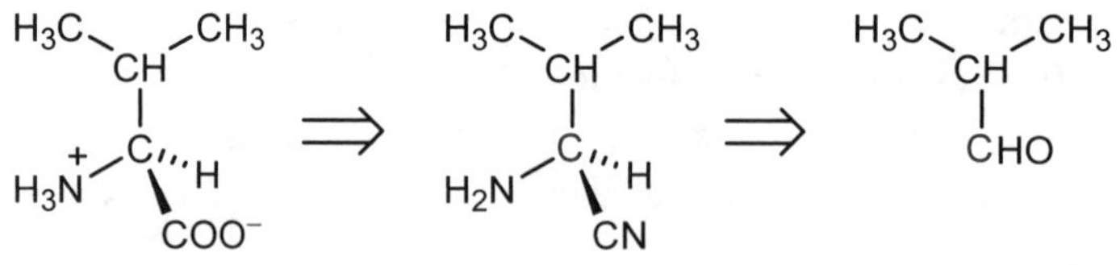

L-Valine

b. The aldehyde needed to make serine may be difficult to work with because it is probably water soluble. It may also undergo self-condensation because both carbonyl and alcohol groups are present. The alcohol group could be protected as its benzyl ether, which would make the molecule more hydrophobic as well as eliminating the presence of the reactive OH group. If the benzyl ether were used, the benzyl group would be removed at the end of the synthesis by means of hydrogenolysis.

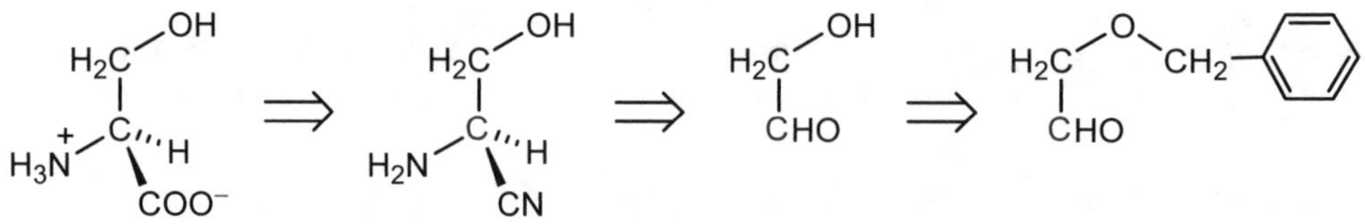

L-Serine

c. The aldehyde needed to make tyrosine is a stable molecule and should present no problem in the Strecker reaction. If the phenol group presented a problem because of its relative acidity, it could be protected as its acetate ester derivative.

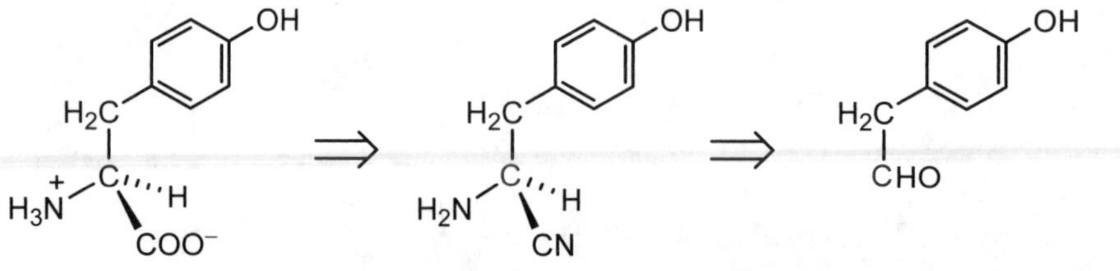

L-Tyrosine

27.6. Each of the amino acids shown in this exercise can be made from the corresponding α-bromo carboxylic acid by its reaction with azide ion followed by hydrogenation.

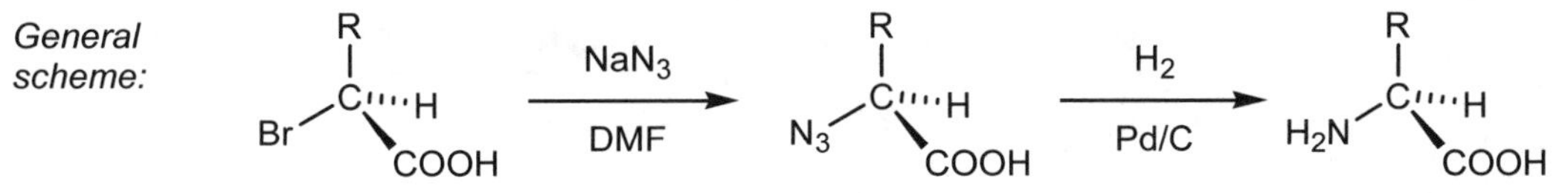

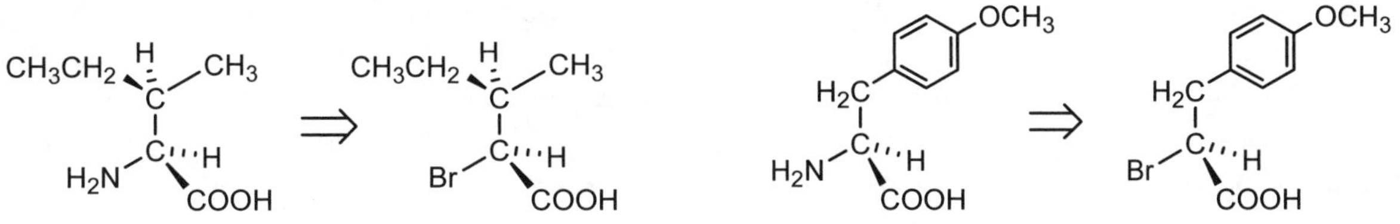

General scheme:

Isoleucine

p-Methoxyphenylalanine

27.7. *tert*-Butyl and benzyl esters are most often prepared by the reaction of the corresponding acid chloride with *tert*-butyl and benzyl alcohol, respectively. This route avoids the use of a strong acid that can generate a carbocation intermediate from the alcohol.

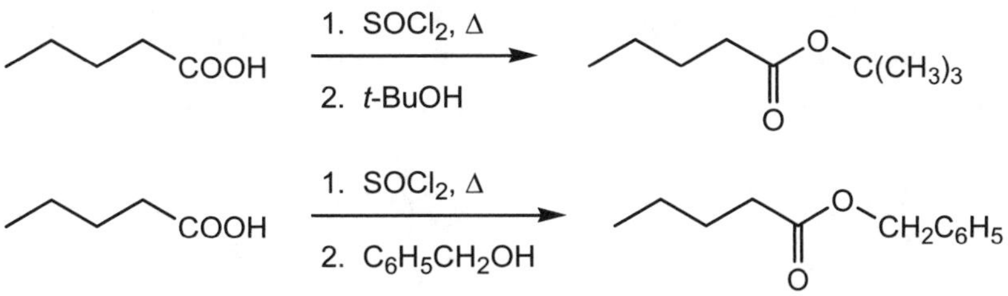

27.8. A *tert*-butyl ester is deprotected (converted to the corresponding carboxylic acid) by its reaction with a strong acid such as trifluoroacetic acid. Protonation of the ester carbonyl group (step 1) yields a cation that dissociates to form the carboxylic acid and the *tert*-butyl carbocation (step 2). The *tert*-butyl carbocation is deprotonated to form isobutylene, which also regenerates the acid catalyst.

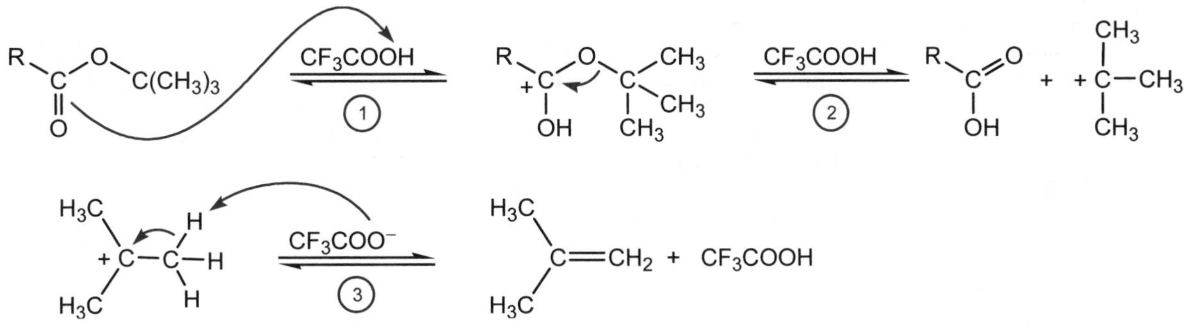

27.9. Converting a chiral glycine enolate ion to an amino acid requires the use of a reactive alkyl halide (methyl, allyl, benzyl, or α–bromo carbonyl). The stereochemistry is set by using the heterocycle shown in the text.

a. This phenylalanine derivative is made by alkylating the enolate derivative of the heterocycle with *p*-bromobenzyl bromide. The heterocycle is cleaved using lithium in liquid ammonia to form the Boc-protected amino acid; the Boc group is removed using trifluoroacetic acid.

27.9. (continued)

b. This α-methyl phenylalanine derivative is made by alkylating the enolate derivative of the heterocycle with benzyl bromide in the first stage, followed by alkylation with methyl iodide in the second. The heterocycle is cleaved with lithium in liquid ammonia to form the Boc-protected amino acid; the Boc group is removed using trifluoroacetic acid. If the opposite stereochemistry were desired, the order of the two alkylation steps would be reversed.

c. This α-methyl aspartic acid derivative is made by alkylating the enolate derivative of the heterocycle with methyl iodide in the first stage, followed by ethyl bromoacetate in the second. The heterocycle is cleaved with lithium in liquid ammonia to form the Boc-protected amino acid as its ethyl ester; both protecting groups (Boc and ethyl ester) are removed with aqueous acid.

27.10. The course of these enantioselective hydrogenation reactions is predicted by comparing the substrate structures with those of the examples given in the text.

a.

b.

27.11. The synthesis of Phe-Ser-Ala is carried out according to the iterative procedure described in the text. First, the starting dipeptide is deprotected by treating it with trifluoroacetic acid. Phenylalanine (protected as its Boc derivative) is then coupled to the dipeptide using DCC. The resulting tripeptide is subsequently deprotected: LiOH and methanol converts the ester to its carboxylate derivative, trifluoroacetic acid removes the Boc group and protonates the carboxylate ion to form the carboxylic acid, and hydrogenolysis removes the benzyl group from the alcohol group of serine.

27.11. (continued)

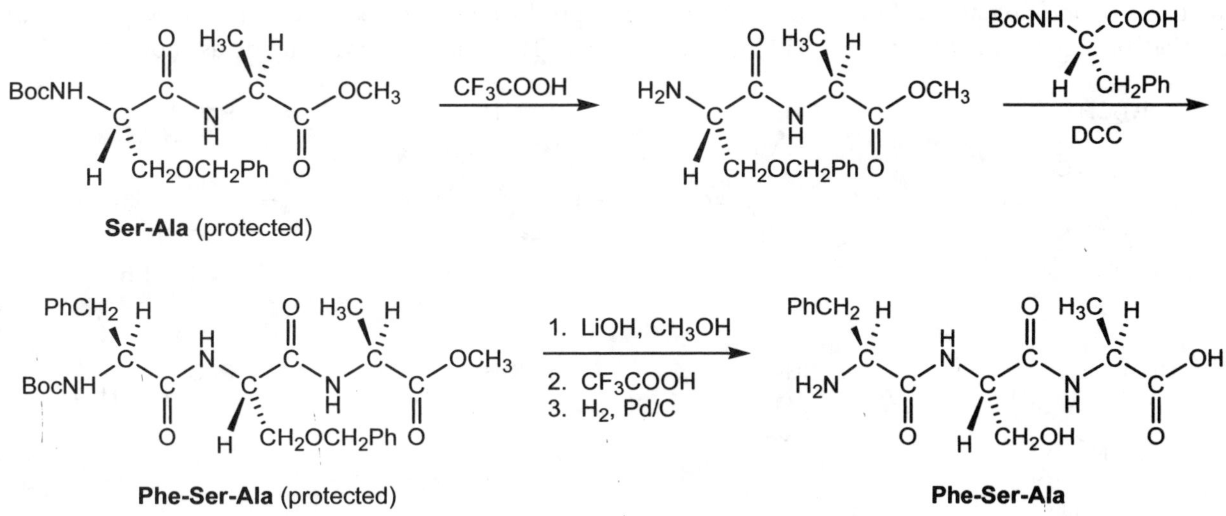

Ser-Ala (protected)

Phe-Ser-Ala (protected)   **Phe-Ser-Ala**

27.12. The overall rearrangement process in the last step of the Edman degradation involves having a nitrogen atom nucleophile replace the sulfur atom attached to the carbonyl group. This requires an addition–elimination mechanism. The carbonyl group is activated by protonation (step 1), and water intercepts the resulting carbocation (step 2). A proton is transferred to the sulfur atom as the C–S bond is broken to form a second carbocation (step 3). This carbocation is then intercepted by the exocyclic nitrogen atom (step 4). Another proton transfer step generates a good leaving group (step 5), and the carbonyl group is regenerated by the loss of a molecule of water (step 6). Finally, a tautomerism process forms the thiocarbonyl group.

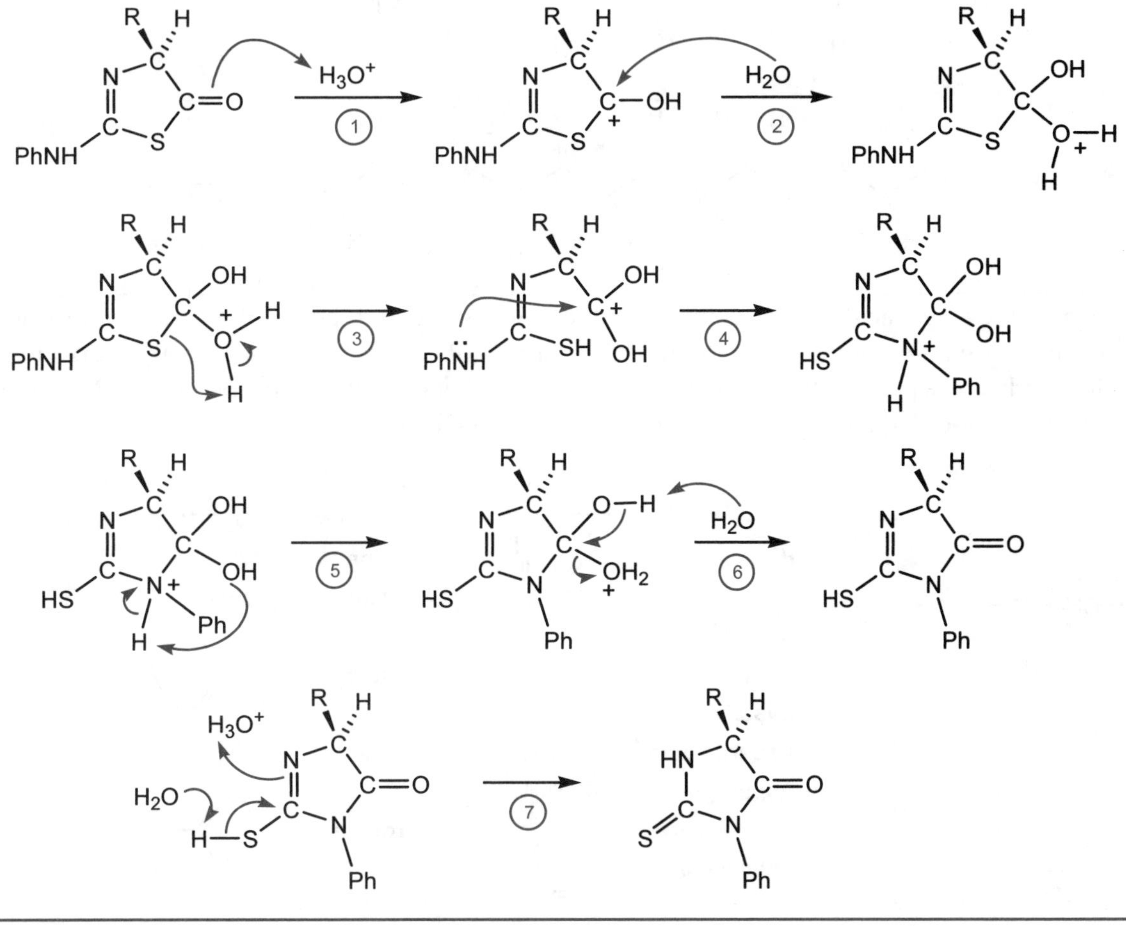

27.13. In the Edman degradation, the structure of each amino acid that is sequentially cleaved is identified by looking at the R group of the heterocycle produced during each round. The R group is the same as the side chain of the amino acid that was cleaved. The assignments are made as shown below:

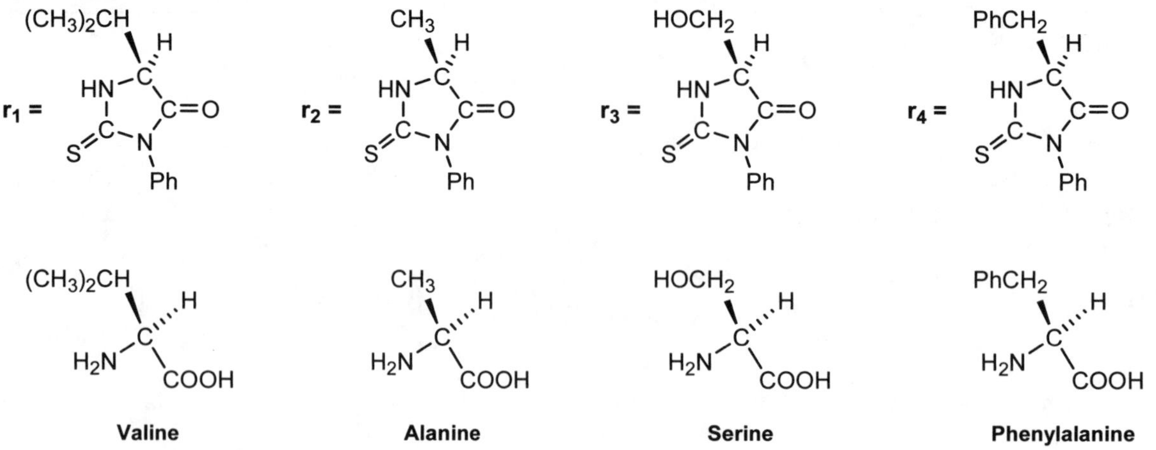

| Valine | Alanine | Serine | Phenylalanine |

27.14. In the antiparallel β-sheet, two strands of protein chains (A and B) run in opposite directions (the N-terminus → C-terminus orientation is indicated by the large arrows in the following structure). This orientation aligns the carbonyl and NH group that form the hydrogen bonds, which are indicated by the dashed lines in color.

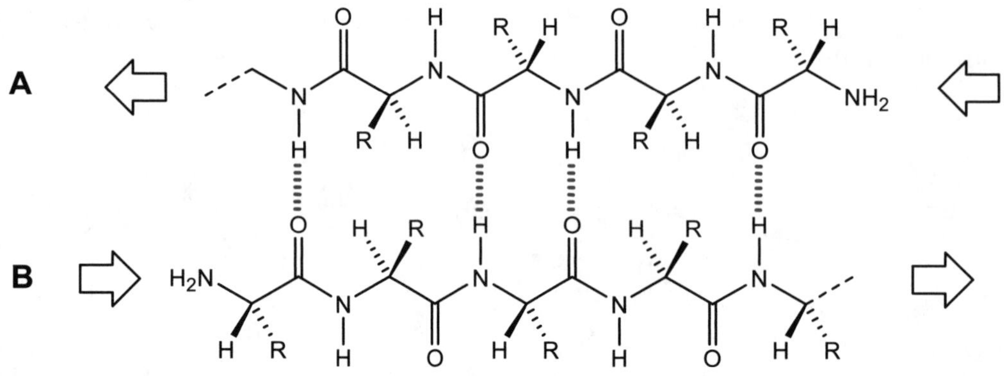

27.15. Learning the amino acid structures in this fashion allows one to recognize the variety of functional groups that are present for a given substructure. The functional groups (or hydrocarbon groups) in this set of amino acids are separated from the alpha carbon atom by a single methylene group. Notice that half of the genetically-coded amino acids have this general side chain structure (that is, $-CH_2X$).

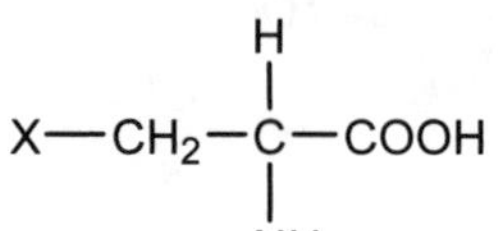

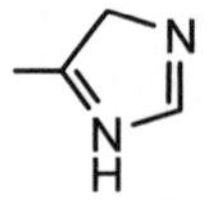

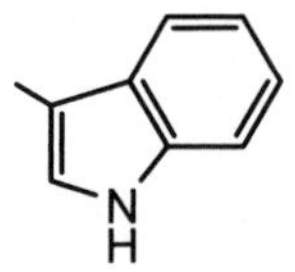

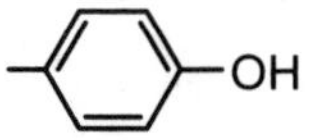

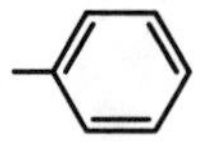

27.16. The functional groups (or hydrocarbon groups) in this set of amino acids are separated from the alpha carbon atom by two methylene groups

	X
Glutamic acid	–COOH
Glutamine	–CONH₂
Methionine	–SCH₃

$$X-CH_2-CH_2-\underset{NH_2}{\overset{H}{\underset{|}{\overset{|}{C}}}}-COOH$$

27.17. The functional groups (or hydrocarbon groups) in this set of amino acids have a branched side chain.

	X
Threonine	–OH
Valine	–CH₃
Isoleucine	–CH₂CH₃

$$CH_3-\underset{|}{\overset{X}{\underset{|}{\overset{|}{CH}}}}-\underset{NH_2}{\overset{H}{\underset{|}{\overset{|}{C}}}}-COOH$$

27.18. Follow the procedure outlined in Example 27.1.

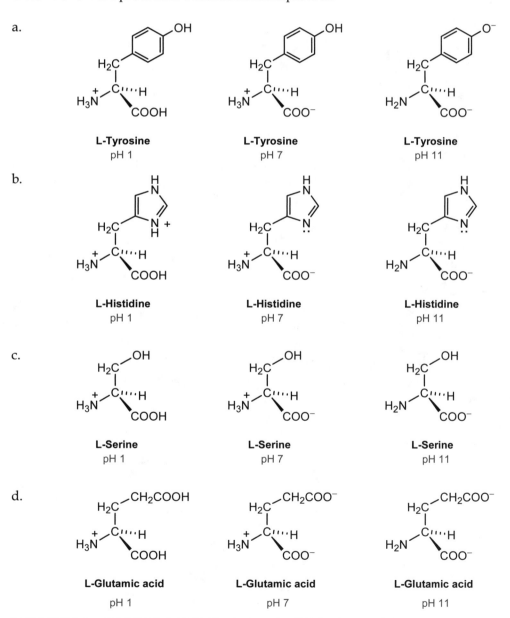

a.

L-Tyrosine
pH 1

L-Tyrosine
pH 7

L-Tyrosine
pH 11

b.

L-Histidine
pH 1

L-Histidine
pH 7

L-Histidine
pH 11

c.

L-Serine
pH 1

L-Serine
pH 7

L-Serine
pH 11

d.

L-Glutamic acid
pH 1

L-Glutamic acid
pH 7

L-Glutamic acid
pH 11

27.19. First, the peptide is drawn with the given sequence, and then the ionization state of each group is evaluated by the procedure outlined in Example 27.1. At pH 7, the carboxylic acid group exists in its deprotonated form, the amino group exists in its protonated form, and a side chain groups exists in the form related to the specific pK_a value of its ionizable group, if present.

a. **Ala-Glu-Val**: only glutamic acid has an ionizable side chain. The carboxy terminus group exists in its conjugate base form, and the amino terminus group is in its conjugate acid form.

b. **Phe-Tyr-Lys**: tyrosine and lysine have ionizable side chains, but only lysine's amino group exists as an ion. The carboxy terminus group exists in its conjugate base form, and the amino terminus group is in its conjugate acid form.

c. **Leu-His-Asn-Ser**: only histidine has an ionizable side chain. The carboxy terminus group exists in its conjugate base form, and the amino terminus group is in its conjugate acid form.

27.20. Nucleophilic aromatic substitution reactions occur by the addition–elimination mechanism. The amino group of isoleucine (in this example), adds to the benzene ring at the position to which the F atom is attached (step 1). Elimination of fluoride ion yields the derivatized amino acid (step 2).

27.21. The amino group of the valine residue at the N-terminus of the polypeptide reacts with DNFB. Complete hydrolysis cleaves the amide groups of the polypeptide, liberating the individual amino acids. The original valine residue is obtained as its DNFB derivative; the other amino acids are obtained in their natural forms.

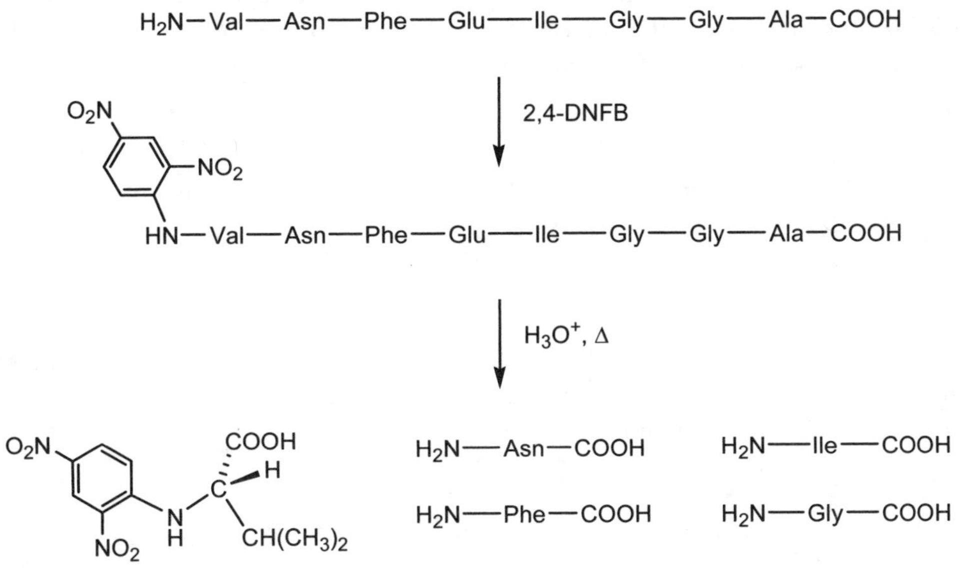

27.22. When the pentapeptide Ile-Ala-Phe-Lys-Ser is treated with 2,4-DNFB, reactions occur between the reagent and any amino group that is present. The structure of the starting peptide, with its side chains, is shown below with the amino groups circled.

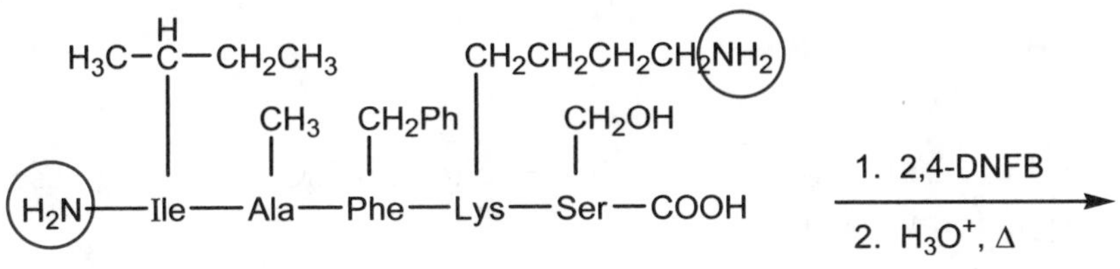

The amino acid that was at the N-terminus is the only one that does not exist as a zwitterion after the hydrolysis step is carried out. Lysine reacts with DNFB at its side-chain amino group (notice that the lysine-DNFB derivative exists as a zwitterion), which will have different properties than the derivative with the DNFB group attached at the α-amino group.

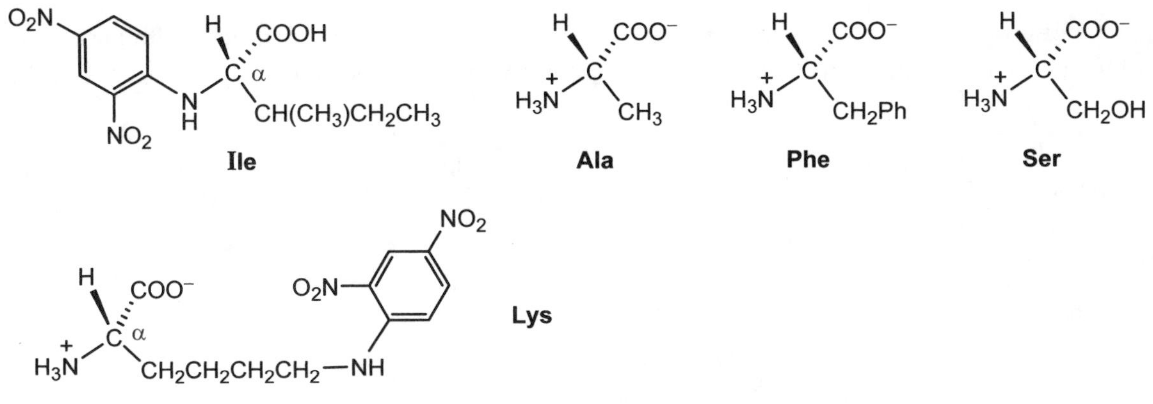

27.23. To predict the structures of the products in each of the following transformations, learn the details of each reaction type, which can be found in the reaction summary section of the chapter. In some of these transformations, reactions from previous chapters are also included.

a. The amino group is acetylated when an amino acid is treated with acetic anhydride. The addition of aqueous acid in the second step protonates the carboxylic acid group, which is deprotonated by the hydroxide ion used in the first step. The starting material is racemic, so the product is also racemic.

b. This transformation is an example of the Strecker synthesis. Use of an isotopically-labeled aldehyde allows one to prepare an isotopically-labeled amino acid. The creation of a new chiral center in the product means that a racemic mixture of amino acids will be obtained.

c. The use of the DuPHOS catalyst allows one to prepare enantiomerically pure amino acids. The second step with aqueous acid removes the Boc group and hydrolyzes the acetal group.

d. The alcohol group is converted to a sulfonate ester upon treatment with TsCl. DBU is a base that is used to promote E2 reactions and form alkenes.

e. The phenol group of a protected tyrosine derivative can be alkylated by treatment with base and an alkyl halide. The starting material is racemic, so the product is also racemic.

27.23. (continued)

f. The use of the DuPHOS catalysts allows one to prepare enantiomerically pure amino acids. The second step is a Suzuki reaction that replaces the Br atom with the phenyl group.

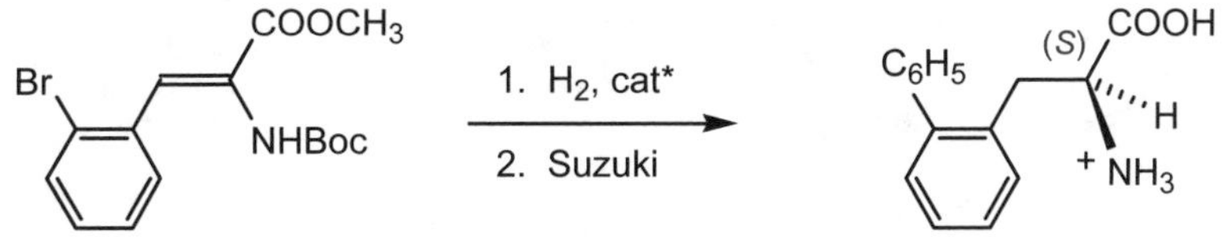

g. An amino acid reacts with phenyl isothiocyanate to form the corresponding thiohydantoin heterocycle. This is the same reaction that the Edman degradation exploits. The starting material is racemic, so the product is also racemic.

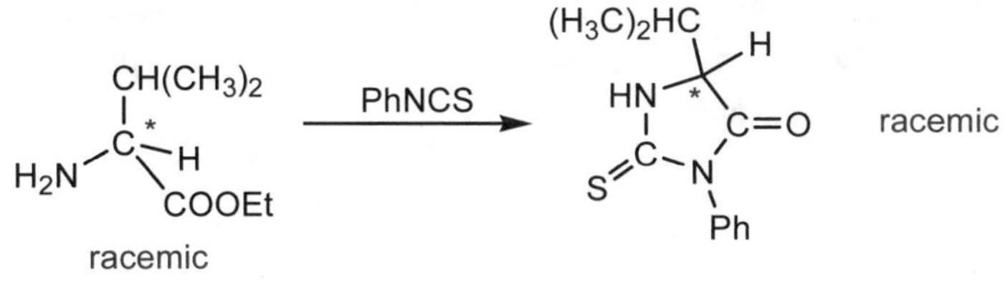

h. Enantiopure amino acids can be made by alkylating glycine enolate ions that are chiral by virtue of their inclusion in a chiral heterocycle.

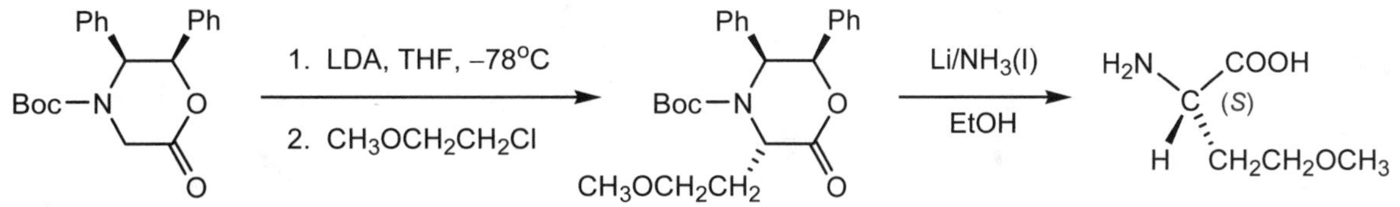

27.24. Any of the methods described in Section 27.3 can be used to prepare the given amino acids because the groups that appear in their side chains are inert to most reagents. Shown below are routes that make use of the chiral glycine enolate described in the text; the starting heterocycle is made from bromoacetic acid. All of the starting compounds are shown in color.

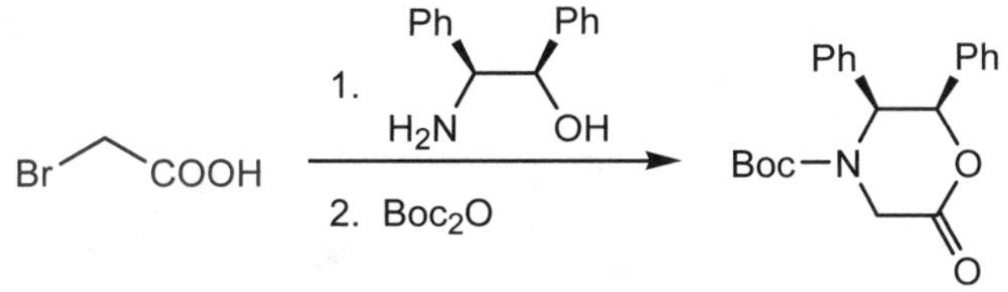

a. The heterocycle is deprotonated and then treated with *p*-bromobenzyl bromide; the amino acid is obtained after reductive cleavage with lithium in liquid ammonia.

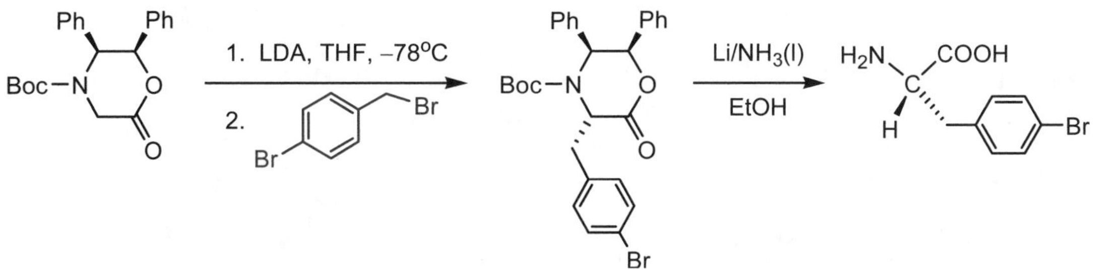

27.24. (continued)

b. The heterocycle is deprotonated and then treated with the bromomethyloxetane; the amino acid is obtained after reductive cleavage with lithium in liquid ammonia.

27.25. The sequence of the dipeptide aspartame has aspartic acid at the N-terminus and phenylalanine at the C-terminus. The carboxy group of the phenylalanine residue is present as its methyl ester. At pH 7, the amino group is in its conjugate acid form, and the carboxy group in the side chain of aspartic acid is in its conjugate base form. There are no other acidic or basic groups in the molecule.

27.26. To prepare the tripeptide Ala-Gly-Ile, one must first synthesize the protected forms of the amino acids. Isoleucine, which is at the C-terminus, is converted to its methyl ester; the other two amino acids are converted to their Boc derivatives.

In the first step of peptide synthesis, the protected forms of Gly and Ile are coupled using DCC.

27.26. (continued)

The amino group of this dipeptide is then deprotected in step 2 using trifluoroacetic acid.

2

BocNH—... + CF₃COOH → H₂N—...

The protected alanine derivative is then coupled to the dipeptide (step 3), and that product is deprotected to form the desired tripeptide.

3

BocNH—CH(CH₃)—COOH + H₂N—... —DCC→ BocNH—...

4

BocNH—... —1. LiOH, CH₃OH / 2. CF₃COOH→ H₂N—...

Ala-Gly-Ile

27.27. At pH 7, the amino groups at the ε-position of the side chain of lysine exist in their protonated forms (below, left). These ammonium ions disrupt the hydrogen bonds that normally stabilize an α-helix. At pH 10, the amino groups are converted to their free-base forms (below, right), and hydrogen bonds form as the stable α-helical structure is formed.

$$\left(\begin{array}{c} \text{C} \\ | \\ \text{H} \end{array}\overset{\text{O}}{\underset{}{\text{C}}}\text{N} \\ | \\ \text{H}\right)_n \quad \xrightarrow{\text{base}} \quad \left(\begin{array}{c} \text{C} \\ | \\ \text{H} \end{array}\overset{\text{O}}{\underset{}{\text{C}}}\text{N} \\ | \\ \text{H}\right)_n$$

H CH₂CH₂CH₂CH₂NH₃⁺   H CH₂CH₂CH₂CH₂NH₂

27.28. γ-Carboxyglutamic acid is like any compound that has two carboxylic acid groups attached to a single carbon atom—it is prone to undergo decarboxylation (Section 22.4c). During amino acid analysis of a protein, the hydrolysis of γ-carboxyglutamic acid under the acidic conditions would be accompanied by decarboxylation.

H₂C—CH(COOH)COOH
H₃N⁺—C(H)—COO⁻

$\xrightarrow[\Delta]{H_3O^+}$

H₂C—CH₂—COOH
H₃N⁺—C(H)—COO⁻   + CO₂

L-γ-Carboxyglutamic acid   **L-Glutamic acid**

27.29. In its protonated form, arginine has the following structure.

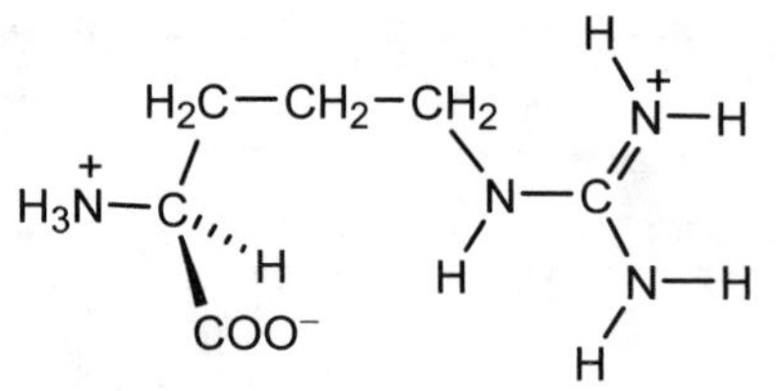

L-Arginine

For the guanidinium ion, three equivalent resonance forms can be drawn. This highly-stabilized group is almost always protonated in biochemical systems.

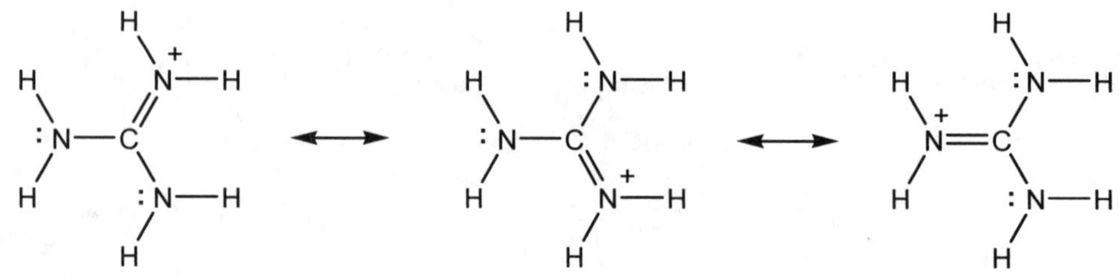

Guanidinium ion

NUCLEIC ACIDS AND MOLECULAR RECOGNITION

28.1. The names of these molecules can be interpreted by comparing their structures with the ones shown in Figures 28.1 and 28.2.

a. Draw the adenine molecule, and then attach the SH group at position 8.

b. Draw the adenosine molecule, and then attach a fluorine atom at C2. Remember that numerals without prime marks are used to designate the positions in the heterocycle; numerals with primes are used to specify the positions in the carbohydrate ring.

c. The "deoxy" prefix means that you remove the oxygen atom(s) from the specified position(s). For this exercise, draw the cytidine molecule without the OH groups at C2′ and C3′ (in the carbohydrate ring).

d. The "deoxy" prefix means that you remove the oxygen atom(s) from the specified position(s). Draw the adenosine molecule without the OH group at C2′. Attach a diphosphate group to the oxygen atom attached to C5′.

e. Draw the uracil molecule, and then attach the trifluoromethyl group at position 5.

28.2. First draw a good Lewis structure for each heterocycle. Then decide if the unshared electron pairs are part of the π system or not: The unshared electron pair on a nitrogen atom that forms a double bond to a neighboring carbon atom will be perpendicular to the aromatic system, but if a nitrogen atom forms only single bonds to its neighbors, then its unshared electron pair will be in a *p* orbital that overlaps to form the π system.

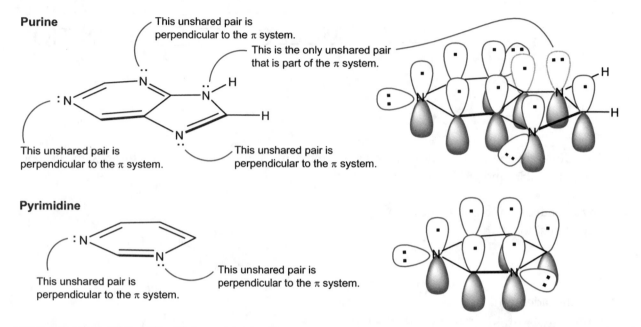

28.3. To draw the tautomeric forms for a given heterocycle, focus on each NH group and its adjacent carbonyl group. A tautomeric structure will have a double bond between the original carbonyl carbon atom and the adjacent N atom, and a hydroxyl group will take the place of the double-bonded O atom.

Uric Acid (11 forms)

28.3 (continued)

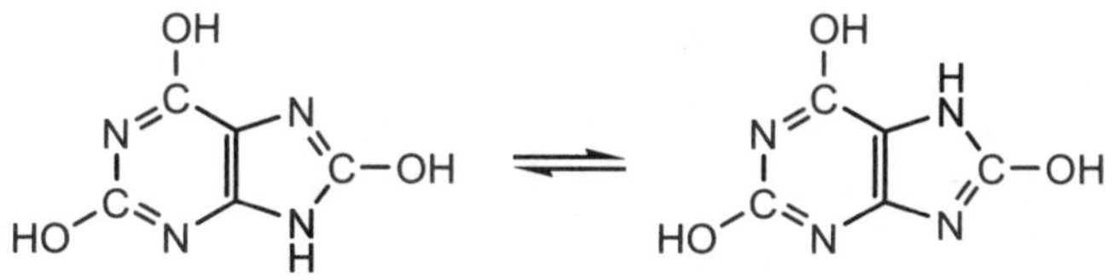

Uracil (5 forms)

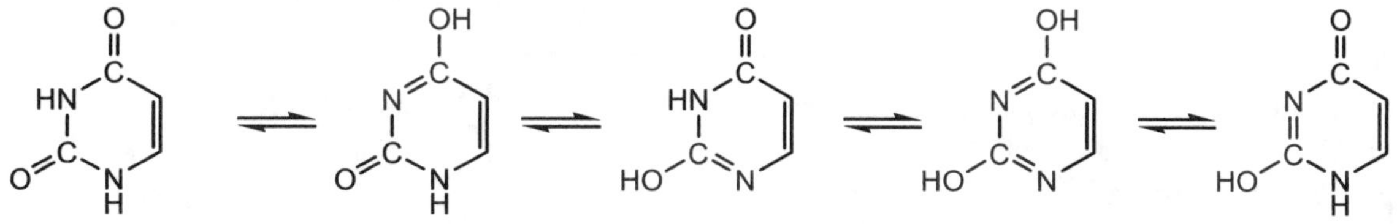

Guanine (2 forms)

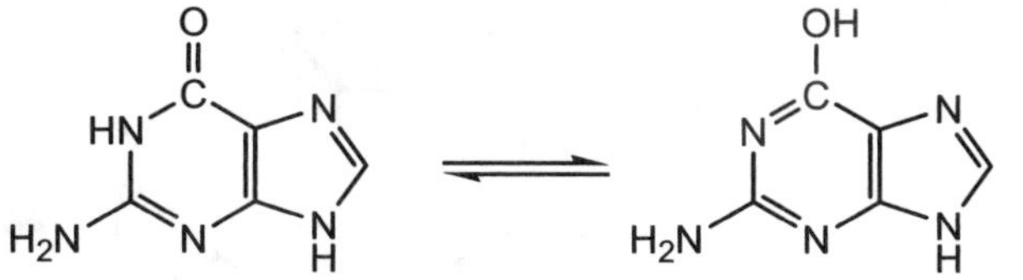

28.4. The synthesis of a pyrophosphate ester requires two steps. The molecule of alcohol first reacts at the middle phosphorus atom of the triphosphate group of ATP, creating a trigonal-bipyramidal, anionic intermediate.

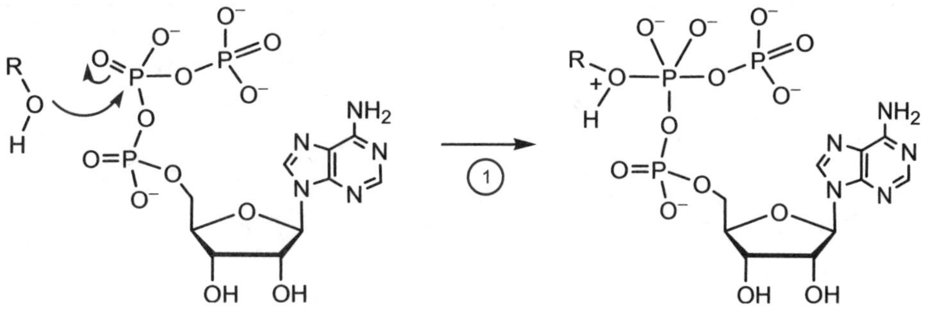

The trigonal-bipyramidal intermediate then collapses to form the pyrophosphate derivative and the protonated form of AMP (step 2), which subsequently loses a proton in step 3 to form AMP.

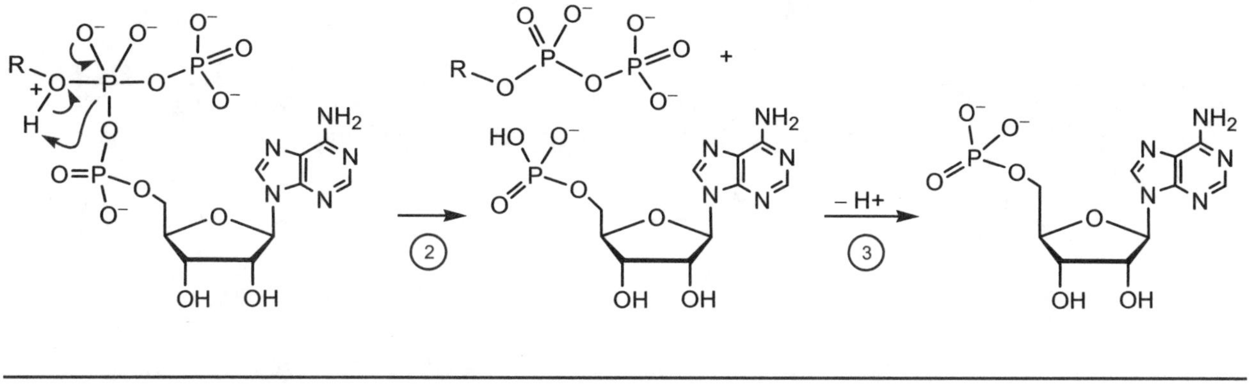

28.5. The structure of the given tetranucleotide is drawn so that the letter at the left end of the sequence (A) corresponds to the nucleoside at the 5′ end of the molecule, which has a phosphate group attached.

The letter T within the sequence tells you that this molecule is DNA, so each carbohydrate ring has an O atom attached at every carbon atom except C2. A phosphate group connects each nucleoside 3′-OH group with the 5′-OH group of its neighbor. The 3′-OH group of the nucleoside at the right end of the sequence (C) has no phosphate group attached.

28.6. The synthesis of dibenzo-18-crown-6 can be accomplished in stepwise fashion, but simply mixing the catechol dianion with the ditosylate derivative of bis(2-hydroxyethyl)ether will also produce the desired compound (in addition to oligomers). Using the dipotassium salt of catechol assists the synthesis of the crown ether product because of the template effect.

28.7. The synthesis of the pyridino-crown ether proceeds in two stages. First, diol **A** is prepared by alkylating bis(2-hydroxyethyl) ether at each hydroxyl group with chloroethanol.

Then, 2,6-lutidine is brominated at its two benzylic positions using NBS in a free radical process.

28.7. (continued)

Dibromo-2-6-lutidine reacts with the dianion of **A** under high dilution conditions and with the template effect of the potassium ion to form the pyridinocrown ether.

The binding strength for a substrate/crown ether complex is maximized when three hydrogen bonds form between the ammonium salt and the crown ether. Thus, the binding constants for the alkylated ammonium salts are expected to decrease in the following order:

can form three hydrogen bonds   can form two hydrogen bonds   can form one hydrogen bond

28.8. The synthesis of cyclotriveratrylene occurs via three consecutive Friedel-Crafts alkylation reactions between an arene ring and a carbocation intermediate. The transformation begins with protonation of the benzylic alcohol (step 1), which forms a carbocation (step 2) that is subsequently trapped by a reaction with a second molecule of 3,4-dimethoxybenzyl alcohol (step 3). Regeneration of the aromatic ring occurs with concomitant formation of a protonated alcohol molecule (step 4).

Dissociation of a molecule of water from this protonated alcohol yields a carbocation (step 5), which reacts with a third molecule of 3,4-dimethoxybenzyl alcohol (step 6). Regeneration of the aromatic ring (step 7), dissociation of a molecule of water (step 8), and intramolecular ring formation (step 9) are followed by regeneration of the aromatic ring (step 10) to create the product molecule.

28.8. (continued)

28.9. The preparation of a urea derivative from the reaction between an amine and an isocyanate takes place in two stages. First, the nitrogen atom of the amine adds to the carbon-nitrogen double bond of the isocyanate to form a zwitterion intermediate (step 1). An acid–base reaction then takes place to yield the neutral product molecule (step 2).

28.10. An alkylphosphate ion can form strong hydrogen bonds with urea. Hydrogen bonds and electrostatic attractions are important for the attractive interactions between an alkylphosphate ion and the guanidinium ion.

28.10. (continued)

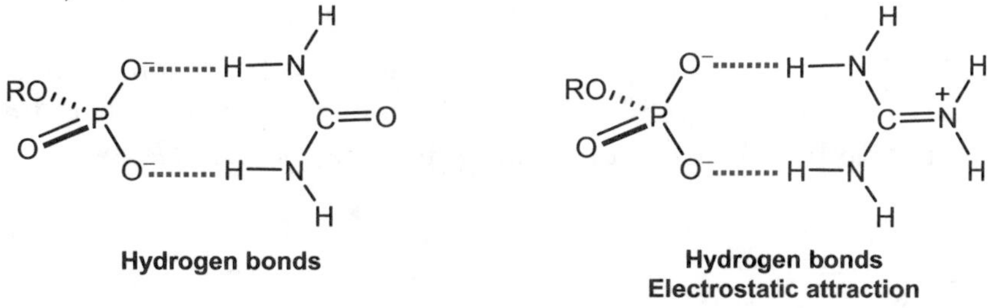

Hydrogen bonds   **Hydrogen bonds**
                            Electrostatic attraction

28.11. Pyridine and pyrazine, respectively, form one and two hydrogen bonds to the convergent carboxylic acid groups of receptor **3**.

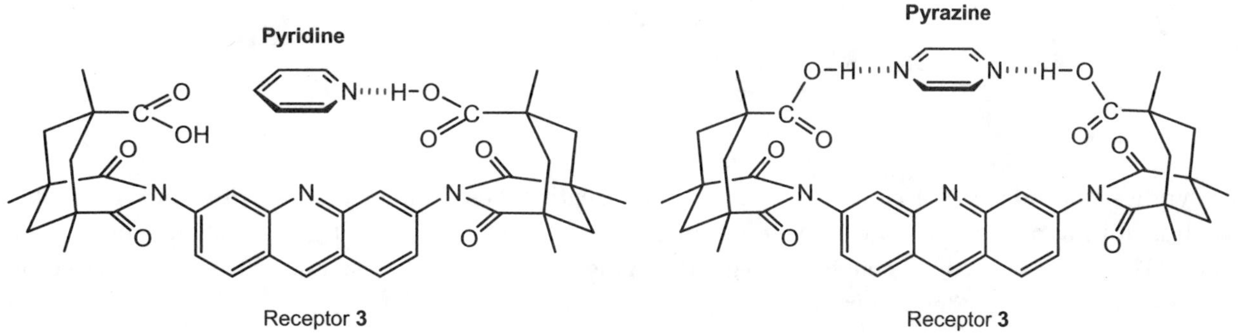

Pyridine

Receptor **3**

Pyrazine

Receptor **3**

Purine forms three hydrogen bonds, and two orientations for these interactions are possible.

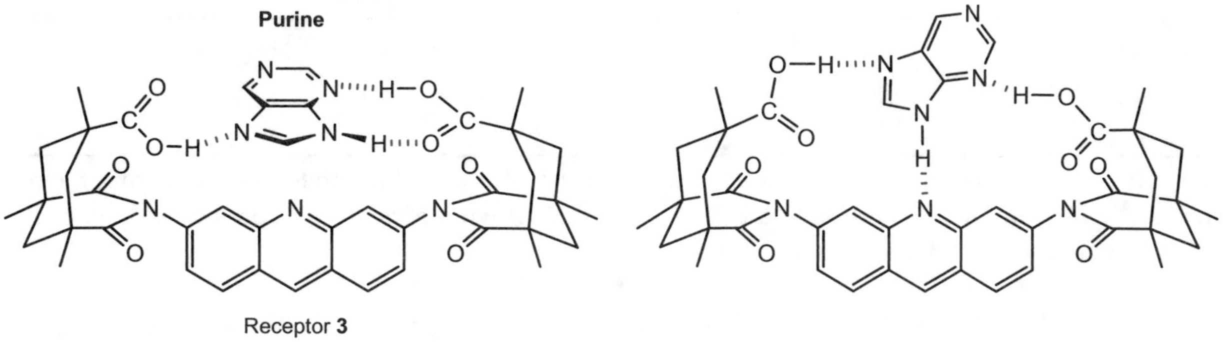

Purine

Receptor **3**

Purine should bind most strongly to the receptor because it forms the greatest number of hydrogen bonds. Pyridine will likely be bound most weakly because only one hydrogen bond can form.

28.12. The hydrogen bond pattern between the given receptor molecule and thymine is apparent if you look at the groups that are involved in forming the hydrogen bonds: The receptor has two N–H groups that are hydrogen-bond donors as well as a nitrogen atom that can act as a hydrogen-bond acceptor. Thymine has one donor group (N–H) flanked by two acceptor groups (the carbonyl oxygen atoms).

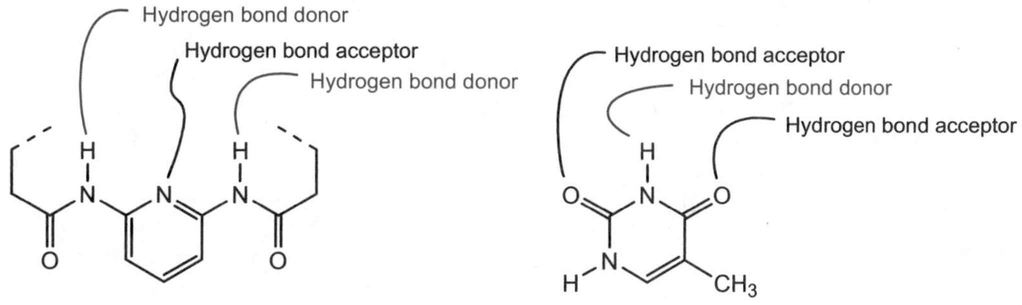

Hydrogen bond donor
Hydrogen bond acceptor
Hydrogen bond donor

Hydrogen bond acceptor
Hydrogen bond donor
Hydrogen bond acceptor

28.12. (continued)

The binding of thymine by the receptor takes place by hydrogen bond formation as well as π-stacking interactions between the two aromatic systems. This stacking interaction is further strengthened by attractive forces between the electron deficient π system of thymine and the electron rich π network of dihydroxynaphthalene.

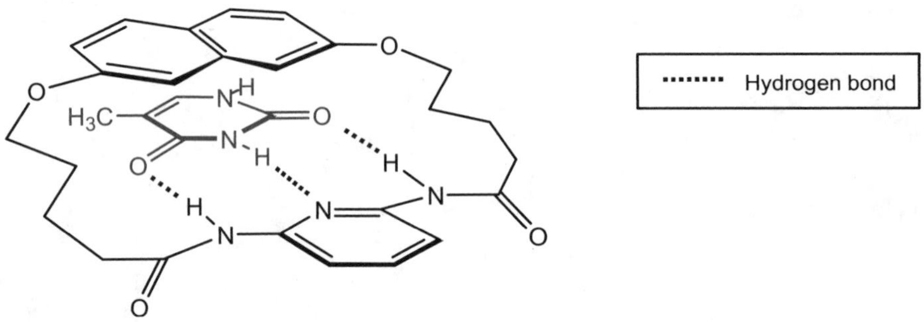

The synthesis of the receptor can be accomplished by treating 2,6-diaminopyridine with γ-butyrolactone. This transformation forms two carboxamide groups and leaves alcohol OH groups at the ends of the chains. These OH groups are converted to good leaving groups (tosylate esters), which are displaced by the nucleophilic dianion derived from 2,7-dihydroxynaphthalene.

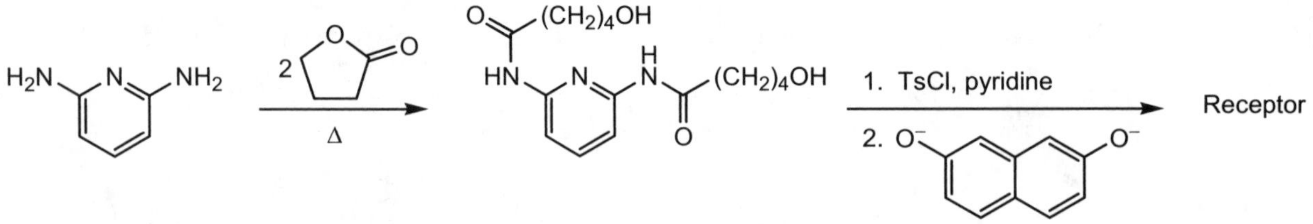

28.13. The names of the given bases, nucleosides, and nucleotides are interpreted by comparing them to structures shown in Figures 28.1 and 28.2. The names of crown ethers are interpreted as described in Section 28.4a and of urea derivatives as in Section 28.6a.

a. Draw the thymine molecule, and then attach a bromine atom at position 6.

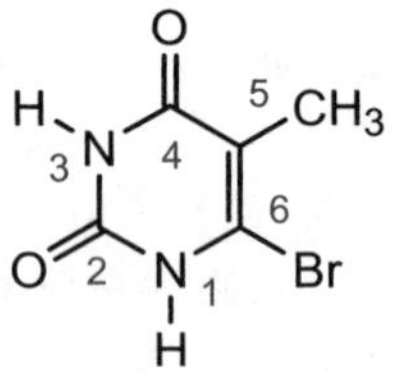

b. Urea is H_2NCONH_2. Any substitution of a proton attached to the nitrogen atoms is designated by the use of the prefixes *N* and *N'* (for the different nitrogen atoms). In this molecule, both benzyl groups are attached to the same nitrogen atom.

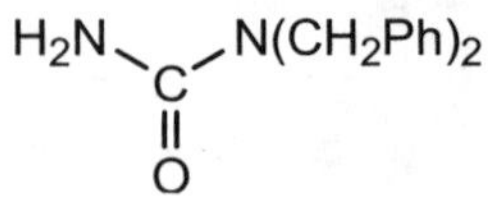

c. The "deoxy" prefix means that you remove the oxygen atom(s) from the specified position(s). Draw the cytidine molecule without the OH group at C2'. Attach a phosphate group to the oxygen atom attached to C5'.

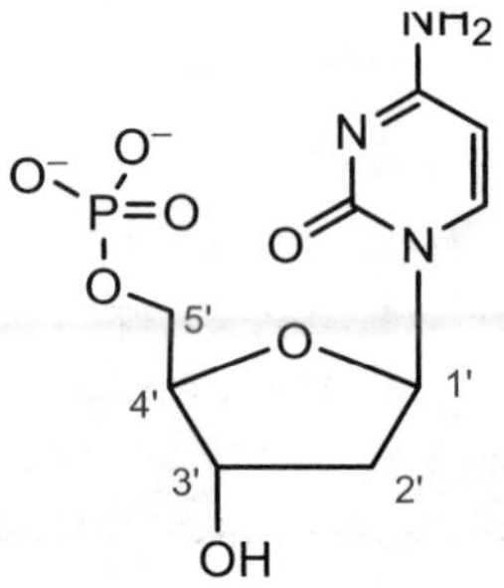

28.13. (continued)

d. Draw the adenosine molecule, and then attach a fluorine atom at position 8 (on the heterocycle). Attach a triphosphate group to the oxygen atom attached to C5'.

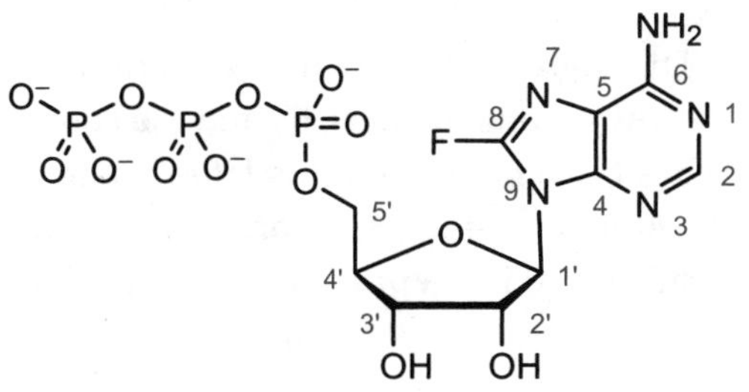

e. The size of a crown ether ring is given by the first numeral (21 atoms), and the number of oxygen atoms (or other heteroatoms) is specified by the second numeral (7).

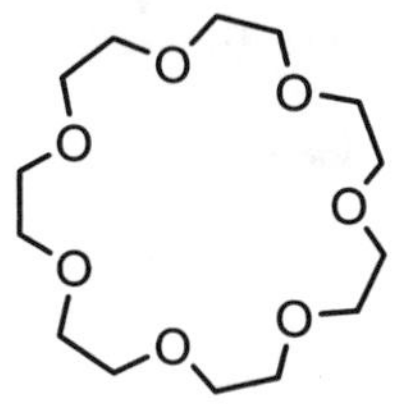

f. Draw the inosine molecule, and then attach a methoxy group at position 2 (on the heterocycle).

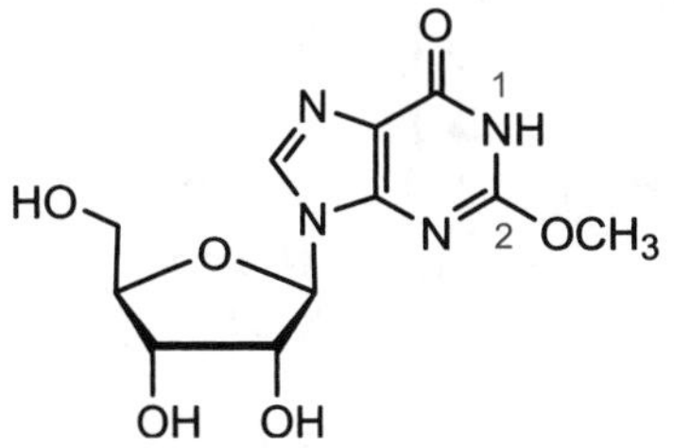

28.14. Use the names of the given bases, nucleosides, and nucleotides given in Tables 26.1 and 26.2 to generate the names of the molecules that correspond to the structures shown in this exercise.

a. The parent compound is guanosine. The C2' position lacks the OH group, and a diphosphate group is attached at C5'. The name of this substance is **2-deoxyguanosine-5'-diphosphate**.

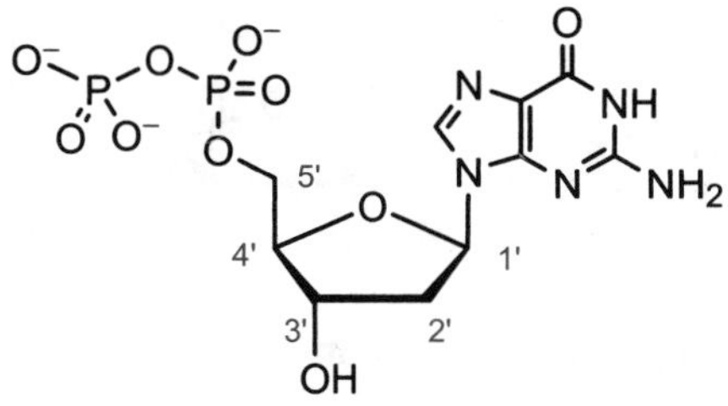

b. The parent compound is guanine. Position 8 has a trifluoromethyl group attached, so the name of this substance is **8-trifluoromethylguanine**.

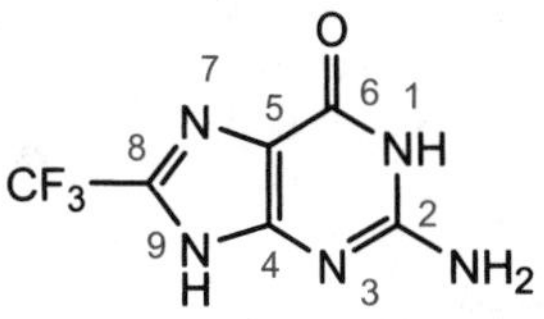

c. The parent compound is urea. One of the nitrogen atoms has a phenyl group attached, so the molecule is **phenylurea** (or *N*-phenylurea).

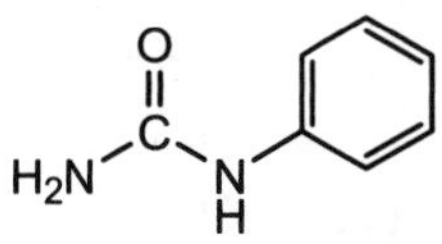

d. The parent compound is adenosine, but the OH groups have been replaced by SH groups. The name must reflect the absence of the OH groups (deoxy) as well as the presence of the SH groups (mercapto). The name of this substance is **2,3-dideoxy-2,3-dimercaptoadenosine**.

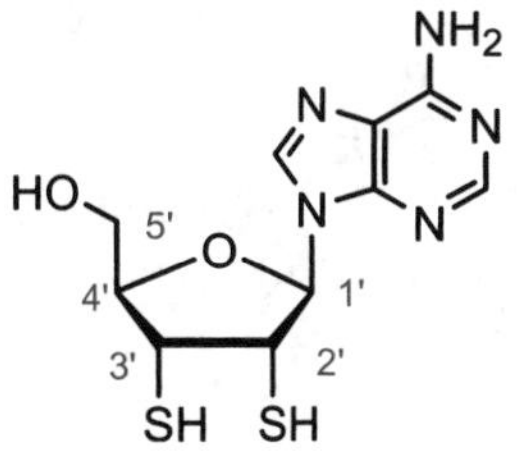

28.15. (continued)

e. The parent compound is uridine. A phosphate group is attached at C5', and the configuration of C2' has been inverted. This latter change is denoted by use of the prefix epi, for *epimer*. The name of this substance is **2'-epiuridine-5'-phosphate**.

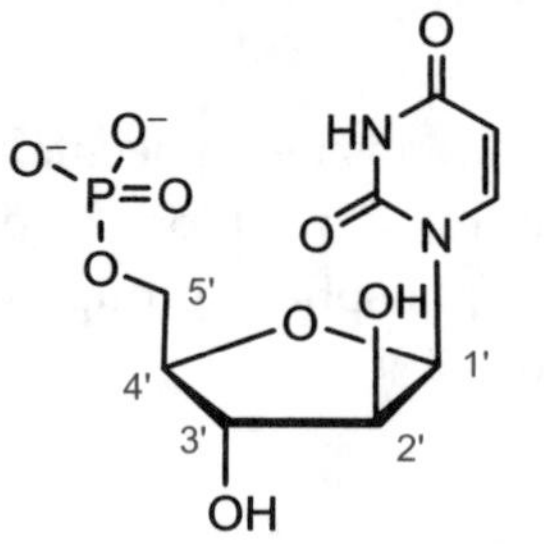

28.16. Follow the procedure outlined in the solution to Exercise 28.5

a. This molecule is DNA because it contains **T**.

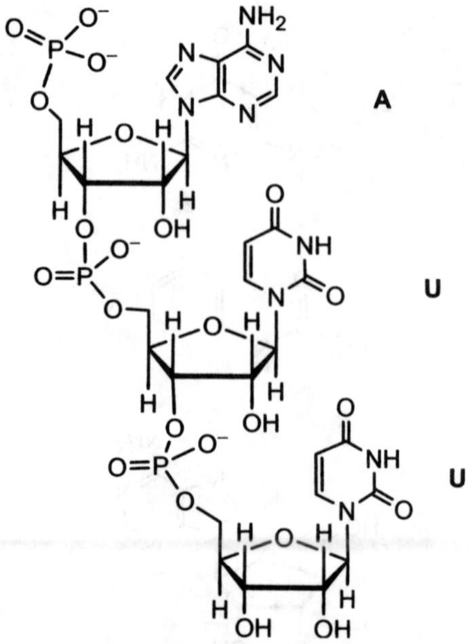

c. This molecule is DNA because it contains **T**.

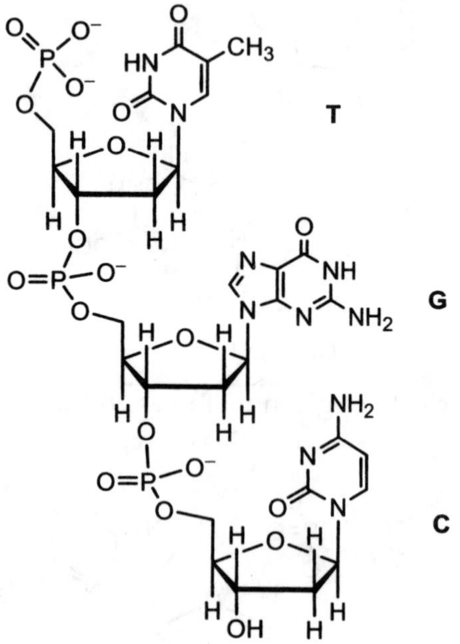

b. This molecule is RNA because it contains **U**.

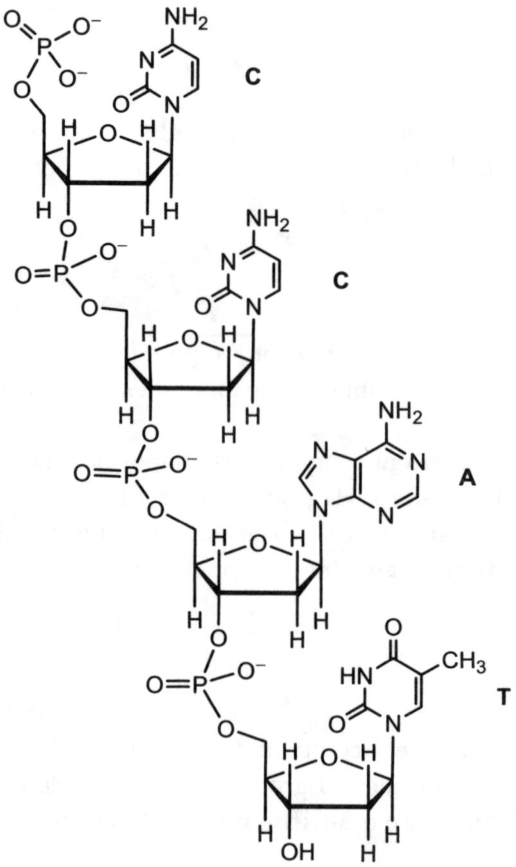

28.16. Inosine has one each of hydrogen bond acceptor and donor groups, and these can form complimentary structures with A, C, or U. Thus, look for the donor and acceptor group on each heterocycle, and form the hydrogen bonds as appropriate.

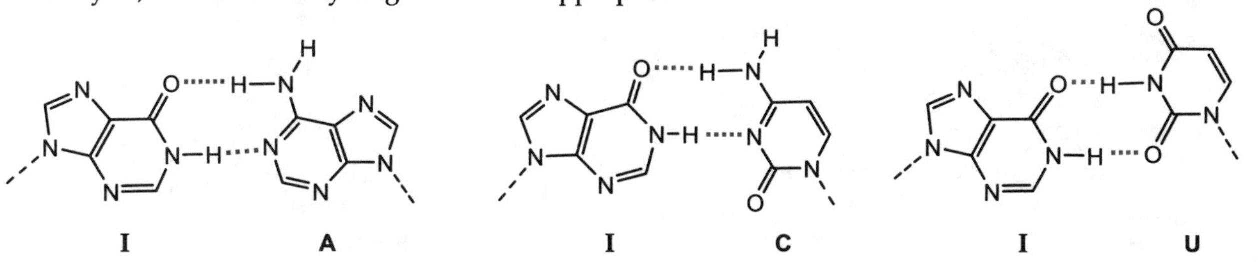

28.17. Pseudouridine forms hydrogen bonds with adenosine in the same way that uridine does. Attaching pseudouridine to the ribose ring through a carbon atom means that two lactam functional groups are present, so two different orientations are possible for a strand of RNA that bears a pseudouridine group.

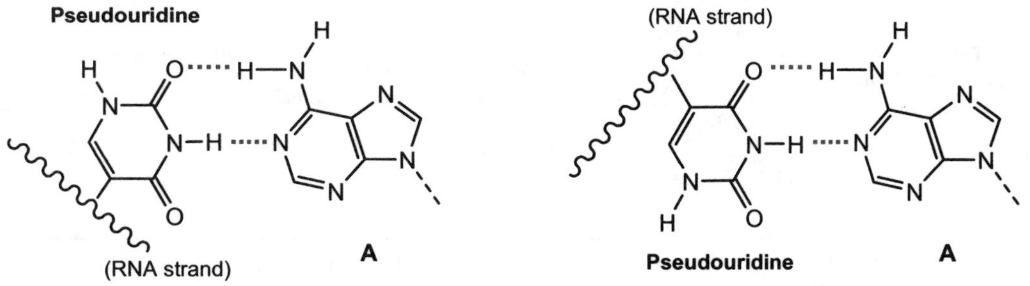

28.18. The structure of 5-bromouracil is similar to the structure of thymine except that a bromine atom is present in place of the methyl group found in T. The nitrogen atom attached to the anomeric carbon atom of the sugar ring requires that the adjacent oxygen-containing functional group exists in the carbonyl form. The NH and its neighboring carbonyl group (circled, below) define the portion of the heterocycle that tautomerizes.   The structures of the lactam and lactim form are as follows:

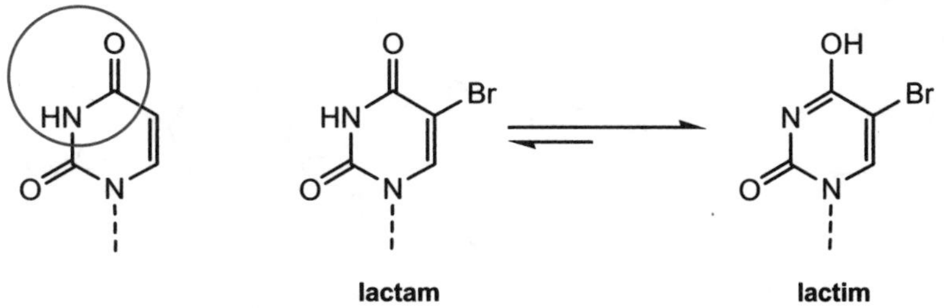

Placing the two forms of 5-bromouracil in equivalent positions relative to the bases of A and G, you can see that the lactam tautomer forms a hydrogen bond with A in the same manner that T would do. The lactim tautomer, however, forms three hydrogen bonds with G.

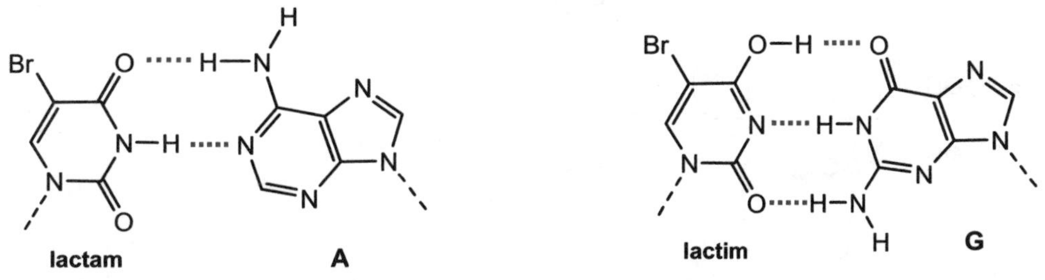

28.19. 7,9-Dimethylguanine exists as a zwitterion in its neutral form. Protonation occurs at the negatively charged oxygen atom, and tautomerism generates the carbonyl group, which is the preferred structural form of the protonated material.

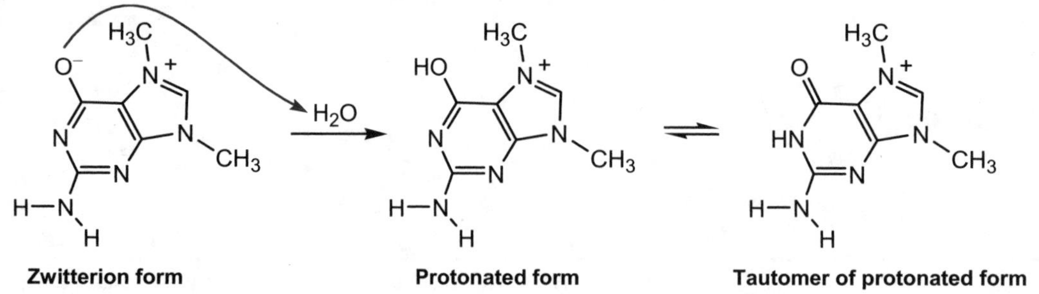

Zwitterion form	**Protonated form**	**Tautomer of protonated form**

The hydrogen bonds that can form are shown below:

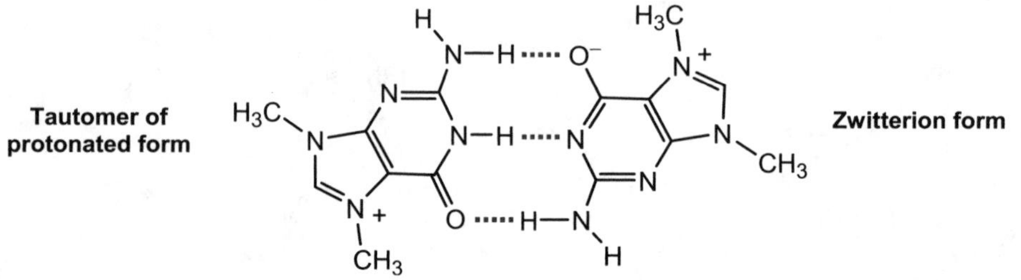

28.20. The synthetic routes used to prepare crown ethers take advantage of procedures involving the S$_{N}$2 reactions between nucleophilic oxyanions (alkoxide or phenoxide ions) and sulfonate ester derivatives of diols. The starting compounds are shown in color.

a.   Alkylation of bis(2-hydroxyethyl)ether is performed at each end using chloroethanol [or a protected form—see the scheme below part (c.) of this exercise]. The diol is converted to its ditosylate derivative, which is treated with the dianion of catechol. Sodium ion provides the needed template to ensure formation of the macrocycle in good yields.

Retrosynthesis

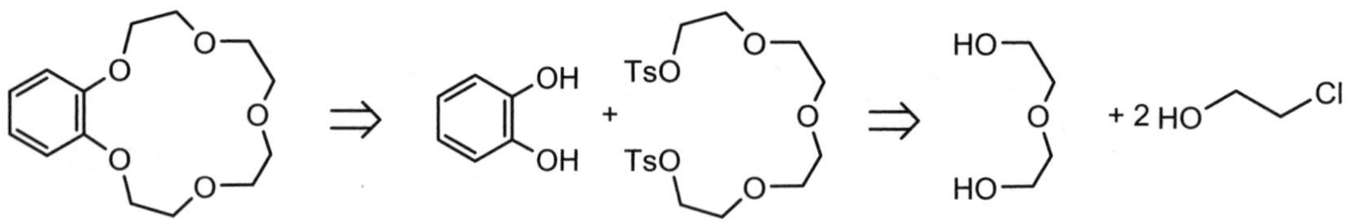

Synthesis

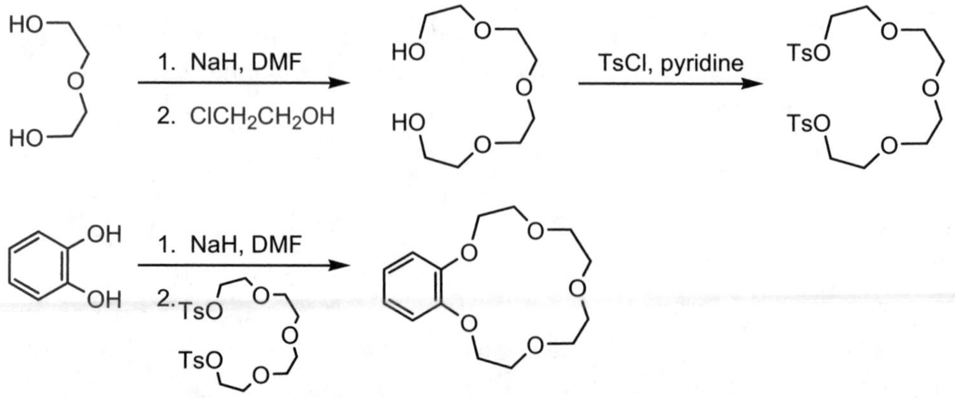

28.20. (continued)

b. The diamino crown ether is made by reducing a bis(carboxamide) derivative with lithium aluminum hydride. The bis(carboxamide) is made from a bis(acid chloride) and a diamine.

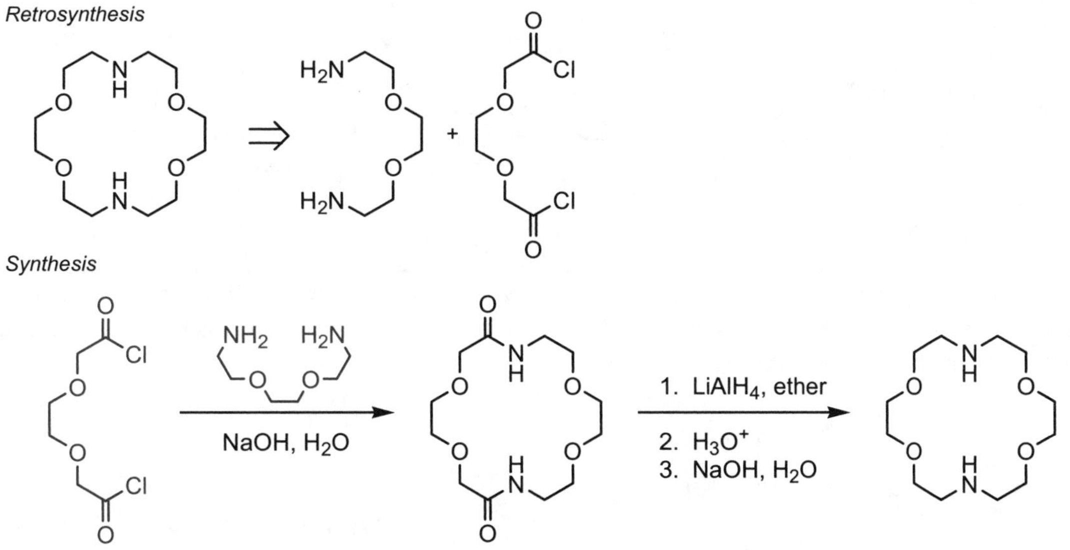

Retrosynthesis

Synthesis

c. The chiral diol is made by catalytic asymmetric dihydroxylation of *trans*-2-butene (Section 16.4d), then the remainder of the carbon skeleton is constructed as in part (a.), starting with bis(2-hydroxyethoxy)ethane. Coupling these two pieces under high dilution conditions makes use of the template effect of the potassium ion to form this chiral 18-crown-6 analog.

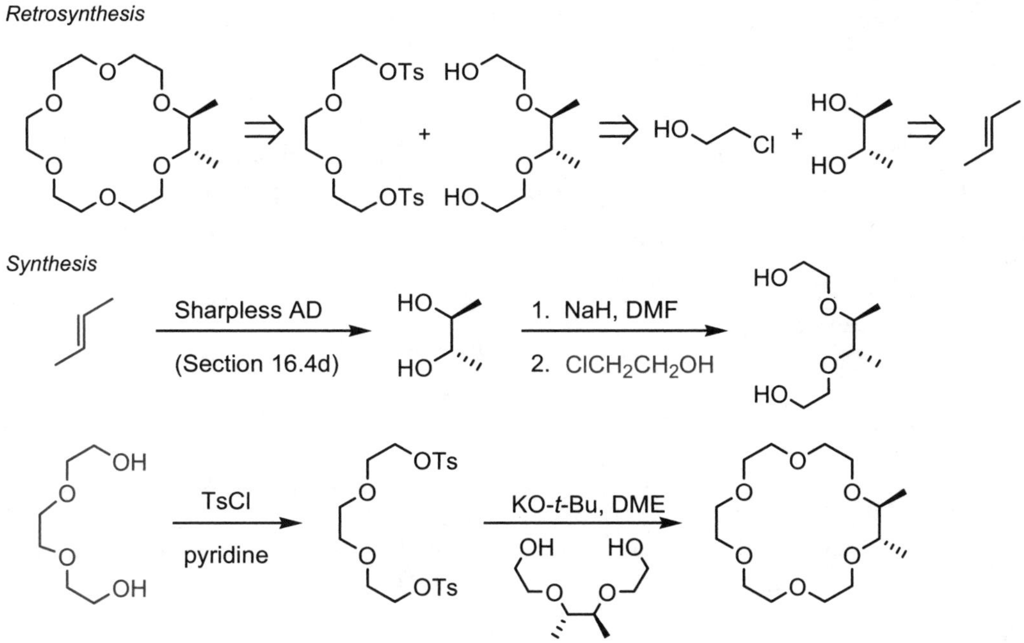

Retrosynthesis

Synthesis

Chloroethanol is sometimes converted with base mostly to ethylene oxide. If this reaction were to occur while attempting the syntheses outlined in parts (a.) and (c.), then protection of the OH group can be accomplished by making the trimethylsilyl derivative. After its use as an alkylating agent, the trimethylsilyl groups are removed using fluoride ion.

28.20. (continued)

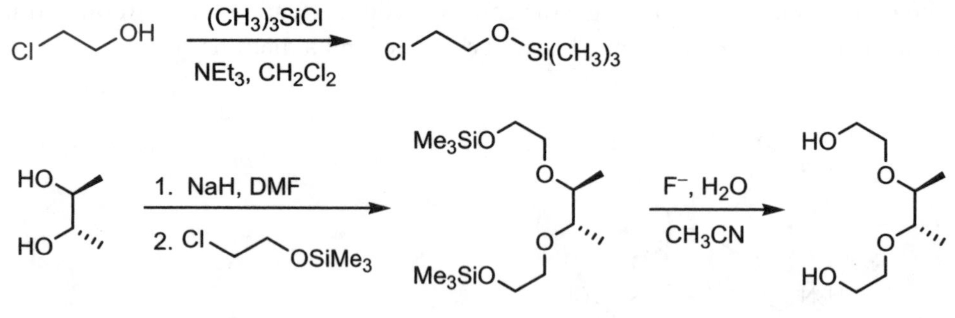

28.21. The condensation reaction that takes place between guanidinium carbonate and dimethyl malonate occurs via two addition–elimination sequences in which the amino groups of the guanidinium ion react with the ester groups of dimethyl malonate.

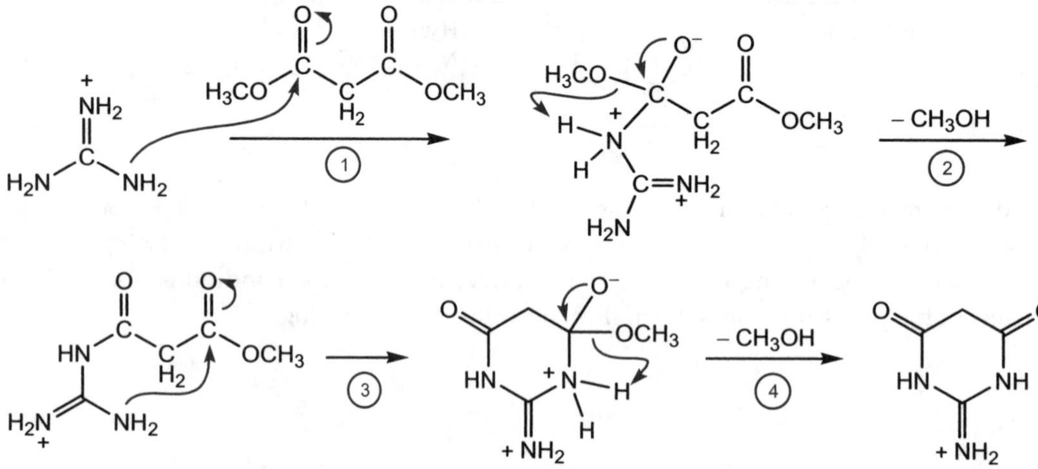

The initially-formed cyclic product tautomerizes twice (step 5), and then carbonate ion acts as a base to remove the proton from the tetrahedral carbon atom to form the aromatic heterocycle (step 6).

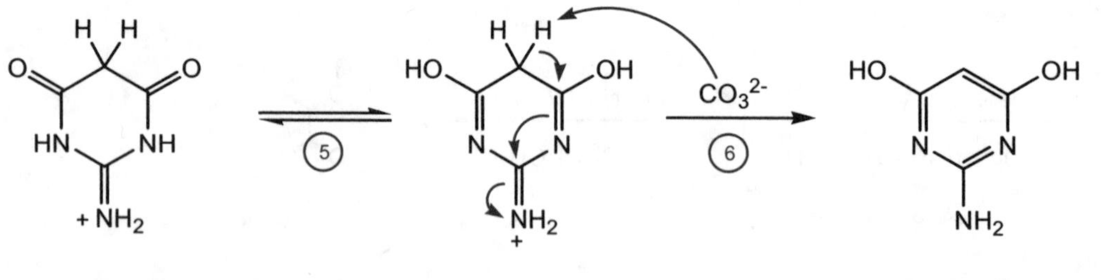

28.22. The first step in the reaction scheme shown in this exercise comprises the alkylation of a heterocyclic nitrogen atom. Sodium hydride deprotonates the NH group (step 1) and this nucleophile reacts with RX by the S_N2 mechanism.

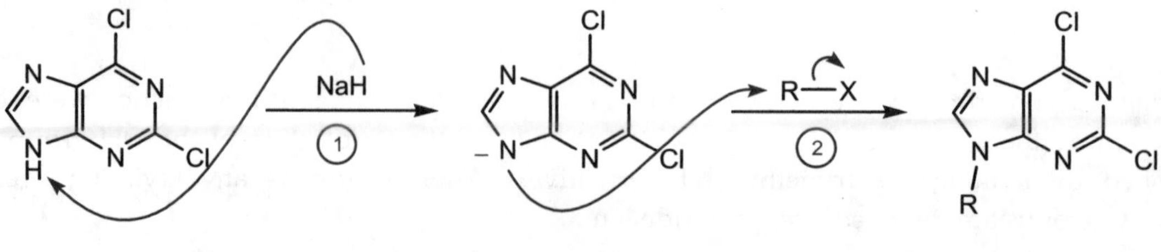

28.22. (continued)

The next step is an addition–elimination reaction that replaces one of the Cl atoms with an amino group.

The amino group undergoes diazotization (step 5—see Section 17.4b), and replacement of the diazonium group by the OH group occurs.

After tautomerism (step 7), another addition–elimination process leads to substitution of the other chlorine atom, and an acid-base reaction finishes the sequence (step 10).

28.23. Both of the receptors shown in this exercise form the same number and types of hydrogen bonds, but receptor **Y** also makes use of the π-stacking interaction between the naphthalene and pyrimidine rings in the recognition process. This stacking interaction is further strengthened by attractive forces between the electron deficient π system of pyrimidine and the electron rich π network of dihydroxynaphthalene.

28.23. (continued)

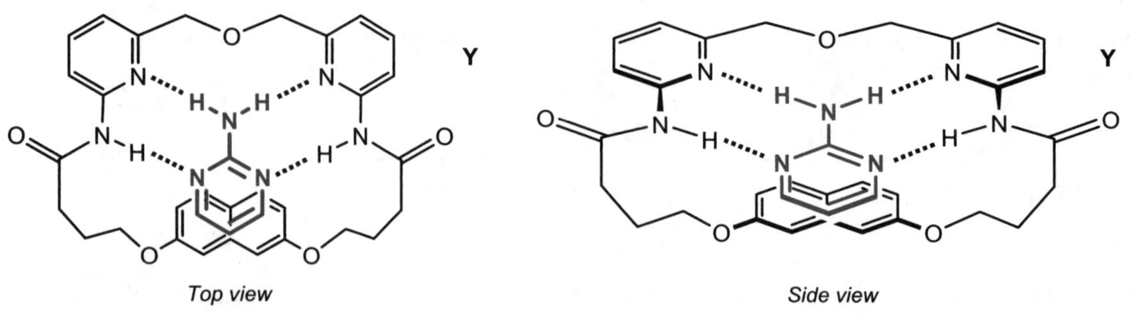

Top view   Side view

28.24. The synthesis of receptor **X** makes use of the amide-forming reactions between the amino groups of the heterocycle and the carbonyl group derived from the specified acid chloride.

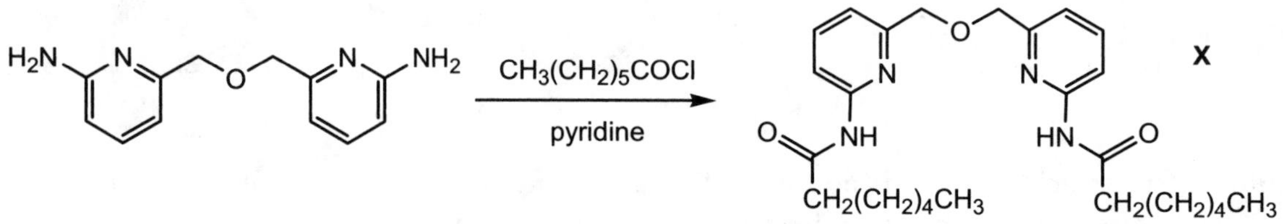

The diamine is prepared by making the benzylic bromide derivative of 2-bromo-6-methylpyridine compound, converting part of that material to the corresponding alcohol derivative via solvolysis (S_N1 reaction), coupling the two pyridine building blocks by means of the S_N2 reaction, and replacing the bromine atoms with amino groups by an addition–elimination reaction.

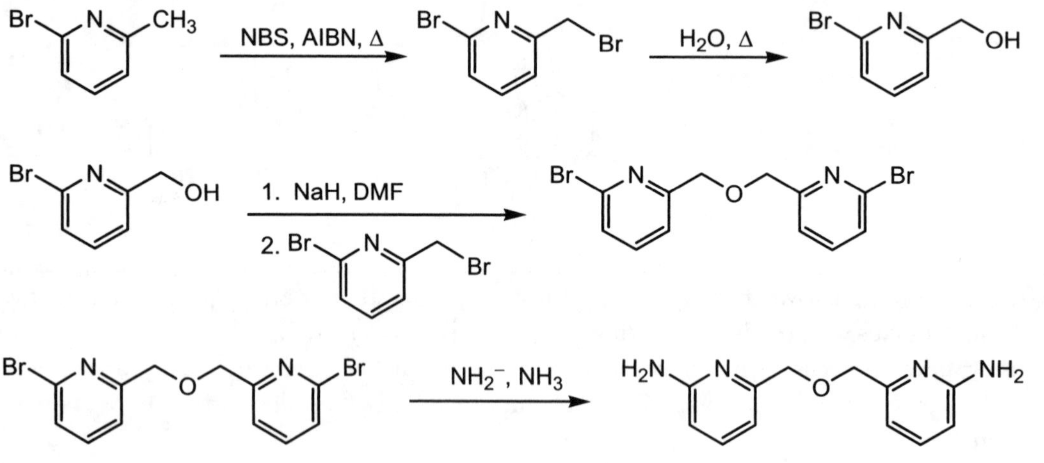

28.25. The three zwitterion forms of receptor **A** are generated by transfer of the phenol proton group to each of the heterocyclic nitrogen atoms.

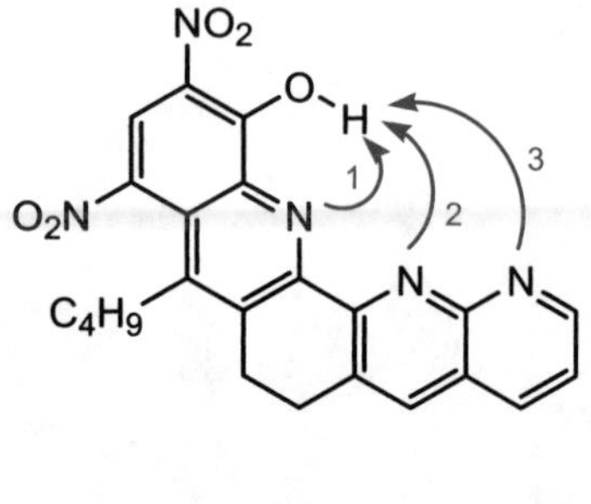

28.23. (continued)

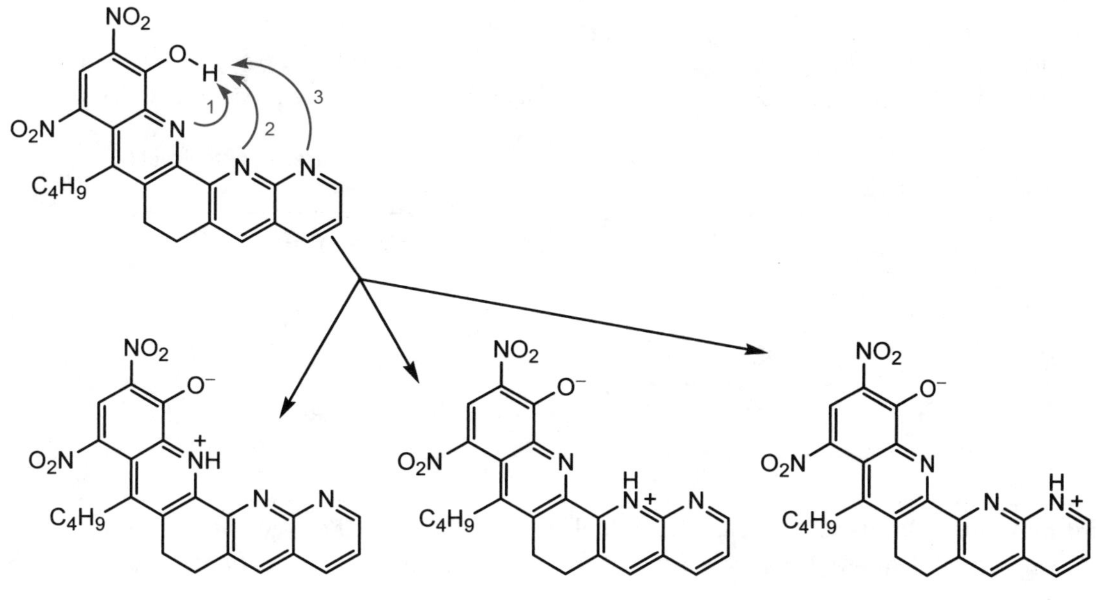

Form **3** can bind creatinine (shown in color) by forming three hydrogen bonds.

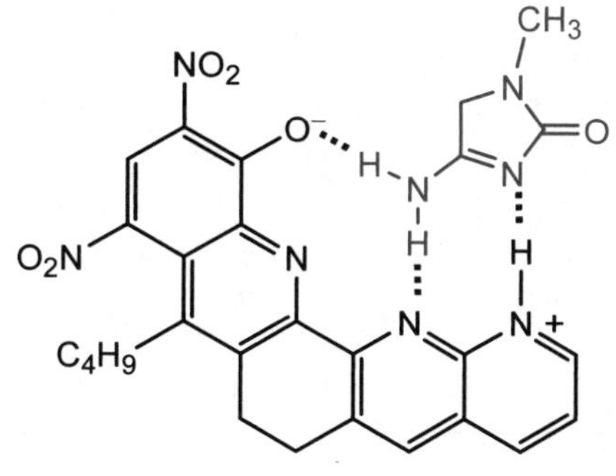

3

28.26. Both receptors shown in this exercise bind the dihydrogen phosphate ion by taking advantage of the fact that the OH groups of $H_2PO_4^-$ act as hydrogen bond donors toward the urea carbonyl groups, and the NH portions of the urea groups act as hydrogen bond donors toward the oxygen atoms of the dihydrogen phosphate ion.

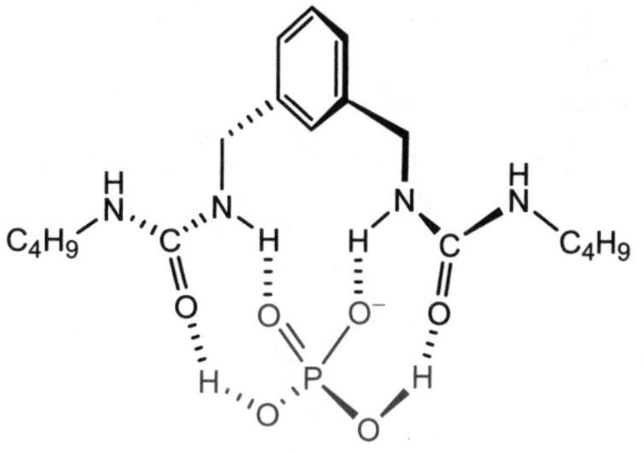

28.27. Derivatives of urea are commonly made by the reaction between an amine and an isocyanate molecule. Receptor **U** in Exercise 28.26 can be made by performing the free radical bromination of *m*-xylene, which forms the α,α'-dibromo derivative. Reaction of this dibromide with potassium phthalimide, followed by hydrolysis (the Gabriel reaction), yields α,α'-diamino-*m*-xylene. This diamine reacts with two equivalents of butyl isocyanate to form the corresponding urea receptor molecule.

28.28. Host molecule **A** binds to both *syn* and *anti* electron pairs on *each* oxygen atom of the acetate ion (four hydrogen bonds). Dimethylurea can form hydrogen bonds only to two unshared electron pairs of the acetate ion.